Handbook for Culturally Competent Care

Eric A. Fenkl • Larry D. Purnell
Editors

# Handbook for Culturally Competent Care

Second Edition

Springer

*Editors*
Eric A. Fenkl
Nicole Wertheim College of Nursing
and Health Sciences
Florida International University
Miami, FL, USA

Larry D. Purnell
College of Health and Nursing Sciences
Excelsior University
Albany, NY, USA

ISBN 978-3-031-70491-8 ISBN 978-3-031-70492-5 (eBook)
https://doi.org/10.1007/978-3-031-70492-5



This Springer imprint is published by the registered company Springer Nature Switzerland AG
The registered company address is: Gewerbestrasse 11, 6330 Cham, Switzerland

# Preface

The usefulness of the Purnell Model for Cultural Competence, its organizing framework, and comprehensive assessment guide has been established globally. They are being used worldwide for individual and family health-care assessment and planning and intervention in health promotion and wellness; illness, disease, and injury prevention; and health maintenance and restoration.

In the second edition of this handbook, chapters from the first edition were updated to reflect the most current information available on each cultural heritage. Many of the chapters were revised and updated by new corresponding authors, providing a fresh perspective of the chapter content. The chapter on "People of Hindu Heritage" was expanded to specifically include people who identify as Sikh, thus resulting in a new chapter title: "People of Hindu and Sikh Heritage." The second edition also includes two completely new chapters not previously published: "People of Jamaican Heritage" and "People of Ukrainian Heritage." Both of which are beneficial to readers seeking culturally competent information for those two groups. Additionally, three consolidated chapters from the more comprehensive *Textbook for Transcultural Health Care: A Population Approach* were added to the handbook including "People of Greek Heritage," "People of Italian Heritage," and "People of Thai Heritage."

The reader will find it helpful that the second edition is organized in the same manner as the first edition. For this second edition, as in the first edition, the aggregate data, a research principle, in culturally specific chapters is meant as a guide. The content may not apply to every individual in that culture depending on individual preferences, acculturation, and assimilation into the dominant culture. The variant cultural characteristics on the model help prevent stereotyping, assuring patients are individuals within their culture.

As in the first edition, each chapter is organized by the model's 12 domains of culturally competent care and provides key approaches and interventions highlighted in bold type, providing a quick reference for working with each individual. Thus, recommended approaches and interventions may need to be adapted based on patient's and family's personal perspectives and circumstances. The health-care provider must make a concerted effort to set aside any ingrained personal biases and prejudices.

With increased worldwide migration, health-care providers are caring for patients whose cultures are foreign to them, making care more complex. Given the worldwide diversity, and the diversity within cultural groups, it is impossible to include the beliefs and practices of all cultures and subcultures.

Specific criteria are used for identifying the groups represented in the guide and were selected based on a variety of the following criteria:

- The group has a large population in North America, such as people of African American, Appalachian, Cuban, Jamaican, Jewish, and Mexican heritages.
- The group is relatively new in its migration status, such as people of Arab, Bosnian, Hmong, Somali, Ukrainian, and Vietnamese heritages.
- The group is widely dispersed throughout the world, such as people of Chinese, Cuban, Greek, Filipino, Hindu, Iranian, Italian, Puerto Rican, and Thai heritages.
- The group has little written about it in the health-care literature, such as people of Brazilian, Haitian, Japanese, Korean, Russian, Sikh, and Turkish heritages.
- The group was of particular interest to the readers, such as people from Amish, Hmong, and Somali heritages.

Miami, FL, USA Eric A. Fenkl
Albany, NY, USA Larry D. Purnell

# Introduction

Caring for culturally diverse patients and families has become commonplace throughout most of North America and worldwide. Geographical areas that previously were largely homogenous, with the exception of possibly one or two culturally diverse groups, no longer exist. This multicultural mosaic being created is adding richness to our communities and societies at large.

Many cultural groups fall into the category of vulnerable populations with increased health and health-care disparities. Even when disparities are not a concern, patient satisfaction may be of concern. If we are to decrease health and health-care disparities and increase patient satisfaction, more attention must be paid to cultural values, beliefs, and practices, which include spirituality and complementary and alternative therapies. If health-care providers and health-care delivery systems do not adapt and include diverse practices, patient and family care are compromised, resulting in not only poorer health outcomes but also an increase in cost. Recent efforts to promote global recognition of multiculturalism have encouraged health-care providers to become more aware of their own cultures and to the cultural differences among their patients in order to provide effective, acceptable, and safe health care. Each provider and organization must make a strong commitment to providing culturally competent care, regardless of the setting in which the care is provided and the backgrounds of the participants in care. As a result of this commitment, health-care organizations and educational programs have begun to recognize the need to prepare health-care professionals with the knowledge, skills, and resources essential to the provision of culturally competent care.

Culturally competent providers value diversity and respect individual differences regardless of one's race, religious beliefs, or ethnocultural background. The goal of this guide is to provide a framework to promote culturally competent care that respects each person's right to be understood and treated as a unique individual.

Educational programs and access to resources necessary for cultural competence are not always readily available in a format that is concise and easy to use. This guide has been developed to be used in a manner that can be easily accessed and understood. This approach to gaining essential knowledge using aggregate data is designed to help health-care professionals provide the highest quality care for patients and families of all backgrounds. Respect is essential to the development of dynamic interpersonal relationships that promote a positive influence on each person's interpretation of, and responses to, health care in a multicultural environment.

The real challenge for health-care providers is gaining timely access to concise information that facilitates an accurate understanding of cultural beliefs from the perspective of both the patient and the provider. This concise guide uses the Purnell Model as a framework to simplify health assessments and interventions accurately while enabling culturally relevant care. Although the primary focus of the handbook is on the needs of health-care providers in the United States, the organizing framework and major concepts presented in each chapter are relevant to multicultural health care throughout the world. The small size of this guide and the outline format of its presentation with recommended highlighted culturally relevant interventions make it an essential resource for all health-care providers who are committed to providing culturally competent care, regardless of their practice setting.

# Contents

# Chapter 1
# Transcultural Diversity and Health Care

**Larry D. Purnell and Eric A. Fenkl**

## 1.1 The Need for Culturally Competent Health Care

The need for cultural competence is increasingly recognized by health-care providers and health-related organizations in the United States and globally. The social ideology of a melting pot has been replaced by recognizing that people deserve respect within their cultural framework and as individuals. The literature on health and health-care disparities across ethnic, social, and economic groups continues to demonstrate compelling evidence for health-care providers and health-care organizations to be attentive to cultural diversity and inclusion as well as cultural competency. Major goals of Healthy People 2030 that impact on health disparities, with culture being one force to help eliminate them, are to (a) attain high-quality, longer lives free of preventable disease, disability, injury, and premature death; (b) achieve health equity, eliminate disparities, and improve the health of all groups; (c) create social and physical environments that promote good health for all; and (d) promote quality of life, healthy development, and healthy behaviors across all life stages.

Culture is defined as the totality of socially transmitted behavioral patterns, beliefs, values, customs, lifeways, arts, and all other products of human work and thought characteristics of a population of people that guide their worldview and decision-making. These patterns may be explicit or implicit, are primarily learned and transmitted within the family, are shared by most members of the culture, and are emergent phenomena that change in response to global phenomena and

L. D. Purnell (✉)
College of Health and Nursing Sciences, Excelsior University, Albany, NY, USA
e-mail: lpurnell@udel.edu

E. A. Fenkl
Nicole Wertheim College of Nursing and Health Sciences, Florida International University, Miami, FL, USA
e-mail: efenkl@fiu.edu



E. A. Fenkl, L. D. Purnell (eds.), *Handbook for Culturally Competent Care*,
https://doi.org/10.1007/978-3-031-70492-5_1

technological advancements. Culture is learned first in the family, then in school, and then in the community and other social organizations. Culture is largely unconscious and has powerful influences on health and illness beliefs and treatment. Health-care providers must recognize, respect, and integrate patients' cultural beliefs and practices into health prescriptions to help eliminate or mitigate health disparities and provide patient satisfaction.

This handbook describes aggregate data on the dominant cultural characteristics of selected ethnocultural groups and provides a guide for assessing cultural values, beliefs, and practices. Based on individuality and the **variant cultural characteristics of culture** listed below, aggregate data will **not fit every individual in a cultural group.** Health-care providers who understand their own cultures and their patients' cultural values, beliefs, and practices are in a better position to interact with their patients and provide culturally acceptable care that increases opportunities for health promotion and wellness; illness, disease, and injury prevention; and health maintenance and restoration. To this end, health-care providers need both cultural general and specific cultural knowledge. The more one knows about a cultural group, the more targeted will be the assessment and interventions. Without specific knowledge of cultural groups for whom they provide care, health-care providers might not know what questions to ask to provide culturally competent care. However, any generalization made from aggregate data about the behaviors of a cultural group or individual is almost certain to be an oversimplification; within all cultures are subcultures, ethnic groups, and individuals who do not adhere to all the values of their dominant culture and vary according to variant cultural characteristics as identified later in this chapter. Subcultures, ethnic groups, and ethnocultural populations are groups of people who have experiences different from those of the dominant culture with which they identify. For this Guide, subcultures are defined as a group of people from several cultures who join together for an array of reasons. For example, a support group for substance abusers, bikers, or veterans may have members from European American, Korean, and Panamanian cultures; they create their own subculture within their dominant culture.

Culture has a powerful unconscious impact on health professionals. Each health-care provider adds a unique dimension to the complexity of providing culturally competent care. The way health-care providers perceive themselves as competent providers is often reflected in the way they communicate with patients. Thus, it is essential for health providers to think about themselves, their behaviors, and their communication styles in relation to their perceptions of different cultures. Before addressing the multicultural backgrounds and unique individual perspectives of each patient, health-care professionals must first address their own personal and professional knowledge, values, beliefs, ethics, and life experiences in a manner that optimizes assessment of and interactions with patients who come from a culture different from that of the health-care provider. Self-awareness in cultural competence is a deliberate and conscious cognitive and emotional process of getting to know oneself: one's own personality, values, beliefs, professional knowledge, standards, ethics, and the impact of these factors on the various roles one plays when interacting with individuals who are different from oneself. The ability to understand oneself sets the stage for integrating new knowledge related to cultural differences into the professional's knowledge base and perceptions of health interventions.

Culture and race are not synonymous. However, the controversial term *race* must be addressed when learning about culture. Race is genetic and includes physical characteristics that are similar among members of the same group, such as skin color, blood type, and hair and eye color. Although there is less than a 1% genetic difference among the races, this difference is significant when conducting health assessments, determining genetic conditions, and prescribing medications and treatments, as outlined in culturally specific chapters that follow. People from a given racial group may, but do not necessarily, share a common culture. Perhaps the most significant aspect of race is social in origin; race can either decrease or increase opportunities, depending on the environmental context.

## 1.2 Variant Characteristics of Culture

Major attributes that shape peoples' worldview and the extent to which people identify with their cultural group of origin are called variant characteristics of culture. Some characteristics may change while others remain stable over life. Variant characteristics include:

- nationality,
- race,
- skin color,
- gender,
- age,
- religious affiliation,
- educational status,
- socioeconomic status,
- occupation,
- military experience,
- political beliefs,
- urban versus rural residence,
- enclave identity,
- marital status,
- parental status,
- physical characteristics,
- sexual orientation,
- gender issues,
- health literacy,
- length of time away from the country of origin, and
- reason for migration (sojourner, immigrant, or undocumented status).

Immigration status influences a person's worldview. For example, people who voluntarily immigrate generally acculturate and assimilate more easily. Sojourners who immigrate with the intention of remaining in their new homeland for only a short time or refugees who think they may return to their home country may not have the need or desire to acculturate or assimilate. Additionally, undocumented individuals (illegal immigrants) may have a different worldview from those who have arrived legally. Some variant characteristics of culture may be fluid and should not be seen as categorically imperative.

The literature reports many definitions for the terms *cultural awareness, cultural sensitivity*, and *cultural competence*. Sometimes, these definitions are used interchangeably. However, cultural awareness has more to do with an appreciation of the external or material signs of diversity, such as the arts, music, dress, or physical characteristics. Cultural sensitivity has more to do with personal attitudes and not saying things that might be offensive to someone from a cultural or ethnic background different from that of the health-care provider. Increasing one's consciousness of cultural diversity improves the possibilities for health-care providers to provide culturally competent care. Cultural competence, as used in this book, means:

1. Developing an awareness of one's own existence, sensations, thoughts, and environment without letting them have an undue influence on those from other backgrounds.
2. Demonstrating knowledge and understanding of the patient's culture, health-related needs, and culturally specific meanings of health and illness.
3. Continuing to learn about cultures of patients to whom one provides care.
4. Recognizing that the variant cultural characteristics determine the degree to which patients adhere to the beliefs, values, and practices of their dominant culture.
5. Accepting and respecting cultural differences in a manner that facilitates the patient's and family's abilities to make decisions to meet their needs and beliefs.
6. Not assuming that the health-care provider's beliefs and values are the same as the patient's.
7. Resisting judgmental attitudes such as "different is not as good."
8. Being open to cultural encounters.
9. Being comfortable with cultural encounters.
10. Adapting care to be congruent with the patient's culture.
11. Engaging in cultural competence is a conscious process and not necessarily a linear one.
12. Accepting responsibility for one's own education in cultural competence by attending conferences, reading professional literature, and observing cultural practices.

As of April 2020, the census recorded a resident population of 331,449,281 in the 50 states and the national capital of Washington, D.C., reflecting an increase of 7.4%, or 22,703,743, over that of 2010 (U.S. Department of Commerce: U.S. Census Bureau 2022). Also, as of 2022, nearly 46.2 million immigrants lived in the United States, the most in U.S. history.

Table 1.1 presents the diversity of the U.S. population.

The census collects data on two categories: race and ethnicity; therefore, these numbers total more than 100%. Race and Hispanic origin are two separate and distinct categories. Race categories as used in Census 2020 include the following:

1. *White* refers to people having origins in any of the original peoples of Europe, the Near East, the Middle East, and North Africa. This category includes Irish, German, Italian, Lebanese, Turkish, Arab, and Polish.
2. *Black* and *African American* refer to people having origins in any of the black racial groups of Africa and include Nigerians and Haitians and any person who self-designated this category regardless of origin.
3. *American Indian* and *Alaskan Native* refer to people having origins in any of the original peoples of North, South, and Central America and who maintain tribal affiliation or community attachment.
4. *Asian* refers to people having origins in any of the original peoples of the Far East, Southeast Asia, and the Indian subcontinent. This category includes the terms *Asian Indian, Chinese, Filipino, Korean, Japanese, Vietnamese, Burmese, Hmong, Pakistani*, and *Thai.*
5. *Native Hawaiian* and other *Pacific Islander* refer to people having origins in any of the original peoples of Hawaii, Guam, Samoa, Tahiti, the Mariana Islands, and Chuuk.
6. *Some other race* was included for people who are unable to identify with the other categories. Additionally, the respondent could identify, as a write-in, with two races (U.S. Department of Commerce: U.S. Census Bureau 2022).

**Table 1.1** Diversity of the U.S. population as of 2021

| | |
|---|---|
| White | 59.3% |
| Hispanic or Latino origin | 18.9% |
| Black and African American | 12.6% |
| Asian | 5.9% |
| Persons with two or more races | 2.3% |
| American Indian/Alaskan Native | 0.7% |
| Some other race | 0.5% |
| Native Hawaiian and other Pacific Islanders | 0.2% |

Data from "US population by year, race, age, ethnicity, & more". USAFacts. May 18, 2023. Retrieved June 4, 2024 from https://usafacts.org/data/topics/people-society/population-and-demographics/our-changing-population/

## 1.3 Reflective Exercises

1. What changes in ethnic and cultural diversity have you seen in your community over the last 5 years? Over the last 10 years? Have you had the opportunity to interact with newer groups?
2. What health disparities have you observed in your community? To what do you attribute these disparities? What can you do as a professional to help decrease these disparities?
3. Who in your family had the most influence in teaching you cultural values and practices? Mother, father, grandparent?
4. What activities have you done to increase your cultural competence?
5. Given that everyone is ethnocentric to some degree, what do you do to become less ethnocentric?
6. How do you distinguish a stereotype from a generalization?
7. How have your variant characteristics of culture changed over time?
8. Distinguish between a stereotype and a generalization.
9. What ethnic and racial groups do you encounter on a regular basis? Do you see any racism or discrimination among these groups?
10. What does your organization do to increase diversity and cultural competence?

## Bibliography

U.S. Department of Commerce: U.S. Census Bureau (2022) https://www.commerce.gov/bureaus-and-offices/census

U.S. Department of Health and Human Services (2024) https://health.gov/healthypeople

US population by year, race, age, ethnicity, & more. USAFacts. 18 May 2023. https://usafacts.org/data/topics/people-society/population-and-demographics/our-changing-population/. Accessed 4 Jun 2024

# Chapter 2
# The Purnell Model for Cultural Competence

**Larry D. Purnell**

## 2.1 The Purnell Model

The Purnell Model for Cultural Competence and its organizing framework can be used in all practice settings and by all health-care providers. Moreover, it has been classified a holographic and complexity grand theory. In fact, even mediators with the American Bar Association have found the model valuable. The model is a circle, with an outlying rim representing global society, a second rim representing community, a third rim representing family, and an inner rim representing the person (Fig. 2.1). The interior of the circle is divided into 12 pie-shaped wedges depicting cultural domains (constructs) and their associated concepts. The dark center of the circle represents unknown phenomena. Along the bottom of the model is a jagged line representing the nonlinear concept of cultural consciousness. The 12 cultural domains and their concepts provide the organizing framework. Each domain includes concepts that need to be addressed when assessing patients in various settings. Moreover, health-care providers can use these same concepts to better understand their own cultural beliefs, attitudes, values, practices, and behaviors. An important concept to understand is that no single domain stands alone; they are all interconnected. The 12 domains are overview/heritage, communications, family roles and organization, workforce issues, biocultural ecology, high-risk health behaviors, nutrition, pregnancy and the childbearing family, death rituals, spirituality, health-care practices, and health-care practitioners.

L. D. Purnell (✉)
College of Health and Nursing Sciences, Excelsior University, Albany, NY, USA
e-mail: lpurnell@udel.edu


E. A. Fenkl, L. D. Purnell (eds.), *Handbook for Culturally Competent Care*,
https://doi.org/10.1007/978-3-031-70492-5_2

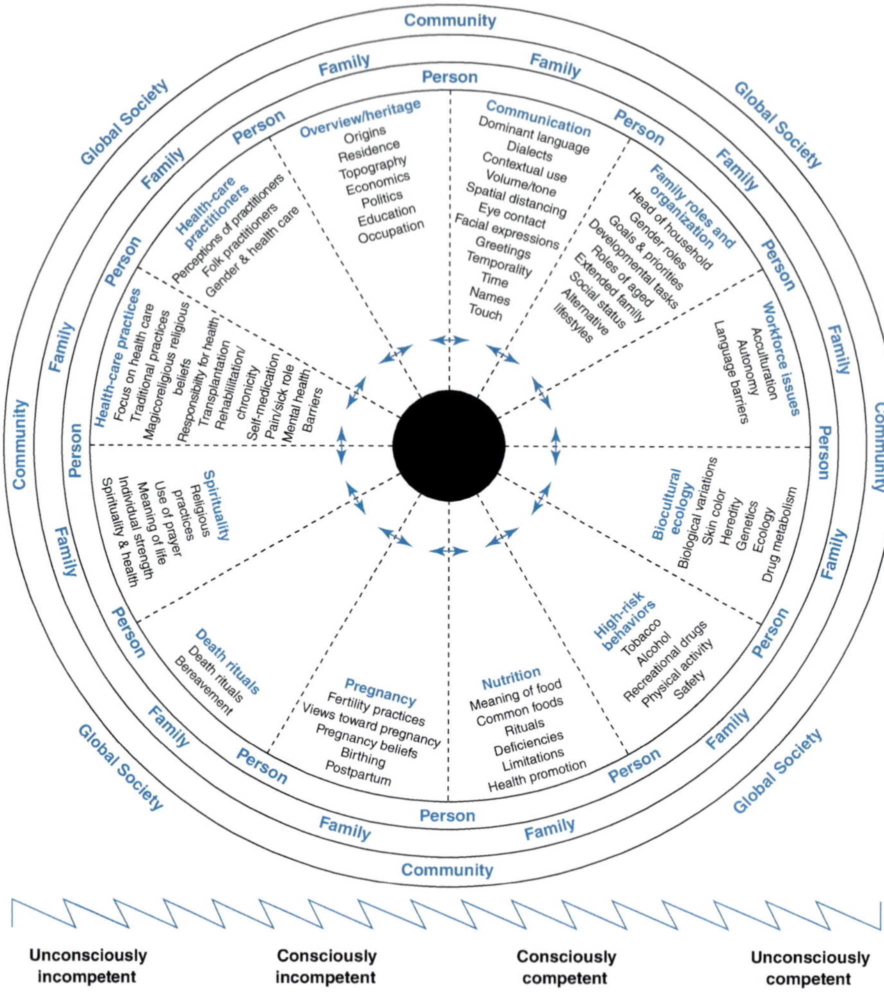


Primary characteristics of culture: age, generation, nationality, race, color, gender, religion

Secondary characteristics of culture: educational status, socioeconomic status, occupation, military status, political beliefs, urban versus rural residence, enclave identity, marital status, parental status, physical characteristics, sexual orientation, gender issues, and reason for migration (sojourner, immigrant, undocumented status)

Unconsciously incompetent: not being aware that one is lacking knowledge about another culture
Consciously incompetent: being aware that one is lacking knowledge about another culture
Consciously competent: learning about the client's culture, verifying generalizations about the client's culture, and providing culturally specific interventions
Unconsciously competent: automatically providing culturally congruent care to clients of diverse cultures

Model created by Larry D. Purnell, PhD, RN, FAAN. Reprinted with permission.

**Fig. 2.1** The Purnell model for cultural competence

### 2.1.1 Assessment Guide

Following each of the domains and concepts is a box that includes suggested questions to ask and observations to make when assessing patients from a cultural perspective. It is recognized that clinicians are not able to complete a thorough cultural assessment for every patient. The list of questions is extensive; thus, the clinician must determine which questions to ask according to the patient's presenting symptoms, teaching needs, and the potential impact of culture. Furthermore, cultural values and beliefs might change over time.

### 2.1.2 Domains and Concepts

#### 2.1.2.1 Overview and Heritage

Includes concepts related to the country of origin and current residence; the effects of the topography of the country of origin and the current residence on health, economics, politics, reasons for migration, educational status, and occupations both present and past. See table below.

#### 2.1.2.2 Communications

Includes concepts related to the dominant language, dialects, and the contextual use of the language; paralanguage variations such as voice volume, tone, intonations, inflections, and willingness to share thoughts and feelings; nonverbal communications such as eye contact, gesturing and facial expressions, use of touch, body language, spatial distancing practices, and acceptable greetings; temporality in terms of past, present, and future orientation of worldview; clock versus social time; and the amount of formality in use of names. Differences between the language spoken by the health-care provider and the patient, educational level, and health literacy are additional communication add to the communication difficulties. Effective communication is the first and probably the most important aspect of obtaining an accurate assessment. See table below.

#### 2.1.2.3 Family Roles and Organization

Includes concepts related to the head of the household, gender roles (a product of biology and culture), family goals and priorities, developmental tasks of children and adolescents, roles of the aged and extended family, individual and family social status in the community, and acceptance of alternative lifestyles such as single parenting, same sex partnerships and marriage, childless marriages, and divorce. See table below.

#### 2.1.2.4 Workforce Issues

Includes concepts related to autonomy, acculturation, assimilation, gender roles, ethnic communication styles, and health-care practices of the country of origin. See table below.

#### 2.1.2.5 Biocultural Ecology

Includes physical, biological, and physiological variations among ethnic and racial groups such as skin color (the most evident) and physical differences in body habitus; genetic, hereditary, endemic, and topographical diseases; psychological makeup of individuals; and the physiological differences that affect the way drugs are metabolized by the body. In general, most diseases and illnesses can be divided into three categories: lifestyle, environment, and genetics.

*Lifestyle* causes include cultural practices and behaviors that can generally be controlled: for example, smoking, diet, and stress.

*Environment* causes refer to the external environment (e.g., air and water pollution) and situations over which the individual has little or no control (e.g., presence of malarial mosquitos, exposure to chemicals and pesticides, access to care, and associated diseases and illnesses). See table below.

*Genetic* conditions are caused by genes.

#### 2.1.2.6 High-Risk Health Behaviors

Includes substance use and misuse of tobacco, alcohol, and recreational drugs; lack of physical activity; increased calorie consumption; nonuse of safety measures such as seat belts, helmets, and safe driving practices; and not taking safety measures to prevent contracting HIV and sexually transmitted infections. See table below.

#### 2.1.2.7 Nutrition

Includes the meaning of food, common foods and rituals, nutritional deficiencies and food limitations, and the use of food for health promotion and wellness, and illness and disease prevention. Multiple diseases and illnesses are a consequence of this major cultural component. See table below.

#### 2.1.2.8 Pregnancy and Childbearing Practices

Includes culturally sanctioned and unsanctioned fertility practices; views on pregnancy; and prescriptive, restrictive, and taboo practices related to pregnancy, birthing, the postpartum period, and family involvement. See table below.

#### 2.1.2.9 Death Rituals

Includes how the individual and the society view death and euthanasia, rituals to prepare for death, burial practices, and bereavement behaviors. Death rituals are slow to change. See table below.

#### 2.1.2.10 Spirituality

Includes formal religious beliefs related to faith and affiliation and the use of prayer; behavior practices that give meaning to life; and individual sources of strength. See the following table.

#### 2.1.2.11 Health-Care Practices

Includes the focus of health care (acute versus preventive); traditional, magico-religious, and biomedical beliefs and practices; individual responsibility for health; self-medicating practices; views on mental illness, chronicity, and rehabilitation; acceptance of blood and blood products; and organ donation and transplantation. See table below.

#### 2.1.2.12 Health-Care Practitioners

Includes the status, use, and perceptions of traditional; magico-religious, and biomedical health-care providers; and the gender of the health-care provider. See table below.

## 2.2 Purnell Cultural Assessment Tool

An extensive cultural assessment is rarely completed in the clinical setting because of time and other circumstances. A seasoned clinical practitioner will know when further assessment is required. Thus, this tool should be used as a guide. Furthermore, cultures change over time. *Shaded items in italics are part of any standard assessment. Other items may also be part of a standard assessment, depending on the organization, setting, and clinical area.*

| Cultural assessment tool | Comments |
|---|---|
| **Overview, inhabited localities, and topography** | |
| *Where do you currently live?* | |
| What is your ancestry? | |
| Where were you born? | |

| Cultural assessment tool | Comments |
|---|---|
| How many years have you lived in the United States (or other country, as appropriate)? | |
| Were your parents born in the United States (or other country, as appropriate)? | |
| What brought you (your parents/ancestors) to the United States (or other country, as appropriate)? | |
| Describe the land or countryside where you live. Is it mountainous, swampy, etc.? | |
| Have you lived in other places in the United States/world? | |
| What was the land or countryside like when you lived there? | |
| What is your income level? | |
| Does your income allow you to afford the essentials of life? | |
| Do you have health insurance? | |
| Are you able to afford health insurance on your salary? | |
| What is your educational level (formal/informal/self-taught)? | |
| *What is your current occupation? If retired, ask about previous occupations* | |
| Have you worked in other occupations? What were they? | |
| Are there (were there) any health hazards associated with your job(s)? | |
| Have you been in the military? If so, in what foreign countries were you stationed? | |
| **Communications**<br>How do you wish to be greeted? | |
| *What is your full name?* | |
| *What is your legal name?* | |
| *By what name do you wish to be called?* | |
| *What is your primary language?* | |
| Do you speak a specific dialect? | |
| What other languages do you speak? | |
| Do you find it difficult to share your thoughts, feelings, and ideas with family? Friends? Health-care providers? | |
| Do you mind being touched by friends? Strangers? Health-care workers? | |
| How do wish to be greeted? Handshake? Nod of the head, verbal greeting only etc.? | |
| Are you usually on time for appointments? | |
| Are you usually on time for social engagements? | |
| Observe the patient's speech pattern. Is the speech pattern high- or low-context? Note: Patients from highly contexted cultures place greater value on silence | |
| Observe the patient when physical contact is made. Does he/she withdraw from the touch or become tense? | |
| How close does the patient stand when talking with family members? With health-care providers? | |
| Does the patient maintain eye contact when talking with the nurse/physician/etc.? | |
| **Family roles and organization** | |
| What is your *marital*/partner status? | |
| *How many children do you have?* | |
| Who makes most of the decisions in your family? | |
| What types of decisions do(es) the female(s) in your family make? | |
| What types of decisions do(es) the male(s) in your family make? | |

| Cultural assessment tool | Comments |
|---|---|
| What are the duties of the women in the family? | |
| What are the duties of the men in the family? | |
| What should children do to make a good impression for themselves and for the family? | |
| What should children *not* do to make a good impression for themselves and for the family? | |
| What are children forbidden to do? | |
| What should adolescents do to make a good impression for themselves and for the family? | |
| What should adolescents *not* do to make a good impression for themselves and for the family? | |
| What are adolescents forbidden to do? | |
| What are the priorities for your family? | |
| What are the roles of the older people in your family? Are they sought for their advice? | |
| Are there extended family members in your household? Who else lives in your household? | |
| What are the roles of extended family members in this household? What gives you and your family status? | |
| Is it acceptable to you for people to have children out of wedlock? | |
| Is it acceptable to you for people to live together and not be married? | |
| Is it acceptable to you for people to admit being gay or lesbian? | |
| What is your sexual preference/orientation? (ask only if appropriate and later in the assessment after a modicum of trust has been established) | |
| **Workforce issues** | |
| Do you usually report to work on time? | |
| Do you usually report to meetings on time? | |
| What concerns do you have about working with someone of the opposite gender? | |
| Do you consider yourself a "loyal" employee? | |
| How long do you expect to remain in your position? | |
| What do you do when you do not know how to do something related to your job? | |
| Do you consider yourself to be assertive in your job? | |
| What difficulty does English (or another language) give you in the workforce? | |
| What difficulties do you have working with people older (younger) than you? | |
| What difficulty do you have in taking directions from someone younger/older than you? | |
| What difficulty do you have working with people whose religions are different from yours? | |
| What difficulty do you have working with people whose sexual orientation is different from yours? | |
| What difficulty do you have working with someone whose race or ethnicity is different from yours? | |
| Do you consider yourself to be an independent decision maker? | |
| **Biocultural ecology** | |
| *Are you allergic to any medications?* | |
| *What problems did you have when you took over-the-counter medications?* | |

| Cultural assessment tool | Comments |
|---|---|
| *What problems did you have when you took prescription medications?* | |
| *What are the major illnesses and diseases in your family?* | |
| *Are you aware of any genetic diseases in your family?* | |
| What are the major health problems in the country from which you come (if appropriate)? | |
| *With what races and ethnic groups(s) do you identify?* | |
| *Observe and document skin coloration and physical characteristics* | |
| *Observe for and document physical handicaps and disabilities* | |
| **High-risk health behaviors** | |
| How many cigarettes a day do you smoke?<br>Do you smoke a pipe (or cigars)? | |
| Do you chew tobacco? | |
| *For how many years have you smoked/chewed tobacco?* | |
| *How much alcohol do you drink each day? Ask about wine, beer, spirits?* | |
| How many energy drinks do you consume each day | |
| *What recreational drugs do you use?* | |
| *How often do you use recreational drugs?* | |
| *What type of exercise do you do each day?* | |
| Do you use seat belts? | |
| What precautions do you take to prevent getting a sexually transmitted infections or HIV/AIDS? | |
| **Nutrition** | |
| *Are you on a special diet?* | |
| Are you satisfied with your weight? | |
| *Which foods do you eat to maintain your health?* | |
| *Do you avoid certain foods to maintain your health?* | |
| Why do you avoid these foods? | |
| Which foods do you eat when you are ill? | |
| Which foods do you avoid when you are ill? | |
| Why do you avoid these foods (if appropriate)? | |
| For what illnesses do you eat certain foods? | |
| Which foods do you eat to balance your diet? | |
| Which foods do you eat every day? | |
| Which foods do you eat every week? | |
| Which foods do you eat that are part of your cultural heritage? | |
| Which foods are high-status foods in your family/culture? | |
| Which foods are eaten only by men? Women? Children? Teenagers? Elderly? | |
| How many meals do you eat each day? | |
| What time do you eat each meal? | |
| *Do you snack between meals?* | |
| *What foods do you eat when you snack?* | |
| What holidays do you celebrate? | |
| Which foods do you eat on particular holidays? | |
| Who is present at each meal? Is the entire family present? | |

| Cultural assessment tool | Comments |
|---|---|
| Do you primarily eat the same foods as the rest of your family? | |
| Where do you usually buy your food? | |
| Who usually buys the food in your household? | |
| Who does the cooking in your household? | |
| How frequently do you eat at a restaurant? | |
| When you eat at a restaurant, in what type of restaurant do you eat? | |
| Do you eat foods left from previous meals? | |
| Where do you keep your food? | |
| Do you have a refrigerator? | |
| How do you cook your food? | |
| How do you prepare meat? | |
| How do you prepare vegetables? | |
| What type of spices do you use? | |
| What do you drink with your meals? | |
| Do you drink special teas? | |
| *Do you have any food allergies?* | |
| *Are there certain foods that cause you problems when you eat them?* | |
| How does your diet change with each season? | |
| Are your food habits different on days you work versus when you are not working? | |
| **Pregnancy and childbearing practices** | |
| *How many children do you have?* | |
| What do you use for birth control? | |
| What does it mean to you and your family when you are pregnant? | |
| What special foods do you eat when you are pregnant? Looking for pica | |
| What foods do you avoid when you are pregnant? | |
| What activities do you avoid when you are pregnant? | |
| Do you do anything special when you are pregnant? | |
| Do you eat non-food substances when you are pregnant? | |
| *Who do you want with you when you deliver your baby?* | |
| In what position do you want to be when you deliver your baby? | |
| What special foods do you eat after delivery? | |
| What foods do you avoid after delivery? | |
| What activities do you avoid after you deliver? | |
| Do you do anything special after delivery? | |
| Who will help you with the baby after delivery? | |
| What bathing restrictions do you have after you deliver? | |
| Do you want to keep the placenta? | |
| What do you do to care for the baby's umbilical cord? | |
| **Death rituals** | |
| What special activities need to be performed to prepare for death? | |
| What special activities need to be performed after death?<br>Who do you want to be with you when dying? | |
| Would you want to know about your impending death? | |
| What is your preferred burial practice? Interment, cremation? | |

| Cultural assessment tool | Comments |
|---|---|
| How soon after death does burial occur? | |
| How do men grieve? | |
| How do women grieve? | |
| What does death mean to you? | |
| Do you believe in an afterlife? | |
| Are children included in death rituals? | |
| **Spirituality** | |
| *What is your religion?* | |
| Do you consider yourself deeply religious? | |
| How many times a day do you pray? | |
| What do you need in order to say your prayers? | |
| Do you practice meditation such as TM, mindfulness, meditation?<br>What gives strength and meaning to your life? | |
| In what spiritual practices do engage for your physical and emotional status? | |
| **Health-care practices** | |
| *In what prevention activities do you engage to maintain your health?* | |
| Who in your family takes responsibility for your health? | |
| Who takes care of family members when they are sick? | |
| *What over-the-counter medicines do you use?* | |
| What herbal teas and folk medicines do you use? | |
| For what conditions do you use herbal medicines? | |
| What do you usually do when you are in pain? | |
| How do you express your pain? | |
| How are people in your culture viewed or treated when they have a mental illness? | |
| How are people with physical disabilities treated in your culture? | |
| What do you do when you are sick? Stay in bed, continue your normal activities, etc.? | |
| What are your beliefs about rehabilitation? | |
| How are people with chronic illnesses viewed or treated in your culture? | |
| *Are you averse to blood transfusions?* | |
| *Is organ donation acceptable to you?* | |
| *Are you listed as a potential organ donor?* | |
| Would you consider having an organ transplant if needed? | |
| Are health-care services readily available to you? | |
| Do you have transportation problems accessing needed health-care services? | |
| Do you feel welcome when you see a health-care professional? | |
| What traditional health-care practices do you use? For example, mineral baths, sweating, acupuncture, acupressure, *cai gao*, coining, *moxibustion*, aromatherapy, etc.? | |
| What home difficulties do you have that might prevent you from receiving health care? | |

| Cultural assessment tool | Comments |
|---|---|
| **Health-care practitioners** | |
| What health-care providers do you see when you are ill? Physicians, nurses? | |
| Do you prefer a same-sex health-care provider for routine health problems? For intimate care? | |
| What healers do you use beside physicians and nurses? | |
| For what conditions do you use healers? | |

## 2.3 Reflective Exercise

Communication is probably the most import aspect of obtaining a good assessment. The onus of effective communication primarily lies with the health-care provider. Thus, providers must understand their own communication styles. When your personal communication practices differ from what is in the scholarly literature, posit why. See the variant cultural characteristics as a guide. Speak to each one of the following points:

1. Identify your cultural ancestry. If you have more than one cultural ancestry, chose the one with which you most closely associate.
2. Explore the willingness of individuals in your culture to share thoughts, feelings, and ideas. Identify any area of discussion that would be considered taboo?
3. Explore the practice and meaning of touch in your culture. Include information regarding touch between family members, friends, members of the opposite sex, and health-care providers.
4. Identify personal spatial and distancing strategies used when communicating with others in your culture. Discuss differences between friends and families versus strangers.
5. Discuss your culture's use of eye contact. Include information regarding practices between family members, friends, strangers, and persons of different age groups.
6. Explore the meaning of gestures and facial expressions in your culture. Do specific gestures or facial expressions have special meanings? How are emotions displayed?
7. Are there acceptable ways of standing and greeting people in your culture?
8. Discuss the prevailing temporal relation of your culture. Is the culture's worldview past, present, or future oriented? Temporality also include punctuality.
9. Discuss the impact of your culture on your nursing and/or health care. Be specific, not something that is very general.

# Chapter 3
# Barriers to Culturally Competent Health Care

**Larry D. Purnell and Eric A. Fenkl**

## 3.1 Introduction

Barriers to culturally competent health care are defined as obstacles or hindrances that make it difficult to obtain or access good health or health care. Despite efforts and goals in the United States and many other countries around the world to reduce or eliminate obstacles, significant disparities continue, especially with what the professional literature refers to as vulnerable populations. Vulnerable populations include the following:

- the economically disadvantaged
- racial and ethnic minorities (including those of mixed race or ethnicities)
- the uninsured and underinsured
- older adults
- children
- the homeless
- people with human immunodeficiency virus (HIV)
- people who are stigmatized for whatever reason, which includes substance misuse
- people with chronic health conditions and severe mental illness
- rural populations
- people with low acculturation and low levels of education
- gender disparities

L. D. Purnell (✉)
College of Health and Nursing Sciences, Excelsior University, Albany, NY, USA
e-mail: lpurnell@udel.edu

E. A. Fenkl
Nicole Wertheim College of Nursing and Health Sciences, Florida International University, Miami, FL, USA
e-mail: efenkl@fiu.edu



E. A. Fenkl, L. D. Purnell (eds.), *Handbook for Culturally Competent Care*,
https://doi.org/10.1007/978-3-031-70492-5_3

- refugees and undocumented immigrants
- distance to a health-care facility
- language difficulties, and an interpreter is not available.

Overall, these disparities are due to environmental, psychological, and social factors; maldistribution or shortage of human and fiscal resources; and lack of access to health-care systems. From this list, it seems clear that under the right set of circumstances, any group or individual can be vulnerable.

Barriers can be categorized in several domains that are presented in this chapter: language and health literacy, availability, accessibility, affordability, appropriateness, accountability, adaptability, acceptability, awareness, attitudes, approachability, alternative practices, and additional services. It should be noted that these domains are not categorically imperative or perfectly distinct categories; they can overlap.

## 3.2 Language and Health Literacy

Language and health literacy barriers include the medical jargon used by health-care providers, inadequate reading level of the patient, or lack of fluency in English or in the patient's mother tongue. Several studies in the United States have identified that a lack of fluency in language is the primary barrier to receiving adequate health care, and not just for people for whom English is a second language.

Language barriers may involve both oral and written language. Interpretation refers to oral communication, while translation refers to written communication. Interpretation requires more than word-for-word substitution. Professional interpreters must not only demonstrate bilingual proficiency, but they must also be knowledgeable of idioms, slang, and colloquialisms as well as medical terminology. Sign language may be a concern for some because there is no standard worldwide sign language. Besides American Sign Language, there are multiple Arabic sign languages. Language and literacy barriers may result in the following:

- Language barriers can negatively affect perceptions of patient and provider care, resulting in patients not returning for follow-up care.
- Language barriers may result in a patient leaving against medical advice. A patient who perceives that he or she is not understood by the health-care provider or who does not understand a provider who is using medical technology may simply leave.
- Limited English proficiency (LEP) can result in unsafe situations caused by poor patient comprehension of their medical condition, treatment plan, or discharge instructions; an inaccurate and incomplete medical history being obtained by the health-care provider; ineffective or improper use of medications or serious medication errors; improper preparation for tests and procedures; and poor or inadequate informed consent.

- **Be alert to comments and behaviors from patients and family members that might mean they do not understand the provider. Such cues include "Can I take these forms home to complete?" or "I forgot my glasses. I'll wait until (friend or family member) comes to help me."**
- LEP patients who do not receive professional interpretation services at admission and discharge have longer lengths of stay and higher readmission rates compared with patients who receive professional interpretation services.
- Just because someone speaks the language does not necessarily mean he or she has the skills to be successful at interpretation.
- Improper interpretation is prone to additions, substitutions, omissions, volunteer opinions, and sematic errors.
- Compared with professional medical interpreters, ad hoc interpreters such as patients' family members or house staff frequently make medical interpretation errors. These errors are significantly more likely to have clinical consequences.
- **Use only certified interpreters when at all possible.**
- **Use dialect specific interpreters whenever possible.**
- **Create a system for interpretation and translation.**
- **Compile a list of commonly used words in the dominant languages of the patients using the facility. This list does not replace the need for an interpreter but may be helpful if a professional interpreter is not available.**
- **Allow time for interpretation.**
- **Be aware that social class differences can affect interpretation.**
- **Maintain eye contact with both the patient and the interpreter.**
- Non-professional interpretation and using children or other family members can lead to a breach of confidentiality. Also, patients might not speak freely, especially regarding issues such as sexuality, drug and alcohol misuse, and domestic violence.
- **Do not place the patient in a precarious situation by asking the patient in front of the family if it is acceptable for a family member or friend to interpret.**
- **Do not use family members except in nonconfidential situations such as obtaining demographic and admitting data and teaching about medication administration and dietary requirements.**
- **Do not use individuals known to the family as interpreters due to confidentiality issues.**
- Language barriers may involve written as well as oral communication. It is therefore important to provide written materials in the patient's native language.
- **Translate satisfaction surveys into the languages of the population served.**
- **Translate treatment plans and medication requirements into the languages of the patients who come to the facility.**
- **Translate pamphlets on common illnesses and diseases into the languages of the patients who come to the facility.**
- **Ensure that translated material is at a grade 6 reading level.**
- **Create pamphlets on high-risk behaviors and occupations that are high risk.**

## 3.3 Availability

Availability is defined as a health-related system offering a needed service and doing so at a time that is reasonable. Patients are routinely referred to Emergency Departments for care when health-care providers' offices and clinics are closed, adding to patient dissatisfaction and increased cost to the patient and health-care system. Pharmacy hours can be severely restricted in the evening and at night, especially in high crime unsafe neighborhoods, making their services unavailable.

- **Create fast-track systems with expanded hours of services in Emergency Departments that can be staffed by advanced practice nurses.**
- **Open satellite urgent care centers making services more readily available and taking some of the workload off the Emergency Department.**
- **Provide patients with a temporary supply of medications until the local pharmacy is open.**

## 3.4 Accessibility

Accessibility is defined as the ability of the patient to actually get to a service when needed. Transportation services may not be available, or challenging road conditions and rivers and mountains may make it difficult for people to obtain needed health-care services when no health-care provider is available in their immediate region. Although federal law requires public facilities to be accessible to people with disabilities, there are still private clinics where accessibility may be a problem. Extremely rural hospitals have critical access facilities receive cost-based reimbursement but access can still be a concern when the closest facility is 50+ miles away.

- **Make vans available at convenient locations and at specified times to transport patients to clinics.**
- **Provide guidance for accessing services and navigating the health-care delivery system by advertising in local newspapers, in newsletters, and on public transportation systems. Local convenience stores and restaurants are additional sites to advertising services.**
- **Provide telephone triage systems where patients and families can talk with a nurse to determine the acuity of a health concern.**

## 3.5 Affordability

Affordability is defined as having the financial resources to access and use health-care services. The service might be available, but if patients do not have financial resources, they may wait until their health condition is more severe, resulting in longer recovery times and an increase in cost.

- **Partner with local philanthropic agencies to provide financial services.**
- **Involve social workers to get patients funds on a temporary basis.**

## 3.6 Appropriateness

Appropriateness is defined as having a service that is needed and congruent with the patient's and family's cultural belief system. Maternal and child services might be available, but there might be a greater need for geriatric and psychiatric services. Perhaps the critical access hospital does not have an infectious disease specialist.

- **Partner with larger organizations that have full-service capabilities to offer hours in rural areas on a regular basis.**
- **Offer full-service health promotion and wellness services, testing, and education a few days each week on a regular basis.**

## 3.7 Organizational Accountability

Organizational accountability is defined as health-care providers' seeking resources for their own education and learning about the cultures of the people they serve.

- **Provide cultural competence in orientation programs.**
- **Obtain local, regional, or national organizations and personnel to provide cultural competence workshops.**
- **Update staff annually on the practices, values, and health beliefs of the populations the organization serves.**
- **Provide staff with assessment forms and guides for culturally competent assessment and interventions.**
- **Incorporate cultural data on intake assessment forms.**
- **Provide print and online resources for staff.**

## 3.8 Adaptability

Adaptability is defined as the health-care environment being able to change and to offer additional services when needed. For example, a mother brings her child to the clinic for an immunization. Can she get a mammogram at the same time or must she make another appointment?

- **Provide several times during the week when full-service health promotion and wellness, services, testing, and education are available. Advertise these hours in local stores such as grocery stores, houses of worship, banks, post offices, and restaurants—including fast food restaurants.**

## 3.9 Acceptability

Acceptability is defined as being able to meet the requirements and provide satisfaction to a patient or family needing a health-care service. Are services and patient education offered in a language preferred by the patient?

- **Match sex of the patient and health-care provider whenever requested and if available.**
- **Provide ethnic concordant providers if requested whenever possible.**

## 3.10 Awareness

Awareness is defined as the patient and family being cognizant of needed health-care services. Is the patient aware that needed services exist in the community? The service may be available, but if patients are not aware of it, the service will not be used.

- **Advertise in local newspapers, in newsletters, and on public transportation systems services that are available in the community. Include convenient stores, grocery stores, houses of worship, libraries, and restaurants. Include advertisements in foreign languages where appropriate.**
- **Partner with local and regional radio and television stations (including foreign-language stations) that advertise the services that are available to patients in the community.**

## 3.11 Attitudes

Attitude is defined as a state of mind or feeling about a patient's or family's health beliefs and values. Adverse subjective beliefs and attitudes from caregivers increase the chance that the patient will not return for needed services until the condition is more severe. Do health-care providers have negative attitudes about patients' home-based traditional practices?

- **Do not tell patients that they are wrong for using home-based treatments. Explain problems that might occur.**
- **Do not let judgmental attitudes interfere with acceptance of patients and families.**

## 3.12 Approachability

Approachability is defined as the patient or family feeling it is easy to talk or deal with health-care providers. Do patients feel welcomed? Do health-care providers and receptionists greet patients in the manner in which they prefer?

- **Greet patients formally until told to do otherwise.**
- **Introduce yourself by name and position.**
- **Maintain eye contact without staring when greeting patients and families. Note that in some cultures, people do not maintain eye contact with people in higher social positions.**

## 3.13 Alternative Practices

Alternative practices are defined as complementary and alternative health-care practices and traditional and folk practices. While some practices are not harmful and may be helpful, some patients may overuse folk practices and folk practitioners before seeking allopathic care.

- **Incorporate nonharmful alternative practices into treatment plans.**
- **Partner with local folk healers and educate them on when referral to allopathic care is advisable.**
- **Refer patients to alternative folk practitioners when the condition (for example: evil eye) warrants it.**

## 3.14 Additional Services

Additional services are defined as value-added benefits that improve health-care access and adaptability. For example, are child- and adult-care services available if a parent must bring children or an aging parent with them for an appointment?

- **Network with local organizations and churches that can volunteer sitter services with professional oversight.**
- **Assist patients with navigating the intricacies of the health-care system.**
- **Assist patients and families with completing bureaucratic forms.**
- **Conduct focus groups to determine community needs.**

## 3.15 Reflective Exercises

1. What barriers do you see to culturally competent care in your organization? School, work, etc.
2. What language barriers to you see in your community such as English as a second language, educational levels, health literacy, etc.?
3. What affordability concerns for health care do you see in your community?
4. What complementary/alternative health-care practices do you use?
5. What complementary/alternative health-care practices are available in your community?
6. How do you want to be addressed? First name, last name with a title?
7. How do you address older people in your community? First name, last name with a title?
8. Is public transportation readily available to health-care services in your community? What might be done to improve them?
9. What does your organization do when a patient needs an interpreter?
10. In addition to English, what other languages are health-care instructions provided in your organization?

## Bibliography


Agency for Healthcare Quality and Research (2016) TeamSTEPPS enhancing safety for patients with limited English proficiency module. www.ahrq.gov/teamsteppstools/

Betancourt J, Green A, Carrillo JE, Ananeh-Firempong A (2003) Defining cultural competence: a practical framework for addressing racial/ethnic disparities in health and health care. Public Health Rep 118:293–302

Center for Health Care Strategies (2016) Reducing barriers to health care: practical strategies for local organizations. www.chcs.org

Flores G (2005) The impact of medical interpreter services on the quality of health care: a systematic review. Med Care Res Rev 62(3):255–299

Flores G, Laws MB, Mayo SJ, Zuckerman B, Abreu M, Medina L, Hardt EJ (2003) Errors in medical interpretation and their potential clinical consequences in pediatric encounters. Pediatrics 111(1):6–14

Giger J, Davidhizar R, Purnell L, Harden J, Phillips J, Strickland O (2007) American Academy of Nursing Expert Panel Report: developing cultural competence to eliminate health disparities in ethnic minorities and other vulnerable populations. J Transcult Nurs 18(2):95–102

Institute of Medicine of the National Academies (2016) Crossing the quality chasm. http://nationalacademies.org/hmd/reports/2001/crossing-the-quality-chasm-a-new-health-system-for-the-21st-century.aspx

Jacobs EA, Agger-Gupta N, Hm Chin A, Piorowski A, Hardt A (2016) Language barriers in healthcare settings. California Endowment. www.calendow.org

Office of Minority Health (2016) National Standards on Culturally and Linguistically Appropriate Services (CLAS). https://minorityhealth.hhs.gov/cultural-and-linguistic-competency

Purnell L (2013) Transcultural health care: a culturally competent approach. F.A. Davis, Philadelphia

Purnell L, Davidhizar R, Fishman D, Strickland O, Allison D (2011) A guide to developing a culturally competent organization. J Transcult Nurs 22(1):5–14

The Joint Commission (2010) Advancing effective communication, cultural competence, and family centered care: a roadmap for hospitals. https://www.jointcommission.org/-/media/tjc/documents/resources/patient-safety-topics/health-equity/aroadmapforhospitalsfinalversion727pdf.pdf

The Joint Commission (2011) Patient-centered communication standards and EPs. http://www.imiaweb.org/uploads/pages/275.pdf

# Chapter 4
# People of African American Heritage

**Mattie Geneva Caldwell and Myra Michelle DeBose**

## 4.1 Overview and Heritage

African Americans (Black People) comprise 45.4 million residents in the United Stated (U.S. Census Bureau 2022). They are the descendants of Africans who were unwillingly brought to America as designated enslaved people to serve a large population of the white race. This event caused the separation of many families who never saw their relatives again.

Their color identity was African, Negro, colored, and less respectful names, such as niggers, and darkies. The external labels transitioned into various shades by the sexual mixing of the races, which produced a wide variation in skin colors (mullato) but was still viewed as beautiful.

Owners usually assigned lighter-skinned persons duties in the "House." Naturally, these enslaved persons learned some of the eclectic ways. The darker-skinned person was assigned duties outside in the fields. The native names of the enslaved persons were abolished and renamed. These names were unfamiliar, and they could not pronounce them. Their status in America was considered the lowest. They were separated, even more so, by a language they could not speak, nor was it interpreted. However, they demonstrated tremendous courage and pride to survive such treatment.

Today, African Americans primarily identify themselves as Black people because they were born in America. Although they are descendants of Africans, they do not know the African landscape or cultural practices. These African Americans built this country from strength based on faith that God would make it possible to free them. The many inventions that Americans enjoy today were ideas conceived by the

M. G. Caldwell (✉) · M. M. DeBose
College of Nursing and School of Allied Health, Northwestern State University of Louisiana, Shreveport, LA, USA
e-mail: caldwellm@nsula.edu; debosem@nsula.edu


E. A. Fenkl, L. D. Purnell (eds.), *Handbook for Culturally Competent Care*,
https://doi.org/10.1007/978-3-031-70492-5_4

Africans. In terms of culture and their ways of living, they transformed America in many ways.

Most families place a high priority on self-identification. To deny the African heritage or self as of the Black race is most offensive. The experiences of the Black race have improved in that more of them are college educated, employed in white collar positions, business owners, and live in the upper echelon of neighborhoods. Despite these improvements, African Americans remain in the "blue collar" segment of the working class. Nevertheless, they are mindful of where they came from by celebrating their African American (Black) heritage every year in February. But if you ask a Black person about the celebration, they will say, "Every day of the year should be celebrated by each person who is of the Black race" (Satterfield 2023).

*All information presented in this chapter represents aggregate information and may not represent anyone identifying as an African American or Black.*

### 4.1.1 Communications

- African Americans have been plagued with labels to identify themselves as well as labels over language. Many scholars who research language have determined Black English, Ebonics, African American English (AAE), African American Vernacular English (AAVE), and most recently, African American Language (AAL), are means by which African Americans may communicate.
- Many believe there is a common vernacular and language among US African Americans regardless of where they live. However, the vocabulary and pronunciation of words can be very different. Speech patterns are fundamental to family culture and history. Storytelling has been a means by which African Americans pass on their history. This means of communication helps people learn and hear the stories of African American ancestors.
- Many agree that African-American Vernacular (AAVE) is the most commonly spoken language of many middle-class African Americans.
- The dominant language of African Americans is English. However, many refer to an informal language called Black or African American English (AAE) or Ebonics. For example, some may pronounce "th" as "d". Therefore, the word these may be pronounced **dese**. Do not stereotype African Americans as speaking only in Black English because most African Americans are articulate and competent in formal English.
- African American English is a credible authentic dialect suitable for respect. Its origins can be found in the Atlantic slave trade. The development of dialects created "first and second language varieties" developed by African Americans. The vibrant English language is mainly due to the distinct features of African American language.
- What transpires within the family is considered private and inappropriate for discussion with strangers. A common phrase that reflects this perspective is, "Don't air your dirty laundry."

- An example of when withholding "dirty laundry" may harm families is when someone in the family becomes unemployed. Historically, men have traditionally been the providers and if they lose their jobs, it can be detrimental to the family. Families are reluctant to discuss job loss, as it is considered a private family matter. Churches often encourage prayer to help resolve these situations. If unemployment goes for too long, additional family struggles may arise.
- Speech is dynamic and expressive and the volume is often loud compared with other cultural groups. Body movements are usually involved when communicating with others and facial expressions can be very demonstrative. This audible voice volume does not necessarily reflect anger; it is merely a dynamic expression.
- Most individuals are comfortable with a close personal space. However, be aware that maintaining prolonged direct eye contact may be misinterpreted as aggressive behavior by some. A few older African Americans may avoid eye, a carryover from the past where direct eye contact was avoided.
- Nonverbal communication can be seen as a respectful greeting. Tilting the head up acknowledges another's presence, and tilting the head down shows respect. Hugs can be given when you are familiar with the person; however, asking if it is okay to hug may go further than assuming it is OK.
- Body language can send direct messages where words are not required. Therefore, it is essential to determine what messages a person tries to convey through body movement, facial expressions, and gestures.
- Generally, most individuals are more present- than past- or future-oriented. Younger and middle-aged people may be more future-oriented, as indicated by the value placed on education. Some are more relaxed about time, believing that making an appointment is more important than being on time. Flexibility in timing appointments may be necessary for those with a circular rather than a linear sense of time.
- African Americans require effective communication and this begins with building trust. Healthcare providers must be aware of their own biases and prejudices. If providers are not consciously aware of them, this can hinder positive healthcare outcomes for this population.
- Showing respect with formality in names may be a first step in gaining trust. Most prefer to be greeted formally as Mr., Mrs., Ms., or Miss with the use of their surname because the "family name" is highly respected and connotes pride in their family heritage. Greet patients formally as Mr., Mrs., Ms., Dr., or Rev. and the last name until told to do otherwise.

### *4.1.2 Family Roles and Organizations*

African Americans have faced a long history of racism in laws, policies, and practices throughout the United States. This has led to the creation of racist institutions and a system of structural racism that favors White people while systematically disadvantaging Black Americans. The impact of this inequality is felt deeply by

African Americans in nearly every aspect of their lives, from where they live, the quality of education they receive to their job opportunities. Access to childcare, health care, and mental and physical health outcomes can be affected. Furthermore, their political standing, power, and treatment in law and justice systems have been affected. The color of their skin has influenced virtually every aspect of African Americans' lives, regardless of their socioeconomic status or social standing.

- The African American community has a distinctive family structure that prioritizes the significance of extended families. Grandmothers, aunts, and other relatives often play a pivotal role in providing emotional and physical support to children. This approach ensures that children receive the necessary care, even when nuclear families are unable to provide it.
- Education and self-reliance are highly valued in the community, and parents work hard to shield their children from the outside world. Because of discrimination and violence in some areas, parents act as a buffer, protecting their children from danger. They instill respect, obedience, and good behavior to ensure children grow up in a safe and secure environment. Adolescents are assigned household chores to teach them responsibility and survival skills.
- Older individuals, particularly grandmothers, are respected for their wisdom and are frequently essential in childcare. Healthcare professionals, such as doctors, nurses, and dentists are highly regarded and social status is significant within the community.
- However, the community faces challenges in some areas, such as teen pregnancy. While premarital teenage pregnancy is not approved, it is accepted after the fact. The extended family provides a robust support system and the adolescent mother assumes primary responsibility for her child. Sometimes, someone other than the mother may become the primary caregiver.
- It is crucial to recognize that acceptance of same-sex relationships is different among families. Revealing these relationships may have unfavorable consequences. Therefore, it is best not to disclose same-sex relationships to others.
- In conclusion, the African American community has a unique family structure emphasizing the importance of extended families, education, and self-reliance. Despite difficulties, the community values respect, obedience, and good behavior, ensuring children grow up in a secure and nurturing environment. Acknowledging and honoring these values when working with the community is essential.

### *4.1.3 Workforce Issues*

- Acknowledging the struggles many African Americans face in the workforce is essential. Unfortunately, prejudicial attitudes and cultural tension can create a hostile workplace atmosphere that makes it difficult for them to succeed. Employers must address these issues and create an inclusive environment.

- The racial wealth gap is another challenge that many African American families face. The persistent disparity in income and wealth between African American and white families highlights the ongoing impact of discrimination and inequality. Addressing this wealth gap could not only benefit African American families but also have a positive effect on the economy.
- It is essential to recognize that African American employees are often underrepresented in leadership positions and overrepresented in low-wage entry-level roles. Seeing African American leadership in the workplace can be encouraging, but there is still work to provide equal opportunities.
- Ignoring African American culture in the work environment can have significant adverse health consequences. Microaggressions and implicit biases can lead to physical and mental health challenges that can result in increased absences from work and decreased productivity. Employers must commit to addressing these issues and implementing policies and processes that promote a healthy and inclusive work environment for all employees.

### *4.1.4 Biocultural Ecology*

- Research and genetics companies, such as **23andMe**, have genetic profiles that provide historical records tracing African Americans as far as the transatlantic slave trade. It shows tens of thousands of genetic profiles tracing modern-day African American ancestry.
- DNA was extracted from the remains of individuals at the historic site of Catoctin Furnace in Maryland and individuals clustered into numerous family groups. These enslaved individuals carried out domestic and agricultural tasks and operated the furnace. These enslaved individuals brought with them skills of iron technology and iron artistry.
- African Americans' gene pool of racial strains is attributed to their skin color which may vary from light to very dark. Lighter-skinned people appear more yellowish brown, whereas darker-skinned African Americans appear ashen.
- Pallor in dark-skinned people can be observed by the absence of the underlying red tones that give brown and black skin its "glow" or "living color."
- To assess conditions such as inflammation, cyanosis, jaundice, and petechiae, palpate the skin for warmth, edema, tightness, or induration.
- To assess for cyanosis in dark-skinned people, observe the oral mucosa or conjunctiva.
- Jaundice is assessed more accurately by observing the sclera of the eyes, the palms of the hands, and the soles of the feet which may have a yellow discoloration.
- Keloid formation, common among dark-skinned people, is one example of the tendency toward overgrowth of connective tissue.
- African Americans have a greater bone density than European Americans, Asians, and Hispanics, resulting in a lower incidence of osteoporosis.

- Because of the darker skin color of most African Americans, the early stages of pressure ulcers are more difficult to observe.
- Pseudofolliculitis ("razor bumps") is more common among males, whereas melasma ("mask of pregnancy") is more common among darker-skinned females during pregnancy.
- Many experience a disproportionate amount of pigment discoloration with vitiligo being the most common. This is an autoimmune disease that causes skin discoloration and is associated with diabetes and thyroid disorders. If left untreated, it can cause skin cancer. Some erroneously believe that dark-skinned people are not at risk for skin cancer or sunburn because of their higher concentration of melanin.
- Dispel myths that darker-skinned people are not at risk for skin cancer and sunburn.
- Birthmarks are more common in this population. Mongolian spots, found more often in newborns, disappear over time. Reassure the family that most Mongolian spots will fade over time.
- Some women have alopecia related to the use of chemicals to straighten/relax the hair or braid. Provide information on the causes of alopecia.
- The pathophysiology of hypertension in African Americans is most frequently related to volume expansion, decreased renin, and increased intracellular concentration of sodium and calcium, making African Americans more genetically prone than whites to retain sodium.
- There is evidence of breast cancer disparities among African Americans; however, there is little consideration of the genomic risk of recurrence by breast cancer subtype. Women with early high-risk breast cancer where there is no evidence of residual invasive cancer have similarly good outcomes regardless of race. However, African American women with HR-positive/*ERBB2*-negative tumors without surrogate biomarkers may have worse outcomes than White women.
- Gene variants and alcohol consumption have been associated with an increased risk of breast cancer in women. There is a correlation between the consumption of over seven or more drinks per week and invasive breast cancer as well as ER (estrogen-positive) and PR (progesterone-receptor) or breast cancer. **Researchers must develop and test biomarker therapy in African American women.**
- Although there has been a decline in leading causes of death such as accidents, cancer, infant mortality, and cardiovascular diseases, the adjusted death rates for African Americans are higher than for whites. Other causes of death are homicide, cirrhosis, malnutrition, chemical dependency, and diabetes.
- African Americans are at a higher risk of misdiagnosis for psychiatric disorders and therefore may be treated inappropriately with drugs.
- African Americans respond to or metabolize alcohol, antihypertensives, beta blockers, psychotropic drugs, and caffeine differently than European Americans. Psychiatric patients experience a higher incidence of extrapyramidal effects with haloperidol decanoate than European Americans. They are more susceptible to

tricyclic antidepressant delirium, showing higher blood levels and a faster therapeutic response.

- As a result, they experience more toxic side effects. **It is crucial to observe African American patients closely for side effects related to tricyclics and other psychotropic medications.**
- Finally, it is essential to keep in mind that light eyes dilate wider in response to mydriatic drugs than dark eyes and must be taken into consideration when treating African Americans.

### *4.1.5 High-Risk Health Behaviors*

- High-risk health behaviors are causing significant challenges for African Americans. These behaviors have been linked to structural and economic disparities that are rooted in racism. The increased risk of disease or injury that comes along with these behaviors can be detrimental and often lead to disability, death, or social problems. Addressing these issues and creating a more equitable and just society for all is crucial.
- Addressing health equity, particularly in the African American community, should involve addressing behaviors such as alcoholism, eating disorders, risky sexual behaviors, tobacco use, and violence. These severe issues affect individuals and families, and we must prioritize efforts to prevent and address them. By raising awareness and providing support and resources, we can work toward healthier communities for all.
- The prevention of high-risk health behaviors is essential to improve the overall health and well-being of African Americans who are affected by chronic diseases. The COVID-19 pandemic has highlighted the lack of access to healthcare for many African Americans living in poor, inner-city neighborhoods. In such situations, families had to rely on each other to survive, and the fear of the healthcare system prevented many African Americans from seeking care. Unfortunately, this mistrust in the healthcare system still exists, and the COVID-19 pandemic has only added to it. Healthcare professionals must address these issues and build trust in the healthcare system.
- It is essential to acknowledge that some health conditions disproportionately impact African Americans. For instance, they have a higher rate of mortality and morbidity in cardiovascular disease, all cancers, diabetes, obesity, and HIV.
- Cardiovascular disease is the top cause of death among all adults in the United States. Among the most common conditions related to cardiovascular disease are heart disease and stroke. Heart disease has several disorders; two briefly discussed here are hypertension and heart attack. Although a stroke is like a heart attack, its damage is caused in the brain. It can be a prominent factor in disability. Unfortunately, these risks are even higher for African Americans due in part to historical and systemic factors that continue to play a significant role in current health outcomes. For instance, adverse social determinants of health—such as

lack of access to healthcare and healthy foods, among other societal issues—are significant contributors to this disparity.

- Hypertension (high blood pressure), if uncontrolled in African Americans, can significantly increase their risk of heart disease and stroke. Irreversible heart damage can even lead to sudden death. That's why early screening for heart disease is crucial for promoting longevity and overall health among this population. Knowing your family history of heart disease is vital. Younger African Americans are developing some form of heart disease.
- Heart attacks can be seen differently in many patients. Some people will experience tightness between their shoulders, pain in their arms or chest, and pain in their upper abdomen. The key to surviving a heart attack is knowledge. They are knowing the symptoms of a heart attack, not being alone, being with someone who knows CPR, and understanding heart disease risk factors.
- Interestingly, despite smoking about the same amount as other racial and ethnic groups, African Americans have an increased risk for cancer deaths overall. Unfortunately, the survival rate following a 5-year diagnosis is lower for African Americans and they are likely to be diagnosed with all cancers at a later disease stage. This makes it more difficult to treat cancer, mainly if it has already spread to other parts of the body.
- To address this issue, it is crucial to inform African Americans that many cancers are preventable. Healthcare providers should make cancer prevention information a standard part of their care plan during every visit. Community engagement is also essential for educating African Americans about their health. Churches are a common place for educating African Americans about their health and they can be used as a platform to share information about cancer prevention. Limiting alcohol use, maintaining a healthy weight, and staying away from tobacco can lower cancer risk. Additionally, committing to regular diagnostic screening for cancer is another way to catch cancer early.
- Sixty percent of African Americans are more likely to be diagnosed with diabetes than white Americans. It is essential to be aware of the risks associated with diabetes, especially for African Americans who are more likely to receive a diabetes diagnosis than white Americans. Diabetes can lead to a narrowing and hardening of blood vessels, which can cause serious complications such as stroke and heart disease. However, some steps can be taken to manage diabetes and reduce the risk of complications such as adopting a healthy diet, engaging in regular exercise, and being knowledgeable about the symptoms of diabetes.
- Obesity has a disproportionate impact on African Americans. This condition of being overweight can lead to hypertension, diabetes, and high levels of cholesterol in the blood, all of which can have serious health consequences. Food deserts contribute to this issue, as people may not have access to healthy food options. Additionally, a lack of exercise and understanding of the value of physical activity can contribute to obesity as well. Raising awareness about these issues and promoting healthy habits is essential to improve the African American community's overall health and well-being.

- Sexually transmitted infections are another high-risk health behavior that has been shown to impact life expectancy among African Americans. Therefore, it is recommended that health screening programs in the community and church activities be initiated where the entire family is present. Health prevention strategies and safe sex practices should be taught at every opportunity with individual patients and families during community events and at churches. It is unfortunate but true that sexually transmitted infections (STIs) can have a significant impact on the life expectancy of African Americans. It is essential to practice safe sex and get tested regularly to prevent the spread of STIs and maintain good health.

### *4.1.6 Nutrition*

- Many foods that African Americans consumed were called "soul foods." African Americans commonly used this term to describe the foods they enjoyed from their African homeland and later became staples of the American culture. Many of the fruits and vegetables from Africa grew well in America so the enslaved Africans planted them. America is known as the "melting pot," and many foods from new and old American crops and European and Native American cultures have influenced African food preparation traditions. The cuisine became known as "soul food." African Americans had to adapt to the foods/spices that were available and therefore compromised on some nutritional properties that is referred to as unhealthy dietary patterns.
- A large majority of these unhealthy adaptations have persisted through many generations of African Americans which contributed to the chronic diseases that are present in African Americans today. Today's foods originated from the south (New Orleans, LA.) which contained okra and was used in gumbo.
- The story of the heirloom rice has been lost to history—the rice, which was grown in the Southern uplands after it was brought from West Africa more than 200 years ago, disappeared from the South by World War I in large part because black farmers who grew it found themselves pushed from the lands they farmed. The rice's legacy in the United States was erased like so many other contributions that the millions of enslaved African Americans made to southern history, culture, and, of course, cuisine.
- Many of the deviations from African food preparation practices decreased nutritional value. They consumed meat and parts of vegetables that Americans did not want. These foods were frequently called "scraps" (fat trimmed from meats, animal organ meat, liver, gizzards, pig intestines, etc.). Some green leafy vegetables had bulbs that were boiled/seasoned to provide a substance that was filling to them and diminished hunger. Yams were a staple in the African culture and can be prepared in various ways. Sweet potatoes and yams are different. Sweet potatoes are primarily used to make pies. They are baked, boiled, and seasoned with extra sugar and butter which can be very filling but is unhealthy.

- Rice and wheat were challenging to grow in some areas (because they were pushed from their farms). The enslaved Africans preferred foods such as millet, sorghum, and cereal grain plants rich in nutrition. Millet and sorghum lacked gluten and unleavened substances and were called "flatbread." Hot water cornbread originated from these products and was used to scoop up food which replaced utensils. Early African Americans used their fingers to eat food because enslaved individuals were forbidden to use utensils for cooking and eating. They were afraid they would use them as weapons.
- Nutrition is a critical social issue as well as a cultural tradition. African Americans came from a rich history of creating cuisine that had great taste and has transformed into the idea of food that promoted health and healing. Access to healthy, affordable, and culturally significant foods for African Americans is now a means for helping them to become healthier and more productive within their communities. Since the Black Panther days (in the 1960s), lack of access to fresh and nutritious foods has become a social justice issue. They believed African Americans (Blacks) had a right to food and started the first "Feed the Children Breakfast Movement "in Los Angeles, CA.
- Although many African Americans view access to food as a right, the lack of access for some people below the poverty level remains an issue. To minimize this problem, people plant gardens, plant seeds in flowerpots, get food from food pantries and receive food stamps to purchase fresh fruits, vegetables, meats, and dairy products.

- Other sources may include food pantries/voluntary groups in which donated foods are free. Hunger and chronic diseases are never-ending problems for most Americans. However, political parties struggle to provide food to those in poverty. Republicans would abolish this measure if they could garner political support.

### *4.1.7 Pregnancy and Childbearing Practices*

- African American women's preparations for childbirth come from a long history of older women teaching their young girls how to be wives and mothers. This teaching comes from the biblical influence of older women who teach and set an example for younger women *(Titus 2:4–5). "The older women teach the younger women to be sober, love their husbands, and their children."* Also, refraining from premarital sex, although not often followed, is one of the many guidelines that African Americans regard as an essential rule to follow to avoid becoming pregnant. Women who live in poverty usually feel they are unable to prevent pregnancy and or see this to procure food and better living conditions through support from the government or the father of the child. They may also experience low self-esteem and surrender to the male who promises to care for them. Over the last two decades, African American women have become more independent and open to being a single parent and perceptions of what others think of them have fallen by the wayside.

- African American females rarely seek abortions. Biblical and societal teachings guide the moral compass in deciding to abort a child. Biblical beliefs are that one is killing a human being, which is one of the great sins against God. "Thou shall not kill" is the sixth commandment of God. When others find out they aborted a child, family and community members will view abortion as shameful for the family. However, over the past decade, abortions have increased in the African American population, accounting for 39% of abortions nationally.
- Preparing for childbirth is a celebrated time in life for most African American women. The new mother transitions from being obsessed with herself to being obsessed with caring for her child. This transition is an essential phase of maturity and a significant motherhood priority.
- In recent years, African American women have become more comfortable in preventing pregnancy by taking oral contraceptives. However, contraceptives are known to serve as precursors for developing breast, ovarian, and uterine cancers. And for that reason, they only take them for a short time until they feel ready to have a baby.
- Once a young woman becomes pregnant, the African American family is thrilled that a new member will join them. However, depending on the age of the female, it may be perceived as an adverse event if the person is less than 18 years of age, unmarried, financially poor, and in a less-than-optimal environment to rear a child. Nevertheless, older African American women remain close to advising and assisting the expectant mother.
- The family network guides many of the childbirth practices. The grandmother (if still alive) is the most influential, followed by the mother and other females who have had a child. Grandma dictates the daily routines, including eating healthy foods and drinking plenty of water. Over many decades, the baby has been believed to dictate what the mother should eat. Cravings for pickles, ice cream, corn starch, and nonfood-white clay (an "iffy today"). Eating clay in substantial amounts can cause constipation and dirt may be contaminated with pollutants from chemical plants and waste disposal systems; this practice has diminished in the culture.
- Preparing for the child's arrival is another essential and fun activity. The sex of the child determines what items may be purchased. Decorating the baby's room and buying clothes and other care items follows a specific color scheme.
- Very little information is documented in the literature about African American women's preparation for childbirth. Most often, they are made to feel marginalized and stereotyped because of the racism against them. So, their perspective on preparing for childbirth is based on what they see Caucasian women do. Participating in childbirth classes is also one of the things only a few of them join in due to a lack of funds and other activities that leave very little time to do so.

### 4.1.8 Death Rituals

- The ideas and rituals of African Americans about death stem from their Christian education and beliefs about God the Father and Creator of all humanity. Most have affiliations with Baptist, Methodist, Catholic, and Muslim churches and beliefs with **Baptist and Methodist being the most preferred**. Baptists and Methodists have the same spiritual beliefs; however, their organizational structures/processes may differ. Others may not be members of either of the traditional spiritualities. It is also believed that life does not end when death occurs. The decedent transitions to another phase of life, waiting to be awakened again to live in Heaven or hell.
- African Americans display their emotions/feelings upon hearing about the death of a loved one by fainting or lowering their bodies to sit on the floor or chair and, most often, crying out loud. This emotional response to the death may not require medical attention. However, if the family member has a medical condition when a severe emotional reaction occurs, they may need medical attention. Episodes of crying/grieving that can last for an undetermined amount of time. Some people don't show their grief in public, which can sometimes result in a severely mentally delayed response, requiring treatment. Family members may need education and counseling at any time following the death of family members or friends.
- African American families have a variety of rituals that comprise their way of mourning and grieving for their loved ones. The death and burial of a family member do not end the connection with that person. And it is not just a funeral, but rather a "homegoing." The dead person is believed to have returned home to God, the Father and Creator of life. Most quote the scripture: "We are confident, I say and willing, to be absent from the body is to be present with the Lord (II Corinthians 5:8). When the person dies, their physical body should be buried but the soul returns to God. It is believed that when Jesus Christ, the Son of God, returns, the physical body and the soul will be joined together and be transformed to look like Jesus Christ. They also believe that the body should be kept intact and embalmed for preservation and are less likely to volunteer as an organ donor. Some situations require autopsies to be performed for legal reasons. Family members will need education on rationales for autopsies. However, the family sometimes requests autopsies if they believe something went wrong during hospitalization or foul play. Understanding "why" the decedent can no longer live ushers in closure and acceptance of their demise.
- African Americans don't rush to bury their loved ones. The funeral is customarily held within 7 days after death. They give family members and friends time to prepare for the funeral. COVID-19 pandemic deaths from 2020 to 2022 changed this practice because Mortuaries were overloaded with many deaths, and funerals were scheduled as far out as 2–3 weeks.
- There is a variation in what occurs during the "homegoing celebration." The ceremony must reflect the love and emotions of the family members. So, they select the most elaborate casket and floral arrangements and plan the homegoing

celebration to provide a healing and calming effect to the family. It also communicates love and respect for the decedent. The most significant pain associated with loss can last for a while. Still, the homegoing celebration brings closure to the fact that the decedent is in a better place and no longer suffers, which lessens the pain.

- The actual homegoing celebration follows a specific cultural order. The processional order follows in a two-by-two fashion: the Minister, Church Leaders, Pallbearers, Family (the closest relative is first in line), and all others follow.
- The casket is usually left open for attendees to view the decedent before the service begins. Persons in the processional may view the decedent as they come in and sit down. Once everyone has entered the chapel, the casket is closed. The casket is usually not left open unless it is a graveside service but may be opened upon request.
- The order of the funeral service is culturally driven. It is as follows: musical prelude, processional, scriptures for the Old and New Testaments (read by a minister), prayer, a song by the choir or soloist and words of the song encouraging family/friends or it may provoke public display of grief. The obituary is read in silence. Special tributes by family/friends and coworkers or a video presentation of the decedent's life may be displayed. A church member reads the resolution. The choir or soloist performs another song. The Pastor eulogizes the deceased. At the end of the service, the Recessional allows the persons in the processional and attendees to leave the church in the same order as they entered the church.
- The descendant's casket is moved to the burial site in a processional style consisting of a fleet of cars. The burial site may be in the ground or a mausoleum, symbolizing that the body has been returned to the Earth. Cremations have increased and the ashes are contained in a pleasant urn and are given to the family.
- The repast is when family and friends gather to remember, reflect, and rejoice in positivity, strength, and support. It focuses on food and fellowship, which signifies the realization of life anew for their loved one without the physical presence of the decedent.

### *4.1.9 Spirituality*

- Most African Americans (Blacks) identify as Christians from religious traditions or beliefs. Most worshippers see the role of religion in fighting racial injustice and gaining respect within society. This belief is changing as racial and injustice events continue to occur. The younger generation views it as a way to fight back and gain equality/equity and respect for diversity through nonviolent events. They are less likely to grow up in mostly Black congregations. Three percent of black Americans identify as atheists or agnostics.
- Fifteen attributes of God encourage African Americans to believe in God's deity, to serve Him, and to believe they will eventually return to Him in heaven.

- They believe in the attributes of God with earthly affairs and three dimensions of deity: God the Father (resides in Heaven), the Son (resides in Heaven but walked the earth for 33 years and is now residing in Heaven), and the Holy Spirit (reside in each believer and is on earth). God presents as an apparition of a human being (ghost-like), but man is not allowed to look upon Him, or he will die. The Trinity, as it is called, has 15 attributes: (1) God is infinite, self exists and without origin (all three forms of God exist from the beginning of time), (2) God is immutable—He never changes, (3) God is self-sufficient—He has no needs, (4) God is Omnipotent—He is all-powerful, (5) God is Omniscient—He is all-knowing, (6) God is Omnipresent—He is always everywhere, (7) God is Wise—He is perfect, unchanging wisdom, (8) God is Faithful—He is infinitely and unchangingly true, (9) God is Good—He is infinitely, unchangingly, kind, and full of Good Will, (10) God is Just—He is infinitely, unchangeably right, and perfect in All He does, (11) God is Merciful—He is infinitely, unchangeably compassionate and kind. (12) God is Gracious—Infinitely inclined to spare the guilty, (13) God is Loving-God is infinitely unchanging and loves humans, (14) God is Holy—He is infinitely, unchangingly perfect, (15) God is Glorious—He is infinitely beautiful and great. These attributes have shaped the moral compass for medicine, nursing, and other business operations.
- The Son was conceived by the works of the Holy Spirit overcoming Mary, not by a human being. Jesus presented Himself as a human without sin. The Holy Spirit is the third dimension of God, who resides on earth (in each believer) to comfort and serve as man's consciousness to do good in life.
- Regarding Christian affiliation, most African Americans affiliate with Baptist and Methodist denominations. Other religious denominations include African Episcopal, Jehovah's Witness, Church of God in Christ, Seventh Day Adventist, Pentecostal, Apostolic, Presbyterian, Lutheran, Roman Catholic, Nation of Islam, and other Islamic sectors. Determine the religious affiliation during the intake interview and respecting their beliefs is prudent.
- Most people take their spiritual beliefs seriously and expect to receive messages through preaching, prayer, and waiting on God's providential intervention. Speaking in tongues is primarily done in the Church of Christ congregations. Baptist churches adhere to the biblical scripture that says if a person speaks in a tongue, it must be to God or where two or at most three persons are present. Also, there must be someone who can interpret it. The Holy Spirit will give them an utterance. If not, the person speaks to himself and not God. If there is no interpreter, he builds himself up, not the church, because no one understands what is being said.
- Prayer is an essential part of their daily routines. It gives them hope and belief that God will be with them and see them through tough times. He may not remove the current situation but will strengthen them to overcome it. Support and allow time for praying for individuals who desire to do so.
- Laying on hands while praying to facilitate healing is discouraged in today's church practices. Biblical teachings explain that healing occurs through God's divine interventions by answering the person's prayer. Healing does not come

from the person who touches the sick. It is a symbol of God touching the sick. Healing is also facilitated by trained groups who are considered an extension of God or His chosen people to perform interventions that promote healing. This healing process can be explained by scientific knowledge. But remember, there are still miraculous healings that no one can explain. The reaction is, "Oh, but God." This phrase does not mean that individuals do not care about their health or are unwilling to practice illness-prevention measures. Sickness can occur through heredity and exposure to agents present in the environment. Recognize and support spiritual beliefs and practices that provide comfort, coping, and support.

- Since biblical times, illness has represented sin. However, in current society, biblical teachings of this belief are supported by scriptures that say sinful acts can cause their family to experience an illness over many generations. Jesus healed the sick when He was present on earth, and He explained that some were not sick because of sin and did not provide a reason. His concern was to demonstrate His powers through healing. Partnering with hospital chaplains or community church clergy is an excellent strategy to promote the patient's beliefs about healing. Most churches have healthcare ministries and emergency response teams to encourage a healthy lifestyle through preventive education.

### *4.1.10 Healthcare Practice*

- Healthcare practices among African Americans have changed over the past decade. Many improvements have been achieved by increasing access to care, making insurance more affordable, maximizing government insurance coverage (Medicare/Medicaid), increasing outpatient clinics in rural areas, increasing nurse practitioner clinics, telehealth medicine, and collaborative practices with family practice physicians. Therefore, better access means people can seek healthcare more often when needed.
- Although African Americans continue to have suspicions about healthcare providers, it has improved over the past two decades. The reasons are that more information about the best healthcare screenings, diagnosis, and treatment is available via television, online, mailing reminders, community health fairs, health fairs at Black churches, healthcare ministries, and family support. Much emphasis has been placed on the most common health problems in the Black patient population, so preventive, maintenance, and treatment guidelines are readily available to patients via online and medical offices.
- African Americans most often experience more than two chronic illnesses simultaneously. This usually creates a stressful situation regarding the cost of medical care.
- Thirty percent of African Americans believe that their health is dependent upon fate or destiny, and only about 50% feel that health is a high priority. Reception

and utilization of health information are well-known significant factors in disease prevention.

- Sometimes African Americans are less likely to follow prescribed medication directions due to a lack of understanding of how medications work and the importance of taking them for a specified time to control or cure medical disorders. For example, when prescribed antihypertensives, the individual may stop taking the medication as soon as the blood pressure normalizes.
- African Americans are sometimes reluctant to follow prescribed medication regimes related to unwillingness to tolerate side effects. For example, men usually deviate from prescribed directions when taking antihypertensives because of decreased libido and sexual functioning. More and more of them are beginning to take this matter more seriously due to the number of heart attacks that occur among family members. Unlike any other tactic, this phenomenon will likely make believers out of them. This has been the number one thing that increased their participation in getting annual physical exams. Ask about the family history of diseases commonly occurring in their family and encourage them to seek screening, preventive care, and medical treatment as early as possible.
- Although access to healthcare has improved among African Americans, sharing medications is a common practice among males and females. For example, blood pressure medications, prescribed analgesics, and antibiotics are frequently given to people with similar complaints, especially if they forget to take their medications or are experiencing symptoms. They are also likely to discontinue medications once the discomforts subside. This habit could explain why some diseases become resistant to treatment or worsen. Discourage patients from sharing medications because they can be more dangerous and cause side effects in one person that may not occur in another.
- African Americans once believed that some diseases/ailments caused by unnatural forces could not be mitigated by routine medical procedures and resorted to seeking help from Voo Doo doctors or people who practiced witchcraft. However, since many people who grew up being exposed to these practices have died, much of this philosophy and belief has been lost.
- Homeopathic medical care has been around for years since the enslaved populations. They did not have access to medical care and, therefore, had to use whatever was thought to relieve discomfort. For example, a paste of clay and vinegar treated cuts and injuries from rusty metals. Some of these practices remain in use today when someone cannot afford healthcare.
- Apple cider vinegar is used to treat other health conditions. African American women frequently use apple cider vinegar to treat yeast infections and vaginal irritations. Most African Americans, male and female, believe apple cider vinegar can control high blood pressure by mixing a teaspoon of vinegar in a small amount of water. Vinegar mixed with sugar, lemon, and honey is used for weight loss. White vinegar is often used to treat skin irritations and discomforts.
- Baking soda, mixed in water, is ingested to treat indigestion or stomachache. Some people mix baking soda, syrup of Black Draught, aspirin, or vinegar. It is also used topically to relieve skin irritations such as burns, insect bites, and rashes.

- Lemons frequently treat sore throat, colds, and flu-like symptoms. They boil the lemons to make tea by adding honey and peppermint. Some gargle to relieve symptoms and others suck on a slice of lemon. Mixing lemon and syrup is also used as a cough syrup and is believed to be as effective as commercial cough syrup. Any combination of these ingredients is believed to relieve symptoms of a cold.
- Salt is used primarily as a topical agent to treat mouth, throat, and sinus tissues. Commercial products using salt are available to treat sinusitis. Salt mixed with water is also used to treat bleeding gums, sore throats, and mouth sores after dental procedures.
- Mental health issues have existed in the African American patient population. However, their perspective of mental health stigmatizes their existence. It is a significant reminder of being made to feel "less than" and is a reminder of the effects of racism (limited mobility and progress). They do not trust the healthcare system for fear of being put away and deprived of a "normal life." Also, there is a fear of being abused by family members and caregivers. Encourage patients and family members to seek professional help as soon as possible. Select a family member the patient trusts to engage in therapy and care.

### *4.1.11 Healthcare Providers*

- Black Americans have mixed ideas about the progress made in improving healthcare experiences and outcomes for Black patients. Forty-seven percent say it has been enhanced over the last 20 years, 31% say it stayed about the same, and 20% feel it is worse.
- When asked about their personal experiences, about 55% said they had a positive experience. They contribute to better care by speaking up and advocating for appropriate treatment. Still, there are perceptions of being treated with less respect than other patients.
- Social determinates contributing to lower quality health outcomes include less access to quality care, living in communities with more environmental problems, healthcare providers being more likely to provide advanced care, working in jobs with risks for health problems, hospitals, and medical centers giving lower priorities to well-being, and more communication issues such as language and cultural differences.
- Although most African Americans say healthcare has improved overall, there are some negative aspects to interacting with providers. Examples include having to speak up and ask for care, the pain they were experiencing not being taken seriously, feeling rushed by healthcare providers, feeling they were treated with less respect, and feeling they received lower quality of care. They were looked down on because of their weight.
- Black adults with higher education are more likely to point to healthcare disparities than those with lower educational levels.

- Thirty-one percent of Blacks prefer to see black providers, 14% somewhat prefer, and 17% see a black doctor or other healthcare provider for routine health care. Encourage patients to discuss their concerns with healthcare providers. If they continue to feel uncomfortable when interacting with the provider, they can choose to find someone else.

## Bibliography


Lanehart S (2023) Language in African American communities. Routledge, New York

Weldon T (2021) Middle-class African American English. Cambridge University Press

Weldon T (2022) Visibly invisible: the study of middle class African American English. Lang Linguist Compass 16(11):e12477. https://doi.org/10.1111/lnc3.12477

Cross CJ, Taylor RJ, Chatters LM (2018) Family social support networks of African American and Black Caribbean adolescents. J Child Fam Stud 27(9):2757–2771. https://doi.org/10.1007/s10826-018-1116-2

Lloyd C, Alvira-Hammonds M, Carlson J, Logan D (2021) Family, economic, and geographic characteristics of Black families with children. Black Children & Families, ChildTrends

Moras A, Shehan C, Berardo FM (2018) African American families: historical and contemporary forces shaping family life and studies. In: Batur P, Feagin J (eds) Handbook of the sociology of racial and ethnic relations. Handbooks of sociology and social research. Springer, Cham. https://doi.org/10.1007/978-3-319-76757-4_6

Aliprantis D, Carroll DR (2019) What is behind the persistence of the racial wealth gap? Econ Commen (2019-03)

Derenoncourt E, Kim CH, Kuhn M, Schularick M (2024) Wealth of two nations: the US racial wealth gap, 1860–2020. Q J Econ 139(2):693–750

Sullivan L, Meschede T, Dietrich L, Shapiro T (2015) The racial wealth gap. Institute for Assets and Social Policy, Brandeis University, DEMOS

Harney É, Micheletti S, Bruwelheide KS, Freyman WA, Bryc K, Akbari A et al (2023) The genetic legacy of African Americans from Catoctin furnace. Science 381(6657):eade4995. https://doi.org/10.1126/science.ade4995

Jackson FL (2023) Community-initiated genomics. Science 381(6657):482–483

Kyalwazi B, Yau C, Campbell MJ, Yoshimatsu TF, Chien AJ, Wallace AM et al (2023) Race, gene expression signatures, and clinical outcomes of patients with high-risk early breast cancer. JAMA Netw Open 6(12):e2349646

Young KL, Olshan AF, Lunetta K, Graff M, Williams LA, Yao S, Zirpoli GR, Troester M, Palmer JR (2023) Influence of alcohol consumption and alcohol metabolism variants on breast cancer risk among Black women: results from the AMBER consortium. Breast Cancer Res 25(1):66. https://doi.org/10.1186/s13058-023-01660-1

Beech BM, Ford C, Thorpe RJ Jr, Bruce MA, Norris KC (2021) Poverty, racism, and the public health crisis in America. Front Public Health 9:699049. https://doi.org/10.3389/fpubh.2021.699049

Thakur N, Lovinsky-Desir S, Bime C, Wisnivesky JP, Celedón JC (2020) The structural and social determinants of the racial/ethnic disparities in the U.S. COVID-19 pandemic. What's our role? Am J Respir Crit Care Med 202(7):943–949. https://doi.org/10.1164/rccm.202005-1523PP

Ford A (2022) Food from the soul: A history of African American culture and nutrition. https://thedo.osteopathic.org/columns/food-from-the-soul-a-history-of-african-american-culture-and-nutrition/. Accessed 24 Feb 2024

Korhonen V (2023) Population of the U.S by race and ethnic group. https://www.statista.com/statistics/183489/population-of-the-us-by-ethnicity-since-2000/. Accessed 22 Feb 2024

Satterfield S (2021) High on the hog: how African American cuisine transformed America season 1. Netflix

Satterfield S (2023) High on the hog: how African American cuisine transformed America season 2. Netflix. https://www.smithsonianmag.com/arts-culture/food-historian-reckons-black-roots-southern-food-180964285/

Abbyad C, Robertson T (2011) African American women's preparation for childbirth from the perspective of African American healthcare providers. J Perinat Educ 20(1):45–53. https://doi.org/10.1891/1058-1243.20.1.45. https://blackchildlegacy.org/wp-content/uploads/2017/04/AA-Womens-prep-fopr-childbirth.pdf

Kortsmit K, Nguyen AT, Mandel MG, Clark E, Hollier LM, Rodenhizer J, Whiteman MK (2022) Abortion surveillance—United States, 2020. MMWR Surveill Summ 71(10):1–27. https://doi.org/10.15585/mmwr.ss7110a1

Dignity Memorial (2024) Black funeral traditions. https://www.dignitymemorial.com/memorial-services/funeral-traditions/black-funeral-traditions. Accessed 3 Feb 2024

Fletcher K (2021) 7 Elements of African American mourning practices & burial traditions-talk deaths. https://www.talkdeath.com/7-elements-of-african-american-mourning-practices-burial-traditions/. Accessed 20 Feb 2024

BST & Crosswalk Staff (2024) 15 Amazing Attributes of God: What they mean and why they matter. https://www.biblestudytools.com/bible-study/topical-studies/15-amazing-attributes-of-god-what-they-mean-and-why-they-matter.html. Accessed 19 Mar 2024

King James Bible. I Corinthians 14:2

Pew Research Center (2023) Faith among Black Americans. https://www.pewresearch.org/religion/dataset/faith-among-black-americans/. Accessed 20 Feb 2024

Purnell L, Fenkl E (2019) Handbook for culturally competent care. Springer, Cham. ISBN 978-3-030-21945-3; ISBN, 978-3-030-21946-8. (eBooks). https://doi.org/10.1007/978-3-030-21946-8

Ellis KR, Hecht HK, Young TL, Oh S, Thomas S, Hoggard LS et al (2020) Chronic disease among African American families: a systematic scoping review. Prev Chronic Dis 17:190431. https://doi.org/10.5888/pcd17.190431

Lang SM (2019) J Health Serv Res 54(2). https://doi.org/10.1111/1475-6773.131

Quant S, Sandberg JG, Altizer K, Arcury T (2015) Home remedy use among African American and white older adults. https://www.ncbi.nlm.nih.gov/pmc/articles/PMC4631220/pdf/nihms709035.pdf. Accessed 18 Mar 2024

Funk C (2022) Black Americans' views about health disparities, experiences with health care. https://www.pewresearch.org/science/2022/04/07/black-americans-views-about-health-disparities-experiences-with-health-care/#:~:text=A%20majority%20of%20Black%20adults,it%20is%20not%20a%20reason. Accessed 22 Mar 2024

# Chapter 5
# People of American Indian/Alaskan Native Heritage

**Kathy Prue-Owens**

## 5.1 Overview and Heritage

This chapter provides a general overview of American Indians and Alaskan Natives (AI/AN) followed by a more detailed description of the Navajo Nation as one exemplar. Space does not permit an extensive discourse on each tribe. The 2020 census reports 9.7 million people identified as American Indian and Alaska Native alone or in combination with another race. This is a significant increase from the 5.2 million in 2010. Around 20% live in American Indian areas/reservations or Alaska Native villages. The census report is tabulated based on an AI/AN write-in response to the 2020 census race question indicating that there is a changing demographics among AI/AN. National findings indicate that the Alaska Native alone population grew 10.9% to 133,311 while the AN alone or in any combination population increased 45.6% to 241,797. A similar increase is seen in the American Indian alone population which grew 11.6% to 2,159,802 while the alone or in any combination population nearly doubled to 6,363,796.

There are more than 574 federally recognized sovereign AI/AN tribes, of which 231 are Alaskan Natives and are divided into five groups: Aleuts, Northern Eskimos (Inupiat), Southern Eskimos (Yuit), Interior Indians (Athabascans), and Southeast Coastal Indians (Tlingit and Haida). In addition, there are approximately 400 non-federally recognized tribes, each with its unique customs and beliefs. Some Alaskan Natives do not mind being called Eskimos or Indians; however, some do not like to be called Native Americans or Eskimos and prefer Alaskan Native, Alaska Indian, or Inuits. The amount of Indian blood or blood quantum necessary to include enrolment requirements is dependent on each tribe. For example, the Cherokee Nation

---

K. Prue-Owens (✉)
Helen and Arthur E. Johnson Beth-El College of Nursing and Health Sciences, University of Colorado Colorado Springs, Colorado Springs, CO, USA
e-mail: kprueowe@uccs.edu



E. A. Fenkl, L. D. Purnell (eds.), *Handbook for Culturally Competent Care*,
https://doi.org/10.1007/978-3-031-70492-5_5

laws are based on legal documentation that demonstrates direct ties to the tribe with no blood quantum. However, the Navajo Nation requires at least one biological parent to be enrolled with the Navajo Nation, the applicant must meet the minimum ¼ Navajo blood quantum requirement. Many of the Sioux Nations from the great plains area require ¼ blood quantum and three generations of lineal descent.

Indian Health Service (HIS): The Indian Health Service (IHS) was designed to fulfill the U.S. government's role in providing health care to American Indians and Alaskan Natives (AI/AN); however, inadequate funding has left those who depend on the IHS with only limited health care.

- Overall, AI/ANs are extremely diverse groups of people. Some are very traditional, some are highly acculturated, and some selectively combine tradition with the European American culture. Many share similar, but not exact, traditions, beliefs, and values.
- AI/ANs are a particularly vulnerable population in terms of accessing health care.
- High poverty rates, increasing suicides, individual high-risk health behaviors, social factors, low educational levels, chronic conditions and greater health disparities make it particularly important for accessing and receiving ongoing supportive and preventive care.
- One-third of AI/ANs are in families with incomes below the federal poverty line. The median household income for American Indian and Alaska Natives is $49,906.
- One in five adults has not graduated from high school; 65 % earn a high school diploma compared with 75.2% of the U.S. population. The college graduation rate is also much lower, with 9.3% compared to the national average of 20.3%.
- Almost half of uninsured adults do not have a usual source of care, making it difficult to receive preventive services and timely care for acute health problems.

The IHS is divided into 12 regions, each with either hospital, clinics, or outreach areas. The regions are listed in Table 5.1. Federally recognized tribes are considered sovereign nations, giving them independent authority over their territories with the power to rule and make laws.

- Although the IHS includes access to some specialty care and has established urban Indian health clinics, shortages in funding and a lack of facilities and health-care professionals have left many eligible AI/ANs struggling to obtain timely, high-quality health care.
- The AI/AN population is a young population with a median age of 31 years compared with 37 years for the US overall.

### *5.1.1 Communications*

- AI/ANs are collective cultures and avoid conflict and confrontation.
- Each tribe has its own language, but some languages are similar to each other.

**Table 5.1** Regions of the Indian Health Service

| Region | Area(s) served |
|---|---|
| Great plains area | North Dakota, South Dakota, Iowa, and Nebraska |
| Alaska area | State of Alaska |
| Albuquerque area | New Mexico, Colorado, and Texas |
| Bemidji area | Indiana, Minnesota, Michigan, and Wisconsin |
| Billings area | Montana and Wyoming |
| California area | California and Hawaii |
| Nashville area | Eastern US |
| Navajo area | Arizona, New Mexico, and Utah |
| Oklahoma city area | Oklahoma, Kansas, and portions of Texas |
| Phoenix area | Arizona, California, Nevada, and Utah |
| Portland area | Oregon |
| Tucson area | Southern Arizona |

- Younger people speak English. Some older people may know only their traditional tribal language or perhaps Spanish in the southwestern part of the United States.
- If needed, use professional interpreters whenever possible. Confidentiality is taken seriously. Often the younger family members will serve as interpreters.
- Trust and privacy are major elements that facilitate a clearer communication.
- Greeting with Mr., Mrs., or Miss shows respect and is always safe. Even the use of sir or ma'am is appropriate. While use of the given name is preferred by some; wait for permission to use it.
- Asking for help with pronunciation of a surname is appreciated.
- Greet everyone in the room and establish relationships. A light handshake is appropriate with a simple nodding of the head gestures an acknowledgment of the person.
- Depending on the situation, punctuality may not be valued. Lateness is not meant as disrespect but is the result of a different sense of time. Explain the importance of timeliness if required and be flexible with appointment times whenever possible. Use of a watch is not common. Be aware that many do not have reliable transportation.
- Silence is a means of respect and thinking before speaking. Interrupting someone while he/she is talking is considered rude.
- Strong emotions are not usually displayed verbally or nonverbally.
- Direct questioning expecting an immediate answer is frowned upon. Ask questions in an indirect way, using a phrase such as "Tell me about…" or "Can you give more information…"
- One does not have the right to speak for someone else.
- Directly telling someone that he/she is wrong is considered rude.
- Maintaining continued eye contact with someone during a conversation is considered disrespectful. Do not maintain direct and sustained eye contact during an assessment.

- Singling out a specific individual for an answer is disrespectful because if the person is wrong, he/she will meet peer disapproval.
- A long pause in a conversation does not necessarily mean the person is finished talking. Wait for a nonverbal gesture that says the person has finished.
- Do not use idioms, colloquialisms, or abstract ideas; those unacculturated to the dominant culture might not understand them.
- Greet AI/ANs with a *light* handshake, not a firm, pumping handshake.
- A nod of the head may denote "I hear you," not necessarily agreement with what has been said.
- Loudness may be interpreted as aggressiveness.
- Most appreciate a space greater than 2 ft during a conversation, especially with strangers, and for some maybe up to 6 ft. Take cues from the patient.

### 5.1.2 Family Roles and Organization

- The extended family is the primary social unit in most AI/AN culture.
- Most define family as members made up of fictive (blood-related) and nonfictive (non-blood-related) kin, extended family, and tribal community. One is rarely alone or without family.
- Some tribes are patriarchal and patrilocal (a social system where a married couple lives with the husband's family); others are matriarchal and matrilocal (a social system where a married couple lives with the wife's family) in structure. Given the individuality in any family system, determine the family spokesperson ahead of time before a decision must be made.
- Traditional women are responsible for child rearing, domestic tasks, and the overall concerns of the family. In a more modernized version women's roles are more varied. Determine in a "simple" how and where the women's role is. Many tribes value the matriarchal role of the women especially if she is the primary caregiver.
- The goal of family and parental support, within the context of the family of origin, is to foster interdependence.
- Family and parental support encompasses cultural and spiritual maintenance and satisfaction of physical and emotional needs.
- Lifelong interdependence among members is fostered.
- In traditional families, children frequently attend adult events, but they are expected to remain silent unless invited to speak.
- Child rearing is a family affair that includes aunts, uncles, and cousins. While the mother bears the primary child-rearing responsibility in a traditional family, fathers, and siblings are active participants.
- Rather than learning from direct verbal instruction, children are expected to learn through observation and from examples.
- It is acceptable for children to be disciplined in public.

- Children demonstrate their skills and knowledge by what they can do rather than what they say.
- Gay and lesbian lifestyles have long been accepted in many tribes, although there are individual variations. Alternative lifestyles are accepted readily in most tribes.
- Older people are held in high respect and are a good source for family care.

### 5.1.3 Workforce Issues

- Punctuality may not be valued. Explain the importance of timeliness in reporting to work if required.
- Understanding of family dynamics as it relates to transportation, child/parent care, and work is important to understand.
- Do not use idioms, colloquialisms, or abstract ideas; those unacculturated with the dominant American culture might not understand them.
- Greet AI/ANs with a *light* handshake, not a firm, pumping handshake. Introduction of others in the room is not always done so acknowledging others is important.
- Do not single out a specific individual for an answer; if the person is wrong, he/she will meet peer disapproval.
- Actions speak louder than words; if something is promised, make sure the promise is carried out or give an explanation as to why it could not be done.

### 5.1.4 Biocultural Ecology

- While some American Indians are well over 6 ft tall, others are well under 6 ft, especially Eskimos.
- Most have little body hair, brown skin, and epicanthal folds; however, with intermarriage, many AI/ANs have no distinct characteristic features.
- Eskimos are susceptible to narrow-angle glaucoma because they have a shallow chamber and thicker lenses.
- AI/ANs have the highest rate of poverty of any racial group in the nation. They also have the highest rates of crime, suicide, alcoholism, gang membership, and sexual abuse. Suicide rates among AI/AN persons increased nearly 20% from 2015 (20.0 per 100,000) to 2020 (23.9), compared with a <1% increase among the overall U.S. population (13.3 and 13.5, respectively)
- Nearly half of all Native American women (46%) have experienced rape, physical violence, or stalking by an intimate partner. One in three Indian women will, at some point in her life, experience the violence and trauma of rape.
- On some reservations, Native American women are murdered at a rate more than ten times the national average. The National Crime Information Center (2016)

reported that 5712 reports of missing American Indian and Alaska Native women and girls (MMIWG) through the U.S.

- This vulnerability state indicates that NA women make up a significant portion of missing and murdered cases. The murder rates are ten times higher than the national average for women living on reservations, but murder is the 3rd leading cause of death for NA women.
- The rate of child abuse is twice as high as the national average. For every 1000 of these cases being reported, an average of 15.2 of these cases were for American Indian/Alaska Native children.
- The prevalence of diabetes mellitus is at least twice that of any other racial and ethnic group. AI/ANs have high rates of obesity.
- The leading causes of death are cancer, chronic liver disease, unintentional injuries, diabetes, and heart problems. AI/ANs die at higher rates than other Americans from tuberculosis, alcoholism, diabetes, cardiovascular disease, unintentional injuries, homicide, and suicide.
- Domestic violence rates are also alarming, with more than 4 in 5 American Indian and Alaska Native adults (83%) experiencing some form of violence in their lifetime. This translates to almost three million people who have experienced psychological aggression or physical violence by intimate partners, stalking, or sexual violence. Explain abuse laws at every opportunity. Educate teachers and tribal leaders on the signs and symptoms of suicide ideology.
- While bullying does not cause suicide, a stressful environment and persistent, emotional victimization can increase a person's risk of suicide. Incorporate anti-bullying recommendations of **stopbullying.gov**.

### *5.1.5 High-Risk Behaviors*

- More than one-quarter smoke. Partner with schools beginning in elementary school for anti-smoking campaigns.
- One in five is a binge drinker. Work with tribal leaders to develop support groups for alcohol and drug misuse. Begin teaching about alcohol and drug misuse beginning in elementary schools.
- Incorporate in elementary schools in conjunction with tribal leaders "I Strengthen My Nation," an IHS campaign that empowers Native youth to resist drugs and alcohol and motivates parents to talk openly to their children about drug and alcohol use. In addition, weRnative.org offers young people information on puberty, sexual violence, suicide, cultural preservation, and other important health topics.

### 5.1.6 Nutrition

- Although great variation exists, most AI/AN diets are high in fat and sugar.
- **Include a dietary assessment as part of the intake interview.**
- Availability of fresh fruits and vegetables may be severely limited, especially during winter in cold climates and on reservations.
- **Ask if any particular food is taboo, what foods are consumed for health promotion, and what foods are consumed during illness.**

### 5.1.7 Pregnancy and Childbearing Practices

- Birth control practices are usually an individual choice.
- Some do not discuss birth control with providers until significant trust is established.
- Each tribe has specific foods believed to be beneficial for pregnancy. **Ask about specific foods at each stage of the pregnancy.**
- Women in many tribes remain active throughout their pregnancies.
- Some believe that wearing a necklace may cause the baby to be born with the umbilical cord wrapped around its neck.
- Some believe that if a pregnant woman walks through a door backward, she will have a breech delivery.
- Some believe that blowing up balloons may cause premature rupture of the membranes.
- Some believe that leaving projects unfinished may prolong labor.
- Stoicism during labor and delivery is valued by some. **Ask frequently if pain medicine is needed.**
- Younger, more urban generations may not adhere to any traditional birthing practices.
- For some tribes, the father is not usually present during the birth. **Do not make the father feel guilty if he does not want to participate until after the delivery.**
- Delivering in a kneeling or squatting position is preferred by some.
- In some tribes, the placenta is saved and buried, returning it to mother earth.
- Postpartum, the mother may be isolated according to the sex of her child; 1 month for a boy and 2 months for a girl. However, this is not a widespread belief.
- Many women resume their normal activities within a week and are expected to care for their babies as well as other family members.
- Women often turn to their mothers for advice after birth.
- Some tribes do not name the baby for a week or longer after the delivery and may include an English name along with a tribal name. **Do not assume that not naming the baby means the parents do not care about or do not want the baby.**
- Some tribes often name a newborn after a recently deceased respected relative.
- Both sexes are seen as gifts from nature.

### 5.1.8 Death Rituals

- In general, AI/AN religions do not have precise beliefs about life after death.
- Some believe in reincarnations with rebirth as a human, as a ghost, as an animal, or as a combination of these.
- Burial ceremonies vary by tribe. For example, the burial ceremony of the Zuni requires individuals to take 3 days off from work.
- Autopsy is generally acceptable. **Explain legal requirements if necessary.**
- **Expect and respect a wide variation on bereavement.**

### 5.1.9 Spirituality

- Some tribes have preserved their spiritual beliefs in writing or as an oral tradition.
- Some adhere to the Native North American Church. Others may be more traditional, practice mostly Christian religions, or combine the two.
- The Native North American Church practices have four requirements for good health: food, sleep, cleanliness, and good thoughts. The four divisions of nature are also a holistic part of spirituality: spirit, mind, body, and life.
- Routes of spiritual connectedness include singing, drumming, dancing, and using sweat lodges. **Warn about dehydration when using sweat lodges.**
- Table 5.2 lists the major values between AI/AN tribes and the European American culture.

### 5.1.10 Health-Care Practices

- Religion, medicine, and healing are inseparable among AI/AN populations.
- Diseases can be attributed to natural or unnatural forces, the scientific theories of illness and disease, or a combination of both.
- Natural remedies are preferred, although the specific herbal remedies vary significantly by tribe and region of the country and depend on environmental influences for what is available.
- **Ask patients about specific herbal and tribal remedies so they do not conflict with or potentiate allopathic medicines and treatment.**
- **Incorporate IHS programs that give general directions for health-care providers: Breastfeeding Promotion and Support, Diabetes Treatment and Prevention, Epidemiology, Elder Care Initiative, Health Weight for Life, Influenza, and HIV/AIDS (www.IHS.gov).**
- **Family conferences are appreciated with serious illness.**
- Mental illness may be seen as violation of tribal taboos or loss of harmony with the environment.

**Table 5.2** Tribal values versus European American values

| Tribal traditional cultural values | European American values |
|---|---|
| Group, clan, or tribal emphasis | Individual emphasis |
| Present oriented | Future oriented |
| Time, always with us | Time, use every minute |
| Age | Youth |
| Cooperation | Competition |
| Harmony with nature | Conquest of nature |
| Giving, sharing | Saving |
| Pragmatic | Theoretical |
| Mythological | Scientific |
| Patience | Impatience |
| Mystical | Skeptical |
| Shame | Guilt |
| Permissiveness | Social coercion |
| Extended family and clan | Immediate family |
| Nonaggressive | Aggressive |
| Modest | Over confident |
| Silence | Noise |
| Respect others' religion | Convert others to religion |
| Religion, way of life | Religion, a segment of life |
| Land, water, forest | Land, etc. |
| Belong to all | Private domains |
| Beneficial, reasonable | Avarice, greedy |
| Use of resource | Use of resource |

- Organ donation is not usually practiced, but it is usually an individual choice.

### *5.1.11 Health-Care Providers*

- "Shaman" is often used refer to traditional AI/AN healer.
- Spirits may be encouraged to occupy the shaman's body during public lodge ceremonies. Drum beating and chanting aid this process.
- Tribal spiritual healers may be used in combination with allopathic providers.
- Ask for disclosure of all treatments and explain possible complications.
- Partner with traditional healers.
- Similar to identifying over the counter, explore possible alternative treatments.

## 5.2 People of Navajo Heritage

The Navajo Nation claim the distinction of being the largest tribe: at least one-fourth Navajo blood is required to be considered a member of the tribe. The Navajo Indians are nomadic and wander great distances, searching for adequate grazing grounds for their sheep. Because of severe economic conditions and high unemployment rates, significant migration occurs into and out of the reservations. Commercial activities on reservations are limited to businesses owned by Navajo Indians or partnerships in which a Navajo must own the controlling interest. Many Native Americans who leave their reservations experience culture differences, and many return to their reservation on a regular basis to refresh and renew themselves through Blessingway ceremonies/dances.

Rug-weaving and jewelry making are the most common forms of art among the Navajo. They are also noted for sand paintings, which were traditionally used in healing ceremonies by medicine people and not originally intended for sale.

### *5.2.1 Communications*

- The Navajo language was not reduced to writing until the 1970s; consequently, many older people speak only their native language, and few are literate in the English language. The younger populations are usually bilingual, with their native tongue spoken primarily in the home. Be extremely careful when attempting to use the Navajo language, because minor variations in pronunciation may change the entire meaning of a word or phrase. Differences in pronunciation, particularly while speaking with an older Navajo, may cause a misunderstanding. Such misunderstandings make subsequent caring for the individual difficult. It is often safer to use an interpreter.
- Talking loudly among Navajo Indians is considered rude. Voice tones are quiet but not monotone. Their language is full of inflections with different meanings, making the language melodious with a quiet volume.
- Most generally do not share inner thoughts and feelings with anyone outside their clan. It may take nontribal health-care providers a long time to build trust.
- Most Navajo are comfortable with long periods of silence. Interest in what an individual says is shown through attentive listening skills. One may be considered immature if answers are given quickly or one interrupts another who is forming a response. Failing to allow adequate time for processing information may result in an inaccurate response or no response. Allow time to respond to questions. Follow-up with a clarification to ensure comments are understood.
- Touch among the Navajo is unacceptable unless one knows the person very well. Observation body language for determining cues related to the permissibility of touch. Ask permission and explain the reason for touch. Never assume that touch is ok even after establishing a relationship.

- The acceptable personal space for American Indians is greater than that of American cultures. Do not take offense if the patient stands at a greater distance from what you are accustomed.
- Shaking hands is the traditional greeting. The handshake is light, more of a passing of the hands. Shaking hands with the Navajo is an acceptable greeting even gesturing a slight head nod is acceptable.
- Pointing with the finger is considered rude. Rather than pointing a finger to indicate a direction, individuals shift their lips toward the desired direction. Do not point with your finger. Rather shift your lips to or verbally confirm the direction.
- Sustained direct eye contact is rude and possibly confrontational. Even close friends do not maintain eye contact. Not maintaining eye contact does not mean the patient or family is being less than truthful.
- The time sequence is present, past, and future, in that order. Very little planning is done for the future because the Navajo view is that many things are outside of one's control and may affect or change the future. In fact, the Navajo language does not have a future-tense verb. To plan for the future is sometimes viewed as foolish. Events do not always start on time, but rather time starts when the group gathers. To help prevent frustration in scheduling events, time factors need to be taken into consideration and the speaker made aware of these unique time perceptions. Appointments may not always be kept, especially if someone else in the clan needs help.
- Older people are addressed as grandmother or grandfather or as mother or father by their clan. Otherwise, they are called by a nickname. A health-care provider can call an older Navajo "grandmother" or "grandfather" as a sign of respect.

### 5.2.2 *Family Roles and Organization*

- Whereas men are considered important, grandmothers and mothers are at the center of Navajo society. When providing family care, note that no decision is made until the appropriate older woman is present. Find the appropriate gatekeeper; otherwise, time is lost and the problem must be addressed again later.
- The relationship between brother and sister is often more important than the relationship between husband and wife.
- Even though children may be named at birth, their names are not revealed until their first laugh, when they are considered to officially have a soul and self-identity. This protects the children and keeps them in tune with the Holy People.
- Infants are kept in the cradleboard until they begin walking. However, hip dysplasia may be exacerbated by cradleboards. The use of diapers has decreased the incidence of hip dysplasia because diapers bind the hips in a slightly abducted position. Encourage the use of diapers, and explain the hazard of using a cradleboard.
- The use of formula has become popular, resulting in an increased incidence of bottle caries because many babies go to bed with a bottle of juice, soda or pop.

This practice causes children to lose their teeth by the age of 4 years. Educate parents about dental caries and the hazards of sweet drinks and the importance of breastfeeding.

- A primary social premise is that no person has the right to speak for another. Thus, children are allowed to make their own decisions; for example, children may be allowed to decide if they want to take their medicine. Explain the importance of taking medicines as prescribed.
- An important ceremonial ritual for teenage girls is the onset of menarche, which is celebrated with special foods that symbolize passage into adulthood, with only aunts and grandmothers participating.
- The family unit consists of the nuclear family and relatives such as sisters, aunts, and their female descendants.
- Family goals do not center on wealth or the attainment of possessions. In fact, if one person has more wealth than other relatives, the member who has more has a responsibility to assist relatives who have less.
- A sister's children are considered the same as her own children. If a mother dies or for some other reason cannot care for her children, the grandmother or sister will raise the children as her own.
- Older people are looked on with clear deference. Younger adults are faced with the responsibility of caring for relatives.
- There are few nursing homes, and hospitals are forced to keep patients until nursing home placement is found. When nursing home placement is found, it may be at a great distance from the family, making it difficult for family visits. If a nursing home becomes the healthcare option, it is important to carry on traditional practices as this can make the transition easier.
- Alternative lifestyles are not discussed but are quietly accepted. Special individuals exist who are not looked on with disfavor but rather are accepted as being different.

### *5.2.3 Workforce Issues*

- Navajo are compelled to attend ceremonies and often must take time from work to do so. Balance the needs of the individual against organizational requirements for taking time off work. Allow annual leave or allow employees to earn religious compensatory time.
- Many American Indians avoid people with whom they are in conflict. Persistence in dealing immediately with the conflict causes additional ill feelings. If one tribal member has a disagreement with another, then that member does not address the second party face to face. Rather the member tells a third person, who then relays the information to the second party. Resolve conflicts through third-party compromise.
- One individual should not be singled out to answer a question because one's mistakes are generally not forgotten by the group. Improved success is achieved

if the American Indian is allowed to observe a task several times before being asked to demonstrate it. Their first effort at completing the task in front of a group should occur without error. This is especially true when delivering care to their people. Mistakes that have been made are discussed in the community. Allow adequate time for observation for a greater chance of success the first time a procedure is performed.

- Educational levels may be low. Work assignments should be made in clear, concrete terms. Input from the individual is at the discretion of that person.
- Younger supervisors may not be respected because they are not perceived as possessing the life experiences necessary to lead. Major decisions are made by the group with the assistance of the group leader, who is generally the senior woman. Thus, a young male manager on the reservation may face resistance when attempting to direct a work group.
- Indian Health Service is required to hire a Native American when possible. This law is referred to as the Indian Preference Law.
- English may not be their primary language. Allow extra time for a verbal response to think about a response and translate it into English. Then confirm the context of the statement.
- Most American Indian students are not good test takers. Although the individual may have the knowledge necessary to complete an examination, the translation of the knowledge into written form is especially difficult. When tested verbally on the same material, students tend to score higher on the examination. To the extent possible, examinations that consist of return demonstration and that do not use abstract terminology are preferred. Provide special classes that teach multiple-choice test taking.

### *5.2.4 Biocultural Ecology*

- Skin color varies from light brown to very dark brown. To assess for oxygenation in darker-skinned people, examine the mucous membranes and nail beds for capillary refill. Anemia is detected by examining the mucous membranes for pallor and the skin for a grayish hue. To assess for jaundice, examine the sclera rather than relying on skin hue. Newborns commonly have Mongolian spots on the sacral area. Do not mistake these spots for bruises and suspect child abuse.
- The Navajo appear Asian, with epithelial folds over the eyes. They are generally taller and thinner than other American Indian tribes. The Navajo have traditionally been good runners and excel in relay races and long-distance running. Remember that these characteristics are not seen with everyone; variations in this population do exist. Good history gathering is important and must be done in a sensitive, calm, nonthreatening manner.
- Table 5.3 lists commonly occurring health conditions in the Navajo in three categories: lifestyle, environment, and genetics.

**Table 5.3** Health conditions common to Navajos

| Health conditions | Causes |
|---|---|
| Shigellosis | Environment, lifestyle |
| Salmonella | Environment, lifestyle |
| Respiratory illnesses | Environment, lifestyle |
| Otitis media | Environment, lifestyle |
| Tuberculosis | Environment, lifestyle |
| Plague | Environment, lifestyle |
| Tick fever | Environment, lifestyle |
| Muerto Canyon Hanta virus | Environment, lifestyle |
| Diabetes mellitus | Genetic, environment, lifestyle |
| Combined immunodeficiency syndrome (SCIDS) | Genetic |
| Navajo neuropathy | Genetic |
| Albinism | Genetic |
| Alcohol misuse | Genetic, environment, lifestyle |
| Suicide | Environment, lifestyle |
| Homicide | Environment, lifestyle |
| Unintentional injuries | Environment, lifestyle |
| Gastrointestinal disease | Environment, lifestyle |
| Infant mortality | Environment, lifestyle |
| Heart disease | Genetic, environment, lifestyle |
| Cirrhosis | Environment, lifestyle |
| Fetal alcohol syndrome | Environment, lifestyle |

- The water on areas of the Navajo reservation is often impure and unchlorinated, making those who drink it susceptible to water-borne diseases such as shigellosis. Salmonella is common because of the lack of refrigeration and hypothermia because of frequent snowstorms and conditions that limit Navajo ability to gather wood. Teach patients how to protect themselves from rodent-borne diseases. At every opportunity teach about the combination of environmental and lifestyle factors to decrease the incidence of parasitic and infectious diseases and illnesses and injury prevention. Good handwashing techniques based on resources is important.
- Lidocaine reactions occur in 29% of the Navajo population as compared with 11–15% of European Americans. Carefully observe for lidocaine reactions particularly when providing emergent care such as for numbing areas like for providing sutures to an individual.

### 5.2.5 *High-Risk Health Behaviors*

- Alcohol use is more prevalent than any other form of substance abuse. Health problems related to alcoholism include motor vehicle accidents, homicide, suicide, and cirrhosis.
- Spousal abuse is common and is frequently related to alcohol use. The wife is the usual recipient of the abuse, but occasionally the husband is abused.
- Suicide is a problem among the adolescent population. Whereas space does not permit an extensive review of the American Indian/Alaskan Native National Suicide Prevention Strategic Plan at https://www.cdc.gov/suicide/pdf/suicideprevention_strategicplan-508.pdf health-care. Providers are encouraged to access this plan. Educate elementary school teachers on suicide ideology and make referrals to health-care providers as necessary.
- Noncompliance with seat-belt use is high. Teach patients the importance of using child safety seats, helmets, and seat belts.

### 5.2.6 *Nutrition*

- Life events and religious ceremonies are celebrated with food. Food is not generally associated with promoting health or illness. Herbs are used in the treatment of many illnesses to cleanse the body of ill spirits or poisons. Encourage and teach about healthy food choices and preparation practices. Access the link between what families traditionally use and provide nutritional support to what is used in the home.
- Sheep are a major source of meat, and sheep brains are a delicacy. Fry bread and mutton are cooked in lard. Access to fresh fruits or vegetables is minimal except during the fall. Squash is common at harvest time. Corn is an important staple. Corn pollen is used in the Blessingway and many other ceremonies.
- Diets may be deficient in vitamin D because many individuals suffer from lactose intolerance or do not drink milk. Recommend nondairy foods that are high in calcium or making stews and purees using animal bones.

### 5.2.7 *Pregnancy and Childbearing Practices*

- Traditional Navajos do not practice birth control and, thus, do not limit the size of their families. Large families are considered favorably because, in times past, many children died at an early age. Encourage prenatal care, carefully explaining the detrimental effects on the mother and infant if prenatal care is not sought.
- Twins are not considered favorably and are frequently believed to be the work of a witch, in which case one of the babies must die. Sometimes the mother may

have difficulty caring for two infants. Twins may be readmitted to the hospital for neglect and failure to thrive. Culturally sensitive counseling related to adoption. Encourage tribal members to adopt the children.

- It is especially important to adhere to the many prescriptive and restrictive taboo practices related to pregnancy, which involve both husband and wife. See Table 5.4, Navajo Taboos Regarding Expectant Women, and Table 5.5, Taking Care of Yourself During Pregnancy: Navajo Rules for Expectant Couples.
- During labor, the mother wears birthing necklaces made of juniper seeds and beads to assist with a safe birth. Woven belts or sashes are used to help push the baby out. This practice is also used by Navajo midwives in caring for their patients.
- Many Navajo are reluctant to deliver their babies in hospital settings. They know that people have died in hospitals and thus perceive that pregnant women should not be around the dead or in a place where people have died.
- A taboo practice among the Navajo is purchasing clothes for an infant before birth. Do not interpret not buying clothes for the infant as not wanting the baby. Simply provide emotional support to the expecting mother.

**Table 5.4** Navajo Taboos regarding expectant women

| |
|---|
| 1. Do not wear two hats at once; you will have twins (or two wives) |
| 2. Do not hit babies in the mouth; they will be stubborn and slow to talk |
| 3. Do not have a weaving comb (rug) with more than five points; your baby will have extra fingers |
| 4. Do not have a baby cross its fingers; its mother will have another one right away |
| 5. Do not swallow gum while you are pregnant; the baby will have a birthmark |
| 6. Do not kill animals while your wife is pregnant; the baby will look like a bird |
| 7. Do not stand in the doorway when a pregnant woman is present |
| 8. Do not make a slingshot while you are pregnant; the baby will be crippled |
| 9. Do not go to ceremonies while pregnant; it will have a bad effect on the baby |
| 10. Do not eat a lot of sweet stuff while you are pregnant; the baby won't be strong |
| 11. Do not sleep too much when you are about to have a baby; the baby will mark your face with dark spots |
| 12. Do not look at a dead person or animal while you are pregnant; the baby will be sickly because of bad luck |
| 13. Do not jump around if you are pregnant or ride a horse; it will induce labor |
| 14. Do not cut gloves off at the knuckles, the baby will have short, round fingers |
| 15. Do not cut a baby's hair when it is small; it won't think right when it gets older |
| 16. Do not put on a Yei mask while your wife is pregnant; the baby will have a big head and look strange |
| 17. Do not let a baby's head stay to one side in the cradleboard; it will have a wide head |
| 18. Do not watch or look at an accident while your wife is present; it will affect the baby |
| 19. Do not sew on a saddle while your wife is pregnant; it will ruin the baby's mouth |

https://www.aastec.net/wp-content/uploads/2020/10/MaternalChildHealth.V7.HiRez_.Sept_.30.20.pdf

**Table 5.5** Taking care of yourself during pregnancy: Navajo rules for expectant couples

| **During prenatal period** | |
|---|---|
| *Mind/soul* | |
| Do's<br>• Keep the peace<br>• Keep thoughts good<br>• Talk with "corn pollen sprinkled" words<br>• Say morning (dawn) prayers<br>• Have shielding prayers if you have nightmares | Don'ts<br>• Argue with partner or others<br>• Scold children<br>• Allow bad thoughts to occupy mind for long period of time<br>• Talk negatively or with criticism |
| *Body* | |
| Do's<br>• Eat foods good for baby<br>• Get up early and walk around<br>• Have a Blessingway ceremony for a safe delivery | Don'ts<br>• Drink milk or eat salt or foods taken away by Navajo ceremonies<br>• Lie around too much<br>• Tie knots<br>• Attend funerals or look at body of deceased person<br>• Be with sick people for long or go to crowded place<br>• Attend healing ceremonies for sick people like "Yei Bei Chai Dance"<br>• Look at dead animals or taxidermic trophies<br>• Look at eclipse of moon or sun<br>• Make plans for baby or prepare layette sets until after birth<br>• Lift heavy things<br>• Kill living things or cut a sheep's throat<br>• Weave rugs or make pottery |
| **During labor** | |
| *Mind/soul* | |
| Do's<br>• Think about a good delivery<br>• Have medicine people do "Singing Out Baby" chants<br>• Have medicine person perform "Unraveling" songs if necessary | Don'ts<br>• Let too many people observe labor; only people who are helping you in some way |
| *Body* | |

(continued)

**Table 5.5** (continued)

| Do's | Don'ts |
|---|---|
| • Loosen your hair<br>• Drink cornmeal mush<br>• Wear juniper seed beads<br>• Burn cedar<br>• Hold onto sash belt when ready to push<br>• Have someone apply gentle fundal pressure during pushing effort<br>• Get in squatting position for pushing<br>• Drink herbal tea to relax if necessary<br>• Drink herbal tea to strengthen contractions if necessary | • Braid or tie hair in a knot<br>• Tie knots |
| **After birth of baby (postpartum period)** | |
| Do's | Don'ts |
| • Bury the placenta<br>• Drink juniper/ash tea to cleanse your insides<br>• Drink blue cornmeal mush<br>• Smear baby's first stool on your face<br>• Breast-feed your baby<br>• Wrap sash belt around waist for days after delivery | • Drink cold liquids or be in cold draft<br>• Smell afterbirth blood for too long<br>• Show signs of displeasure if baby soils on you or during diaper change<br>• Have sexual intercourse for 3 months after delivery |

https://www.aastec.net/wp-content/uploads/2020/10/MaternalChildHealth.V7.HiRez_.Sept_.30.20.pdf

- Immediately after birth, the placenta is buried as a symbol of the child being tied to the land. Sometimes it is burned in a fire. This is considered a safe place because fire is sacred and protects the baby against evil spirits.
- After birth, the baby is given a mixture with juniper bark to cleanse its insides and rid it of mucus. In addition, a ceremonial food of corn pollen and boiled water is given. Corn symbolizes healthy nutrients and an enduring nature.

### *5.2.8 Death Rituals*

- One death taboo involves talking with patients concerning a fatal disease or illness. Effective discussions require that the issue be presented in the third person, as if the illness or disorder occurred with someone else. Never suggest that the patient is dying. To do so would imply that the provider wishes the patient dead. If the patient does die, it would imply that the provider might have evil powers.
- The body must go into the afterlife as whole as possible. The body is not buried for approximately 4 days after death.
- A cleansing ceremony must be performed after an individual dies or the spirit of the dead person may try to assume control of someone else's spirit. Family members are reluctant to deal with the body because those who work with the dead must have a ceremony to protect themselves from the deceased's spirit.

- If the person dies at home, the home/dwelling must be abandoned, or a ceremony must be held to cleanse it.
- Before burial, a ring is placed on the index finger of each hand, and the shoes are put on the wrong feet. This allows living relatives to recognize individuals if they come back and present themselves at ceremonial dances. Do not wear rings on your index fingers, as older people may not want to be around them.
- Excessive displays of emotion are not considered favorably among some tribes. Support survivors and permit family bereavement and grieving in a culturally congruent and sensitive manner that respects the beliefs of the tribe.

### *5.2.9 Spirituality*

- The American Indian religion predominates in many tribes. Sometimes hospital admissions are accompanied by traditional ceremonies and consultation with a pastor. Even if people are strong in their adopted beliefs, they honor their parents and families by having a traditional healing ceremony.
- Many Navajo start the day with prayer, meditation, corn pollen, and running in the direction of the sun.
- Spirituality for most is based on harmony with nature. The meaning of life is derived from being in harmony with nature. The individual's source of strength comes from the inner self and also depends on being in harmony with one's surrounding.
- Prayers ask for harmony with nature and for health and invite blessings to help the person exist in harmony with the earth and sky. Prayer helps the Navajo to attain fulfillment and inner peace with themselves and their environment.
- Spirituality cannot be separated from the healing process in holistic ceremonies. Illness results from not being in harmony with nature, from the spirits of evil people such as witches, or from violation of taboos. Healing ceremonies restore an individual's balance mentally, physically, and spiritually. Assist with arrangements for healing ceremonies and practicing traditional religious beliefs and prayer.
- The core concepts of traditional Indian medicine are shown in Table 5.6.

### *5.2.10 Health-Care Practices*

- Many older people do not understand the theory of germs.
- Asking patients questions in order to make a diagnosis fosters mistrust. This approach is in conflict with the practice of traditional medicine men who tell people what is wrong without their having to say anything.
- The medicine man is expected to diagnose the illness and prescribe necessary treatments for regaining health. In Western health care, the health-care provider

**Table 5.6** Basic concepts of traditional Indian medicine

| |
|---|
| • Indians believe in a Supreme Creator |
| • Each person is a threefold being composed of mind, body, and spirit |
| • All physical things, living and non-living, are part of the spiritual world |
| • The spirit existed before it came into the body, and it will exist after it leaves the body |
| • Illness affects the mind and the spirit as well as the body |
| • Wellness is harmony |
| • Natural unwellness is caused by violation of a taboo |
| • Unnatural wellness is caused by witchcraft |
| • Each individual is responsible for his or her own health |

asks the patient what he or she thinks is wrong and then prescribes a treatment. This practice is sometimes interpreted by American Indians as ignorance on the part of the white healer.

- Often great distances must be traveled to reach hospitals or health-care facilities. Many families do not have adequate transportation and must wait for others to transport them into town. Immunizations may be missed because parents do not have transportation. Pay close attention to the immunization status of patients on their arrival at the emergency department or clinic. If patients are not current with immunizations, scheduling an appointment may be a waste of time because they may not be able to return until a ride is found. Take time to administer the immunization on the spot or make a referral to the public health nursing office. It is important to address as many needs as possible during one visit as to limit the need for another visit.
- Pain control may be ineffective because the actual intensity of the individual's pain is not obvious to the health-care provider and because patients do not request pain medication.
- Pain is viewed as something that is to be endured. Herbal medicines may be used without the knowledge of the health-care provider. Establishing trust will encourage patients to fully disclose herbal treatments used for pain control. Offer pain medicine and explain that it will promote healing.
- Mental illness results from witches or witching (placing a curse) on a person. In these instances, a healer who deals with dreams or a crystal-gazer is consulted. Individuals may wear turquoise to ward off evil; however, a person who wears too much turquoise is sometimes thought to be an evil person and, thus, someone to avoid.
- Rehabilitation as a concept is relatively new. Explain concepts of rehabilitation particular how it can link with the culture.

- Those with physical or mental handicaps are not considered different; rather, the limitation is accepted, and a role is found for them within the society.
- Cultural perceptions of the sick role for the American Indian are based on the ideal of maintaining harmony with nature and with others. Older people frequently work even when they are seriously ill and often must be encouraged to rest.
- Autopsy and organ donation are unacceptable practices to traditional American Indians. Explain that autopsy is a legal requirement in some instances.

### *5.2.11 Health-Care Provider*

- Native healers are divided primarily into three categories: those working with the power of good, the power of evil, or both. Generally, these healers are divinely chosen and promote activities that encourage self-discipline, self-control, and acute body awareness. Some healers are endowed with supernatural powers, whereas others only have knowledge of herbs and specific manipulations. Acceptance of Western medicine is variable, with a blending of traditional health-care beliefs.
- Many are suspicious of American Indian physicians. Many health concerns of American Indians can be treated by both traditional and Western healers in a culturally competent manner when these health-care providers are willing to work together and respect each other's differences.
- Western health-care providers, traditional medicine men, and herbal healers receive respect on the reservation. However, not all individuals accord equal respect to these groups, and many prefer one group over the other or use all three.
- Navajo tribal health-care providers divide their knowledge into preventive measures, treatment regimens, and health maintenance. An example of a preventive measure is carrying an object or a pouch filled with objects prescribed by a medicine man that wards off the evil of a witch. Do not remove medicine pouches from patients.
- The various types of traditional health-care providers are given in Table 5.7.
- Male health-care providers are generally limited in the care they provide to women, especially during menses. Women are generally modest and wear several layers of slips. This practice is very common among elderly women. Provide a same-gender health-care provider for intimate care ensuring a family member is available. Not necessarily right at the examination table but within the room.

**Table 5.7** Traditional American Indian health-care providers

| |
|---|
| 1. People who can use their power only for good can transform themselves into other forms of life and can maintain cultural integration in times of stress |
| 2. People who can use their powers for both evil and good are expected to do evil against someone's enemies. People in this group know witchcraft, poisons, and ceremonies designed to afflict the enemy |
| 3. The diviner diagnostician, such as a crystal-gazer, can see what caused the problem but cannot implement a treatment. Another example of this type is a hand trembler. These people, instead of using crystals, practice hand trembling over the sick person to determine the cause of an illness |
| 4. Specialist medicine people treat the disease after it has been diagnosed and specialize in the use of herbs, massage, or midwifery |
| 5. Those who care for the soul send guardian spirits to restore a lost soul |
| 6. Singers, who are considered to be the most special, cure through the power of their song. These healers use laying on of hands and usually remove objects or draw disease-causing objects from the body while singing |

## 5.3 Reflective Exercises

1. How does suicide, violence, and rape in AI/AN communities compare with the national average? What can health-care providers do to help decrease these conditions.
2. What are the major health conditions of AI/ANs? On which of these can the health-care provider have a major impact.
3. What are the major high-risk health behaviors among AI/ANs? What are some major tactics to help decrease these behaviors?
4. List the major values between AI/AN tribes and the European American culture.
5. Some AI/ANs lack trust in Indian physicians. Why might this happen? What can you do to help gain their trust in Indian physicians?
6. What are the names of traditional healers? What conditions do they treat.
7. What tactics can be used to incorporate traditional healers with allopathic providers?
8. List several birthing practices of the Navajo.
9. What are the tenets of the Native American Church?
10. How might you approach explaining to a patient and family of an impending death of a Navajo?

## Bibliography

2020 Navajo Nation Maternal and Child Health Needs Assessment (2020) https://www.aastec.net/wp-content/uploads/2020/10/MaternalChildHealth.V7.HiRez_.Sept_.30.20.pdf

Adakai M, Sandoval-Rosario M, Xu F, Aseret-Manygoats T, Allison M, Greenlund KJ et al (2018) Health disparities among American Indians/Alaska Natives—Arizona, 2017. Morb Mort Wkly Rep 67(47):1314–1318

American Indian Religious Freedom Act of 1978 (AIRFA) (1978) 42 U.S.C. § 1996

Anderson RW (2016) Native American reservation constitutions. Const Polit Econ 27:377–398. https://doi.org/10.1007/s10602-016-9225-7

Blood Quantum (2019) American Indian Enterprise & Business Council-CDC, NGO, 501C3. http://www.aiebc.org

Blood Quantum and Sovereignty: A Guide (2022) Native Governance Center. https://nativegov.org/

Breathett K, Sims M, Gross M, Jackson EA, Jones EJ, Navas-Acien A et al (2020) Cardiovascular health in American Indians and Alaska Natives a scientific statement from the American Heart Association. Circulation 141:e948–e959

Bureau of Indian Affairs (BIA) (2023). https://www.bia.gov/bia

Carlson LA, Roberts MA (2006) Indian lands, "Squatterism," and slavery: economic interests and the passage of the Indian removal act of 1830. Explor Econ Hist 43:486–504. https://doi.org/10.1016/j.eeh.2005.06.003

Cherokee Tribal Policy (2023) http://www.cherokee.org.

Childbearing Practices of Native Alaskans (2020)

Clark Callister L (2016) Integrating cultural beliefs and practices when caring for childbearing women and families. http://nursekey.com/integrating-cultural-beliefs-and-practices-when-caring-for-childbearing-women-and-families/

Cooper LA, Ortega AN, Ammerman AS, Buchwald D, Paskett ED, Powell LH et al (2015) Beyond health equity: achieving wellness within American Indian and Alaska Native communities. Am. J. Public Health 105(S3):S376–S379

Dennis MK, Brewer JP (2017) Rearing generations: Lakota grandparents commitment to family and community. J Across Cult Gerontol 32:95–113. https://doi.org/10.1007/s10823-016-9299-8

Department of Interior Bureau of Indian Affairs Federal Register (2019) Indian entitles recognized by and eligible to receive services from the United States Bureau of Indian Affairs. 84(22). 1 Feb 2019/Notices

Diabetes and American Indians/Alaska Natives. https://minorityhealth.hhs.gov/diabetes-and-american-indiansalaska-natives

Drug Abuse & Native American Indian Youth (2014) http://dodgingbullets.org/drug-abuse-native-american-indian-youth/. Accessed 6 Oct 2014

Ficklin E, Tehee M, Killgore RM, Isaacs D, Mack S, Ellington T (2022) Fighting for our sisters: community advocacy and action for missing and murdered Indigenous women and girls. J Soc Issues 78:53–78

Gaffney A, McCormick D (2017) America: equity and equality in health 2. The Affordable Care Act: implications for health-care equity. Lancet 389:1442–1452

Gampa V, Smith C, Muskett O, King C, Sehn H, Malone J et al (2017) Cultural elements underlying the community health representative-client relationship on Navajo Nation. BMC Health Serv Res 17:19. https://doi.org/10.1186/s12913-016-1956-7

Garrow CE (2015) Tribal criminal law and procedure, 2nd edn. Rowman & Littlefield, London

Gone JP, Blumstein KP, Dominic D, Fox N, Jacobs J, Lynn RS et al (2017) Teaching tradition: diverse perspectives on the pilot urban American Indian Traditional Spirituality Program. J Community Psychol 59(3-4):382–389. https://doi.org/10.1002/ajcp.12144

Greenleaf AT, Roessger KM, Williams JM, Motsenbocker J (2019) Effects of a rite of passage ceremony on veterans' well-being. J Couns Dev 97:171–181. https://doi.org/10.1002/jcad.12248

Greensky C, Stapleton MA, Walsh K, Gibbs L, Abrahason J, Finnie DM et al (2014) A qualitative study of traditional healing practices among American Indians with chronic pain. Pain Med 15:1795–1802

Halabi S (2019) The role of provinces, states, and territories in shaping federal policy for indigenous peoples health. Am Rev Can Stud 49(2):231–246. https://doi.org/10.1080/02722011.2019.1613797

Hanson JD (2012) Understanding prenatal health care for American Indian women in a northern plains tribe. J Transcult Nurs 23(1):29–37. https://doi.org/10.1177/1043659611423826

Haozous EA, Strickland CJ, Palacios JF, Arambula Solomon TG (2014) Blood politics, ethnic identity, and racial misclassification among American Indians and Alaska Natives. IJERPH 2014:1–9. https://doi.org/10.1155/2014/321604
Health Care for Urban American Indian and Alaska Native Women (2015) Number 515 (Reaffirmed 2021). https://www.acog.org/clinical/clinical-guidance/committee-opinion/articles/2012/01/health-care-for-urban-american-indian-and-alaska-native-women
Indian Affairs Manual (2019) https://www.bia.gov/policy-forms/manual
Indian Health Service (2023) https://www.ihs.gov
Indian Health Service "Fact Sheets" (2023) https://www.ihs.gov/newsroom/factsheets/
Indian Health Service Indian Health Disparities (2019) https://www.ihs.gov/aboutihs
Indian Preference Law (2000) Indian Health Service. https://www.opm.gov/
Indian Removal Act: 1830 (2023) Library of Congress. http://www.loc.gov/rr/program/bib/ourdocs/Indian.html
Johansson P, Schober DJ, Leon MA, Brueggemann G, Tutsch SF, Lyden E et al (2019) Adapting an evidence-based cardiovascular disease risk reduction intervention to urban American Indians. J Health Care Poor Underserved 30(2):618–636. https://doi.org/10.1353/hpu.2019.0046
Johnson-Jennings M, Billiot S, Walters K (2020) Returning to our roots: tribal health and wellness through land-based healing. Genealogy 4(91):1–21
Joseph G (2021) The role of sovereignty in Indigenous community-based health interventions: a qualitative metasynthesis. Am J Community Psychol 2023:1–18
Ka'apu K, Burnette CE (2019) A culturally informed systematic review of mental health disparities among adult indigenous men and women of the USA: what is known? Br J Soc Work 49:880–898
Leavitt RA, Ertl A, Sheats K, Petrosky E, Ivey-Stephenson A, Fowler KA (2018) Suicides among American Indian/Alaska Natives—national violent death reporting system, 18 states, 2003-2014. Centers for Dis Control Prev Morb Mort Wkly Rep 67(8):237–242. https://doi.org/10.15585/mmwr.mm6708a1
Lewis JP, Boyd K, Allen J (2018) "We raise our grandchildren as our own:" Alaska Native grandparents raising grandchildren in Southwest Alaska. J Cross Cult Gerontol 33:265–286. https://doi.org/10.1007/s10823-018-9350-z
Lopez JH, Bobroff KL (2019) Rooted indigenous core values: culturally appropriate curriculum and methods for civic education reflective of Native American culture and learning styles. Multicult Perspect 21(2):119–126. https://doi.org/10.1080/15210960.2019.1606640
Masotti P, Dennem J, Banuelos K, Seneca C, Valerio-Leonce G, Inong CT et al (2023) The Culture is Prevention Project: measuring cultural connectedness and providing evidence that culture is a social determinant of health for Native Americans. BMC Pub Health 23(1):741–751. https://doi.org/10.1186/s12889-023-15587-x
Mihesuah DA (2015) Sustenance and health among the five tribes in Indian territory, postremoval to statehood. Ethnohistory 62:263–284. https://doi.org/10.1215/00141801-2854317
Mitchell FM (2019) Water (in)security and American Indian Health: social and environmental justice implications for policy, practice, and research. R Soc Public Health 176:98–105. https://doi.org/10.1016/j.puhe.2018.10.010
National Archives and Records Administration (2019) U.S. Department of the Interior. A guide to tracing American Indian & Alaska Native ancestry. http://www.archives.gov/research/native-americans/dawes/intro.html
Native American Heritage Month (2023) https://www.nativeamericanheritagemonth.gov/. Accessed 30 Nov 2023
Native Knowledge 360° Framework for Essential Understandings about American Indians. Smithsonian National Museum of the American Indian. https://americanindian.si.edu/nk360/pdf/NMAI-Essential-Understandings.pdf
Nava LT, Zambrano JM, Arviso KP, Brochetti D, Becker KI (2015) Nutrition-based interventions to address metabolic syndrome in the Navajo: a systematic review. JCN 24:3024–3045. https://doi-org.libproxy.uccs.edu/10.1111/jocn.12921

Navajo Nation Government Reform Project Prepared By Dine Policy Institute (2013) https://ongd.navajo-nsn.gov/

Navajo Nation Human Rights Commission 2014 Shadow report, regarding the United States of America report submitted by states parties under Article 9 of the International Convention on the Elimination of All Forms of Racial Discrimination. Navajo Nation, AZ

Navajo Tribal Policy (2023) http://www.cherokee.org/Services/146/Page/Default.aspx.

Ng AE, Adjaye-Gbewonyo D, Vahratian A (2023) Health conditions and health care use among American Indian and Alaska Native adults by tribal land residential status: United States, 2019–2021. U.S. Department of Health and Human Services CDC National Center for Health Statistics No. 185. https://www.cdc.gov/nchs/products/index.htm

O'Keefe VM, Reger GM (2017) Suicide among American Indian/Alaska Native military service members and veterans. Psychol Serv 14(3):289–294. https://doi.org/10.1037/ser0000117

Office of Indian Education (2023). http://www.ade.state.az.us/asd/indianed. Accessed 1 Dec 2023

Office of Program Planning and Development Navajo Area Indian Health Service: Area Profile (1988) Navajo Area Indian Health Service

Official Native American Church of North America (2024) https://nativeamerican.church/

Ornelas IJ, Osterbauer K, Woo L, Bishop SK, Deschenie D, Beresford SAA (2018) Gardening for health: patterns of gardening and fruit and vegetable consumption among the Navajo. J Community Health 43:1053–1060. https://doi.org/10.1007/s10900-018-0521-1

Pacheco JA, Pacheco CM, Lewis C, Williams C, Barnes C, Rosenwasser L et al (2015) Ensuring healthy American Indian generations for tomorrow through safe and healthy indoor environments. Int J Environ Res Public Health 12:2810–2822. https://doi.org/10.3390/ijerph120302810

Payne HE, Steele M, Bingham JL, Sloan CD (2018) Identifying and reducing disparities in mental health outcomes among American Indians and Alaskan Natives using public health, mental healthcare and legal perspectives. Adm Policy Ment Health Ment Health Serv Res 45:5–14. https://doi.org/10.1007/s10488-016-0777-7

Pollock NJ, Naicker K, Loro A, Mulay S, Colman I (2018) Global incidence of suicide among Indigenous peoples: a systematic review. BMC Med 16(1):145–152. https://doi.org/10.1186/s12916-018-1115-6

Public Law 67-85 (1921) The Snyder Act of 1921. https://www.bia.gov/

Public Law 93-638 (1975) Indian Self-Determination and Education Assistance Act of 1975. https://www.bia.gov/

Purnall LD, Fenkl EA (2021) Textbook for transcultural health care: a population approach cultural competence concepts in nursing care, 5th edn. Springer, Cham

Ross RE, Garfield LD, Brown DS, Raghavan R (2015) The Affordable Care Act and implications for health care services for American Indian and Alaska Native individuals. J Health Care Poor Underserved 26(4):1081–1088. https://doi.org/10.1353/hpu.2015.0129

Sanchez-Rivera AI, Jacobs P, Spence C (2023) Detailed data for hundreds of American Indian and Alaska Native tribes. A look at the largest American Indian and Alaska Native tribes and villages in the nation, tribal areas and states. https://www.census.gov/library/stories/2023/10/2020-census-dhc-a-aian-population.html

Schmidt RW (2011) American Indian identity and Blood Quantum in the 21st century: a critical review. Journal of Anthropology 2011:1–9

Sequist TD (2021) Improving the health of the American Indian and Alaska Native Population. JAMA 325(1):1035–1036

Shaeffer Riley N (2016) The new trail of tears. Encounter Books, New York

Sharma G, Keliher A, Deen J, Parker T, Hagerty T, Choi EE et al (2023) AHA scientific statement status of maternal cardiovascular health in American Indian and Alaska Native individuals: a scientific statement from the American Heart Association. Cir Cardiovasc Qual Outcomes 2023:443–455. http://www.ahajournals.org/journal/circoutcomes

Spirituality (2002) Alaskan Native Knowledge Network. http://www.ankn.uaf.edu/NPE/spirituality.html

Stopbullying.gov (2023) https://www.stopbullying.gov/

Substance Abuse and Mental Health Services Administration (2019) About us. http://www.samhsa.gov/about-us

Traditional Native American Values and Behavior (2023) http://nwindian.evergreen.edu/curriculum/ValuesBehaviors.pdf

Tribal Enrollment Process (2023) U.S. Department of the Interior. https://www.doi.gov/tribes/enrollment. Accessed 1 Dec 2023

Trude ACB, Kharmats A, Jock B, Liu D, Lee K, Martins PA et al (2015) Patterns of food consumption are associated with obesity, self-reported diabetes and cardiovascular disease in five American Indian communities. Ecol Food Nutr 54:437–454. https://doi.org/10.1080/03670244.2014.922070

U.S. Census Bureau (2020) Census Report for 2020. http://www.census.gov

U.S. Department of the Interior: Bureau of Indian Affairs (2016) https://www.bia.gov/

U.S. Department of the Interior: Bureau of Indian Education (2016) https://www.bie.edu/

U.S. Government Accountability Office GAO (2012) Report to the honorable Dan Boren, house of representatives Indian issues federal funding for non-federally recognized tribes GAO-12-348

United Nations Human Rights Office of the high commissioner indigenous peoples and the United Nations human rights system, Fact Sheet No. 9 Rev. 2

Villarroel MA, Clarke TC, Norris T (2020) Health of American Indian and Alaska Native adults, by urbanization level: United States, 2014–2018. U.S. Department of Health and Human Services CDC National Center for Health Statistics No. 372. https://www.cdc.gov/nchs/products/index.htm

Walters KL, Johnson-Jennings M, Stroud S, Rasmus S, Charles B, John S et al (2018) Growing from our roots: strategies for developing culturally grounded health promotion interventions in American Indians, Alaska Native, and Native Hawaiian communities. Prev Sci 21:1–11. https://doi.org/10.1007/s11121-018-0952-z

Wang H (2021) Why the Navajo Nation was hit so hard by coronavirus: understanding the disproportionate impact of the COVID-19 pandemic. Appl Geogr 134:1–10. https://doi.org/10.1016/j.apgeog.2021.102526

WERNATIVE. https://www.wernative.org/

Yeager Washington E, van Hover S (2011) Teaching about Navajo citizenship and sovereignty. Soc Stud 102(2):80–87. https://doi.org/10.1080/00377996.2010.497177

Zamora-Kapoor A, Sinclair K, Nelson L, Lee H, Buchwald D (2019) Obesity risk factors in American Indians and Alaska Natives: a systematic review. Pub Health 174:85–96. https://doi.org/10.1016/j.puhe.2019.05.021

# Chapter 6
# The Amish

**Christine Nelson-Tuttle**

## 6.1 Overview and Heritage

**Today's Amish live in rural areas in a band of more than 20 states, stretching westward from Pennsylvania, Ohio, and Indiana to as far west as Montana, with some scattered settlements in Florida and the province of Ontario, Canada. Recently, Amish have started new communities in six states that had not had settlements prior: Colorado, Nebraska, New Mexico, South Dakota, Vermont, and Wyoming. Population estimates of Amish in 2022 were 373,620 with an increase of approximately 12,150 since 2021. Amish communities in Canada include the provinces of Ontario, Manitoba, New Brunswick, and Prince Edward Island. As affordable land has decreased, some Amish have migrated to Argentina and Bolivia. The Old Order Amish, the so-called for their strict observance of traditional ways that distinguish them from other, more progressive "plain folk," are the largest and most notable group. The Amish emerged after 1693, when they parted ways with the Anabaptist movement that originated in Switzerland in 1525. After experiencing severe persecution and martyrdom in Europe, the Amish and related groups immigrated to North America in the seventeenth and eighteenth centuries. Some variant groups are named after their factional leaders (for example, Egli and Beachy Amish); some are called conservative Amish Mennonites; and others, New Order Amish. No Amish live in Europe today after leaving in the nineteenth and twentieth centuries due to religious persecution.**

C. Nelson-Tuttle (✉)
College of Nursing, Niagara University, Lewiston, NY, USA
e-mail: tuttlec@dyc.edu



E. A. Fenkl, L. D. Purnell (eds.), *Handbook for Culturally Competent Care*,
https://doi.org/10.1007/978-3-031-70492-5_6

Even though the Amish have persisted in relative social isolation, they continue to adapt and change at their own pace, accepting innovations selectively. Although most Amish homes do not have electricity and electronic labor-saving devices and appliances, that does not preclude their openness to using state-of-the-art medical technology necessary for health promotion. The Amish selectively adapt to technology when it is necessary; selected use telephones for business and health-care purposes is more commonly permitted in the more progressive orders of Amish. They also may use other technology such as electrically powered ventilators for home care when required. With the advent of technology and not being able to order from such catalogs as Sears, they have been forced to order online. In many Amish communities, local stores that cater to the Amish will place electronic orders for them.

### 6.1.1 Communications

- English is the language of school, of written and print communications, and of contact with most non-Amish outsiders.
- At home and in the immediate Amish communities, **Deitsch/Deutsch**, or Pennsylvania German, is used. In this highly contextual culture, less overt verbal communication is required, and more reliance is placed on implicit, often unspoken, understandings. Much of what passes for "general knowledge" in the information-rich popular culture is screened, or filtered, out of Amish awareness.
- **Expect all Amish patients of school age and older to be fluently bilingual. They can readily understand spoken and written directions and answer questions presented in English, although their own terms for some symptoms and illnesses may not have exact equivalents in Deitsch and English.**
- Most Amish have severely restricted their own access to print media, permitting only a few newspapers and periodicals. Most have also rejected electronic media such as radios, television, and entertainment and information applications of film and computers. Since the decline of mail order catalogs, many Amish consider alternative methods of securing necessary supplies. In some communities, the Amish are able to go to community stores where personnel will order supplies and equipment for them online.
- The Amish are clearly not outwardly demonstrative or exuberant. Fondness and love of family members is held deeply but privately. ***Demut***, humility, is a priority value, the effects of which may be observed in public as a modest and unassuming demeanor. ***Hochmut***, pride or arrogance, is avoided.
- **The expression of joy and suffering is not entirely subdued by dour or stoic silence.**
- The Amish present in an unpretentious, quiet manner, with modest outward dress in plain colors lacking any ornament, jewelry, or cosmetics.
- They are unassertive and nonaggressive and avoid confrontational speech styles and public displays of emotion.

- Amish self-perception is grounded in the present. In public, Amish avoid eye contact with non-Amish, but in one-on-one clinical contacts, patients can be expected to express openness and candor with unhesitating eye contact.
- They are generally punctual and conscientious about keeping appointments.
- Using first names with Amish people is appropriate because there are only a limited number of surnames. For example, it is preferable to use John or Mary during personal contacts rather than Mr. or Mrs. Miller. Within Amish communities, individuals are identified further by nicknames, residence, linked with description of occupation or a spouse's given name.
- **Greet Amish patients with a handshake and a smile.**
- **Ask Amish if they have access to a telephone where messages may be left if needed.**
- **If educational videos are shown in the health-care organization, ask Amish if they wish to view them or if they prefer printed material. If preoperative or postoperative/discharge care is usually delivered via online video, they may be willing to view them in the office or health-care facility if no other reasonable options are available. If this is necessary, offer them paper and pencil to take notes for review at home.**

### *6.1.2 Family Roles and Organization*

- Amish society is patriarchal, but women are accorded high status and respect. Practically speaking, husband and wife may share equally in decisions regarding the family farming business. In public, the wife may assume a retiring role, deferring to her husband, but in private they are typically partners. **Do not assume that the spokesperson for the family is the primary decision-maker. Verify who the health-care provider has permission to share information with to facilitate health-care decision-making for the family. This should be documented in the patient chart.**
- The Amish family pattern is the three-generational family. This kinship network includes consanguine relatives consisting of the parental unit and households of married children and their offspring. It is not uncommon for multi-generations to reside together.
- The highest priority for parents is child rearing, a charge given them by the church.
- Young people older than 16 years may experiment with non-Amish dress and behavior, but the expectation is that they will be baptized Amish before marriage. Unmarried children live in the parents' home until marriage. Single adults are included in the social fabric of the community.
- Families are the units that make up church districts. The size of church districts is measured by the number of families rather than by the number of church members. When the church district increases to a size that is not manageable, a new district may be formed to splinter off.

- Grandparents have respected status as elders; they provide valuable advice, material support, and services that include child care to the younger generation. Family emotional and physical proximity to older adults facilitates elder care.
- **Include the extended nuclear family in health educational activities.**

### 6.1.3 Workforce Issues

- The Amish place a high value on hard work, with little time off for leisure or recreation.
- In the past, the Amish have worked almost exclusively in agriculture and farm-related tasks. Amish avoidance of compromising associations with "worldly" organizations, such as labor unions, restricts them to non-union work, which often pays lower hourly rates. Work settings that rely heavily on modern technology are not open to the Amish.
- Some Amish with carpentry and other businesses have landlines and or cell phones that are located in the shop or outside the building in a weather protected box. Several families may share one telephone. **Ask if there is a telephone where messages may be left and what medical information is permissible to leave in a message (or would they prefer a message instructing them to call the health provider back)**.
- Amish may avoid eye contact with non-Amish, but in one-on-one contacts in the work environment, they can be expected to express openness and candor with unhesitating eye contact. **Do not take offense at sustained eye contact.**
- Because English is the language of instruction in schools and is used with business contacts in the outside world, there is generally no language barrier for the Amish in the workplace. **Do not use slang or idiomatic expressions that the Amish may not understand.**

### 6.1.4 Biocultural Ecology

- The Amish are essentially a closed population, with exogamy occurring very rarely. Most are of German and Swiss descent; therefore, their physical characteristics differ, with skin variations ranging from light to olive tones. Hair and eye colors vary accordingly.
- Statistics do not exist for Amish cardiovascular and respiratory diseases and illnesses or for the incidence of cancer and diabetes mellitus. Obesity is common, especially among women.
- Table 6.1 lists commonly occurring health conditions for the Amish. A description of the common genetic disorders is in Table 6.2.

**Table 6.1** Health conditions common to the Amish

| Health conditions | Causes |
|---|---|
| Limb-girdle muscular dystrophy | Genetic |
| Ellis-van Creveld syndrome | Genetic |
| Dwarfism | Genetic, unknown |
| Polydactylism | Genetic |
| Cartilage hair hypoplasia | Genetic |
| Glutaric aciduria | Genetic |
| Manic-depressive disorder | Genetic |
| Pyruvate kinase deficiency | Genetic |
| Hemophilia B | Genetic |
| Obesity | Lifestyle |

**Table 6.2** Common genetic disorders

| |
|---|
| • Dwarfism |
| • *Ellis-van Creveld syndrome* is characterized by short stature and an extra digit on each hand, congenital heart defects, and nervous system involvement resulting in some amount of mental deficiency |
| • *Cartilage hair hypoplasia*, also a dwarfism syndrome, is characterized by short stature; fine, silky hair; and deficient cell-mediated immunity that increases susceptibility to viral infections |
| • *Pyruvate kinase anemia*, a rare blood-cell disease with jaundice and anemia |
| • *Hemophilia B* |
| • *Phenylketonuria* (PKU) results in an inability to metabolize the amino acid phenylalanine, evidenced as high blood levels of the substance and eventually severe brain damage if the disorder is untreated |
| • *Glutaric aciduria*, a progressive neurologic disease, which can be prevented by screening individuals at risk, restricting dietary protein, and thus limiting protein catabolism, dehydration, and acidosis during illness episodes |
| • Health-care providers need to plan for family and community education about genetic counseling and screening of newborns for genetic conditions |

## *6.1.5 High-Risk Health Behaviors*

- Farm and traffic accidents are major health concerns because of horse-drawn vehicles. Transportation-related injuries involving farm animals are the largest group. Falls from ladders and down hay holes result in numerous orthopedic injuries. **Encourage close monitoring of children who operate farm equipment and transportation vehicles.**
- Alcohol and recreational drug use occur at low rates among the Amish, although few statistics as to the frequency can be found. **Teach about safety factors. Ask about alcohol and recreational drug use in a nonjudgmental approach and when family members are not present. It is important to remember that when young adults may choose to try non-Amish activities during rumspringa, they may be at increased risk for injuries due to lack of familiarity.**

### 6.1.6 Nutrition

- Most Amish families grow their own produce. Typical meals include meat; potatoes or noodles or both; a cooked vegetable; bread; something pickled, such as red beets; cake or pudding; and coffee.
- At mealtimes, all members of the household are expected to be present.
- Snacks and meals tend to be high in fat and carbohydrates. Common snacks are large, home-baked cookies about 3 in. in diameter, ice cream, pretzels, and popcorn.
- **When asking about weight control, suggest reducing portion size, decreasing the amount of sugar used in baking, limiting fatty meats, offering more fresh fruits and vegetables and altering food preparation practices.**

### 6.1.7 Pregnancy and Childbearing Practices

- Children are considered gifts from God. The average number of live births per family is seven. Birth control is viewed as interfering with God's will and thus should be avoided. Nevertheless, some Amish women do use intrauterine devices, but this practice is uncommon. **Approaching the subject of birth control obliquely may make it possible for an Amish woman or man to sense the health-care provider's respect for Amish values and thus encourage discussion. "When you want to learn more about birth control, I would be glad to talk to you" is a suggested approach. It is also important for health-care providers to offer education if using hormonal medication for non-contraceptive use. An Amish woman may not be aware of medicinal uses other than birth control for these medications.**
- Women participate in prenatal classes, often with their husbands. Women are interested in learning about all aspects of perinatal care, but they may choose not to participate in sessions when videos are used. **Prenatal class instructors should inform them ahead of time when videos or films will be used so they can decide whether to attend.**
- Most women prefer home births and choose to use Amish or non-Amish lay midwives. The Amish have no major taboos or requirements for birthing. Men may be present, and most husbands choose to be involved; however, they are likely not to be demonstrative in showing affection verbally or physically.
- The laboring woman cooperates quietly, seldom audibly expressing discomfort. Women sometimes use herbal remedies to promote labor. **Knowledge about and respect for Amish health-care practices alert health-care providers to the possibility of simultaneous treatments that may or may not be harmful.**
- The postpartum mother resumes her family role managing, if not doing, all the housework, cooking, and child care within a few days after childbirth. Grandmothers or other female relatives often come to stay with the new family

for several days to help with care of the infant and give support to the new mother. Older siblings are expected to help care for the younger children and to learn how to care for the newborn.
- When hospitalized, the family may want the patient to spend the least allowable time in the hospital. **Include extended nuclear family in care of the newborn. Ask the father how extensively he wants to be involved with labor and delivery.**

### *6.1.8 Death Rituals*

- Families are expected to care for the aging and the ill in the home. However, when hospitalization is required, **make private arrangements for family members to stay overnight in the hospital.** A wake-like "sitting up" through the night is expected for the seriously ill and dying.
- While it is unusual for Amish to use cremation, many will utilize local funeral services for embalming. This would enable family and friends to travel for viewing and participation in funeral services. The viewing and services may occur in a family home or barn building depending on the size of gathering.
- The funeral ceremony is simple and unadorned, with a plain wooden coffin and no flowers. Although grief and loss are keenly felt, verbal expression may seem muted as if to indicate stoic acceptance of suffering. No eulogy is offered.
- The Amish community will assist the family by preparing daily meals, assisting in performing farm work, business obligations or household management during the period of bereavement.
- **Accept varied means of grieving from Amish families. Just because grief is not overtly demonstrated does not mean that family members do not care about their loved one.**

### *6.1.9 Spirituality*

- Amish religious and cultural values include honesty, order, personal responsibility, community welfare, humility, non-resistance or nonviolence, and obedience to parents, church, and God.
- Amish are socialized to sustain injuries, grieve, and move on without fixing blame or seeking punishment or redress from perpetrators.
- Amish settlements are subdivided into church districts similar to rural parishes, with 30–50 families in each district. Local male leaders are chosen from their own community and are generally untrained and unpaid. No regional or national church hierarchy exists to govern internal church affairs. To maintain harmony within a group, individuals often forgo their own wishes. In addition to Sunday

services every other week, silent prayer is always observed at the beginning of a meal. A prayer may also end the meal.

- **When choosing among health-care options, families usually seek counsel from religious leaders, friends, and extended family, but the final decision resides with the immediate family.**

### *6.1.10 Health-Care Practices*

- The body is considered to be the temple of God, and human beings are stewards of their bodies. Medicine and health care should always be used with the understanding that it is God who heals. Nothing in the Amish understanding of the Bible forbids them from using preventive or curative medical services. They are highly involved in the practices of health promotion and wellness; illness and disease prevention, and health restoration. Men are involved in major health-care decisions and often accompany the family to the chiropractor, physician, or hospital. Health-care decisions are influenced by three factors: (1) type of health problem, (2) accessibility of health-care services, and (3) perceived cost of the service. Grandparents are frequently consulted about treatment options. **Include the extended nuclear family in health-care decision-making.**
- Many Amish do not carry health insurance. Some have formalized mutual aid, such as the Amish Aid Society. **Ask Amish families if they have access to the Amish Aid Society and assist with contacting them if desired. Some health-care systems or hospitals in areas where there is a large Amish population, have established reduced set costs for common medical procedures when paid in cash, as preferred by the Amish.**
- **Providers must be aware that some individuals may withhold important medical information from medical providers by neglecting to mention folk and alternative care being pursued at the same time. Health-care providers must remain open to hearing and learning more about these types of care choices.**
- **When the Amish use professional health-care services, they want to be partners in their health care and want to retain their right to choose from all culturally sanctioned health-care options.**
- Care is expressed in culturally encoded expectations, which the Amish best describe in their dialect as ***abwaarde***, meaning "to minister to someone by being present and serving when someone is sick in bed."
- Some accept medical advice regarding the need for high-technology treatments such as transplants, ventilator support, or other high-cost interventions.
- The patient's family seeks prayers and advice from the bishop and deacons of the church, the extended family, and friends, but the decision is generally a personal family choice. Family members may also seek care from Amish healers and other alternative care practitioners, who may suggest nutritional supplements.

- Herbal remedies include those handed down by successive generations of mothers and daughters.
- Common folk illnesses are described in Table 6.3.
- Health-care knowledge is passed from one generation to the next through women.
- **Inquire about the full range of remedies being used. For the Amish patient to be candid, the provider must develop a context of mutual trust and respect.**
- The Amish are unlikely to display pain and physical discomfort. **Remind Amish patients that medication is available for pain relief if they choose to accept it.**
- There are no cultural or religious rules or taboos prohibiting blood transfusions or organ transplantation and donation. Some may opt for organ transplantation after the family seeks advice from church officials, extended family, and friends, but the patient or immediate family generally makes the final decision.
- Children with mental or physical differences are sometimes referred to as "hard learners" and are expected to go to school and be incorporated into classes with assistance from other student "scholars" and parents. The mentally ill are generally cared for at home whenever possible.

### *6.1.11 Health-Care Providers*

- Amish hold all health-care providers in high regard. Health is integral to their religious beliefs, and care is central to their worldview. They tend to place trust in people of authority when they fit Amish values and beliefs.
- Amish usually refer to their own healers by name rather than by title, although some say *brauch-doktor* or ***braucher***. In some communities, both men and women provide these services. Amish folk healers use a combination of treatment modalities, including physical manipulation, massage, ***brauche***, herbs and teas, and reflexology. Most prefer health-care providers who discuss health-care options, giving consideration to cost, need for transportation, family influences, and scientific information.
- **Because Amish are not sophisticated in their knowledge of physiology and scientific health care, health-care providers should be prepared to use educational materials that may include anatomic models or illustrations. Providers should bear in mind that the Amish respect authority and that they may unquestioningly follow orders.**

**Table 6.3** Common Amish folk illnesses

| |
|---|
| • Brauche, sometimes referred to as sympathy curing or powwowing, is referred to as "warm hands" and includes the ability to feel when a person has a headache or a baby has colic |
| • **Abnemme** is a condition in which the child fails to thrive and appears small, ill, and weak. Specific treatments for the child may include incantations |
| • **Aagwachse**, or livergrown, meaning "hide-bound" or "grown together," includes crying and abdominal discomfort that is believed to be caused by jostling in rough buggy rides |

## 6.2 Reflective Exercises

1. What are your concerns regarding access and quality care for an Amish family?
2. During your assessment of a 60-year-old Amish woman with a breast lump, she reinforces that she does not want to be "cut upon" at her age. What health teaching and advice might you give to her?
3. How might health-care providers use the Amish values of the three-generational family and their visiting patterns in promoting health in the Amish community?
4. What elements would you include for an Amish patient wanting to lose weight and prepare healthier meals for her family?
5. A 42-year-old Amish woman has just delivered her tenth baby. She tells the nurse that with ten children, she just does not want any more. What approach might the nurse take to discuss contraception with her?
6. An Amish family comes to the clinic with a chronically sick 4-year-old boy. They tell you that they have been seeing a folk healer but the child has not improved over the last year. What approach should the nurse take in determining care?
7. Name at least three common genetic disorders among the Amish. Describe conventional treatment. How might an Amish person react if you encouraged genetic counseling?
8. Distinguish the terms *aagwachse*, *agnemme*, and *abwarde*.
9. What language would you use to provide discharge instruction to an Amish family?
10. What is meant by "warm hands"?

## Bibliography

Amish Cultural Practices: Mutual Aid (2023) https://groups.etown.edu/amishstudies/cultural-practices/mutual-aid/

Amish Population Profile (2022) https://groups.etown.edu/amishstudies/statistics/amish-population-profile-2022/#:~:text=The%20estimated%20population%20of%20the,states%20and%20four%20Canadian%20provinces

Amish Social Organization: Family (2023) https://groups.etown.edu/amishstudies/social-organization/family/

Amish Technology (2019) https://www.ohiosamishcountry.com/articles/amish-technology

Johnson-Weiner KM (2020) The lives of Amish women (young center books in Anabaptist and Pietist studies). Johns Hopkins University Press, Baltimore

Kraybill DB, Johnson-Weiner KM, Nolt SM (2018) The Amish. Johns Hopkins University Press, Baltimore

Nelson-Tuttle C (2021) People of Amish heritage. In: Purnell L, Fenkl EA (eds) Textbook for transcultural health care: a population approach, 5th edn. Springer, Cham

Nolt SN (2015) A history of the Amish, 3rd edn. Good Books, New York

Stavisky J (2022) In plain view: the daily lives of Amish women. Herald Press, Harrisonburg, VA

# Chapter 7
# People of Appalachian Heritage

**Mary Lou Clark Fornehed**

## 7.1 Overview and Heritage

The term *Appalachian* describes people born in the Appalachian mountain range and their descendants who live in or near Appalachia. Appalachia comprises 25 million people in 423 counties in 13 states—Georgia, Alabama, Mississippi, Virginia, West Virginia, North Carolina, South Carolina, Kentucky, Tennessee, Ohio, Maryland, New York, and Pennsylvania. Thus, one can see the tremendous diversity that exists is such a large region. Most of the region is rugged, mountainous terrain that is partially responsible for its residents' values and traditions. Substandard secondary and tertiary roads as well as limited public bus, rail, and airport facilities prevent easy access to much of the area.

Most Appalachians have always been a racially and ethnically diverse population with strong family ties to Native American Indians, as well as the Spanish, French, and Eastern European people. Many of those who live in Appalachia can trace their family roots back 200 years or more and it is common to find whole communities comprising of extended, related families.

Appalachians are loyal, caring, family-oriented, religious, hardy, independent, honest, patriotic, and resourceful. The concept of "home" is associated with the land and the family, not a physical structure. Although many are well educated, for some education beyond elementary levels is not considered important because it is not viewed as necessary to earning a living in their traditional occupations. High unemployment and low wages necessitate that many Appalachians migrate throughout the region for better economic opportunities.

M. L. C. Fornehed (✉)
Whitson-Hester School of Nursing, Tennessee Technological University, Cookeville, TN, USA
e-mail: mcfornehed@tntech.edu



E. A. Fenkl, L. D. Purnell (eds.), *Handbook for Culturally Competent Care*,
https://doi.org/10.1007/978-3-031-70492-5_7

### *7.1.1 Communications*

- The dominant language among Appalachians is English, with many words derived from the sixteenth-century Elizabethan English. A small population may still speak Elizabethan English in their home environment. This can cause communication difficulties with health-care providers who are not familiar with the dialect. Many drop the *g* on words ending in *ing*. For example, *writing* becomes *writin'*, *reading* becomes *readin'*, and *spelling* becomes *spellin'*. Consonants may be added, and vowels may be pronounced with a diphthong that can cause difficulty to one unfamiliar with this dialect; hence, *poosh* for *push*, *boosh* for *bush*, *warsh* for *wash*, *hiegen* for *hygiene*, *deef* for *deaf*, *welks* for *welts*, *whar* for *where*, *hit* for *it*, *hurd* for *heard*, and *your'n* for *your*. **If unfamiliar with the exact meaning of a word, ask the patient to elaborate.**
- Appalachians practice the *ethic of neutrality*, which helps shape communication styles, their worldview, and other aspects of the culture.
- Some Appalachians may be less precise in describing emotions and are more concrete in conversations; many answer questions in a direct manner. **Use more open-ended questions when obtaining health information and eliciting opinions and beliefs about health-care practices.**
- Many Appalachians do not easily trust or share their thoughts and feelings with **outsiders** and are sensitive to direct questions about personal issues. **Sensitive topics are best approached with indirect questions and suggestions. Establish trust before asking questions that are sensitive in nature if possible.**
- Because of past experiences with large mining and timber companies, many dislike authority figures and institutions that attempt to control behavior. **If time permits, "sit a spell" and "chat" before getting down to the business of collecting health information. To establish trust, it is necessary to demonstrate an interest in the patient's family and other personal matters, drop hints instead of giving orders, and solicit patients' opinions and advice.**
- More traditional individuals may stand at a distance when talking with people in health-care situations. Direct eye contact from strangers may be considered as aggression or hostility.
- Most Appalachians are comfortable with silence and when talking are likely to speak without emotion, facial expression, or gestures.
- **Calling a person by his or her first name with the title Miss** (pronounced "miz," similar to "Ms.," when referring to women, whether single or married) **or Mr. (for example, Miss Brenda or Mr. Erbie) denotes familiarity with respect.**
- **Some individuals may respond better to verbal instructions and education, with reinforcement from videos rather than from printed communications.**
- Based on fatalistic views, some individuals believe they have little or no control over nature and that the time of death is "predetermined by God." Thus, one frequently hears expressions such as "I'll be there, God willing, or if the crick (creek) don't rise."

- Nonverbal behaviors should be assessed when speaking with Appalachians, considering the context with which they are communicating.
- When individuals are not seen because they are late for an appointment and are asked to reschedule, they may not return because of difficult transportation or feelings of rejection. **Be flexible with appointments.**
- Appalachians are comfortable with silence. **Allow a modicum of silence before initiating conversations.**

### *7.1.2 Family Roles and Organization*

- Rural Appalachian persons often refer to their family as "kin."
- The traditional household continues to be patriarchal, however, gender roles are usually well defined.
- Grandparents becoming the head of a household is a rapidly growing part of the family unit.
- Many of the people who live in Appalachia can trace their family roots back 200 years or more and it is common to find whole communities comprising extended, related families.
- Young children and adolescents in Appalachia and across the United States do experience adverse childhood experiences (ACEs) (Centers for Disease Control and Prevention).
- One's obligation to extended family outweighs the obligations to school or work.
- Children, single or married, may return to their parents' home where they are readily accepted whenever the need arises.
- **The family network can be a rich resource when health teaching and assistance with personalized care are needed. The family rather than the individual must be considered as the basic treatment unit.**
- Although alternative lifestyles are usually readily accepted in the Appalachian culture, **Do not disclose same-sex relationships to others.**

### *7.1.3 Family Decision Making at End of Life*

- Caring for persons with life-limiting illness has changed in the last 20 years, creating complex situations when families must make decisions at end of life.
- In rural Appalachia, family plays an integral role in these decisions. **Supporting family care at home is very important.**
- Communication is essential to rural Appalachians when making decisions at end of life. **Maintaining a trusting and caring nurse practice with families and persons dying is vital.**
- Knowledge and economics play a large role in decision making at end of life.

- Rural appalachians value combining folk care with traditional medicine at end of life. **Preserving their folk care practices are vital.**
- Rural Appalachian families may choose **not** to receive hospice care based on the desire to continue curative care. **Becoming active with your state legislator to propose both adult and pediatric palliative and hospice care is provided in your area is important.**
- Pediatric palliative and hospice care is very limited in rural Appalachia.
- Rural Appalachian families may only decide to use hospice when caregivers are exhausted or when pain and symptom management is no longer possible using folk care methods. **Negotiations with community resources (family church) to support families in their homes at end of life can be helpful.**

### *7.1.4 Workforce Issues*

- Because many Appalachians value family above all else, reporting to work may become less of a priority when a family member is ill or other family obligations are pressing. For some, the preferred work pattern is to work for an extended period of time, take some time off, and then return to work. **Liberal leave policies for funerals and family emergencies are a necessary part of the work environment.**
- More traditional individuals may stand at a distance when talking with people in the workforce. **Do not take offense if Appalachians stand at a greater distance than to what you are accustomed.**
- A harmonious environment that fosters cooperation and agreement in decision making is valued. Health-care workers who come from outside the area may have some difficulty establishing rapport in the workplace if they lack an understanding and appreciation for Appalachian workplace etiquette. **Familiarize yourself with the Appalachian culture of neutrality when working with Appalachian staff.** (See Sect. 7.1.1.)
- Most Appalachians are comfortable with silence, and when talking are likely to speak without emotion, facial expression, or gestures. **Do not mistake silence and lack of facial expressions as not caring or not listening.**
- The Appalachian ethic of neutrality and the values of individualism and non-assertiveness, with a strong people orientation, may pose a dichotomous perception at work for outsiders who may not be familiar with the Appalachian way of life.
- Because many Appalachians align themselves more closely with horizontal rather than hierarchical relationships, they are sometimes reluctant to take on management and leadership roles. Many Appalachians prefer to work at their own pace, devising their own work rules and methods for getting the job done.

### 7.1.5 *Biocultural Ecology*

- Sometimes, the influence of American Indians can be observed in olive-toned skin.
- **Those of Scots-Irish background and others with light skin tones are at increased risk for skin cancer. Take precautions and teach patients to protect themselves from the harmful effects of the sun.**
- Predominant occupations, particularly farming, textile manufacturing, mining, furniture-making, and logging, place residents at increased risk for accidents and respiratory diseases such as black lung, brown lung, and emphysema.
- Children are at greater risk for sudden infant death syndrome, congenital malformations, and infections. Only 70% of children are immunized, compared with 90% for the nation as a whole. The area also has an incidence of childhood injuries due to burns, trauma, poisoning, child neglect, and abuse that is higher than the national average. Poor oral hygiene has high rates in Appalachia.
- **Work with community organizations and schools to promote immunizations, oral hygiene, and prevention of childhood injuries.**
- **Dental care is challenging, with rural Appalachians experiencing the worst oral in health in the United States.**
- Table 7.1 lists commonly occurring health conditions for Appalachians in three categories: lifestyle, environment, and genetics.

### 7.1.6 *High-Risk Health Behaviors*

- High smoking rates continue throughout Appalachia. Underage use of alcohol is widespread among teens.
- Rural Appalachians have a disproportionately higher rate of substance misuse disorder.

**Table 7.1** Health conditions common to people of Appalachian heritage

| Health conditions | Causes |
|---|---|
| Black lung, brown lung | Environment, lifestyle |
| Emphysema | Environment, lifestyle |
| Tuberculosis | Environment, lifestyle |
| Hypochromic anemia | Environment, lifestyle |
| Cardiovascular disease | Genetic, lifestyle |
| Sudden infant death syndrome | Genetic, lifestyle |
| Diabetes mellitus | Genetic, lifestyle |
| Otitis media | Environment, lifestyle |
| Parasitic infections | Environment, lifestyle |
| Cancer | Genetic, environment, lifestyle |
| Suicide | Environment, lifestyle |

- Suffer from reduced mental health care.
- **Initiate anti-smoking campaigns in elementary schools and promote the benefits of not being overweight.**

### *7.1.7 Nutrition*

- Common foods are sweet potato pie; molasses candy; apple beer; gooseberry pie; pumpkin cake; and pickled beans, fruit, corn, beets, and cabbage, usually high in sodium.
- Bone marrow is used to make sauces, and stomach, intestines (chitlins or chitterlings), pigs' feet, tail, and ribs are also commonly eaten.
- Many recipes contain lard, and meats are commonly preserved with salt. Low-fat game meat is usually breaded and fried with lard or animal fat, negating the overall gains from these low-fat meat sources.
- Most diets include sweet pre-packaged drinks, Kool-Aid with added sugar, very sweet iced tea, and soda.
- **Explain the benefits of low-fat preparation practices and the potential harmful effects of excess salt and refined sugar.**
- Many children replace meals with snacks. The most common snacks are candy, salty foods, desserts, and carbonated beverages. Many adolescents skip breakfast and lunch entirely, preferring to eat snack foods. This pattern of snacking can result in deficiencies in vitamin A, iron, and calcium. **Educate students on healthy eating and food preparation practices beginning in elementary school.**

### *7.1.8 Pregnancy and Childbearing Practices*

- Fertility control methods include birth control pills, condoms, and tubal ligation; abortion is an individual choice.
- A disproportionate number of teenage pregnancies occur at a younger age among Appalachians when compared with non-Appalachians. **Fertility practices and sexual activity, both sensitive topics for many teenagers, are areas in which outsiders unknown to the family may be more effective than health-care providers who are known.**
- Beliefs about pregnancy are given in Table 7.2.
- The birthing mother is expected to accept childbirth as a short, intense, natural process that will bring her closer to the Earth and must be endured. **Let the birthing mother know that pain medicine is available if she desires.**

**Table 7.2** Beliefs about pregnancy

| |
|---|
| • Boys are carried higher and the mother's belly appears pointy, whereas girls are carried low |
| • Picture-taking can cause a stillbirth |
| • Reaching over one's head can cause the cord to strangle the baby |
| • Wearing an opal ring during pregnancy may harm the baby |
| • Being frightened by a snake or eating strawberries or citrus fruit can cause birthmarks |
| • If the mother experiences a tragedy, a congenital anomaly may occur |
| • If the mother craves a particular food during her pregnancy, then she should eat that food or the baby will have a birthmark similar to the craved food |

## *7.1.9 Neonatal Abstinence Syndrome*

- **With the disproportionate misuse of opioids, babies are born suffering from neonatal abstinence syndrome (NAS).**
- **Approximately every 15 min a baby is born addicted to opioids or alcohol.**
- **In rural Appalachia, these women have limited access to medication assisted treatment for opioid addiction.**

## *7.1.10 Death Rituals*

- When death is expected, family and friends may stay through the night. **Make accommodations for some family members to stay in the health-care facility.**
- **Diversity in these practices can also include accepting death by some family while others continue to pray for a miracle.**
- Funeral services can last for 3 h. The amount of time for a service varies according to the age of the deceased. For example, the service for an elderly person is usually longer than for a younger person. The body is displayed for hours, either in the home or at the church, so that all those who wish to view the body may do so.
- At the end of the service, all who wish to view the body again may do so, with the closest relative being the last to view the body.
- The deceased is usually buried in his or her best clothes. A common practice is to bury the deceased with personal possessions.
- Burial sites can vary. For example, the author's "daddy" was buried in a cemetary less than a mile from his home where family and community have been buried for over 200 years. Cremation is becoming more popular, however, with the movement or outmigration of children and other family members.
- After the funeral services are completed, elaborate meals are served either in the home or at the church. Services are accompanied by singing before, during, and after the service.

- Clergy help families through the grieving process by providing counseling and support to family members. **Ask family if they wish to see a chaplain and help them with contacting one.**

### *7.1.11 Spirituality*

- Prayer for many Appalachians is a primary source of strength. Prayer is personally designed around specific church and religious beliefs and practices, which vary widely throughout the region and between and among churches of similar faith.
- There is a great diversity of religious practices throughout Appalachia, such as Roman Catholic, Eastern Orthodox, Luthern, Jehovah's Winess, Jewish, Muslim, Mornon, Episcopalian, Mennonite, Quaker, Hindu, and Unitarian as well as people of other faiths all live and worship in Appalachia. Do not make assumptions regarding religious practices.
- **Churches in many parts of Appalachia serve as the social centers of the community and are a good location for health teaching.**
- Many small churches have lay preachers instead of trained ministers. Most believe that to be a preacher, a person must have a divine calling.
- Some free-will churches (for example, The Holiness Church) preach against attending movies, ball games, and social functions where dancing occurs. Other sects believe in handling poisonous snakes, although this is rare, the practice continues; it is believed that the snake will not bite those who have faith.
- Religious beliefs may also be of a spiritual nature, not tied to the tenets of any singular faith, and reflect the harmony of the mountains and being at one with life.
- Some free-will churches—for example, the Holiness Church—preach against attending movies, ball games, and social functions where dancing occurs.
- Another practice, the ingestion of strychnine in small doses during religious services, is believed to increase sensory stimuli. Needless to say, this practice can precipitate convulsions if ingested in large enough amounts.
- Meaning in life comes from the family and "living right," which is defined by each person and usually means living right with God and in the beliefs of a chosen church.
- **Forming partnerships between health-care providers and faith-related organizations for health promotion and wellness, and illness and disease prevention has strong potential for improving the health of Appalachians. Respect the spiritual beliefs of Appalachians without expressing negative comments.**

### 7.1.12 *Health-Care Practices*

- Because self-reliance activities and nature predominate over people, many believe that it is best to let nature heal.
- For some who do not believe in owing money, seeing a health-care provider may be postponed until the condition is severe.
- Many may not see formal biomedical health-care providers until self-medicating and folk remedies have been exhausted. When they finally seek formal health care, the condition has become severe, takes longer to treat, and has a less favorable outcome.
- Individuals may feel powerless regarding their own health and abdicate self-responsibility in favor of high expectations and unrealistic dependence on the health-care system, with the physician taking charge of their care completely.
- Folk Medicine is a traditional part of the Appalachian culture with a wide variation in use of folk remedies that honor their elders.
- When older people see a health-care provider, many expect immediate help. Physicians who dispense medications in their offices are considered helpful.
- A strong belief in folk medicine is a traditional part of the culture. Using herbal medicines, poultices, and teas is common practice among individuals of all socioeconomic levels. Although many of these home remedies are not harmful, some may have deleterious effects when used to the exclusion of, or in combination with, prescription medications.
- **Ascertain if individuals intend to use folk medicines simultaneously with prescription medications and treatment regimens so that these remedies can be incorporated into the plan of care and dialog can be undertaken to prevent adverse effects.**
- **Integrating folk medicine into allopathic prescriptions has a greater chance of improving patients' compliance with health prescriptions and interventions.**
- Bureaucratic forms foster fear and suspicion of health-care providers. **Help patients complete bureaucratic forms if needed.**
- **Be aware that patients may be especially sensitive to criticism. If the provider uses language that the patient does not understand, the provider may be perceived as "stuck up." Decrease language barriers by decoding the jargon of the health-care environment.**
- Individuals with mental impairments or physical handicaps are readily accepted and not turned away. Those with a mental handicap are not perceived as crazy but rather as having "bad nerves" or being "quite turned" or "odd turned."
- The traditional belief is that disability is a natural and inevitable part of the aging process. **Provide factual information about disability and explain that it is not an inevitable part of the aging process.**
- For many, pain is something that is to be endured and accepted stoically. When a person becomes ill or has pain, personal space collapses inward and the person

expects to be waited on and cared for by others. **Explain that self-care activities and taking pain medicine will hasten the healing process.**

- **Rural children with life-limiting illness continue to follow with their primary care and nonhospice providers for symptom relief at end of life.**
- A 10-step pattern of health-seeking behaviors was identified see Table 7.3:

### *7.1.13 Health-Care Providers*

- Folk practitioners are primarily older women but may be men. Grannies and herb doctors are trusted and known to individuals and the community for giving more personalized care.
- **For patients to become more accepting of biomedical care, it is important for health-care providers to approach individuals in an unhurried manner and engage patients in decision making and care planning.**
- Generally, there is no problem providing care to opposite-gender patients.
- There are significant shortages of specialty providers throughout Appalachia. Barriers to seeing these specialty providers in larger urban areas include economics, long distances, lack of family or community resources for travel.

**Table 7.3** Health-seeking behaviors of rural Appalachian

| |
|---|
| 1. At the onset of symptoms, self-care practices that are usually learned from mothers are implemented |
| 2. When the symptoms persist, they call their mother, if she is available |
| 3. If the mother is unavailable, they call the female in their kin network that is perceived as knowledgeable regarding health. If a nurse is available, they may seek the nurse's advice |
| 4. If relief is not achieved, they use over-the-counter (OTC) medicine they have seen advertised on television for symptoms that most closely match their own |
| 5. If that is ineffective, they use some of a neighbor's medicine |
| 6. Next, they ask the local pharmacist for a recommendation; this usually marks the first encounter with a professional health-care provider. (Of course, they usually do not tell the pharmacist that they tried the neighbor's medicine.) The pharmacist may strongly suggest that they see a health-care provider; however, on their insistence, the pharmacist may recommend another OTC medication |
| 7. When no relief is achieved, they seek a local health-care provider. The provider treats them to the best of his or her ability |
| 8. If the condition does not resolve, the local health-care provider refers them to a specialist in the area |
| 9. The specialist treats the condition to the best of his or her ability |
| 10. If unsuccessful, the specialist refers them to the closest tertiary medical center. These ten steps may not always follow the sequence presented here; some steps may be skipped, and not all steps are always completed. The time around these ten steps may be several years. Often by the time they are referred for definitive treatment, compensatory reserves have been depleted, and they die at large medical centers. It is essential to inquire in a nonjudgmental way about all treatments used for an illness |

- As a large area of Appalachia is medically and health-care provider underserved, "outsider health-care providers" may be seen as disrespectful of the Appalachian culture and therefore not trusted to provide care. Trust is gained slowly.

## 7.2 Reflective Exercises

1. A patient wants to incorporate ginseng with prescriptive medication. What is your advice?
2. What evidence do you see with the "ethic of neutrality" among Appalachians or any other group or person?
3. Identify at least three health conditions common among Appalachians. What might you do to help decrease these conditions?
4. What tactics might you use to help decrease the high smoking rates among Appalachians?
5. How might you work with local community churches to teach health food practices and prevention of sexually transmitted infections?
6. How is *fatalism* displayed among Appalachians?
7. An overweight teenager is drinking 8–10 glasses of "sweet tea" on a daily basis. How might you encourage her to decrease the amount of "sweet tea" she drinks on a daily basis?
8. A close family friend whose spouse is dying of metastatic breast cancer tells you "I am wore out and don't know how much more I can do." What might you do to help your friend get some respite from this situation?

## Bibliography

Appalachian Regional Commission (2023) About the Appalachian region. https://www.arc.gov/about-the-appalachian-region/

Center for Oral Health Research in Appalachia (2020) https://www.dental.pitt.edu/research/center-oral-health-research-appalachia

Centers for Disease Control and Prevention (2019a) Adverse childhood experiences (ACEs). https://www.cdc.gov/aces/about/index.html

Centers for Disease Control and Prevention (2019b) Preventing adverse childhood experiences. https://stacks.cdc.gov/view/cdc/82316

Fornehed MLC, Mixer SJ, Lindley LC (2020) Families' decision making at end of life in rural Appalachia. J Hosp Palliat Nurs 22(3):188–195

Fornehed M, Svynarenko R, Keim-Malpass J, Cozad M, Qualls K, Stone W, Lindley L (2022) Comparison between rural and urban Appalachian children in hospice care. South Med J 115(3):192–197. https://doi.org/10.14423/SMJ.0000000000001365

Hatcher J, Voigts K, Culp-Roche A, Adegboyega A, Scott T (2018) Rural grandparent headed households: a qualitative description. Online J Rural Nurs Health Care 18(1):40–62

Hufford M (2006) Folklore and folklife. In: Abramson R, Haskell J (eds) Encyclopedia of Appapachia. The University of Tennessee Press, Knoxville, pp 844–909

March of Dimes (2020) Neonatal abstinence syndrome (NAS) https://www.marchofdimes.org/complications/neonatal-abstinence-syndrome-(nas).aspx

Mixer SJ, Fornehed MLC (2021) People of Appalachian heritage. In: Purnell LD, Fenkyl EA (eds) Textbook for transcultural health care: a population approach. Cultural competence concepts in nursing care, 5th edn. Springer, Cham

Mixer SJ, Fornehed ML, Varney J, Lindley LC (2014) Culturally congruent end-of-life care for rural Appalachian people and their families. J Hosp Palliat Nurs 16(8):526–535. https://doi.org/10.1097/NJH.0000000000000114

National Alliance on Mental Illness [NAMI] (2018) Mental health by the numbers. https://www.nami.org/Learn-More/Mental-Health-By-the-Numbers. Accessed 20 Jun 2024

National Institute on Drug Abuse [NIDA] (2018a) Drugs, brains, and behavior: the science of addiction. https://www.drugabuse.gov/publications/drugs-brains-behavior-science-addiction/drug-misuse-addiction

National Institute on Drug Abuse [NIDA] (2018b) Pregnant women in Appalachia face barriers to opiod treatment. https://pubmed.ncbi.nlm.nih.gov/29949454/

Tennessee Department of Health (2021) Neonatal abstinence syndrome surveillance summary week 35: August 29–September 04, 2021. https://www.tn.gov/health/health-program-areas/fhw/neonatal-abstinence-syndrome-/data-and-reports.html

U.S. Department of Health and Human Services (USDHHS) (2020a) Administration shortage designation advisor. https://datawarehouse.hrsa.gov/tools/analyzers/hpsafind.aspx

U.S. Department of Health and Human Services (USDHHS) (2020b) Loan repayment. https://nhsc.hrsa.gov/loan-repayment/nhsc-loan-repayment-program

Webb A (2013) African Americans in Appalachia. https://oxfordaasc.com/page/2527. Accessed 26 Jan 2020

Yao N, Hector EA, Anderson R, Balkrishnan R (2017) Cancer dispartities in rural Appalachia: incidence, early detection and survivorship. J Rural Health 33:375–381

# Chapter 8
# People of Arab Heritage

**Suha Ballout**

## 8.1 Overview and Heritage

Exploring the richness of the Arab heritage showcases a journey through time, tracing back to the nomadic desert tribes that roamed the Arabian Peninsula. Rooted in a shared language, **Arabic**, and united by the teachings of **Islam**, a faith that blossomed in seventh-century Arabia. Arab heritage is a mix of vibrant cultural heritage, literature, cuisine, traditional crafts, and resilience in the face of political challenges and conflicts. Despite common bonds of language and history, the Arab residents of each country defy homogeneity, exhibiting diversity of religion, thoughts, attitudes, and behaviors. These cultural variations weave a vibrant narrative within and across countries and regions. From the tradition-bound farmers of rural Yemen and fertile agriculture in Palestine to the enterprising small business owners of Beirut and Syria, each facet of Arab identity reflects a unique blend of history, geography, and circumstance. Significant cultural differences exist not only within individual countries and regions but also across them. It is important to recognize that the individual differences among Arabs exceed the boundaries set by cultural, religious, and societal norms. Immigrant Arab communities, on the other hand, may display considerable cultural variations influenced by factors like religion, country of origin, refugee status, length of residence, ethnic background, education, economic situation, employment status, social networks, and proficiency in English.

Interestingly, despite this diversity lies a unifying thread—a profound respect for individuality that transcends cultural, religious, and social norms. Arabs, whether native or immigrant, embody a spectrum of experiences shaped by a multitude of factors including religion, country of origin, refugee status, and more. Each layer

S. Ballout (✉)
Manning College of Nursing and Health Sciences, University of Massachusetts,
Boston, MA, USA
e-mail: Suha.Ballout@umb.edu



E. A. Fenkl, L. D. Purnell (eds.), *Handbook for Culturally Competent Care*,
https://doi.org/10.1007/978-3-031-70492-5_8

adds complexity, enriching the collective heritage and highlighting the dynamic nature of Arab identity. In this chapter, we discuss the Arab heritage to explore its complexities, celebrate its diversity, and honor its enduring legacy.

## 8.2 Understanding Arab American Identity, Perceptions, and Successes

**Tapestry of Landscapes, Cultures, and Societies**
The Arab countries span the Middle East and North Africa and encompass a diverse array of landscapes, cultures, and societies. The term "Arab countries" typically refers to the nations where Arabic is the primary or official language and where Arab culture and heritage are predominant. The majority of Arab countries include Algeria, Bahrain, Comoros, Djibouti, Egypt, Iraq, Jordan, Kuwait, Lebanon, Libya, Mauritania, Morocco, Oman, Palestine, Qatar, Saudi Arabia, Somalia, Sudan, Syria, Tunisia, United Arab Emirates, and Yemen. It is worth noting that some countries, such as Sudan and Somalia, have many non-Arab populations and have multiple official languages. In addition, there are Arab populations and cultural influences in other countries outside of this list, however, they may not be considered predominantly Arab nations. Most Arab immigrants to the United States arrived during the late nineteenth and early twentieth centuries. This period witnessed significant waves of immigration from various regions of the Arab world, particularly from present-day Lebanon, Syria, Palestine, and Jordan. Many Arab immigrants during this time sought economic opportunities, escaping political instability, religious persecution, and poverty in their homelands. They settled in urban areas such as New York, Detroit, and Chicago where they contributed to the cultural and economic fabric of American society. Subsequent waves of Arab immigration occurred in the mid-twentieth century and continue to the present day with immigrants arriving for various reasons: including family reunification, education, and employment opportunities.

**The Legacy of Arab Immigration**
Arab immigrants in the United States contribute to the nation's rich culture with an estimated population of around three million persons. Arab immigrants trace their ancestry to various countries across the Arab world, including Lebanon, Syria, Egypt, Morocco, Iraq, Jordan, Palestine, Yemen, Tunisia, Algeria, and numerous Gulf countries. Their immigration history is marked by waves with the first wave of immigrants arriving between 1887 and 1913, driven by a quest for economic opportunities and the desire to secure financial stability potentially enabling them to invest in their ancestral villages or establish businesses upon their return. These first-wave Arab Americans engaged in unskilled labor with a majority being male and a large proportion being undereducated coming from mountainous or rural regions. Currently, Lebanon accounts for 32% of Arab American immigrants while 11% originate from Egypt (Arab American Institute Demographics 2024). The second wave of immigrants came to the United States after World War II with an increase after the Palestinian-Israeli conflict and the enactment of the Immigration

Act of 1965. Unlike the economically driven Lebanese Syrian Christians of the first wave, the majority of second-wave immigrants were refugees fleeing war-torn and politically unstable nations primarily occupied Palestine, Jordan, Iraq, Yemen, Lebanon, and Syria. From 2008 to 2018, immigrants from the Arab world originated from countries including Iraq, Egypt, Somalia, Morocco, and Lebanon with a large portion including professionals and individuals pursuing higher education who have chosen to remain in the United States (Arab American Institute Demographics 2024). These migration patterns highlight the diverse socioeconomic factors influencing Arab immigration to the United States, ranging from economic aspirations to escaping conflict and seeking educational opportunities. Geographically, Arab immigrants reside in metropolitan areas with approximately 94% residing in urban centers. Large Arab American populations can be found in states such as Michigan, California, New York, Illinois, and Florida with cities like Dearborn (Michigan) and Los Angeles (California) hosting substantial Arab American communities (Arab American Institute Demographics 2024).

**Perceptions of People of Arab Heritage and Arab Countries**

Sentiments toward Arabs and Arab countries around the world vary based on several influences such as historical events, political dynamics, media portrayals, cultural exchanges, and personal experiences. Appreciation for Arab culture, history, and contributions to various fields such as literature, science, and architecture is widespread. Arab civilization has left a lasting mark on history through its advancements in mathematics, astronomy, medicine, and philosophy, among other disciplines. From the poetry of Al-Mutanabbi to the architectural works of Alhambra, Arab creativity and innovation continue to inspire admiration. The hospitality and generosity associated with Arab societies evoke respect and admiration. Arab countries, with their rich cultural heritage, ancient landmarks, and natural beauty are recognized as premier tourist destinations attracting visitors eager to explore historical sites like the pyramids of Egypt, the ancient city of Petra in Jordan, or the bustling markets of Morocco. There is widespread support for Arab causes such as the Palestinian struggle for statehood among individuals and groups advocating for human rights and social justice reflecting a global commitment to fairness and equity.

On the other hand, unfortunately, negative sentiments toward Arabs and Arab communities persist and are fueled by prejudice, discrimination, and stereotypes perpetuated in media portrayals that depict Arabs in a negative light (Alsaidi et al. 2023; Hashem and Awad 2021; Kader et al. 2020). The association of Islam with Arab culture leads to Islamophobia, which can result in bias and discrimination against Arabs compounded by geopolitical tensions between Arab-majority countries and Western nations (Hashem and Awad 2021; Kader et al. 2020). Countries like Iraq, Syria, and Iran are often met with hostility due to perceived conflicts and tensions further exacerbating negative perceptions. Also, fear and distrust toward Arab immigrants or refugees prevail, particularly in the context of debates surrounding immigration policies and national security concerns contributing to social divisions and barriers to integration. These negative sentiments not only perpetuate harmful stereotypes but also hinder efforts toward promoting understanding and cooperation between diverse communities. Healthcare professionals must recognize that many Arab Americans do not align with the ideological anti-Western movement. Arab individuals should not be stereotyped based on their cultural background. Research conducted by Kulwicki and colleagues examining the impact of 9/11 on Arab American nurses in Detroit showed instances of discrimination against Muslims and Arabs including name-calling, intimidation, verbal assaults regarding appearance and religion, as well as unwarranted suspicion and questioning often supported by the media. Similar perceptions of Arab Americans related to anti-Muslim sentiments among healthcare providers have been reported in other studies.

Mixed sentiments toward Arab countries are also prevalent, reflecting the complexities of their political landscapes and socioeconomic transformations. In countries like Egypt or Saudi Arabia, admiration for their rich cultural heritage often coexists with criticism of human rights records and political repression leading to ambivalence and conflicting views. Similarly, skepticism or uncertainty is observed toward Arab countries undergoing rapid social and economic changes such as Saudi Arabia and the Gulf states where modernization efforts sometimes clash with traditional values creating a sense of unease among some observers. Overall, sentiments toward Arabs and Arab countries are multifaceted influenced by individual experiences, cultural backgrounds, and geopolitical realities. Efforts aimed at promoting cross-cultural understanding, challenging stereotypes, and fostering dialog can play an essential role in cultivating more nuanced and empathetic attitudes toward Arabs and Arab countries on a global scale.

**The Arab Identity and the Healthcare Profession**

The diverse nature of Arab culture poses a challenge in presenting a comprehensive account of Arab Americans compounded by limited research literature on Arabs in the Americas. Early Arab immigrants, primarily Christians from present-day Lebanon and Syria, arrived during the late nineteenth and early twentieth centuries, emphasizing assimilation and easily integrating into mainstream American society. However, Arab Americans tend to be overlooked in national studies as they are often classified as White in census data rather than as a distinct ethnic group. To provide a more inclusive portrayal of Arab Americans incorporating literature describing Arabs in their home countries supplements research on Arab Americans residing in various states such as Michigan, Illinois, New York, Ohio, and the San Francisco Bay area of California. While it is assumed that first-generation immigrants may share some attitudes and behaviors with their counterparts in the Arab world, caution is advised against making assumptions about Arab or Arab American clients without thorough individual assessment.

Healthcare professionals need to understand Arab culture, history, religion, and the diverse experiences of Arab communities around the world. This knowledge is key for providing culturally competent care and fostering positive patient–provider relationships. Healthcare professionals must recognize and challenge negative stereotypes and prejudices that may impact the care of Arab patients, particularly in the context of Islamophobia and discrimination. Understanding the complex socio-political landscapes of Arab countries can inform discussions around health disparities, access to care, and the impact of geopolitical tensions on individual and community health. By promoting cross-cultural understanding, challenging stereotypes, and fostering dialog, healthcare professionals can contribute to creating more inclusive and equitable healthcare environments for Arab patients and communities worldwide.

**Educational and Occupational Success Among Arab Americans**
People of Arab heritage, in general, prioritize professional occupations, emphasizing the importance of education as a pathway to white-collar careers. Recent data indicates that both US-born and foreign-born Arab Americans exhibit higher educational attainment compared to the national average. Over 90% of Arab Americans have completed high school, a contrast to 32% of the total U.S. population. Around 20% of Arab Americans hold post-graduate degrees. In comparison to European Americans, Arab Americans are more likely to be self-employed and hold managerial and professional specialty occupations. Around 45% are employed in managerial or professional roles while 29% work in sales and 16% in service positions. Arab American households in the United States have a mean annual income of $60,398, surpassing the national average of $52,029 for all households (Arab American Institute Demographics 2024). These data show the strong emphasis on education and professional success within the Arab American community contributing to their overall economic prosperity and higher-than-average household income.

### *8.2.1 Communication*

**Dominant Language and Dialects**
Arabic is the official language across the Arab world, including both modern and classical forms. Modern Standard Arabic is used in formal contexts such as written communication, radio broadcasts, and academic lectures while colloquial Arabic specific to each community is utilized for everyday spoken interactions (Suleiman 2019). Arab individuals often blend Modern Standard Arabic and colloquial Arabic depending on the context and level of formality. The diversity of dialects, characterized by variations in accent, intonation, and vocabulary, can pose challenges in communication between Arab immigrants from different regions such as Syria and Lebanon compared to those from Iraq and Yemen (Suleiman 2019). Generally, Arab communication is characterized by repetition and expressive gestures, especially in serious discussions which may be perceived as confrontational or aggressive by external observers. While English serves as a common second language in countries like Egypt, Jordan, Lebanon, Yemen, Iraq, and Kuwait, French is prevalent in Lebanon, Algeria, and Morocco. Literacy rates in the Arab world exhibit significant disparities ranging from 93% for men and 30% for women in Yemen to 95% for men and 85% for women in Jordan.

Despite proficiency in languages other than Arabic, Arab American patients still encounter significant language and communication barriers within American healthcare settings. These challenges often arise from differences in cultural norms, communication styles, and healthcare terminology between Arab American patients and healthcare providers (Abuelezam et al. 2018; Aseeri 2020; Fleischer and Sadek

2024). Studies have highlighted that many Arab American patients face difficulties expressing their healthcare needs and understanding medical instructions from providers resulting in misunderstandings and confusion (Abuelezam et al. 2018; Fleischer and Sadek 2024). Consequently, this miscommunication can lead to delayed appointments, reluctance to disclose personal health information, and overall dissatisfaction with healthcare experiences. These communication barriers have tangible consequences as they delay Arab American patients' access to timely and appropriate healthcare services and compromise the effectiveness of healthcare treatment. Addressing these challenges requires healthcare professionals to adopt culturally sensitive communication strategies, utilize interpreters or bilingual staff when necessary, and foster trust and rapport with Arab American patients to ensure their healthcare needs are adequately met.

**Arab Communication Dynamics**

Arab communication is characterized by a nuanced approach where the context of the situation often carries more significance than the words spoken. Privacy is highly valued among Arabs leading them to resist disclosing personal information to strangers, particularly regarding familial health matters. Within the circle of friends and relatives, Arabs freely express their emotions and feelings. These communication patterns are better understood when viewed through the lens of Arab cultural norms. In close-knit family settings, personal needs are often anticipated without explicit verbalization due to the strong bonds within the family unit. Also, nonverbal cues and unspoken expectations play a significant role in communication within Arab families.

Establishing personal relationships with healthcare professionals is important for Arabs before they feel comfortable sharing personal information. Both the manner and tone of communication are as essential as the actual words spoken, as Arabs are sensitive to the respect and courtesy shown to them (Abuelezam et al. 2018; Suleiman 2019). Social niceties such as greetings, inquiries about well-being, and sharing a cup of tea or coffee precede discussions of business matters. Physical proximity, steady eye contact, and appropriate touch (e.g., limited to members of the same sex) are common in conversations, reflecting the closeness and warmth inherent in Arab interpersonal interactions. Adhering to proper sitting and standing etiquette is also critical as deviations may be interpreted as a sign of disrespect.

In personal relationships, verbal agreements hold greater significance than written contracts as they are seen as a matter of honor. Yet, in business dealings, there is an increasing reliance on written contracts. Arab culture places great emphasis on maintaining harmonious relationships and upholding dignity and honor. When faced with adversity or wrongdoing, Arabs often respond with a sense of resignation expressed through the phrase "*maalesh*," meaning "never mind; it doesn't matter." In challenging situations, individuals may be shielded from unpleasant news for as long as possible and informed gently when necessary. Disagreements are often hinted at indirectly or mediated through intermediaries to preserve social harmony and save face for all parties involved. This mediation process assures the petitioner that their concerns have been addressed with maximum influence and tact (Khosh et al. 2020).

**Suggestions for Communication with People of Arab Heritage or Descent**

1. Healthcare professionals should consider adopting an approach that balances expertise with warmth, aiming to minimize status differences. Arab Americans often feel uncomfortable in the presence of authority figures and are sensitive to their feelings, fearing discrimination or belittlement.
2. Nurses and other healthcare professionals should consider investing time in building rapport before discussing medical matters. Showing genuine interest in the patient's background and adjustment to American life can facilitate open communication and help in understanding potential traumatic immigration experiences.
3. Role responsibilities, especially regarding history taking and physical examinations, should be openly communicated, especially to newer immigrants who may not be familiar with the healthcare system's hierarchy. While some Arab American immigrants may recognize nurses' elevated scope of practice in the United States, they may still perceive them as medical assistants or housekeepers.
4. Nurses should consider conducting comprehensive assessments and explaining the relevance of the information gathered to address physical complaints effectively.
5. Healthcare professionals should try to understand and respect family dynamics, recognizing that a family member may serve as a spokesperson for the patient. Family members may also edit information they deem inappropriate and act as advocates for the patient seeking assistance from influential intermediaries if necessary.
6. Healthcare professionals should try to convey hope and optimism in their communication, understanding that the concept of "false hope" may not resonate with Arab patients who believe in God's infinite power to cure. The amount and type of information shared should be carefully considered to maintain a balance.
7. Respect for the patient's modesty and dignity is paramount. Islamic teachings prohibit unnecessary physical contact between unrelated adults of opposite genders such as shaking hands. Healthcare professionals should be mindful of these cultural norms, especially when interacting with female patients to ensure their comfort and trust in the healthcare setting.

**Predestination and Time Perception Among Arab Immigrants**

First-generation Arab immigrants often adhere to the belief in predestination that asserts that God has predetermined the events of one's life. This belief encourages individuals to strive for the best outcomes while recognizing that ultimate control rests with a higher power. Plans and aspirations are often accompanied by the phrase "*inshallah*," meaning "if God wills," signifying an acknowledgment of divine authority over one's fate. Additionally, both blessings and misfortunes are commonly attributed to the will of God rather than solely to individual actions (Mpofu 2018).

Across the Arab world, there exists a relaxed attitude toward punctuality except in the context of formal business or professional engagements. Life moves at a more leisurely pace compared to Western standards where social gatherings and appointments often lack strict start and end times. While some individuals may prioritize punctuality, it is common for appointments to begin later than scheduled. However, within Arab American communities, especially among those engaged in professional or business occupations, punctuality, and adherence to deadlines and appointments are regarded as essential virtues. This cultural disposition toward time reflects a broader worldview that values interpersonal relationships and flexibility over rigid adherence to schedules. Understanding these nuances in cultural attitudes toward time can facilitate smoother interactions and collaborations between Arab immigrants and individuals from different cultural backgrounds promoting mutual understanding and respect.

**Navigating Social Etiquette in Arab Culture**

Traditional etiquette stresses the exchange of handshakes upon both arrival and departure in Arab culture. However, when an Arab man is introduced to an Arab woman, it is customary for him to wait for the woman to initiate the handshake. In traditional Muslim customs, persons may refrain from shaking hands with members of the opposite gender. In such cases, both women and men often place their right hand over their chest as a respectful gesture in place of shaking hands.

Titles hold significance in Arab culture and are typically used in conjunction with the individual's first name. For instance, it is common to address someone as "Mr. Khalil" or "Dr. Ali." Alternatively, some individuals may prefer to be addressed with familial titles, such as "mother" (*Um*) or "father" (*Abu*), followed by the name of their eldest son (e.g., Abu Khalil, which translates to "father of Khalil"). It is also customary for married women to retain their maiden names reflecting the importance of lineage and familial connections in Arab society (Suleiman 2019).

## *8.2.2 Family Roles and Organization*

**Gender Dynamics and Modesty in Arab Families**

Gender dynamics within Arab families often adhere to traditional patriarchal structures where men typically hold authoritative roles and decision-making responsibilities while women are often perceived as subordinate. Younger family members defer to their elders (Aldossari and Chaudhry 2024; Druker Shitrit et al. 2024; Hurley 2024). Despite the persistence of these conventional roles in some regions, there is a noticeable trend toward more progressive attitudes regarding gender equality, particularly in urban contexts (Hurley 2024). Modesty is highly valued among both genders in Arab families, influencing their attire and behavior with conservative clothing choices prevalent. For Muslim women, the hijab is widely viewed as a symbol of modesty and cultural identity covering the body except for the face and hands. Many choose to wear it as an expression of their faith and cultural adherence (Aldossari and Chaudhry 2024; Almutawaa et al. 2024; Druker Shitrit et al. 2024).

In traditional Arab families, women are seen as subordinate to men. While husbands may hold obvious influence in public, wives often exercise significant power behind the scenes, particularly concerning domestic affairs and childcare. This power dynamic emphasizes the importance of understanding cultural intricacies within Arab families. Within extended families, decision-making authority often falls to older male figures, although women may gain influence with age, particularly after their children reach adulthood. The bond between mothers and sons is typically strong with sons often prioritizing their mother's wishes and preferences. However, as nuclear families replace extended ones, these traditional norms are evolving (Aldossari and Chaudhry 2024).

Gender roles in Arab families are typically delineated, with men fulfilling roles as breadwinners, protectors, and decision-makers while women are responsible for childcare, household management, and maintaining successful marriages. Despite increasing opportunities for women in professional careers in urbanized Arab countries, family and marriage remain primary commitments for many and caregiving often takes precedence. Unfortunately, these traditional gender roles are sometimes misinterpreted in Western contexts fueling stereotypes of dominant Arab males and passive, oppressed females. This misinterpretation can contribute to conservative Arab Americans perceiving criticisms of gender roles as attacks on their culture and family values (Aldossari and Chaudhry 2024; Khosh et al. 2020).

Modesty is a core value among Arabs, influencing attire choices and behavior with many considering the hijab a symbol of Muslim identity and moral integrity. However, while the hijab may be seen as protective within Muslim communities, it is often misinterpreted by some Americans as a symbol of oppression. It is essential to recognize that the hijab is a personal choice with some women within the same family choosing to wear it while others do not depending on individual preferences. Even those who do not wear it during the day often wear it during religious activities like prayer and Quran reading. Understanding these nuances of gender dynamics and modesty within Arab families is crucial for healthcare professionals to provide culturally sensitive care, build trust, and improve health outcomes for Arab patients (Almutawaa et al. 2024; Druker Shitrit et al. 2024).

Gender dynamics and modesty norms influence family dynamics and social interactions within Arab communities reflecting broader cultural values and traditions. While traditional gender roles may persist in some areas, there is a growing acceptance of more egalitarian perspectives on gender within Arab societies with an ongoing emphasis on modesty and cultural identity. Healthcare professionals must understand these gender dynamics within Arab families to foster cultural sensitivity, respect traditional gender roles, and build trust with Arab patients. This understanding facilitates effective communication by enabling healthcare providers to tailor their approach to cultural norms, thereby enhancing rapport and understanding. Also, awareness of gender dynamics enables healthcare professionals to navigate health decision-making processes within Arab families ensuring the inclusion of relevant family members and addressing disparities that may disproportionately affect women or men. Respect for privacy and modesty preferences such as the preference for same-gender healthcare providers contribute to a more dignified care experience. This cultural understanding enables healthcare professionals to deliver culturally competent care that honors patients' values and preferences leading to improved healthcare outcomes and patient satisfaction.

**Parental Roles and Child Rearing in Traditional Arab Families**
In traditional Arab families, the roles of fathers and mothers concerning their children are clearly defined and distinct. Typically, fathers take on the role of disciplinarians while mothers serve as allies and mediators providing unwavering love and kindness. Although some fathers may believe in maintaining a degree of fear, family relationships are generally characterized by affection and sentimentality. Children are deeply cherished, indulged, and actively involved in all family activities (Abdul-Haq 2018). Within Arab culture, the upbringing of children to reflect positively on the family is a significant responsibility. The character and achievements (or shortcomings) of a child are often attributed to the influence of upbringing and parental guidance. Given the emphasis on collective familism over individualism in Arab culture, adherence to adult rules is highly valued, and child-rearing practices are geared toward fostering accommodation and cooperation (Abdul-Haq 2018).

Maintaining family honor is paramount and children are expected to conduct themselves honorably to avoid disgracing their families. Also, respect for parents and elders is instilled from a young age in traditional families with children encouraged not to question elders and to obey older siblings although these norms are gradually evolving to grant children more autonomy. Discipline methods may include physical punishment and shaming with a focus on external repercussions rather than internal reflection. Unlike Western adolescents, which emphasize the acquisition of personal identity and separation from the family, Arab adolescents are encouraged to remain closely connected to their family system. Family interests and opinions often heavily influence career and marriage decisions with academic success being particularly emphasized due to its association with social status.

Behaviors that may tarnish family honor, such as academic underachievement, sexual activity, substance use, and delinquency are actively avoided, especially among girls who are expected to uphold chastity and decency (Abdul-Haq 2018). Adolescence in North America may offer Arab adolescents' greater opportunities for academic achievement and more freedom in career choices compared to their counterparts in Arab countries. However, this career choice freedom can lead to cultural conflicts between American and Arab values causing significant tensions within Arab American families. Parents often express concerns about conflicting values regarding dating, extracurricular activities, alcohol consumption, and substance use among their children, highlighting the importance of understanding these cultural nuances for healthcare professionals.

**Family Dynamics and Social Structures**
In Arab society, the family serves as the fundamental socioeconomic unit where members collaborate to ensure livelihood, raise children, and uphold their status and influence within the community. Family members typically reside close to one

another often engaging in intermarriage such as between first cousins although this practice varies across different countries. Regardless of practicality or capability, there exists a strong expectation among family members to provide support to one another with loyalty to the family superseding personal needs. Preserving family honor is of utmost importance and guides the conduct of all family members (Abdul-Haq 2018).

Within the hierarchical structure of the family, older members command significant respect and it is customary for children, especially sons, to assume the responsibility of caring for elderly parents. Consequently, elderly parents are commonly looked after within the family home until their passing, despite the sacrifices involved. In traditional Arab families, the burden of caring for family members primarily falls on older men, with brothers assuming responsibility for unmarried sisters in the absence of the father. In the unfortunate event of a husband's demise, his family is expected to provide for his widow and children. Family leaders are also expected to leverage their influence to provide support to relatives, although these expectations are shifting as nuclear families increasingly replace extended family structures, leading to challenges in caring for elderly relatives (Abdul-Haq 2018).

While educational achievements, prestigious occupations like engineering, medicine, law, and accumulated wealth contribute to social status in Arab culture. Family lineage remains the foremost determinant of social standing in traditional Arab societies. Certain character traits such as piety, generosity, hospitality, and good manners are honored and can further elevate an individual's social status within the community.

**Navigating Norms and Activism**

Most adults in Arab cultures enter a marital relationship with monogamous unions being the norm despite the option for polygamous marriages according to Islamic law. However, studies show that polygamous marriages are relatively rare among Arab Muslims with only a small percentage opting for multiple spouses. While homosexuality exists across all cultures, it continues to be highly stigmatized within Arab societies (Abboud et al. 2020, 2022). A significant proportion of HIV/AIDS cases among Arabs are attributed to men engaging in sexual activity with other men reflecting the challenges faced by LGBTQ individuals in Arab communities (Saliba et al. 2024). In some Arab countries, homosexual behavior is even considered a criminal offense leading many individuals to conceal their sexual orientation due to fear of familial shame and social ostracism (Abboud et al. 2022). Nevertheless, there has been a notable increase in activism among Arab American LGBTQ individuals with some openly advocating for the rights of the LGBTQ community in Arab countries and among Arab Americans (Abboud et al. 2024a, b). In certain Arab countries like Lebanon, there are emerging LGBTQ organizations that aim to support and advocate for the rights of this marginalized population challenging legal and cultural barriers to alternative lifestyles (Abboud et al. 2020; Hatoum 2018). Recent research in Lebanon has shown positive attitudes and behaviors among mental health providers toward LGBTQ patients suggesting a potential shift toward greater acceptance and support within certain sectors of Arab society (Dandashly 2021; Hatoum 2018; Lebanon's LGBT People

Reclaim Their Power 2020). Similarly, LGBTQ Arab organizations in the United States are actively engaged in raising awareness and combating cultural stereotypes and stigma associated with LGBTQ identities within Arab communities to promote equity and acceptance for all individuals regardless of sexual orientation (Dandashly 2021; Hatoum 2018; Lebanon's LGBT People Reclaim Their Power 2020).

### *8.2.3 Workforce Issues*

**Cultural Influences on Work Dynamics and Mental Health**

Cultural variances that may influence work dynamics include beliefs and customs regarding family structure, gender roles, control over life circumstances, fostering harmonious interpersonal relationships, safeguarding dignity and honor, and the significance placed on upholding one's reputation. Disparities in attitudes toward time management, educational methodologies, cognitive patterns, and the degree of emphasis on objectivity may exist between Arab and American cultures. However, due to the exposure of many second-generation professionals to education in the United States, these distinctions are often more pronounced among less-educated, first-generation Arab Americans.

In recent times, exacerbated by ongoing political unrest in Arab nations and escalating unemployment and poverty rates, stress, depression, and mental health conditions have become increasingly prevalent among first-generation immigrants, refugees, and individuals residing in conflict-ridden Arab regions (Horne 2024). Factors contributing to this mental health burden include separation from family members, challenges adapting to American societal norms, marital discord, and intergenerational tensions, particularly concerning adolescents acculturated to American values through educational and social environments (Abuelezam and El-Sayed 2021; Pampati et al. 2018). Also, concerns about the safety and economic welfare of relatives in their home countries have heightened the susceptibility of Arab immigrants to mental health conditions.

Discrimination-related stressors also significantly impact the work experiences of Arab Americans (Alsaidi et al. 2023; Kader et al. 2020). While major incidents of discrimination such as termination and physical assaults are relatively uncommonly reported, instances of intimidation, suspicion, derogatory remarks about religious practices, and patient refusal for treatment by Arab American nurses have been reported (Hashem and Awad 2021; Saadi et al. 2023). Arab Americans are acutely aware of prevalent misconceptions held by the broader American society regarding Arabs, including perceptions of inferiority, backwardness, sinister motives, and propensity for violence. Additionally, the misrepresentation of mainstream Islam and the stereotyping of Muslims as extremists and confrontational further exacerbate the challenges faced by Arab Americans in navigating their religious practices within a predominantly secular American society (Hashem and Awad 2021). Conflicts often arise concerning matters like marriage, divorce, banking, inheritance, and the observance of religious obligations such as Friday prayers and dietary restrictions, particularly among children balancing Islamic requirements with the norms of the American public school system (Fleischer and Sadek 2024).

**Understanding Work Dynamics in American and Arab Environments**
In contrast to the American workplace culture, which emphasizes meeting deadlines, achieving profit targets, and maintaining a competitive edge, the Arab work environment is characterized by a more relaxed and sociable atmosphere. Business relationships often intertwine with personal friendships, often over cups of sweet tea, blurring the lines between socializing and work-related discussions. Managers in Arab workplaces typically rely on personal influence and persuasion to motivate employees, and performance evaluations often consider not only job skills but also personality and social behavior.

Distinct workplace norms further differentiate American and Arab work cultures. In the United States and some other countries, professional positions are typically earned based on merit with laws applied uniformly and work often taking precedence over familial obligations. Values such as honesty, reliance on facts and logic and direct, constructive criticism are highly regarded. In contrast, in the Arab world, social status within the workplace is often influenced by family connections, although these practices are gradually evolving. Flexibility with rules and regulations is common with family responsibilities often prioritized over job demands. Subjective perceptions frequently guide actions, and criticism is often taken personally, seen as a challenge to one's dignity and family honor.

Supervisors and managers in Arab offices are expected to openly praise their employees to reassure them that their efforts are recognized and appreciated. While such direct praise may sometimes make Americans feel uncomfortable, Arabs typically expect and appreciate acknowledgment when they believe they have earned it. This cultural contrast highlights the importance of understanding and adapting to different workplace norms and communication styles when operating in multicultural environments.

## *8.2.4 Biocultural Ecology*

**Skin Color and Other Biological Variations**
While there is a common stereotype associating Arabs with dark or olive skin tones, it is worth recognizing the diversity within the Arab population (Galaz 2019). While many Arabs do indeed have darker complexions, it is common to find individuals with blonde or auburn hair, blue eyes, and fair skin. This diversity is particularly evident among Arabs from North African countries like Egypt, Morocco, and Tunisia where individuals may have black skin and African features (Galaz 2019; Kader et al. 2020). When assessing health conditions in individuals with darker skin tones such as Arabs, it is important to note that detecting changes in color such as pallor (i.e., pale skin) or cyanosis (i.e., bluish discoloration) can be more challenging. In such cases, healthcare professionals may find it helpful to examine the oral

mucosa (i.e., the inner lining of the mouth) and conjunctiva (i.e., the membrane covering the eye) for signs of these color changes. This comprehensive approach ensures accurate assessment and diagnosis, leading to appropriate healthcare interventions.

**Diseases and Health Conditions**
People of Arab descent encounter various health challenges influenced by genetic factors, cultural practices, socioeconomic status, and disparities in healthcare access. Predominant among these concerns are cardiovascular diseases like coronary artery disease, hypertension, and stroke often exacerbated by sedentary lifestyles, high-cholesterol diets, and tobacco use (Ahmadi et al. 2024; Al-Ajlouni et al. 2024; Ferrari et al. 2024; Rashidi et al. 2024). Also, Type 2 diabetes is increasingly prevalent in Arab communities, driven by factors such as obesity, unhealthy dietary habits, and potential genetic susceptibility (Al-Ajlouni et al. 2024). Cancer, including lung and colorectal cancer, contributes significantly to mortality rates among Arab Americans influenced by tobacco consumption, environmental exposures, and genetic predisposition (Allehebi et al. 2024; AlRabeeah et al. 2024; Al-Shamsi and Musallam 2024; Mansour et al. 2024). Mental health disorders like depression, anxiety, and post-traumatic stress disorder are prevalent, particularly among those affected by conflict and acculturation stress, compounded by stigma and limited access to culturally competent mental healthcare services (Abuelezam and El-Sayed 2021; Pampati et al. 2018; Sweileh 2024). Infectious diseases such as tuberculosis, hepatitis, and parasitic infestations remain significant concerns in rural Arab regions due to inadequate sanitation and healthcare infrastructure (Al Salman et al. 2020; Mostafavi et al. 2022). Genetic disorders like G-6-PD deficiency, sickle cell anemia, and thalassemias are more prevalent due to high rates of consanguineous marriages and genetic predisposition. Maternal and child health issues persist characterized by high rates of infant mortality and pregnancy complications necessitating improved access to prenatal care and maternal healthcare services (Alqifari 2024; Elbarazi et al. 2024; Rezki and Fateh 2024). Respiratory conditions like asthma and COPD are prevalent and exacerbated by environmental factors such as air pollution and tobacco smoke exposure (Almarshoodi et al. 2024; Fleischer and Sadek 2024; Hamid et al. 2024, 2024; Hanafi et al. 2024; Olimat et al. 2024). Unhealthy lifestyle behaviors, including tobacco use and sedentary habits, contribute to chronic diseases like obesity and diabetes among Arab populations. Nurses and healthcare professionals play a crucial role in providing culturally sensitive care, advocating for health equity, and implementing tailored interventions to address the unique healthcare needs of individuals with Arab heritage (Wehbe-Alamah et al. 2024).

The health challenges faced by Arabs in war-torn countries are multifaceted and deeply entrenched within the complex context of ongoing conflict, displacement, and humanitarian crisis (Hamid et al. 2024; Hanafi et al. 2024; Mohsin et al. 2024; Veronese et al. 2024). One primary concern is the prevalence of trauma and injuries resulting from the violence of war, including bombings, gunfire, and armed conflict. Civilians often sustain severe physical injuries such as burns, fractures, and wounds, necessitating immediate medical attention and long-term rehabilitation. Moreover, the psychological impact of war is profound, with many individuals grappling with post-traumatic stress disorder (PTSD), depression, anxiety, and other mental health conditions. Exposure to violence, displacement, and the loss of loved ones exacerbates the prevalence of mental health disorders among war-affected populations (Mahamid et al. 2024; Manzanero et al. 2024; Tarabay and Golm 2024; Veronese et al. 2024). Additionally, the conditions of displacement, overcrowded living environments, and inadequate access to clean water and sanitation facilities heighten the risk of infectious disease outbreaks (Vahidniya et al. 2024). Diseases like cholera, typhoid fever, malaria, and diarrheal illnesses spread rapidly in refugee camps and displaced communities, posing significant public health challenges. Furthermore, conflict disrupts food production, distribution, and access leading to widespread malnutrition and famine. Both acute and chronic malnutrition among children are prevalent with devastating consequences for growth, development and overall health. Access to healthcare is severely impacted by conflict with medical facilities destroyed, shortages of essential medicines and the displacement of healthcare workers. Many individuals struggle to access basic healthcare services, including maternal and child health services resulting in preventable deaths and complications. The plight of displaced individuals further exacerbates health vulnerabilities, exposing them to harsh living conditions, inadequate shelter, and limited access to healthcare. Children are particularly vulnerable facing malnutrition, lack of education, and exposure to violence and trauma. Reproductive health services are also disrupted posing risks for pregnant women and newborns due to limited access to skilled birth attendants and emergency obstetric care. Addressing the health needs of Arabs in war-torn countries requires collaborative efforts from humanitarian organizations, governments, and the international community. Urgent interventions are needed to provide emergency medical care, mental health support, nutrition assistance, and access to clean water and sanitation. Long-term efforts should focus on rebuilding healthcare infrastructure, strengthening health systems, promoting peace and stability, and addressing the root causes of conflict to improve the health and well-being of affected populations (Hamadeh et al. 2024; Manzanero et al. 2024; Tarabay and Golm 2024; Veronese et al. 2024).

**Variations in Drug Metabolism**

Limited information is available regarding drug metabolism and sensitivity among individuals of Arab heritage. It has been reported that approximately 1–1.4% of Arabs may face challenges in metabolizing debrisoquine and similar substances including certain antiarrhythmics, antidepressants, beta-blockers, neuroleptics, and opioid agents. As a result, a small proportion of Arab Americans may experience elevated blood levels and adverse effects when prescribed standard dosages of antidepressants. Conversely, some individuals may not adequately metabolize codeine into morphine, leading to suboptimal analgesic effects even with typical codeine dosages. This variation in drug metabolism highlights the importance of considering individual genetic factors and ethnic differences when prescribing medications to individuals of Arab descent as it may impact treatment efficacy and safety outcomes. Healthcare providers should be aware of these potential differences to optimize medication management and minimize the risk of adverse drug reactions among Arab populations.

### 8.2.5 High-Risk Health Behaviors

High-risk health behaviors among people of Arab descent or heritage encompass a range of practices that pose significant threats to individual and public health. Despite Islamic teachings discouraging tobacco use, smoking remains deeply ingrained in Arab culture, often seen as a gesture of hospitality. This cultural norm is reflected in the higher smoking rates and lower cessation rates among Arab Americans compared to European Americans. Although efforts have been made to reduce tobacco use in many Arab countries, smoking continues to rise in some regions. In addition to conventional cigarette smoking, the use of water pipes, commonly known as hookah, shisha, or narghile, is prevalent among both adults and youth (Atrooz et al. 2024; HARB 2024).

Data from the 2018 Arab and Chaldean Behavioral Risk Factor Survey indicate a slight decrease in the percentage of Arab Americans reporting ever smoking or using hookah from 38.1% in 2013 to 30.8% in 2018. However, there remains a perception among many smokers that hookah use is a safer alternative to cigarette smoking. The prevalence of cigarette smoking among Arab adults is reported to be 17.2%, with a higher prevalence among males than females. Acculturation levels have been linked to nicotine dependence with less-assimilated Arab Americans exhibiting higher smoking rates (Atrooz et al. 2024; HARB 2024; Labibidi et al. 2024).

Alcohol consumption among Arab Americans is influenced by Islamic prohibitions, leading to overall lower rates of alcohol consumption compared to the general population. However, a subset of Arab Americans engages in substance abuse, primarily involving alcohol. Substance abuse treatment centers in areas with large Arab American populations report admissions for alcohol, marijuana, heroin, and crack cocaine abuse, primarily among males. Despite these challenges, a significant portion of Arab Americans abstain from alcohol use, reflecting adherence to cultural norms (Rostam-Abadi et al. 2024; Safadi 2024).

Sexual health practices also contribute to high-risk behaviors among people of Arab descent with limited data available on HIV/AIDS prevalence among Arab Americans (Alsabri et al. 2024; Saliba et al. 2024). Cultural norms of modesty, particularly among Arab women, may inhibit access to reproductive health services, including HIV testing and prevention measures. Despite improvements in breast and cervical cancer screening rates, challenges persist in addressing domestic violence and other reproductive health issues among Arab American women, especially new immigrants.

Socioeconomic factors significantly influence health behaviors and outcomes among Arab Americans with education and income levels serving as strong indicators of health risk and service utilization. Disparities in healthcare coverage exist, with a higher prevalence of uninsured individuals among Arab Americans compared to other ethnic groups. Additionally, physical and mental disabilities among Arab Americans are comparable to those among white Americans, highlighting the need for accessible and culturally competent healthcare services (Abuelezam et al. 2018; Alsabri et al. 2024).

Addressing high-risk health behaviors among people of Arab descent requires comprehensive strategies that consider cultural beliefs, socioeconomic factors, and access to healthcare. Culturally sensitive interventions aimed at tobacco cessation, substance abuse prevention, and sexual health education are essential. Moreover, efforts to improve healthcare access and reduce disparities should prioritize education and outreach initiatives tailored to the needs of Arab American communities. By addressing these challenges holistically, healthcare providers and policymakers can work toward improving health outcomes and promoting well-being among people of Arab heritage.

### *8.2.6 Nutrition*

**Meaning of Food**

Food serves as a cornerstone of Arab culture, deeply intertwined with social customs, familial bonds, and expressions of identity. Across the Arab world, sharing meals with family and friends is not merely a matter of sustenance but a cherished ritual that fosters connection and community. Whether it is a festive celebration, a casual gathering, or a formal dinner party, food acts as a unifying force bringing people together to forge lasting memories and strengthen relationships. One of the striking aspects of Arab cuisine is its remarkable diversity, reflecting the intricate blend of flavors, ingredients, and culinary techniques across different regions. From the aromatic spices of North African dishes to the hearty stews of the Levant and the seafood specialties of the Gulf, each culinary tradition tells a unique story of cultural heritage and regional pride. This diversity not only celebrates the richness of Arab culinary heritage but also serves as a testament to the historical trade routes and cultural exchanges that have shaped Arab cuisine over centuries. Food plays a central role in religious observances, particularly during Ramadan, the holy month of fasting observed by Muslims worldwide. Each day, fasting is broken with a communal meal known as **iftar**, where families and communities come together to share a feast and express gratitude for blessings. This shared experience of breaking fast symbolizes unity, compassion, and spiritual renewal, reinforcing bonds of kinship and solidarity.

In addition to religious rituals, food also serves as a powerful expression of hospitality in Arab culture. The act of preparing and serving food to guests is considered a gesture of warmth, generosity, and respect. Whether welcoming visitors into the home or dining out at a local restaurant, the abundance and variety of dishes offered reflect the host's desire to ensure the comfort and satisfaction of their guests. For Arab women, in particular, the preparation and presentation of food carry deep cultural significance. Traditionally, a woman's culinary skills are seen as a reflection of her love and dedication to her family, with elaborate meals serving as a tangible expression of care and devotion. In social settings, the quality and quantity of food served are indicators of hospitality and esteem with hosts striving to uphold their honor and reputation through the generosity of their offerings. So, food embodies the essence of Arab identity, encapsulating values of community, tradition, and shared heritage. Through the sharing of meals, Arabs reaffirm their connections to family, culture, and faith, celebrating the richness of their culinary traditions while forging bonds of kinship and friendship that transcend borders and generations. Islamic teachings emphasize holistic well-being, encouraging adherents to consume wholesome foods, avoid prohibited substances like pork and alcohol, and maintain moderation in all aspects of life. As Arab communities navigate the complexities of modern life, the preservation and celebration of culinary heritage continue to serve as a source of resilience, unity, and cultural pride.

**Common Foods and Food Rituals**

Common Foods and Food Rituals among people of Arab heritage or descent are not only about sustenance but also about cultural preservation, social bonding, and religious observance. Arab cuisine, characterized by its rich flavors and diverse ingredients, serves as a cornerstone of cultural identity and pride. Rice, bread (especially flatbreads like pita and khubz), lentils, chickpeas, and couscous form the foundation of many traditional Arab dishes. These staple ingredients are versatile and are prepared in various forms across different regions, reflecting both local tastes and historical influences. *Mezze*, a selection of small dishes served as appetizers or snacks is an integral part of Arab dining. *Mezze* platters often feature a colorful array of dips, salads, and finger foods, providing a communal and convivial dining experience. Arab cuisine is renowned for its succulent grilled meats, such as kebabs, shawarma, and kofta. These dishes are often seasoned with a blend of aromatic spices and herbs, adding depth and complexity to the flavors. Rice is a staple in Arab cooking and is often featured in savory dishes such as *biryani*, *maqluba*, and *kabsa*. These rice-based dishes are often cooked with fragrant spices and served as hearty main courses for family meals and festive occasions.

Arab desserts are a celebration of sweetness and indulgence, with treats like *baklava*, *knafeh*, and *halva* tantalizing taste buds with their rich textures and flavors. These confections are often enjoyed during special occasions and religious festivals symbolizing joy and abundance. Coffee and tea hold a special place in Arab culture, serving as symbols of hospitality and social connection. Arabic coffee, infused with cardamom and other spices, is traditionally served in small cups as a gesture of welcome and respect. Mint tea, brewed with fresh mint leaves and sugar, is a refreshing beverage enjoyed throughout the Arab world.

Food rituals among people of Arab heritage or descent are deeply ingrained in daily life and social interactions. Offering food to guests is considered a sacred duty in Arab culture, reflecting the value placed on generosity and hospitality. Hosts take pride in preparing elaborate meals for guests, ensuring they feel welcome and cherished. Family gatherings and celebrations often revolve around food, with meals serving as a focal point for strengthening familial bonds and fostering a sense of belonging. These gatherings are opportunities for sharing stories, laughter, and cherished family recipes. Food also plays a central role in religious rituals, particularly during Ramadan, the month of fasting observed by Muslims. The pre-dawn meal (*suhoor*) and the evening meal (*iftar*) during Ramadan are occasions for communal dining and spiritual reflection reinforcing values of empathy, gratitude, and self-discipline.

Cooking and sharing traditional recipes are cherished traditions passed down through generations, with family members often gathering in the kitchen to bond over food preparation. These culinary traditions serve as a link to the past, preserving cultural heritage and nurturing a sense of continuity and belonging. Arab festivals and celebrations are marked by feasting and culinary indulgence with communities coming together to enjoy food, music, and festivities. Traditional dishes are prepared with care and served in abundance, symbolizing abundance, prosperity, and communal unity. So, food serves as more than just nourishment for people of Arab heritage or descent—it reflects cultural identity, social cohesion, and shared values. By honoring and preserving culinary traditions, individuals and communities celebrate their heritage and forge connections that transcend borders and generations. Also, understanding food-related customs and rituals is essential for healthcare providers to provide culturally competent care, particularly during religious observances like Ramadan where dietary practices may impact medical management and treatment adherence.

**Dietary Practices for Health Promotion**

Dietary practices among people of Arab heritage or descent are deeply rooted in tradition, culture, and culinary heritage. Traditional Arab dietary beliefs often associate good health with eating properly, consuming nutritious foods, and fasting to cure disease. This perspective is influenced by teachings attributed to the prophet Mohammed who emphasized the importance of moderation in eating and the relationship between diet and health. In Arab dietary tradition, maintaining balance among different food types—such as hot, cold, dry, and moist—is considered essential for well-being. According to these beliefs, illness can result from excessive eating, consuming nutritionally deficient food, or mixing opposing types of foods. Conversely, abstinence and moderation are seen as remedies that allow the body to expel disease. The condition of the alimentary tract holds particular importance in Arab perceptions of health with gastrointestinal complaints often being a primary reason for seeking healthcare among Arab Americans. However, dietary-related health issues extend beyond digestive concerns. For example, obesity has become a significant problem among second-generation Arab American women and children, often due to the adoption of American dietary habits, including the consumption of high-fat and calorie-dense snacks.

To promote health and well-being within Arab communities, it is essential to encourage dietary practices that align with cultural traditions while supporting optimal nutrition:

1. Encourage the consumption of whole, minimally processed foods such as fruits, vegetables, whole grains, legumes, nuts, and seeds. These foods provide essential nutrients and fiber while supporting overall health and well-being.
2. Promote mindful eating practices and moderation in portion sizes to prevent overeating and support weight management. Encourage individuals to listen to their hunger and fullness cues and to enjoy meals in a relaxed setting.
3. Advocate for a balanced diet that includes a variety of carbohydrates, proteins, and healthy fats. Incorporate lean sources of protein such as poultry, fish, legumes, and tofu, along with healthy fats from sources like olive oil, avocados, and nuts.
4. Encourage the reduction of highly processed foods, sugary snacks, and high-fat desserts in favor of whole, nutrient-dense options. Educate individuals about reading nutrition labels and making informed choices about packaged foods.
5. Respect cultural preferences and traditions while promoting healthful dietary practices. Explore ways to adapt traditional recipes and meals to align with nutrition recommendations without compromising cultural identity.
6. Emphasize the importance of staying hydrated by drinking water throughout the day. Encourage the consumption of herbal teas and limit sugary beverages to support hydration and overall health.
7. Provide education and resources about nutrition, healthy eating habits, and disease prevention within the context of Arab culture. Offer cooking classes, nutrition workshops, and culturally relevant resources to empower individuals and families to make healthful choices.

By promoting these dietary practices within Arab communities, healthcare providers and educators can support individuals in maintaining cultural traditions while prioritizing their health and well-being. Collaboration between healthcare providers, community leaders, and individuals is essential to address dietary-related health issues and promote positive health outcomes within Arab populations.

**Nutritional Deficiencies and Food Limitations**

Nutritional deficiencies and food limitations among individuals of Arab heritage are influenced by various factors, including cultural practices, dietary habits, socioeconomic status, and access to healthcare. While Arab cuisine is known for its rich flavors and ingredients, certain dietary patterns and cultural norms may contribute to specific nutritional concerns. Iron deficiency is a prevalent concern, particularly among women of childbearing age and children due to insufficient dietary intake (Jalambo et al. 2018). The traditional Arab diet may lack adequate sources of iron, emphasizing the importance of incorporating iron-rich foods such as legumes, fortified cereals, and leafy greens into daily meals to prevent anemia. Limited consumption of dairy products in Arab cultures may contribute to calcium deficiency, impacting bone health. To address this, individuals should be encouraged to consume alternative sources of calcium such as fortified plant-based milk, leafy greens, almonds, and sesame seeds to support optimal bone health. Factors such as limited sun exposure and cultural practices like covering the skin may lead to vitamin D deficiency, especially in regions with minimal sunlight. Dietary sources of vitamin D and supplementation may be necessary to address deficiencies and support bone health and immune function (Masoud et al. 2020). Individuals following a strict vegetarian or vegan diet in Arab cultures may be at risk of vitamin B12 deficiency due to limited intake of animal products. Incorporating fortified foods or considering supplementation can help prevent deficiencies, particularly among those with limited meat consumption.

In regions with low iodized salt intake or limited access to iodine-rich foods like seafood, iodine deficiency disorders may occur. Educating individuals about the importance of iodized salt and promoting the consumption of iodine-rich foods can help prevent deficiencies and support thyroid health. Inadequate folate intake may contribute to neural tube defects in newborns and anemia in adults. Encouraging the consumption of folate-rich foods such as leafy greens, legumes, citrus fruits, and fortified grains is essential for preventing deficiencies and supporting overall health. Food allergies and intolerances, such as lactose intolerance or gluten sensitivity, may affect individuals of Arab heritage, necessitating dietary modifications and medical intervention to prevent adverse reactions and ensure optimal nutrition.

Cultural and religious dietary practices, such as halal food requirements and fasting during Ramadan, may impact nutrient intake and dietary patterns. Healthcare providers should respect and support individuals' cultural and religious beliefs while guiding them on maintaining a balanced diet during periods of fasting or dietary restrictions. Addressing nutritional deficiencies and food limitations among individuals of Arab heritage requires a holistic approach that considers cultural beliefs, dietary preferences and access to resources. Healthcare providers can play a vital role in promoting nutrition education, providing culturally sensitive dietary counseling, and facilitating access to nutritious foods and supplements to support optimal health and well-being within Arab communities. Additionally, understanding and accommodating cultural and religious dietary restrictions are essential for providing culturally competent care to individuals of Arab heritage.

### *8.2.7 Pregnancy and Childbearing Practices*

**Fertility Practices and Views Toward Pregnancy**

Fertility practices and views toward pregnancy among individuals with Arab heritage are influenced by a complex interplay of cultural, religious, social, and familial factors. Arab culture places a high value on fertility and childbearing, often considering it a blessing and a fulfillment of familial and societal expectations (Raghavan et al. 2024; Sharkia and Taubman-Ben-Ari 2024). Islamic teachings emphasize the importance of procreation and raising righteous offspring further reinforcing positive attitudes toward pregnancy and childbirth (Nisar and Benedict 2024). Pregnancy is viewed as a communal event within Arab families and communities. Expectant mothers receive extensive support from family members, friends, and neighbors throughout their pregnancy journey, including emotional support, practical assistance, and advice on prenatal care and childbirth. Couples often engage in preconception care practices to optimize their health and fertility. This may involve adopting a nutritious diet, maintaining a healthy lifestyle, and seeking medical advice to address any underlying health conditions that could affect fertility or pregnancy outcomes.

Some individuals incorporate traditional remedies and practices into their fertility journey, such as herbal supplements, dietary modifications, and rituals believed to enhance fertility and promote a healthy pregnancy. These practices hold cultural and historical significance within Arab communities. In cases of infertility, couples may seek medical interventions such as fertility treatments and assisted reproductive technologies. Despite some cultural reservations, seeking medical assistance for infertility is becoming more common among Arab individuals and couples. Expectant mothers receive prenatal care and education to ensure the health and well-being of both mother and baby. This includes regular check-ups with healthcare providers, prenatal screenings, childbirth education classes, and guidance on nutrition, exercise, and childbirth preparation.

Gender roles influence fertility practices within Arab families, with women often expected to prioritize motherhood (Sharkia Taubman-Ben-Ari 2024). Men also play a supportive role in the fertility journey, providing emotional support and participating in family planning decisions. Childbirth is considered a significant life event, and expectant mothers adhere to cultural customs and traditions surrounding childbirth and postpartum care. This includes preferences for specific birthing practices and the involvement of female relatives or midwives during labor. Despite positive attitudes toward pregnancy and childbirth, individuals may encounter stigma associated with infertility or pregnancy loss within Arab communities. Open discussions about reproductive health and support for those experiencing fertility challenges are essential to address stigma and promote well-being.

While high fertility rates were favored historically, there has been a decline in fertility rates influenced by urbanization, education, and access to healthcare. Contraceptive use has become more prevalent, although cultural, religious, and socioeconomic factors may influence contraceptive decision-making. Overall, fertility practices and views toward pregnancy among individuals with Arab heritage are shaped by cultural traditions, religious beliefs, family dynamics, and socioeconomic factors. By understanding and respecting these diverse perspectives, healthcare providers can offer culturally sensitive care and support to individuals and couples navigating the fertility journey and pregnancy experience within Arab communities.

**Prescriptive, Restrictive, and Taboo Practices in the Childbearing Family**
Prescriptive, restrictive, and taboo practices surrounding childbirth and the postpartum period among individuals of Arab heritage are deeply rooted in cultural, religious, and societal norms.

In Arab culture, childbirth is viewed as a family affair. It is common for extended family members, particularly female relatives, to be actively involved in the pregnancy, childbirth, and postpartum care. They provide emotional support, practical assistance, and traditional remedies to the expectant mother. Arab families often adhere to traditional birthing practices passed down through generations. These may include specific rituals, prayers, or customs believed to ensure safe delivery and protect the mother and baby from harm.

Modesty is highly valued in Arab culture, particularly concerning matters related to childbirth and the postpartum period (Druker Shitrit et al. 2024). Women may be expected to maintain modesty by covering their bodies and refraining from interactions with unrelated males during labor and after childbirth. In some Arab communities, women may be subject to dietary restrictions during pregnancy and postpartum. Certain foods or spices believed to be harmful to the mother or baby may be avoided, while others considered beneficial for recovery and lactation may be encouraged.

There may be a taboo surrounding discussions about childbirth complications or maternal health issues within Arab families. Some individuals may avoid talking openly about potential risks or challenges during pregnancy and labor due to cultural beliefs or fears of stigma. Naming the baby before birth is often considered taboo in Arab culture, as it is believed to attract the evil eye or negative energy. Instead, families may wait until after the baby is born to announce and formally name the newborn. While breastfeeding is encouraged and widely practiced in Arab communities, there may be taboos surrounding public breastfeeding. Some women may feel uncomfortable nursing in public due to cultural norms and expectations regarding modesty.

These practices reflect the cultural values, religious beliefs, and social norms within Arab families. While some practices may provide emotional support and cultural continuity, others may present challenges or limitations for individuals navigating the childbearing experience. It is important for healthcare providers to be aware of these practices and to provide culturally sensitive care that respects the preferences and beliefs of individuals of Arab heritage.

### *8.2.8 Death Rituals*

**Death Rituals and Expectations**

Death rituals and expectations among people with Arab heritage are deeply rooted in cultural, religious, and societal traditions. Islamic traditions heavily influence these rituals, as most individuals of Arab heritage practice Islam. According to Islamic beliefs, death is viewed as a transition to the afterlife, and the deceased is treated with utmost respect and dignity. Following death, the body of the deceased is typically washed and covered in white cloth, a process known as "*ghusl*" and "***kafan***." This ritual purification is performed by family members or designated individuals of the same gender. Muslims believe in performing a special funeral prayer for the deceased, known as "*Salat al-**Janazah***," usually performed in congregation to seek forgiveness and mercy for the deceased. Prompt burial is encouraged in Islam, preferably within 24 h after death. Graves are typically oriented toward the *Qiblah* (the direction of the Kaaba in Mecca) and marked with simple headstones. Family and friends gather to participate in the burial process and offer condolences to the bereaved.

Arab communities observe a mourning period following the death of a loved one, known as "***azza***," lasting 3 days to 1 week. During this time, family members receive visitors who offer condolences and support. Death is considered a communal event in Arab culture and the bereaved receive extensive support from family, friends, and neighbors, providing practical assistance, emotional support, and condolences. Expressing condolences and offering support to the bereaved is a significant aspect of Arab death rituals. Visitors often bring food and gifts to the grieving family as a gesture of sympathy and solidarity. Memorial services may be held on the anniversary of the deceased's death, known as "***Yawm al-Dhikr***" (Day of Remembrance).

Arab Muslims believe in an afterlife where individuals are held accountable for their actions in this world, viewing death as a transition to eternal life. While Islamic traditions provide a framework for death rituals among Arab communities, there may be variations based on regional, cultural, and individual preferences. So, these rituals play a crucial role in providing comfort and honoring the memory of the departed loved ones.

**Responses to Death and Grief**

In Arab heritage communities, responses to death and grief are deeply influenced by cultural, religious, and societal norms. Mourning periods and practices may vary among Muslims and Christians emigrating from different Arab countries. Extended mourning periods may be practiced if the deceased is a young man, a woman, or a child. However, in some cases, Muslims may perceive extended periods of mourning as defiance of the will of God. Family members are asked to endure with patience and good faith in Allah what befalls them, including death. Whereas friends and relatives are to restrict mourning to 3 days, a wife may mourn for 4 months, and in some special cases, mourning can extend to 1 year. Although weeping is allowed, beating the cheeks, or tearing garments is prohibited. For women, wearing black is considered appropriate for the entire period of mourning.

Death is considered a communal event, and the bereaved receive extensive support from family, friends, and neighbors. Community members often rally around the grieving family, providing practical assistance, emotional support, and condolences. This sense of community helps alleviate the burden of grief and provides comfort to those mourning the loss. Expressing condolences and offering support to the bereaved is a significant aspect of Arab culture. Visitors often visit the grieving family, bringing food, and gifts, and offering heartfelt condolences. These gestures of sympathy and solidarity are deeply appreciated and help the grieving family feel supported during their time of loss. Religious rituals and practices play a central role in coping with death and grief among individuals of Arab heritage, especially among Muslims. Islamic traditions provide a framework for mourning rituals, including washing, and covering the deceased, performing funeral prayers, and observing a mourning period. These practices offer solace and spiritual comfort to the bereaved, reinforcing beliefs about the afterlife and divine mercy.

Arab communities typically observe a mourning period following the death of a loved one, known as "*azza*." During this time, family members receive visitors who offer condolences and support. Mourning rituals may include recitation of the Quran, prayers for the deceased, and gatherings to honor the memory of the departed loved one. Cultural customs and traditions also influence responses to death and grief among individuals of Arab heritage. These may vary based on regional, familial, and individual preferences. For example, some families may observe specific rituals or practices passed down through generations while others may incorporate local customs or traditions into their mourning process. While grief is a natural response to loss, individuals of Arab heritage often view death as a natural part of life and accept it as God's will. This perspective provides a sense of peace and resignation helping the bereaved cope with their loss and find meaning amid sorrow. Responses to death and grief among people of Arab heritage are characterized by a strong sense of community support, adherence to religious practices, and cultural customs that provide comfort and solace to the bereaved during their time of mourning.

### *8.2.9 Spirituality*

**Religious Practices and Use of Prayer**

Not all Arab Americans adhere to Islam; various Christian denominations such as the Copts from Egypt, the Chaldeans from Iraq, and the Maronites from Lebanon are prominent within Arab communities. Despite the diversity of religious practices and liturgies among Christians and Muslims, there are shared theological beliefs owing to Islam's roots in Judaism and Christianity. Both Muslims and Christians believe in a singular God, share reverence for many prophets, anticipate the Day of Judgment, acknowledge the existence of Satan, and hold beliefs regarding heaven, hell, and an afterlife. Notably, Islam diverges from Catholicism and Christian Orthodoxy in its absence of a formal priesthood. Instead, Islamic scholars or religious ***sheikhs***, esteemed figures within Islamic communities, fulfill the role of *imam*, or "prayer leader." The imam's responsibilities encompass leading congregational prayers, officiating marriages and funerals, and providing spiritual guidance and interpretation of Islamic teachings. Seeking counsel from a local imam can be beneficial for Arab American Muslims grappling with healthcare decisions (Mpofu 2018).

The degree of religious observance varies among individuals, ranging from nominal adherence to devout practice. However, owing to the intertwined nature of religion and governance in many Arab countries, a certain level of religious engagement is culturally expected. For instance, Islamic principles profoundly influence various facets of life in Jordan, among other Arab countries. Under Islamic law, abortion is considered a criminal act, while foster parenting is encouraged, although adoption is prohibited (Büchler and Schneider Kayasseh 2018). Infertility treatments sanctioned by Islamic jurists are the only options available. Family matters, including marriage, divorce, guardianship, and inheritance, are adjudicated in Islamic law courts, applying *shariah*.

Public schools incorporate Islamic education and provide prayer spaces and schedules are adjusted to accommodate Islamic holidays and the weekly congregational prayer on Fridays. During Ramadan, businesses close during daylight hours, and workdays are abbreviated to facilitate fasting. Friday marks the beginning of the weekend to enable attendance at communal prayers. Also, adhering to Islamic tradition, which mandates respect and protection for followers of other monotheistic faiths, Christians have separate religious institutions and educational facilities. In Saudi Arabia, however, the practice of religions other than Islam is officially prohibited. For Arab American Christians, church involvement holds significant cultural and spiritual importance. Many partake in fasting and participate in ceremonial church services to observe Catholic and Orthodox Christian holidays. Christian symbols, such as crosses and depictions of the Virgin Mary, may be displayed or worn. Also, educational institutions often provide classes on Christianity to reinforce religious teachings and traditions.

**Meaning of Life and Individual Sources of Strength**
For Muslims, Islam embodies the concept of "submission to Allah," representing a profound commitment to the divine will and guidance. Life's purpose revolves around devout worship and spiritual devotion, aimed at preparing for the afterlife by fulfilling religious obligations outlined in the Qur'an, Islam's holy scripture, and the *hadith*, the recorded sayings, and actions of the Prophet Muhammad. Central to Islamic practice are the five pillars, or fundamental duties, which include the declaration of faith, performing the five daily prayers, giving alms to the needy, observing fasting during the holy month of Ramadan, and undertaking the pilgrimage to Mecca, known as *Hajj*.

Despite the strong emphasis on familial ties and interconnectedness in Arab societies, religious faith often holds the utmost significance. Whether adherents of Islam or Christianity, Arabs typically exhibit a profound identification with their religious communities, with religious affiliation constituting a fundamental aspect of personal identity alongside family lineage. In everyday life, the presence and influence of God and His divine power are acknowledged and revered, shaping individuals' beliefs, actions, and interactions within their communities and broader society.

**Spiritual Beliefs and HealthCare Practices**
Many Muslims integrate spiritual practices into their approach to healthcare, often combining spiritual remedies such as prayer and *Quranic* recitation with conventional medical treatment. Devout patients may request specific accommodations, such as facing Mecca during prayers and ensuring cleanliness through ritual ablution before worship. In Islamic tradition, cleanliness is predominant, as prayers are only deemed acceptable when performed in a clean environment, with a clean body and attire. Hygiene is also emphasized, reflecting a belief in the importance of cleanliness for both physical and spiritual health. Also, Muslims are encouraged to support their communities by visiting the sick and assisting those in need, reflecting a sense of social responsibility and compassion. Illness is sometimes perceived as a form of divine punishment for sins, though many Muslims view it as a test or trial that serves to expiate sins and strengthen character. Responses to illness often

include patience, endurance, and unwavering hope in Allah's mercy with devout individuals believing that even seemingly irreversible conditions may be cured if they align with Allah's will. Suffering is seen as a means of earning greater rewards in the afterlife, reinforcing the belief in the sanctity of life and prohibiting practices like euthanasia and assisted suicide. Arab American Christians share similar spiritual beliefs regarding healthcare with Orthodox or Catholic Christians. They prioritize caring for the body and observe similar burial practices. A visit from a priest is customary, especially for Catholic patients who may receive the sacrament of the sick as part of their spiritual care. This intertwining of faith and healthcare underscores the importance of spirituality in providing comfort and support to individuals facing illness or adversity within Arab American communities.

### *8.2.10 Healthcare Practices*

**Health-Seeking Beliefs and Behaviors**

Health-seeking beliefs and behaviors among individuals of Arab heritage are deeply intertwined with cultural, religious, and social factors, shaping their approach to wellness and healthcare utilization (Aseeri 2020; Zimmerman and Shaw Jr 2020). Arab culture emphasizes the significance of familial and community ties, influencing health-seeking behaviors. Seeking advice and support from family members, friends, and community leaders is common practice. Decision-making regarding healthcare often involves consulting with elders or respected community members reflecting the collective nature of decision-making within Arab families. Religious beliefs, particularly within Islam, profoundly impact health-seeking behaviors. Islamic teachings emphasize the importance of maintaining both physical and spiritual well-being. Prayer, Quranic recitation, and adherence to religious rituals are integrated into healthcare practices. Many Muslims believe in the healing power of faith and may combine spiritual practices with conventional medical treatment, viewing illness as a test of faith and an opportunity for spiritual growth.

Traditional healing practices, such as herbal remedies and cupping therapy, are prevalent among individuals of Arab heritage. These practices, rooted in cultural traditions and passed down through generations, are often utilized alongside modern healthcare. While some traditional remedies complement conventional treatment, others may be preferred due to cultural preferences or beliefs, highlighting the importance of holistic approaches to health. Access to healthcare services significantly influences health-seeking behaviors. In regions with limited healthcare infrastructure, individuals may rely more on home remedies or traditional healers. Conversely, in areas with advanced healthcare systems, access to hospitals, clinics, and trained medical professionals shapes health-seeking behaviors, underscoring the role of healthcare accessibility in influencing health outcomes.

The stigma surrounding certain health conditions, particularly mental health disorders, may deter individuals from seeking timely medical attention within Arab communities (Abo-Rass et al. 2024; Sweileh 2024). Cultural taboos or misconceptions about illness can contribute to delays in seeking care or reluctance to discuss health concerns openly highlighting the need for culturally sensitive healthcare practices. Gender roles and expectations impact health-seeking behaviors, with women often assuming the role of primary caregivers. Cultural norms regarding modesty and gender segregation may influence access to certain healthcare services, particularly for women's health issues. Modesty concerns may lead to reluctance in seeking care, particularly if it involves disrobing or examination by male physicians.

Among Arab immigrants or diaspora communities, health-seeking behaviors may evolve due to exposure to different healthcare systems (Abo-Rass et al. 2024; Aseeri 2020). Acculturation may lead to a blend of traditional practices with mainstream healthcare approaches as individuals navigate cultural identity and integration into new societies, highlighting the dynamic nature of health beliefs and practices. Socioeconomic factors, including education level and income influence health-seeking behaviors. Higher education and socioeconomic status correlate with better access to healthcare services and increased health literacy leading to proactive health-seeking behaviors and preventive care practices. However, disparities in access to care may persist among marginalized populations, underscoring the importance of addressing socioeconomic barriers to healthcare access.

While Arab Americans may readily seek care for acute symptoms, preventive care utilization such as well-child visits and health screenings may be limited, particularly among low-income populations. Cultural factors, including present orientation and reluctance to plan, contribute to the underutilization of preventive services. Arab Americans prioritize coping with existing stressors and illnesses over preventive measures, highlighting the need for culturally tailored approaches to preventive care delivery. Understanding these multifaceted influences is crucial for healthcare providers to deliver culturally competent care that respects the beliefs and preferences of individuals from Arab heritage. By addressing cultural, religious, and social factors, healthcare providers can foster trust, enhance communication, and promote equitable access to healthcare services within Arab communities.

**Responsibility for Healthcare**
Responsibility for healthcare among individuals of Arab heritage is deeply rooted in cultural, familial, and religious beliefs. In Arab communities, the family plays a central role with members assuming caregiving roles for ill relatives and seeking guidance from elders in healthcare decision-making. Religious teachings, particularly within Islam, emphasize the importance of physical and spiritual well-being, integrating prayer and religious rituals into healthcare practices. While familial and religious factors shape healthcare decisions, individuals also bear accountability for their health. Arab communities prioritize preventive measures and encourage individuals to seek medical care when needed. Community support further enhances healthcare access with organizations and religious institutions offering assistance and advocacy.

Healthcare providers are respected authority figures in Arab culture, with patients trusting their expertise and guidance. However, differences in healthcare approaches between Arab and American cultures can lead to misunderstandings. American healthcare emphasizes individualism and informed decision-making, while Arab culture values familism and submission to God's will in healthcare matters. Regarding medical disclosure, Arab physicians may prioritize withholding grave diagnoses to preserve patients' hope contrasting with American practices. Healthcare financing also differs with communal resources and government facilities often serving as primary means of healthcare access in Arab countries. Understanding and respecting these cultural nuances are crucial for healthcare providers to deliver effective and culturally sensitive care to individuals of Arab heritage. By acknowledging the multifaceted nature of responsibility for healthcare and bridging cultural gaps, providers can ensure better healthcare outcomes for Arab patients.

**Folk Practices**

Folk practices among people of Arab heritage offer a window into the cultural tapestry that shapes daily life and community interactions (El-Shamy 2018). These practices, deeply rooted in tradition, religion, and social customs reflect a blend of superstitions, healing traditions, and rituals that have been passed down through generations. Despite Islamic teachings discouraging superstition, beliefs in the evil eye, envy, and supernatural entities like jinn persist in Arab folklore (El-Shamy 2018). Protective measures such as wearing amulets, reciting Quranic verses and invoking the name of Allah are common practices to ward off the evil eye. Concerns about jealousy and envy are particularly associated with the misfortune caused by the evil gaze.

Traditional healing practices based on the Prophet's teachings and the theory of humor are prevalent in Arab culture. Remedies often involve treating illnesses with their opposites, such as using hot treatments for cold ailments. Cupping therapy, cautery, and herbal remedies are favored methods alongside spiritual interventions like prayers and dietary adjustments. Rituals seeking divine protection and blessings play a significant role in Arab folk practices. These rituals may involve reciting prayers from the Quran, performing blessings for newborns or homes, or seeking intercession from saints or revered figures for specific needs or challenges. Arab communities mark significant life events with celebratory traditions, including weddings, births, and religious holidays. These occasions are characterized by communal gatherings, music, dance, and feasting, fostering social cohesion and cultural expression.

Cultural taboos and customs shape behavior within Arab communities, guiding interactions and social norms. Practices such as avoiding disrespectful gestures and upholding hospitality and respect for elders are deeply ingrained cultural values. While not universally practiced, some Arab communities engage in astrological or divinatory practices for guidance or insight into the future. Astrology, palmistry, and dream interpretation hold cultural significance for those who adhere to these beliefs. Music and dance are integral components of Arab folk culture, serving as avenues for artistic expression, storytelling, and community bonding. Traditional instruments like the *oud* and *tabla* accompany regional dance styles, fostering a sense of cultural identity and heritage. Solemn ceremonies and collective expressions of grief characterize Arab mourning rituals. Customs such as washing and shrouding the deceased, reciting prayers, and observing mourning periods reflect the cultural significance placed on honoring the departed. These folk practices underscore the richness and complexity of Arab culture, evolving while preserving cultural identity and heritage. Despite variations across regions and communities, these traditions continue to shape the lives and experiences of individuals of Arab heritage bridging the past with the present in a tapestry of shared beliefs and customs.

**Barriers to Healthcare**

Newly arrived and unskilled refugees from poorer parts of the Arab world face significant challenges in accessing healthcare services in their host countries. Factors such as refugee status, recent arrival, cultural differences, language barriers, financial constraints, and lack of awareness about available services compound the stresses of immigration and impact both health status and responses to health problems (Childress et al. 2024; Mohsin et al. 2024). These individuals are at particular risk for increased exposure to ill health and inadequate access to care due to a combination of cultural, linguistic, socioeconomic, and systemic barriers.

Language and communication barriers pose significant challenges for recent Arab immigrants, as limited proficiency in the host country's language impedes effective communication with healthcare providers. Without clear communication channels, misunderstandings may arise, leading to suboptimal care and reluctance to seek medical attention. Also, cultural competence among healthcare providers is essential to address the diverse needs and beliefs of Arab patients. Cultural misunderstandings, stereotypes, and biases may further hinder access to quality care and erode trust in healthcare professionals.

The stigma surrounding mental health issues and seeking psychological support is prevalent in many Arab communities contributing to underutilization of mental healthcare services and untreated mental health conditions. Socioeconomic disparities, including poverty and lack of health insurance coverage, further exacerbate barriers to healthcare access limiting preventive care and specialized treatments for vulnerable individuals. Immigration status and documentation concerns also deter undocumented immigrants or those with uncertain status from seeking medical care, fearing legal repercussions or deportation.

Cultural mistrust and discrimination, compounded by historical biases and negative portrayals of Arabs in media and society create additional barriers to healthcare access (Alsaidi et al. 2023). Arab individuals may fear discrimination or bias from healthcare providers based on ethnicity, religion, or national origin, leading to avoidance of healthcare services or selective disclosure of information. Limited availability of culturally competent care providers proficient in the Arabic language and familiar with Arab customs further delays access to quality care, particularly in regions or healthcare settings where such providers are scarce.

Community and family dynamics, including patriarchal structures and gender roles, also influence healthcare decision-making and access among individuals of Arab heritage. Cultural norms regarding modesty, privacy, and familial obligations may impact individuals' autonomy in seeking healthcare, with women and younger family members often deferring to elder relatives or male guardians. Addressing these complex barriers to healthcare requires a comprehensive approach that prioritizes cultural competence, linguistic accessibility, and equitable access to resources and services. Healthcare systems and providers must actively work to foster inclusive environments, promote diversity in the workforce, and implement policies and practices that recognize and address the unique needs of individuals with Arab heritage. By addressing these barriers, healthcare systems can enhance access to care, improve health outcomes, and promote health equity for individuals of Arab heritage.

**Cultural Responses to Health and Illness**

Arabs regard pain as unpleasant and something to be controlled, often expecting immediate relief postoperatively due to their confidence in medical science. However, this expectation, combined with conserving energy for recovery, may lead to reluctance in adhering to postoperative routines like frequent ambulation. While expressive responses to pain are typically reserved for immediate family, during childbirth or for illnesses accompanied by spasms, Arabs may express pain more openly (Abudoush et al. 2023; Maani et al. 2021; Maki et al. 2022). This differing response between family and healthcare providers can sometimes lead to conflicting perceptions of pain relief adequacy.

The stigma surrounding mental illness is pervasive within Arab communities, often leading to denial, attributions to "bad nerves" or blame on evil spirits. Lebanese families for example, may deny mental illness altogether fearing community reactions, resulting in underreporting and avoidance of professional help (Pampati et al. 2018). Consequently, mental distress may manifest as vague complaints like abdominal pain or lassitude, and patients may prefer somatic treatments over counseling with prayer and herbal remedies being more acceptable than psychiatric intervention. Hospitalization may also be resisted due to fears of abandonment further complicating access to mental healthcare. Similarly, attitudes toward disabilities are generally negative with low expectations for education and rehabilitation. Misconceptions about mental retardation persist due to limited exposure and social stigma resulting in the disabled being kept from public view. Family members typically assume caregiving responsibilities, often indulging and protecting the disabled person reflecting acceptance of dependency within Arab culture. These cultural responses underscore the importance of understanding and respecting cultural nuances in healthcare delivery. By acknowledging and addressing these beliefs and attitudes, healthcare providers can foster trust, improve patient-provider communication, and ultimately enhance healthcare outcomes within Arab communities.

**Blood Transfusions and Organ Donation**

Blood transfusions and organ donation are significant medical procedures within Arab communities, each carrying cultural and religious implications. Blood transfusions are generally accepted among Arabs and Arab Americans, especially in emergencies or for treating severe conditions like anemia or bleeding (Almutairi et al. 2018; Braester et al. 2022). While saving lives often overrides religious concerns in emergencies, conservative Muslim communities may have religious beliefs that limit the acceptance of blood transfusions (Almutairi et al. 2018; Braester et al. 2020). Healthcare providers must ensure individuals understand the necessity and implications of blood transfusions, respecting their autonomy and providing adequate information for informed consent.

Organ donation remains contentious among Arabs and Arab Americans due to religious beliefs, cultural values, and ethical considerations. Views on organ donation vary, influenced by beliefs about death, the afterlife, and body sanctity (Abdulrazeq et al. 2020; Ali et al. 2020; Al-Qerem et al. 2022; Doerry et al. 2022; Janahi et al. 2018; Mekkodathil et al. 2020). Some religious scholars permit it under specific conditions while others oppose it due to concerns about desecration or divine will. Cultural norms, familial expectations, and societal taboos also influence attitudes toward organ donation. Low donation rates are attributed to a lack of public education, misconceptions, and limited support from religious authorities. Providers must approach organ donation discussions sensitively, considering diverse perspectives and engaging in respectful dialog.

In summary, while blood transfusions are generally accepted, organ donation remains complex due to arying religious, cultural, and ethical considerations. Sensitivity and understanding of individual beliefs are crucial for healthcare providers to support informed decision-making and promote acceptance of organ donation within Arab communities.

### *8.2.11 Healthcare Providers*

**Traditional Versus Biomedical Providers**

Arab Americans exhibit a unique blend of traditional and biomedical healthcare practices, often seeking treatment in the United States, especially for complex health conditions. They possess a deep respect for Western medicine, valuing treatments that involve surgery, removal of causative agents, or intravenous therapies aimed at eradicating ailments. However, while they hold high regard for physical medical interventions, there is often a lack of trust or respect for mental health treatments among many Arab Americans. A prevailing sentiment among Arab Americans is that mental health services are not effective and are only necessary for individuals with severe mental disorders. Consequently, mental health services are underutilized within Arab American communities despite a greater need for such services among distressed immigrant populations. This skepticism toward mental health treatment may stem from cultural beliefs and stigmas surrounding mental illness.

Gender and age considerations also play significant roles in healthcare interactions among Arab Americans. In Arab societies, interactions between unrelated males and females are uncommon with healthcare settings often segregated by sex. Shyness in women is valued and traditional Arab customs may lead Muslim men to ignore women out of politeness. Consequently, some Arab Americans may feel embarrassed or stressed when interacting with healthcare providers of the opposite sex. This discomfort may manifest as a reluctance to discuss personal information or undergo physical assessments and hygiene procedures. Arab American women may prefer to be seen by male healthcare providers, potentially excluding or denying men the opportunity to interact with or diagnose health conditions accurately, particularly for high-risk female patients within the Arab American community.

**Status of Healthcare Providers**

Arab Americans hold science and medicine in high regard, often recognizing and celebrating the historical contributions of Arabs to the field of medicine. They view doctors as authoritative figures with significant knowledge and power. When seeking medical care, Arab American patients who lack English proficiency typically prefer to see Arabic-speaking doctors due to a sense of cultural and linguistic affinity. Many believe that Arab American doctors understand their cultural background and communication style better, making them feel more comfortable and at ease during medical consultations. However, among patients who are fluent in English, there is typically no preference for Arab doctors over American doctors. Some English-speaking patients even prefer American doctors, perceiving them as more professional and respectful than their Arab American counterparts.

In Arab society, medicine is highly esteemed, while nursing is often seen as a less desirable profession due to societal norms that dictate certain behaviors for women. However, efforts are being made in many Arab countries to improve the image of nursing and enhance the working environment for nurses (Wehbe-Alamah et al. 2024). American nurses are generally viewed more favorably due to their education, expertise, and ability to perform tasks traditionally reserved for physicians such as physical examinations. Younger Arab American immigrants, particularly those from Lebanon, Iraq, and Jordan tend to have more positive perceptions of nursing as a profession compared to the older generation. The hierarchical structure of Arab society is reflected in the roles and status of physicians and nurses. Physicians, typically older males, are seen as authority figures who expect nurses to defer to them in decision-making and information disclosure. Nurses, often younger females, conform to these expectations and may be relegated to roles as medical assistants and housekeepers rather than being recognized as critical thinkers and health educators. However, nursing is gradually gaining recognition as a profession in many Arab countries, with the establishment of professional organizations resembling the American Nurses Association.

## 8.3 Case Studies

### *8.3.1 Reflective Exercise 8.2*

Suha Abul Mona, an unmarried 23-year-old Arab woman, visits the clinic accompanied by her Middle Eastern parents and older brother. Suha complains of left lower quadrant abdominal pain and spotting, causing concern for her family. Despite feeling visibly uncomfortable and anxious, Suha's family expresses worry about her condition.

Describing a persistent, dull pain in her left lower abdomen, Suha rates it as a 6 out of 10 on the pain scale, noting it began a day ago and has since worsened. She mentions experiencing light vaginal spotting for the past 2 days, which is unusual for her menstrual cycle. Suha denies any history of irregular periods or recent changes in menstrual pattern, attributing her intermittent nausea to the abdominal discomfort. Having no significant medical history, surgeries, or chronic conditions, Suha denies recent illnesses, travel, or exposure to infectious diseases. She is not currently taking any medications, including contraceptives.

The family actively participates in Suha's healthcare decisions, expressing deep concern for her well-being. The nurse observes how traditional gender roles and familial expectations influence communication dynamics with Suha's mother taking the lead in expressing health concerns while her father and older brother provide support. The nurse acknowledges the influence of Suha's unmarried status on discussions about reproductive health and sexual activity within the family, respecting cultural norms and societal expectations.

Performing a thorough physical examination, the nurse evaluates Suha's abdominal pain and vaginal bleeding ordering laboratory tests like a complete blood count and urinalysis to assess for underlying causes. Maintaining confidentiality regarding sensitive health information, particularly related to sexual health and reproductive issues, is essential for Suha's privacy and comfort.

1. What steps should the nurse take in this situation?
2. How can the nurse employ culturally sensitive approaches to effectively address the needs of the AbulMuna family?
3. How can the nurse ensure the best care for Suha?
4. Is it appropriate for the nurse to discuss the possibility of pregnancy with Suha when her parents are not present?

### *8.3.2 Reflective Exercise 8.2.1*

Latif, a 75-year-old man diagnosed with end-stage liver cancer, is undergoing chemotherapy treatment. Experiencing constant nausea and decreased appetite as side effects of his cytotoxic medications, Latif often refuses to eat the food provided to him.

Recognizing Latif's eating difficulties, the nurse arranges for the dietician to assess his dietary preferences and provide personalized recommendations. During the visit, Latif expresses concern about the halal status of the meat in his meals, indicating his discomfort with consuming non-halal food. Despite the nurse's explanation that the food was selected based on his dietary agreement, Latif remains uneasy about the meat's halal status. Also, he voices dissatisfaction with the lack of vegetables and salads in his meals, believing them to be essential for nutrition.

1. Drawing from insights into Arab cultural norms, what steps could the nurse take in response to Latif's concerns?
2. How can the nurse and the dietician collaborate to improve Latif's dietary intake in a culturally sensitive manner?
3. What strategies can the nurse employ to prevent similar situations from occurring in the future?

### *8.3.3 Reflective Exercise 8.2.3*

Ms. Al-Mansoori is a 32-year-old pregnant woman who arrives at the delivery suite experiencing contractions. As the nurses assist Ms. Al-Mansoori to her bed, her husband contacts the nurse and expresses his preference for no men to enter his wife's room. He insists that all nurses, doctors, and other staff attending to his wife be female. Also, he requests that no male personnel enter the room during the night as his wife will have her veil removed while she sleeps. The nurse faces a dilemma as the doctor on duty in the delivery suite is male and Ms. Al-Mansoori is fully dilated and prepared to deliver imminently. In response, the nurse places a note outside the Al-Mansoori family's room, indicating that males are not permitted inside based on the patient's preference.

1. Assess the nurse's reaction based on your nursing education.
2. Interpret the cultural significance behind Mr. Al-Mansoori's actions.
3. What actions can the nurse take in this scenario?

## References

Abboud S, Naal H, Chahine A, Taha S, Harfouch O, Mahmoud H (2020) "It's mainly the fear of getting hurt": experiences of LGBT individuals with the healthcare system in Lebanon. Ann LGBTQ Public Popul Health 1(3):165–185

Abboud S, Veldhuis C, Ballout S, Nadeem F, Nyhan K, Hughes T (2022) Sexual and gender minority health in the Middle East and North Africa region: a scoping review. Int J Nurs Stud Adv 4:100085. https://doi.org/10.1016/j.ijnsa.2022.100085

Abboud S, Chaudhry AB, Pachankis JE (2024a) A qualitative exploration of minority stress, mental health, and sexual health among Arab immigrant sexual minority men in the United States. Cult Divers Ethn Minor Psychol. https://psycnet.apa.org/record/2024-50365-001

Abboud S, Chaudhry AB, Taweh O, Pachankis JE (2024b) "There's a lot of conflict between your queer identity and also your Muslim or Arab identity;" a qualitative exploration of the intersectional minority stressors of US-born Arab sexual minority cisgender men. SSM-Qual Res Health 5:100393. https://www.sciencedirect.com/science/article/pii/S2667321524000027

Abdul-Haq AK (2018) The Arab family: formation, function and dysfunction. In: Caring for Arab patients. CRC Press, pp 77–88. https://doi.org/10.1016/j.socscimed.2012.07.030

Abdulrazeq F, Matsumoto MM, Zourob M, Al-Dobai A, Zeyad K, Marwan N, Qeshta A, Aleyani O (2020) Barriers in knowledge and attitudes regarding organ donation among urban Jordanian population. Saudi J Kidney Dis Transpl 31(3):624–638. https://journals.lww.com/sjkd/fulltext/2020/31030/Barriers_in_Knowledge_and_Attitudes_Regarding.7.aspx

Abo-Rass F, Nakash O, Gelaye B, Khatib A, AboJabel H (2024) Determinants of intentions to seek formal mental health help among Palestinian adolescents in Israel. Int J Soc Psychiatry 70(4):720–729. https://doi.org/10.1177/00207640231224658

Abudoush AN, Ali K, Kiran T, Panagioti M, Poliakoff E, Mo Hom N, Husain N (2023) Exploring chronic pain related attentional experiences, distress and coping strategies among Arabic-speaking individuals in Jordan and the United Kingdom. Front Psychol 14:1268179. https://doi.org/10.3389/fpsyg.2023.1268179

Abuelezam NN, El-Sayed AM (2021) Objective and subjective poor mental health indicators among Arab Americans in Michigan: a population-based study. Ethn Health 26(2):225–234. https://doi.org/10.1080/13557858.2018.1494822

Abuelezam NN, El-Sayed AM, Galea S (2018) The health of Arab Americans in the United States: an updated comprehensive literature review. Front Public Health 6:262. https://www.frontiersin.org/articles/10.3389/fpubh.2018.00262/full

Ahmadi M, Ahadi S, Khadembashiri MA, Khadembashiri MM, Mahalleh M, AziziKia H, Zare HR, Khah ASR, Hekmat H, Daroudi R (2024) Burden of ischemic heart disease in the Middle East and North Africa (MENA) and attributable risk factors: an epidemiological analysis from 1990 to 2019. IJC Heart Vasc 50:101316

Al Salman J, Al Dabal L, Bassetti M, Alfouzan WA, Al Maslamani M, Alraddadi B, Elhoufi A, Enani M, Khamis FA, Mokkadas E (2020) Management of infections caused by WHO critical priority Gram-negative pathogens in Arab countries of the Middle East: a consensus paper. Int J Antimicrob Agents 56(4):106104. https://www.sciencedirect.com/science/article/pii/S0924857920302879

Al-Ajlouni YA, Al Ta'ani O, Shamaileh G, Nagi Y, Tanashat M, Al-Bitar F, Duncan DT, Makarem N (2024) The burden of cardiovascular diseases in Jordan: a longitudinal analysis from the global burden of disease study, 1990–2019. BMC Public Health 24(1):879. https://doi.org/10.1186/s12889-024-18316-0

Aldossari M, Chaudhry S (2024) Gendered precarity in Saudi Arabia: examining the state policies and patriarchal culture in the labor market. Gender Work Organ. https://doi.org/10.1111/gwao.13119

Ali A, Ahmed T, Ayub A, Dano S, Khalid M, El-Dassouki N, Orchanian-Cheff A, Alibhai S, Mucsi I (2020) Organ donation and transplant: the Islamic perspective. Clin Transpl 34(4):e13832. https://doi.org/10.1111/ctr.13832

Allehebi A, Al-Omair A, Mahboub B, Koegelenberg CF, Mokhtar M, Madkour AM, Al-Asad K, Selek U, Al-Shamsi HO (2024) Recommended approaches for screening and early detection of lung cancer in the Middle East and Africa (MEA) region: a consensus statement. J Thorac Dis 16(3):2142. https://www.ncbi.nlm.nih.gov/pmc/articles/PMC11009596/

Almarshoodi K, Echevarria C, Kassem A, Mahboub B, Salameh L, Ward C (2024) An international validation of the "DECAF score" to predict disease severity and hospital mortality in acute exacerbation of COPD in the UAE. Hosp Pharm 59(2):234–240. https://doi.org/10.1177/00185787231209218

Almutairi AF, Salam M, Abou Abbas O, Nasim M, Adlan A (2018) The public's risk perception of blood transfusion in Saudi Arabia. J Blood Med 9:75–82. https://doi.org/10.2147/JBM.S165923

Almutawaa DS, Nuttall P, Mamali E, Al-Mutawa FS, AlJuma DHM (2024) Paradoxes of (un) veiling and the extended self: the experiences of Arab-Muslim women in Kuwait. J Islamic Market 15(1):172–191. https://www.emerald.com/insight/content/doi/10.1108/JIMA-07-2022-0196/full/html

Al-Qerem W, Carter N, Ling J (2022) Attitudes to organ donation and transplantation: an insight from Jordan. Exp Clin Transplant 20(6):602–608. https://www.researchgate.net/profile/Jonathan-Ling-3/publication/361822043_Attitudes_to_Organ_Donation_and_Transplantation_An_Insight_From_Jordan/links/63b342aec3c99660ebc28d97/Attitudes-to-Organ-Donation-and-Transplantation-An-Insight-From-Jordan.pdf

Alqifari S (2024) Antenatal care practices: a population-based multicenter study from Saudi Arabia. Int J Women's Health 16:331–343. https://doi.org/10.2147/IJWH.S452934

AlRabeeah SM, Alzahrani EM, Aldhahir AM, Siraj RA, Alqarni AA, AlDraiwiesh IA, Alqahtani AS, Almqati BS, Alharbi TG, Almuntashiri AA (2024) A population-based study of 15,000 people on knowledge and awareness of lung cancer symptoms and risk factors in Saudi Arabia. Front Oncol 14:1295847. https://www.frontiersin.org/articles/10.3389/fonc.2024.1295847/full

Alsabri MAH, Elsayed SM, Elshanbary AA, Sayed M, Zaazouee BF, Zehra SA, Alkubati SA, Alrubaiee GG (2024) Knowledge, attitude and perceptions of healthcare workers in Arab countries regarding basic life support; a systematic review and meta-analysis. Arch Acad Emerg Med 12(1):e32–e32. https://www.researchgate.net/profile/Mohammed-Alsabri/publication/378567140_Arab_Countries_Regarding_Basic_Life_Support_a_Systematic_Review_and_Meta-Analysis/links/65e04855adf2362b635cd815/Arab-Countries-Regarding-Basic-Life-Support-a-Systematic-Review-and-Meta-Analysis.pdf

Alsaidi S, Velez BL, Smith L, Jacob A, Salem N (2023) "Arab, brown, and other": voices of Muslim Arab American women on identity, discrimination, and well-being. Cult Divers Ethn Minor Psychol 29(2):163. https://psycnet.apa.org/record/2021-26567-001

Al-Shamsi HO, Musallam KM (2024) Not only a Western world issue: Cancer incidence in younger individuals in the United Arab Emirates. CA Cancer J Clin 74:227. https://doi.org/10.3322/caac.21839

Arab American Institute Demographics (2024). Arab American Institute. https://www.aaiusa.org/demographics. Accessed 18 Apr 2024

Aseeri A (2020) Health literacy, social support, and diabetes self care among individuals of Arab descent. Wayne State University. https://search.proquest.com/openview/1ddfcd69fbf34a64d3914affe9a1f8c5/1?pq-origsite=gscholar&cbl=51922&diss=y&casa_token=paUBStI0_1YAAAAA:DBb1EWZ-ZTMT8u1NvERa9sW-V_K-_cz8ZnzIeFdFDla7yt-vRuzPpGI2byUrlJSRHetrocOL4pEY

Atrooz F, Asfar T, Oluwole OJ, Garey L, Salim S, Zvolensky MJ (2024) Insights into cigarette and waterpipe use among Houston-based Arab immigrants and refugees. Open J Soc Sci 12(3):182–195. https://www.scirp.org/journal/paperinformation?paperid=131801

Braester A, Akria L, Yakir O, Shelev L, Barhoum M (2020) Blood myths and ethnic–cultural background as impediments in patient blood management education. J Relig Health 59(6):2918–2927. https://idp.springer.com/authorize/casa?redirect_uri=https://link.springer.com/article/10.1007/s10943-020-01015-6&casa_token=Qm8lA8O8zpQAAAAA:zTmSnU8HI_BSyZmR56_LA6iOiSSg0WYxK3_hV1J3I5FUxWAOCRQwaydLCSmisTDXQrPhD72_3TXd1lgOEQ

Braester A, Shaoul E, Mizrachi O, Akria L, Shelev L, Barhoum M (2022) Patient involvement in the transfusion decision-making can change patient blood management practice for better or worse. Acta Haematol 145(1):5–8. https://karger.com/aha/article/145/1/5/827341

Büchler A, Schneider Kayasseh E (2018) Fostering and adoption in Islamic law-under consideration of the laws of Morocco, Egypt, and the United Arab Emirates. Electron J Islamic Middle Eastern Law (EJIMEL) 6:31–56. https://www.zora.uzh.ch/id/eprint/157736/

Childress S, Shrestha N, Covington E, Stark L, Seff I, Black B, McKay M (2024) Risk and protective factors for well-being and barriers to help-seeking among Arab-speaking MENA immigrants and refugees in North America: a scoping review. Glob Soc Welf:1–20. https://idp.springer.com/authorize/casa?redirect_uri=https://link.springer.com/article/10.1007/

s40609-023-00330-6&casa_token=fTqExXa5Vs8AAAAA:Vw9jEMgjrggYrbfIyecaI-0fAYjiGogd75_v7GbStiRpVuh_Ogji7OmcpN-x-Xr_6UaxMnyUzBn7nGzq

Dandashly A (2021) The EU and LGBTI activism in the MENA—the case of Lebanon. Mediterr Polit 27(4):1–21. https://doi.org/10.1080/13629395.2021.1883287

Doerry K, Oh J, Vincent D, Fischer L, Schulz-Jürgensen S (2022) Religious and cultural aspects of organ donation: narrowing the gap through understanding different religious beliefs. Pediatr Transplant 26(7):e14339. https://doi.org/10.1111/petr.14339

Druker Shitrit S, Ben-Asher S, Ben-Atar E (2024) "The modesty guard" reflected in online comments in mainstream news sites. J Inf Commun Ethics Soc. https://www.emerald.com/insight/content/doi/10.1108/JICES-08-2023-0108/full/html

Elbarazi I, Alam Z, Ali N, Loney T, Al-Rifai RH, Al-Maskari F, Ahmed LA (2024) Health literacy among pregnant women in The United Arab Emirates: the Mutaba'ah study. Womens Health 20:17455057231224179. https://doi.org/10.1177/17455057231224179

El-Shamy H (2018) Folklore of the Arab world. Humanities 7(3):67. https://www.mdpi.com/2076-0787/7/3/67

Ferrari AJ, Santomauro DF, Aali A, Abate YH, Abbafati C, Abbastabar H, Abd ElHafeez S, Abdelmasseh M, Abd-Elsalam S, Abdollahi A (2024) Global incidence, prevalence, years lived with disability (YLDs), disability-adjusted life-years (DALYs), and healthy life expectancy (HALE) for 371 diseases and injuries in 204 countries and territories and 811 subnational locations, 1990–2021: a systematic analysis for the Global Burden of Disease Study 2021. Lancet. https://www.thelancet.com/journals/lancet/article/PIIS0140-6736(24)00757-8/fulltext

Fleischer NJ, Sadek K (2024) Arab, Middle Eastern, and North African health disparities research: a scoping review. J Racial Ethn Health Disparities:1–19. https://idp.springer.com/authorize/casa?redirect_uri=https://link.springer.com/article/10.1007/s40615-024-01972-8&casa_token=wgtoSf-JmvsAAAAA:r0-b77UGrkKjpI9mVb2gB2y7MYSscqkc04ay_WS5TLZl6BuAIAjjvYO9aKj3KpQtIQXBvSAuzqheOirJ5A

Galaz S (2019). The most invisible of the invisibles: skin color and Arab American political ideology. https://polisci.ucsd.edu/undergrad/departmental-honors-and-pi-sigma-alpha/Sydney-Galaz_Senior-Honors_2019.pdf

Hamadeh A, El-Shamy F, Billings J, Alyafei A (2024) The experiences of people from Arab countries in coping with trauma resulting from war and conflict in the Middle East: a systematic review and meta-synthesis of qualitative studies. Trauma Violence Abuse 25(2):1278–1295. https://doi.org/10.1177/15248380231176061

Hamid GA, Muthanna A, Baom N (2024) Lung cancer epidemiology and histopathological classification: a study from Yemen. https://www.researchgate.net/profile/Gamal-Abdul-Hamid/publication/378667299_LUNG_CANCER_EPIDEMIOLOGY_AND_HISTOPATHOLOGICAL_CLASSIFICATION_A_STUDY_FROM_YEMEN/links/65e31f46c3b52a117006acbc/LUNG-CANCER-EPIDEMIOLOGY-AND-HISTOPATHOLOGICAL-CLASSIFICATION-A-STUDY-FROM-YEMEN.pdf

Hamid GA, Muthanna A, Baom N (2024) Lung cancer epidemiology, diagnosis and treatment. https://www.researchgate.net/profile/Gamal-Abdul-Hamid/publication/378310759_Lung_Cancer_Epidemiology_Diagnosis_and_Treatment/links/65d3d7b801325d4652155da2/Lung-Cancer-Epidemiology-Diagnosis-and-Treatment.pdf. Accessed 20 Apr 2024

Hanafi I, Samra DA, Alsaqqa R, Naeem A, Shebli B, Ajlyakin G (2024) Lung cancer diagnoses and outcomes during the Syrian War, 2011-2018. JAMA Netw Open 7(3):e242091–e242091. https://jamanetwork.com/journals/jamanetworkopen/article-abstract/2816029

HARB A 2024. Hookah smoking and aggressive behavior among university Arab students in Israel. https://ejournal.vfu.bg/pdfs/Adel%20Harb,%20HOOKAH%20SMOKING%20AND%20AGGRESSIVE%20BEHAVIOR%20AMONG%20UNIVERSITY%20ARAB%20STUDENTS%20IN%20ISRAEL%20(VFU).pdf. Accessed 20 Apr 2024

Hashem HM, Awad GH (2021) Religious identity, discrimination, and psychological distress among Muslim and Christian Arab Americans. J Relig Health 60(2):961–973. https://idp.springer.com/authorize/casa?redirect_uri=https://link.springer.com/article/10.1007/s10943-020-01145-

x&casa_token=UEZlZ2-fKhAAAAAA:lLVDF6zf_DtZDIy5cf4yfPO-iTmkDshhBRqJeBJ0tVSPiHvLXS93FOb9SngjlECm4dgWyimL_MHIzZ_Euw

Hatoum BA (2018) Refusing to be unnatural—a study of the LGBT rights movement in Lebanon. https://www.duo.uio.no/handle/10852/64441

Horne R (2024) Employment and social trends by region. World Employment and Social Outlook 2024(1):37–60. https://doi.org/10.1002/wow3.204

Hurley Z (2024) Developing feminist-decolonial-capabilities for emancipatory pedagogy: a case of Gulf-Arab women's learner-positionalities. In: Lacković N, Cvejic I, Krstić P, Nikolić O (eds) Rethinking education and emancipation. Springer, Cham, pp 117–136. https://doi.org/10.1007/978-3-031-47044-8_6

Jalambo MO, Karim NA, Naser IA, Sharif R (2018) Prevalence and risk factor analysis of iron deficiency and iron-deficiency anaemia among female adolescents in the Gaza Strip, Palestine. Public Health Nutr 21(15):2793–2802. https://www.cambridge.org/core/journals/public-health-nutrition/article/prevalence-and-risk-factor-analysis-of-iron-deficiency-and-irondeficiency-anaemia-among-female-adolescents-in-the-gaza-strip-palestine/654623F6ACCAD908CCD61ACEA523BFF1

Janahi FK, Al Rais A, Al Rukhaimi M, Khamis AH, Hickey D (2018) Public awareness of knowledge, belief, and attitude regarding organ donation and organ transplantation: a national survey from the United Arab Emirates. Transplant Proc 50(10):2932–2938. https://www.sciencedirect.com/science/article/pii/S0041134518310479

Kader F, Bazzi L, Khoja L, Hassan F, de Leon CM (2020) Perceived discrimination and mental well-being in Arab Americans from southeast Michigan: a cross-sectional study. J Racial Ethn Health Disparities 7:436–445. https://idp.springer.com/authorize/casa?redirect_uri=https://link.springer.com/article/10.1007/s40615-019-00672-y&casa_token=Iipn1SNuD-cYAAAAA:XO3rkUAdnz031z6W7XBPOLxhMyKPeK0EoMl1jgt1JxbGKbgqWBCO9MUNgjqRTjkomv2hIPQ2OJrMwlLEhQ

Khosh NK, Khalil AA, Alhaded HHS (2020) Cultural values and norms of communication: a view from the Middle East. Proceedings of ADVED, 2020 (6th ed). https://www.academia.edu/download/106965242/96.pdf

Labibidi H, Lababidi G, Al Rifai M, Nasir K, Al-Kindi S (2024) Cardiovascular disease in Arab Americans: a literature review of prevalence, risk factors, and directions for future research. Am J Prev Cardiol 18:100665. https://www.sciencedirect.com/science/article/pii/S2666667724000333

Lebanon's LGBT People Reclaim Their Power (2020) Human rights watch. https://www.hrw.org/news/2020/05/07/lebanons-lgbt-people-reclaim-their-power

Maani W, Alrabayah M, Atiyat B, Abed A (2021) The role of Arab and Muslim scholars in defining and treating pain: a review article. Jordan Med J 55(2):55–60. https://archives.ju.edu.jo/index.php/jmj/article/view/105717

Mahamid F, Veronese G, Bdier D (2024) Trauma predicts the desire to migrate and psychological well-being via belongingness and quality of life among Palestinians. Psychol Trauma. https://psycnet.apa.org/record/2024-44918-001

Maki D, Lempp H, Critchley D (2022) An exploration of experiences and beliefs about low back pain with Arab Muslim patients. Disabil Rehabil 44(18):5171–5183. https://doi.org/10.1080/09638288.2021.1928301

Mansour R, Al-Ani A, Al-Hussaini M, Abdel-Razeq H, Al-Ibraheem A, Mansour AH (2024) Modifiable risk factors for cancer in the middle East and North Africa: a scoping review. BMC Public Health 24(1):223. https://doi.org/10.1186/s12889-024-17787-5

Manzanero AL, Aroztegui J, Fernández J, Guarch-Rubio M, Álvarez MÁ, El-Astal S, Hemaid F (2024) War, torture and trauma in preadolescents from Gaza Strip. Two different modalities of PTSD. Anuario de Psicologia Juridica 34(1):1–12. https://journals.copmadrid.org/apj/art/apj2024a6

Masoud MS, Yakout SM, Al-Attas OS, Alokail MS, Al-Daghri NM (2020) The association between iron and vitamin D status in Arab adolescents. Public Health Nutr 23(7):1208–1213. https://www.cambridge.org/core/journals/public-health-nutrition/article/association-between-iron-and-vitamin-d-status-in-arab-adolescents/5F7C822F9FC6EC2A2BACBD957D520 98A

Mekkodathil A, El-Menyar A, Sathian B, Singh R, Al-Thani H (2020) Knowledge and willingness for organ donation in the middle eastern region: a meta-analysis. J Relig Health 59(4):1810–1823. https://doi.org/10.1007/s10943-019-00883-x

Mohsin K, Mula-Hussain L, Gilson R (2024) HealthCare Access Barrier (HCAB) framework for the barriers to cancer care during conflicts: perspective from Iraq. BMJ Oncol 3(1) https://bmjoncology.bmj.com/content/3/1/e000252.abstract

Mostafavi E, Ghasemian A, Abdinasir A, Mahani SAN, Rawaf S, Vaziri MS, Gouya MM, Nguyen TMN, Al Awaidy S, Al Ariqi L (2022) Emerging and re-emerging infectious diseases in the WHO Eastern Mediterranean region, 2001-2018. Int J Health Policy Manag 11(8):1286. https://www.ncbi.nlm.nih.gov/pmc/articles/PMC9808364/

Mpofu E (2018) How religion frames health norms: a structural theory approach. Religions 9(4):119. http://dx.doi.org.ezproxy.lib.umb.edu/10.3390/rel9040119

Nisar M, Benedict SM (2024) Exploring gender equality: a comparative study of contemporary feminism, Islamic perspective, and feminism in Sufism. Islamic Perspective, and Feminism in Sufism (9 Jan 2024). https://papers.ssrn.com/sol3/papers.cfm?abstract_id=4721634

Olimat S, Altawalbeh SM, Maabreh RS, Basheti IA (2024) Asthma–COPD overlap syndrome among smoker subjects with asthma: a cross-sectional study. J Appl Pharm Sci 14(4):72–80. https://japsonline.com/abstract.php?article_id=4194&sts=2

Pampati S, Alattar Z, Cordoba E, Tariq M, Mendes De Leon C (2018) Mental health outcomes among Arab refugees, immigrants, and U.S. born Arab Americans in Southeast Michigan: a cross-sectional study. BMC Psychiatry 18(1):379. https://doi.org/10.1186/s12888-018-1948-8

Raghavan D, Natarajan J, Seshan V, Alsaraireh A (2024) Impact of recurrent pregnancy loss on depression, anxiety and fertility related stress in Omani women: a cross sectional survey. Afr J Reprod Health 28(1):13–21. https://www.ajol.info/index.php/ajrh/article/view/266139

Rashidi M-M, Saeedi Moghaddam S, Azadnajafabad S, Mohammadi E, Khalaji A, Malekpour M-R, Keykhaei M, Rezaei N, Esfahani Z, Rezaei N (2024) Mortality and disability-adjusted life years in North Africa and Middle East attributed to kidney dysfunction: a systematic analysis for the Global Burden of Disease Study 2019. Clin Kidney J 17(1):sfad279. https://academic.oup.com/ckj/article-abstract/17/1/sfad279/7425645

Rezki K, Fateh D (2024). The reality of health development in Algeria in light of the current challenges. https://www.asjp.cerist.dz/en/downArticle/512/7/1/239629. Accessed 20 Apr 2024

Rostam-Abadi Y, Gholami J, Shadloo B, Mohammad Aghaei A, Mardaneh Jobehdar M, Ardeshir M, Sangchooli A, Amin-Esmaeili M, Taj M, Saeed K, Rahimi-Movaghar A (2024) Alcohol use, alcohol use disorder and heavy episodic drinking in the Eastern Mediterranean region: a systematic review. Addiction 119:984. https://doi.org/10.1111/add.16444

Saadi A, Taleghani S, Dillard A, Ryan G, Heilemann M, Eisenman D (2023) Nurses' experiences with racial, ethnic, cultural, and religious discrimination in the workplace: a qualitative study. AJN Am J Nurs 123(5):24–34. https://journals.lww.com/ajnonline/fulltext/2023/05000/Original_Research__Nurses__Experiences_with.18.aspx?casa_token=oJElk1fppTEAAAAA:k-lzMcwd080yHCf4NWqvQNy_qwEIK6xZKimt_jfdzfc1zHy-gIrOmVOHXUGdM1Tj5kCj9KdZyK0RLrB7oA45RNr8

Safadi R (2024) The liver in Greco-Arabic and Islamic medicine. Clin Liver Dis 23(1):e0137. https://journals.lww.com/cld/fulltext/2024/01010/the_liver_in_greco_arabic_and_islamic_medicine.36.aspx

Saliba B, Kang M, Wells N, Mao L, Prestage G, Hammoud MA (2024) Hiding in plain sight: highlighting the research gap on access to HIV and other sexual health services for underrepresented gay men in developed Western countries–insights from a scoping review with a focus on Arab men. Sex Health. https://www.publish.csiro.au/SH/SH23131 21

Sharkia SA, Taubman-Ben-Ari O (2024) Personal growth and life satisfaction among Arab mothers after fertility treatment–the role of stress and optimism. J Happiness Stud 25(1):6. https://idp.springer.com/authorize/casa?redirect_uri=https://link.springer.com/article/10.1007/

s10902-024-00712-3&casa_token=mgfv7uN1X1AAAAAA:rq1XsnFSBlxsrsrTOZlo9tZ2-O3LnT3wJdW3F5Tkg24eD1SA7jtn9RT4mAc6C0Id-F6H7RKVFLFUxuymlA

Suleiman Y (2019) Arabic language and national identity: a study in ideology. Edinburgh University Press. https://books.google.com/books?hl=en&lr=&id=eacxEAAAQBAJ&oi=fnd&pg=PR5&dq=arabic+language&ots=G7hx5uGDbm&sig=0vqKkbLXagVCywGebXAadyzJ1d0

Sweileh W (2024) Stigma research in Arab countries: a rapid review. Ment Health Soc Incl. https://www.emerald.com/insight/content/doi/10.1108/MHSI-10-2023-0111/full/html

Tarabay J, Golm D (2024) The transmission of intergenerational trauma and protective factors in survivors of the Lebanese Civil War and their adult offspring. Int J Intercult Relat 99:101952. https://www.sciencedirect.com/science/article/pii/S014717672400021X

Vahidniya N, Hasani Z, Kamali SM (2024) Effect of warfare and forced displacement on mental health of children. Int J Travel Med Glob Health 12(1) https://www.ijtmgh.com/article_188419.html

Veronese G, Mahamid F, Bdier D, Obaid H, Cavazzoni F (2024) The development and validation of the Palestinian children's traumatic events checklist in a war-torn environment. BMC Psychiatry 24(1):254. https://doi.org/10.1186/s12888-024-05731-1

Wehbe-Alamah H, Doumit MAA, Bernstein J, Aboul-Enein BH (2024) Nursing research in Arab countries: current status, trends, challenges, and opportunities. SAGE Open Nurs 10:23779608241246871. https://doi.org/10.1177/23779608241246871

Zimmerman MS, Shaw G Jr (2020) Health information seeking behaviour: a concept analysis. Health Info Libr J 37(3):173–191. https://doi.org/10.1111/hir.12287

# Chapter 9
# People of Brazilian Heritage

**Thaís Moreira São João**

## 9.1 Overview and Heritage

Brazil, the largest country in South America, borders all South American countries except Chile and Ecuador. Brazilians are not a homogeneous group given the diversity of the size and topography of the country as well as individuality and the variant characteristics of the culture as described in Chap. 1.

There is significant literature on Brazilian health conditions, practices, and beliefs depicting the culture of the various groups in Brazil, including arts, music, dance, and cuisine. Brazilian heritage is rich in its mixture of Portuguese, French, Dutch, German, Italian, Japanese, Chinese, African, Arab, and indigenous Brazilian Indians. Brazilians are not considered Hispanics, despite some similarities between their ethnic features and some Hispanic countries. Because Brazil is located in Latin America, they are also considered Latinxs.

Most Brazilians in the United States are concentrated in communities around Boston; New York; Newark, New Jersey; and Miami. Smaller groups exist in Los Angeles, Detroit, and other locales. Larger Brazilian settlements in the United States have their churches, spiritualists, beauty shops, travel services, and support services. The exact number of Brazilians living in the United States is unknown. Approximately, 550,000 Brazilians are living in the United States, according to the most recent U.S. Census Bureau Data. Many Brazilians subsist in urban slums without privacy and hope to earn enough money to return home. Alternative lifestyles are also typical. They frequently leave children, partners, and family behind to become slaves of work in any type of situation. There are relatively few second- and third-generation Brazilians where both the mother and father have immigrated to the United States. Like other immigrants, many Brazilians are underemployed after

T. M. S. João (✉)
College of Nursing, University of Rhode Island, Kingston, RI, USA
e-mail: thais_sao-joao@uri.edu


E. A. Fenkl, L. D. Purnell (eds.), *Handbook for Culturally Competent Care*,
https://doi.org/10.1007/978-3-031-70492-5_9

immigrating, often giving up their professions to earn money as domestic workers, waiters, cab drivers, and other undervalued positions. Immigrants often move to large cities where many networks help find informal occupations.

### *9.1.1 Communications*

- Brazilian Portuguese is the official language of Brazil and continues to dominate Brazilian communities. Dialects vary among Brazilians according to region. Dialects from Rio de Janeiro, São Paulo, and Northeast are the most frequent. Language is commonly a barrier for Brazilians to have access to culturally competent health care in the United States.
- **Obtain an interpreter for Brazilians when necessary.**
- Many Brazilians continue to be of "proper" old-world orientation in which true feelings are not divulged for fear of hurting the communication receiver. Everything is said to be "*tudo bem*" ("fine"), almost in a stoic sense. However, in the intimate circle of family and compatriots, sharing thoughts and feelings is common. **Establish trust before attempting to obtain sensitive information in the health history.**
- Young adult and adolescent Brazilians in the United States are generally more acculturated because of their desire and need to assimilate into the new culture. Sharing thoughts and feelings is more common among intragenerational groups rather than intergenerational groups.
- Most Brazilians use touch and direct eye contact. They frequently kiss each other on the cheek when they meet and say good-bye. Men shake each other's hands and slap each other on the back with the other hand. This gesture frequently ends in an embrace. Children are kissed, and there is much touching. The kissing of a child frequently includes the combination of a "kiss and smell."
- Spatial distancing is close. Facial expressions and symbolic gestures are commonplace.

  **Do not take offense if Brazilians stand closer than you are accustomed.**

- Most Brazilians in the United States are future-oriented. In general, some are not punctual and tend to arrive "a bit" late—from minutes to hours—especially for social occasions. Everyone seems to know the behavior of tardiness and plans around it. However, those in professional circles are punctual. **Carefully explain the importance of being on time for health-care appointments.**
- Brazilian names are lengthy, but the common trend is to use only the first and last names. Traditionally, names appear as first name, mother's family name, and father's family name. "Junior" is added to a name if the son has been named after the father and "*Neto*" if the son has been named after the grandfather (third generation). When one marries, they may opt to add the partner's last name, drop their mother's maiden name or their father's name, or keep them both. At times "*de,*" "*da/do,*" or "*das/dos*" is added to a name to denote "of" and seems to be done out of tradition. No rigid protocol is apparent.

- There is a growing trend of single-parent families, with a prevalence of female-headed families.
- **Ask the patient his or her full name and the name that is used for legal purposes.**
- In day-to-day relationships, people are called by their first name or *Seu, Senhor* (more respectful, equivalent to Sir or Mr.) preceding the first name of a man, or *Dona* preceding the first name of a woman (equivalent to Ms. or Madam). Mothers, grandmothers, or respected strangers are referred to as *A Senhora,* and fathers, grandfathers, and respected men are called *O Senhor*. Clinicians are addressed as *Doutor* or *Doutora,* and professors are addressed as *Professor* or *Professora*. The latter two are followed by the first name.
- **Address Brazilian patients formally with Mr., Mrs., Ms., or an appropriate title until told to do otherwise.**

### *9.1.2 Family Roles and Organization*

- Brazilian society is mostly sexist, with the middle and upper classes being patriarchal in structure. However, as women assert their equality, more egalitarian relationships are becoming evident. Lower socioeconomic households tend to be more matriarchal.
- **Be sure to identify the family decision-maker for health concerns.**
- Children are important to Brazilian families. If a mother and child face deportation from the United States, the baby is generally accepted into the family and often raised by the maternal grandmother. Wealthier family members may raise the child of a poorer relative, if needed. However, these children often enter the family in a second-class capacity.
- Older people live with one of their children when self-care is a concern, and nursing home placement is rare. Older people are respected, seen as family counselors, and are always addressed as *O Senhor* or *A Senhora*. They are included in family activities and usually accompany their children's families on vacation. The extended family is also very important.
- **Include the older members of the family and extended family in health-care decision-making.**
- Most Brazilian families are catholic. Godparents are derived from this religious view and considered an important family extension. Poor families frequently ask their *patron* and *patrona* (employer and spouse) to be godparents to their children. Godparent may support caring for the child's clothing, schooling, and others. The godmother is called ***comadre*** by the mother. ***Compadre*** is used in reference to the godfather.
- Although historically common in the lower socioeconomic classes, middle-class households with a single-female parent are becoming increasingly common among Brazilians in the United States. Birth certificates may contain one or both names of parents If the father is absent at the birth or denies to recognize the child, the enfant will mostly receive the mother's last name only.

- Social status is very important in Brazilian society, demonstrated in the titles that people use with each other and the practice of listing both parents' surnames. Class separation is maintained discretely by literacy status.
- Brazilians, especially from the south and southeast of Brazil, have become more accepting of LGBTQIA+ relationships. However, for many, same-sex relationships carry a stigma. **Do not disclose same-sex relationships to family members or others.**

### *9.1.3 Workforce Issues*

- Brazilians value diplomacy over honesty even when they promise to attend to something the next day, knowing that it will be impossible.
- **Ask for specific dates and times for when work is to be completed.**
- Professional Brazilians show up for work on time. Less educated Brazilians may find it difficult to adhere to time schedules in the American workplace.
- **Explain the value of punctuality in the American workforce.**
- The English intonation and the pronunciation of certain words are particularly difficult for some Brazilians. Brazilians generally respect authority and are frequently more comfortable in employment situations where rules and job specifications are well defined.
- **Speak clearly and slowly with work directions and ask for clarification to ensure the employee understands the job requirements.**
- Many undocumented Brazilians find employment within the Brazilian community where they may never have to learn the new language. The categorization of Brazilians in the United States under the general category of "Hispanics" adds to their discomfort. **Do not refer to Brazilians as Hispanics.**

### *9.1.4 Biocultural Ecology*

- The "typical" Brazilian is a ***moreno*** with brown skin and eyes and black or brown hair. However, individuals particularly from the southern states of Brazil may have blond hair and blue eyes.
- Table 9.1 lists health conditions commonly occurring for Brazilians in three categories: lifestyle, environment, and genetics. Because intestinal worms are common in Brazilian immigrants, parasitic diseases should be considered during health assessments.
- Interviews have substantiated that the incidence of gastrointestinal diseases increases when Brazilians first move to the United States. Changes in eating habits from the Brazilian long and ample midday dinner to American fast foods have left Brazilians in the United Stateswith gastric complaints. A genetic tendency toward lactose intolerance can contribute to some of these gastric prob-

**Table 9.1** Health conditions common to Brazilians

| Health condition | Cause |
|---|---|
| Malaria | Environment, lifestyle |
| Chagas disease | Environment, lifestyle |
| Dengue fever | Environment, lifestyle |
| Meningitis | Environment, lifestyle |
| Rabies | Environment, lifestyle |
| Yellow fever | Environment, lifestyle |
| Schistosomiasis | Environment, lifestyle |
| Typhoid fever | Environment, lifestyle |
| Hansen's disease | Genetic |
| Hepatitis | Environment, lifestyle |
| Tuberculosis | Environment, lifestyle |
| Parasitic skin infections | Environment, lifestyle |
| Cholera | Environment, lifestyle |
| Cardiovascular diseases | Genetic, environment, lifestyle |
| Lactose intolerance | Genetic |

lems. An increased incidence of allergies, especially in children of Brazilian immigrants, has also been reported.

- Brazil represents the largest number of people living with HIV in Latin America, an endemic infection following Brazilians to the United States, and for which documentation is found. Since 2010, HIV is no longer considered grounds for inadmissibility to the United Statesand since 2013, Brazil has had much success in rolling out a program of highly active antiretroviral therapy (HAART) as well as pre- and post-exposer therapy (PrEP/PEP) for its citizens. Nevertheless, due to socioeconomic constraints and lack of medical insurance in the United States, many Brazilians living in the United States, and particularly those who migrate between the United Statesand Brazil, may be at greater risk for non-treatment of HIV infection or HIV treatment nonadherence.

**Refer patients as needed to social services to help them obtain needed services.**

### *9.1.5 High-Risk Health Behaviors*

- Because Brazilian immigrants frequently settle in Brazilian enclaves in large cities, they are subject to the same risk factors as any socially vulnerable urban subpopulation. The greatest risks are violence, drugs, and crime. Adolescents run the risk of resolving their adolescent identity crises by either banding together or joining gangs.
- **School officials can help decrease violence and crime among Brazilians by instituting after-school programs such as dance, music, sports, and other programs for teens.**

- Smoking is a high-risk behavior among Brazilians living in the United States. Among men, drinking hard liquor is also prevalent. Accessibility and use of street drugs and an individual's desperate search for quick money are other identifiable high-risk behaviors and often involve living in crowded urban conditions where rent is inexpensive. Although many immigrants are aware of the ill effects of smoking and recreational drugs, loneliness and frustration are deterrents to stopping these habits.
- **Encourage smoking cessation and moderation in alcohol consumption. Help patients find community support groups to decrease loneliness and smoking cessation programs in elementary school.**

### *9.1.6 Nutrition*

- Food is important in the celebration of all rites among Brazilians. Food and its counterpart, hunger, are often viewed as symbols that determine social relations. Food has symbolic content, is used as a reward or punishment, and establishes and maintains social relations.
- The mainstay of the Brazilian American's diet continues to be rice, beans, and farina. Roast beef, fresh chicken, and seafood are sought when they are not too expensive.
- *Cafe de manha* (breakfast) typically consists of bread with *cafe com leite* (half coffee and half hot milk). Sometimes *cuscus* (dry cornmeal mush) is served with milk. Fruit, fruit juices, and scrambled eggs, with or without sliced hot dogs, are common special breakfast fares among middle-class families. Sometimes sweet potatoes and yams grace the breakfast table.
- *O almoco* (dinner) is eaten at noon. This heavy meal, consisting of beans, rice, and farina, often includes *puree* (mashed potatoes) and *macarrao* (pasta). Desserts such as *pudim de leite* (custard), various cornmeal pastries, fruit, and *doce* (a sweet paste made by boiling sugar and fruit or fruit pulp) are common. A typical vegetable salad consists of finely cubed carrots, potatoes, and *shushu* (summer squash-like plant). A fruit salad with finely cubed fruits is also common.
- Brazilians living in the U.S. are tipycally vitamin- and health-food conscious. Although this luxury is often not available to those who have immigrated for fast money, documented residents generally become health-food consumers. The preference, especially among young Brazilian women, is to rely on vitamins instead of food consumption to help them remain thin.
- Fruit juices are expensive, and special foods that are common to the Brazilian diet are hard to procure in the United States. Food limitations are imposed by expense and inaccessibility of Brazilian mainstay foods. However, many Brazilian communities in the United States have ethnic markets and restaurants. Large-chain supermarkets often carry a section of ethnic foods, some of which are reasonably priced.
- **Help patients identify low-cost nutritious, culturally acceptable foods.**

### 9.1.7 Pregnancy and Childbearing Practices

- Although Brazil is predominantly a Catholic country, birth control is taught and used. Women are encouraged by their physicians or clinic personnel to have tubal ligations to prevent unwanted pregnancies. Immigrants in the United Statesgenerally practice birth control. Abortions of unwanted pregnancies occur as a matter of personal choice if needed. Thus, fertility practices among immigrant Brazilians are a matter of convenience with a traditional fatalistic overtone.
- **Help women identify acceptable methods of fertility control.**
- Herbal teas are used for bringing on late menstrual periods and for stimulating natural abortions. At times, single women try to become pregnant to facilitate their chance of remaining permanently in the United States. This opportunity is greatly enhanced if the child is born here and has been able to attend school.
- Pregnant women are encouraged not to do heavy work and not to swim. Taboos also warn against having sexual relations during pregnancy.
- **Provide factual information about sexual relations during pregnancy.**
- Some foods are to be avoided, and specific foods are recommended during pregnancy. Taboos generally vary according to geographic region, socioeconomic status, and ethnic background. **Determine taboo and acceptable food choices for pregnant women on an individual basis.**
- Many mothers prefer to give their babies powdered dry milk in place of breastfeeding. Many women wish to regain their pre-pregnancy weight as soon as possible. Some women often feel that their milk is *fraca* (weak). Breastfeeding is linked to a social stigma that a mother who breastfeeds may often be thought of as abandoned or sexually unattractive.
- **Provide factual information on the nutritious effects and immune properties of breastfeeding.**
- A postpartum woman eats chicken soup to help her body return to normal. She is also advised not to eat spicy foods or *repadura* (a molasses candy) and not to drink *garapa* (sugar water) or *caldo de cana* (sugar cane juice) if she breastfeeds her infant.
- **Support postpartum women's nonharmful food choices and practices.**

### 9.1.8 Death Rituals

- The cultural response to the death of a baby can vary widely among Brazilians, as cultural practices and beliefs are influenced by factors such as religion, regional customs, and personal beliefs. In general, Brazilians may experience profound grief and mourning following the loss of a baby, regardless of the circumstances.
- If financially possible, the families of Brazilians who die in the United Statespersonally accompany the body to Brazil for burial in the family vault. If

family members cannot come to the United States, relatives meet at the airport upon the body's arrival in Brazil.
- **Refer to social services and funeral homes if needed to assist with sending the body to Brazil.**
- Responses to death and grief depend on the family. The fatalistic expression, "It was God's will," helps the grieving process among the rich and the poor.
- Older people wear black for various amounts of time depending on their relationship with the family member. Frequently, the final portrait is hung in the family *chaper* or near the family altar, and prayers are recited. An eternal light burns.
- Relatives are honored on the anniversaries of their death, both at home and at masses. Often, the family places an obituary of remembrance with or without a picture of the deceased in the local newspaper on the anniversary of the death.
- **Support bereavement practices within their cultural context.**

### 9.1.9 Spirituality

- Although 90% of Brazilians are Catholic, various Protestant sects exist. A few, including Catholics, incorporate Indian animism, African cults, Afro-Catholic syncretism, and Kardecism, a spiritualist religion embracing Eastern mysticism.
- Aside from the ***curandeiros*** (folk healers), special healers exorcise and pray for the wellness of their patients. Saints are asked for help, and people wear medals or little pouches of special powders around their necks.
- **Accept a wide range of religious practices among Brazilians. Integrate nonharmful religious practices into health prescriptions.**
- The meaning of life is found in religion, economy, fatalism, and reality. For some, life is *uma luta* (a battle). For others, life has an almost hedonistic attitude.
- The greatest source of strength for Brazilians is their immediate and extended families.
- **Include nuclear and extended family in health-care decisions.**
- **Do not remove medals or pouches without asking permission.**

### 9.1.10 Health-Care Practices

- Most Brazilians do not talk about their illnesses unless the illnesses are very serious. Generally, illness is discussed only within the family. Many Brazilians feel that talking about an illness such as cancer negatively influences their condition.
- **Do not disclose the patient's health condition to others. Accept that some patients may not want to talk about their illness.**
- Because many Brazilians tend to shun hospitals, when they are hospitalized their families accompany them and stay around the clock. The patient is often brought

food from home. The family is the nucleus of responsibility for health care and is eager to participate in care. **Include the family in the care of the patient whether in the hospital, long-term-care facility, or at home.**

- Brazilians residing in the United Statesare legally required to have health insurance. However, many cannot afford to pay for medical care and thereby revert to self-care.
- Brazilians are known for their self-medication practices. Antibiotic, neuroleptic, antiemetic, and most other prescription drugs are easily obtained over the counter in Brazilian pharmacies. Once in the United States, it becomes difficult to obtain the many drugs readily available in Brazil. Eventually, incoming Brazilians bring medicines requested by their friends and thus maintain the circulation of medications not available to Brazilians living in the United States.
- **Encourage patients to fully disclose the use of all medications and treatments so they do not conflict with prescription medicines.**
- Because Brazilians tend to self-medicate, the procurement of necessary health care is often avoided or delayed. Consulting with someone who has the same condition or with friends who know someone who has a similar condition may be the first step taken. A trip to the local pharmacist may be the second. A third response may be a telephone call to Brazil asking for a particular medicine.
- The Brazilian culture is rich in folk practices and depends on geographic region, ethnic background, socioeconomic factors, and generation. Many Brazilians prefer to use homeopathic medicines and herbs. Traditional and homeopathic pharmacies are supplemented by ***remedios populares*** (folk medicines) and ***remedios caseiros*** (home medicines). **Ask about the use of herbal remedies.**
- Folk remedies and traditional health-care practices are intermeshed when a serious illness may be best treated by traditional caretakers. Some take homeopathic *"glóbulos"*(little white balls) prepared specifically for certain ailments.
- At times, support services for documented and undocumented Brazilians are hard to find for those who do not have language skills or the self-esteem to become assimilated into the culture of their newly found environments. Language is a major problem for these immigrants. They neglect to learn English and prefer to get by in their enclave community, which may be detrimental to health assessment. **Help patients find legal assistance through social services. Obtain an interpreter when necessary.**
- Another barrier to health care for Brazilians in the United Statesis cost. This, combined with lack of knowledge about the health-care system and facilities. **Refer patients in financial need to social services departments.**
- Brazilians generally do not like to talk about pain. However, once the emotional barrier is removed, they feel relieved to be able to discuss their discomfort. Many pain-relieving medicines are available without a prescription in Brazil. Frequently, a person requiring these on a regular basis can request that friends or friends of friends bring a supply from Brazil.
- **Explain the facts about pain medicine, and encourage disclosure of all pain medicine and treatments being used.**

- Most Brazilians do not work if they are seriously ill. Sickness is a neutral role and is considered socially exempt. This role is free of guilt, blame, and responsibility. Sickness is often seen as something that just happens.
- Better-educated Brazilians accept blood transfusions, organ donation, and organ transplantation. As in the United Statesand other parts of the world, acceptance depends on religious credence and individual preference.

### *9.1.11 Health-Care Providers*

- The folk-health field has many types of health-care providers. *Curandeiros* are divinely gifted, ***rezadeiras*** (praying women) help exorcise illnesses, card readers can predict fortunes, ***espiritualistas*** are able to summon souls and spirits, *conselheiros* are counselors or advisors, and ***catimbozeiros*** are sorcerers. Additionally, the *mae* or *pai de santo* are head priestesses or priests from the African-Brazilian Umbanda or Xango religion. All have the power to heal their believers.
- **Encourage patients to disclose the use of folk healers and treatments prescribed. Incorporate nonharmful practices into prescriptions.**
- Brazilians in the United Statestend to respect physicians and nurses. Medical education is prestigious and highly sought by aspiring university students.

## 9.2 Reflective Exercises

1. Explain the extensive name format for Brazilians? How would you address a woman and a man?
2. What are some common health conditions of Brazilians in the United States?
3. Under what conditions can Brazilians with AIDS come to the United States?
4. Identify at least three barriers to health care for Brazilians in the United States.
5. Identify at least three high-risk health behaviors common among Brazilians. What might you do to help decrease these barriers?
6. Identify three Brazilian fold practitioners. What are the tenets of each?
7. Identify taboo practices for pregnant Brazilian women.
8. Describe communication practices of Brazilians with intimates. With non-intimates.
9. What are *comadres* and what are their roles?
10. What is the greatest source of strength among Brazilians?

## Bibliography

Boccolini CS, Lacerda EMA, Bertoni N, et al. (2023). Trends of breastfeeding indicators in Brazil from 1996 to 2019 and the gaps to achieve the WHO/UNICEF 2030 targets. BMJ global health, 8(9), e012529. https://doi.org/10.1136/bmjgh-2023-012529.

Coler MS, Coler MA (2013) People of Brazilian heritage. In: Purnell L (ed) Transcultural health care: a culturally competent approach, 4th edn. F. A Davis, Philadelphia

Garcia-Cerde R, de Medeiros PFP, Silva LF, et al (2023) Use of integrative and complementary health practices by Brazilian population: results from the 2019 National Health Survey. BMC Public Health 23, 1153 https://doi.org/10.1186/s12889-023-16083-y

Gomes MR, da Cunha MS, de Souza SCI et al (2023) Reproduction of parental occupations, income and poverty in Brazil. Soc Indic Res 167:339–362. https://doi.org/10.1007/s11205-023-03103-y

Roman A (2023) A Closer Look Into Brazil's Healthcare System: What Can We Learn? Cureus, 15(5), e38390. https://doi.org/10.7759/cureus.38390

Silva DFD, Paiva CE, Paiva BSR (2022) Cross-cultural adaptation and translation of the pediatric intensive care unit-quality of dying and death into Brazilian Portuguese. Adaptação transcultural Para o português brasileiro do questionário pediatric intensive care unit-quality of dying and death. Rev Bras Ter Intensiva 33(4):592–599. https://doi.org/10.5935/0103-507X.20210086

# Chapter 10
# People of Chinese Heritage

**Hsiu-Min Tsai, Tsung-Lan Chu, Wen-Pin Yu, and Ruoh-Lih Lei**

## 10.1 Overview and Heritage

China is the second most populous country in Asia and the second most populous country in the world, with a population of 1.409 billion, dispersed over 3.7 million square miles of land (Wikipedia 2024). Ninety-two percent of the population is Han; the remaining 8% is a mixture of 56 different nationalities, religions, and ethnic groups. Chinese people who live in the "Chinatowns" in the United States maintain many of their cultural beliefs and values and insist that health-care providers respect these values and beliefs with their interventions. In the United States, over two million people identify themselves as Chinese, with the largest communities residing in California, New York, Hawaii, and Texas. This aggregate profile of Chinese in the United States may not fit any one person due to individuality, acculturation, and other variant characteristics of culture (see Chap. 1).

A university education is highly valued; however, few have the opportunity to achieve this life goal because of limited enrollment opportunities. Often, young

H.-M. Tsai (✉)
College of Nursing, Hungkuang University, Taichung City, Taiwan
e-mail: hmtsai2024@hk.edu.tw

T.-L. Chu
Administration Center of Quality Management Department, Chang Gung Medical Foundation, TaoYuan, Taiwan
e-mail: jec75@cgmh.org.tw

W.-P. Yu
Nursing Department, Chang Gung Memorial Hospital, TaoYuan, Taiwan
e-mail: wenping@cgmh.org.tw

R.-L. Lei
Nursing Department, Hungkuang University, Taichung City, Taiwan
e-mail: lilylei@hk.edu.tw



E. A. Fenkl, L. D. Purnell (eds.), *Handbook for Culturally Competent Care*,
https://doi.org/10.1007/978-3-031-70492-5_10

adults come to Western countries to attend universities seeking more advanced prestigious educations. Many newer immigrants are professionals from Hong Kong who moved to Canada, the United States, and other Western countries to avoid the repatriation in 1997.

Ideals based on the teachings of Confucius play an important part in their values and beliefs, emphasize the importance of family, and reinforce the idea that all relationships embody power and rule. Other important values are filial piety, industry, patriotism, deference to those in hierarchal positions, tolerance of others, loyalty to superiors, respect for rites and social rituals, knowledge, benevolent authority, thrift, patience, courtesy, and respect for tradition.

### *10.1.1 Communications*

- The official language of China is Mandarin (***pu tong hua***), which is spoken by about 70% of the population. Other distinct dialects such as Cantonese, Fujianese, Shanghainese, Toishanese, and Hunanese are so different that often two groups cannot understand one another verbally. However, the written language is the same throughout the country, and consists of over 50,000 ideographs (about 5000 common ones).
- **If unable to obtain a dialect-specific interpreter, the interpreter should try writing the question.**
- Most Chinese people speak in a moderate to low voice tone and consider Americans to be loud. **Be cautious about tone of voice when interacting with Chinese patients.**
- When asked whether they understand what was just said, the Chinese invariably answer yes to avoid loss of face. **Obtain information in a manner that does not require a "yes" or "no" answer.**
- Negative queries are difficult to understand. **Place instructions in a specific order, such as "First, at nine o'clock every morning get the medicine bottle. Second, take two tablets out of the bottle. Third, get your warm water. Fourth, swallow the pills with the water."**
- **Do not use complex sentences with "ands" and "buts." Have patients demonstrate instructions to ensure that they are understood.**
- Despite their reputation for not openly displaying emotion with strangers, among family and friends they are open and demonstrative. The Chinese share information freely once a trusting relationship has developed.
- Most individuals maintain a formal distance with each other as a form of respect.
- **When touching is necessary, provide an explanation and ask permission.**
- Many Chinese people are uncomfortable with face-to-face communications, especially when there is direct eye contact. Most prefer to sit next to others. **Arrange seating to promote positive communication.**

- Titles are important to Chinese people. The family name is stated first and then the given name. Calling an individual by any name except his/her family name is impolite.
- Women do not use their husband's name after they get married. Therefore, unless the woman is from Hong Kong or Taiwan or has lived in a Western country for a long time, do not assume that her last name is the same as her husband's. Her family name comes first, followed by her given names and finally by her title. Many take an English name as an additional given name because Chinese names are often difficult for Westerners to pronounce. Some give permission to use only the English name. In addition, some switch the order of their names to be the same as those of Westerners, with the family name last. This practice can be confusing. **Ask the patient which is his/her family name and which is the given name. Address patients by their family name and title. Ask persons how they wish to be addressed.**

### *10.1.2 Family Roles and Organization*

- Kinship has traditionally been organized around the male lines. Each family maintains a recognized head who has great authority and assumes all major responsibilities for the family.
- Chinese culture, based on Confucianism, is known for its patriarchal traditions. Traditionally, women have often taken on the role of homemakers, overseeing all domestic responsibilities, while husbands have assumed the roles of family leaders and primary breadwinners (Zhong and Guo 2018). In recent times, some men include housework, cooking, and cleaning as their responsibilities when their spouses work. Most believe that the family is most important and, thus, each family member assumes changes in roles to achieve this harmony.
- Children are highly valued in China because of the government's past mandate that each married couple may only have one child, although there are exceptions to this rule: e.g., if the first born is a female in rural areas or where the only child of a couple is disabled or dies in infancy. Resources are lavished on the child. Independence is not fostered. The entire family makes decisions for the child even into young adulthood.
- Children born in Western countries tend to adopt the Western culture easily while their parents and grandparents may to maintain traditional Chinese culture in varying degrees. Adolescents maintain their respect for elders even when they disagree with them. Teenagers value a strong and happy family life and seldom do things that jeopardize that unanimity. Adolescents question affairs of life and make great efforts to see at least two sides of every issue.
- Children feel pressure to succeed to help improve the future of the family; thus, most children and adolescents value studying over playing and peer relationships. Children are taught to curb their expression of feelings because individuals who do not stand out are successful. **With an increase in cultural adaptation,**

**children are becoming more outspoken, and they engage in more reading of media from the Western world, as well as watching television and movies.**

- The perception of family is developed through the concept of relationships. Each person is identified in relation to others in the family. The individual is not lost, just defined differently from individuals in Western cultures.
- Young men and women enter the workforce immediately after high school if they are unable to continue their education. Many continue to live with their parents and contribute to the family even after marriage (in their 20 s) and the birth of a child (in their 30 s).
- Extended families are important. Children may live with their grandparents or aunts and uncles so individual family members can obtain a better education or reduce financial burdens.
- Teenage pregnancy is not common among the Chinese but it is increasing among Chinese in America.
- Older people are venerated and viewed as very wise. Children are expected to care for their parents when self-care becomes a concern; in China, law mandates this.
- **When Chinese immigrants need additional assistance, call on local Chinese organizations to obtain help.**
- Maintaining reputation is accomplished by adhering to the rules of society. True equality does not exist in the Chinese mind; if more than one person is in power, then consensus is important. If the person in power is not present at decision making meetings, barriers are raised, and any decisions made are negated unless the person in power agrees. **Ensure that key decision-makers within the family attend patient conferences or other times when health-care decisions need to be made.**
- The word for "privacy" in Chinese has a negative connotation and means something underhanded, secret, and furtive. People grow up in crowded conditions; they live and work in small areas. The value of group support does not place a high value on privacy.
- **Offer to make arrangements for the family to remain with the patient 24 h a day if needed.**
- The Chinese may ask many personal questions about salary, life at home, age, and children. Refusal to answer personal questions is accepted as long as it is done with care and feeling. The one subject that is taboo is sex and anything related to sex. This may create a barrier for a Western health-care provider who is trying to assess a Chinese patient with sexuality concerns. **Approach issues of sexuality tactfully and indirectly and only after trust is obtained.**
- In China, homosexuality has been legal since 1997. In 2019, same-sex marriage officially became legal in Taiwan (Executive Yuan 2022), making it the first country in Asia and the 27th country in the world to recognize same-sex marriage. This granted equal marriage rights to both same-sex and heterosexual couples. However, the LGBT community still faces stigmatization in society, with gay relationships largely criticized at the national level and continuing to bear a stigma. **Do not disclose same-sex relationships.**

### 10.1.3 *Workforce Issues*

- Chinese adapt to the culture in the workplace quickly. They frequently call on other Chinese people to teach them and to discuss how to fit into the new culture more quickly. **Assign a Chinese as a mentor if available.**
- The Chinese are accustomed to giving co-workers small gifts of appreciation for helping them acculturate and adapt to the American workforce. Often, Americans seek opportunities to reciprocate with a gift, such as at a birthday party, farewell party, or other occasion. Some gifts are not appropriate. **Do not give gifts that are culturally inappropriate; for example, giving an umbrella means that you wish to have the recipient's family dispersed; giving a gift that is white in color or wrapped in white could be interpreted as meaning the giver wishes the recipient dead; and giving a clock could be interpreted as never wanting to see the person again or wishing the person's life to end. If other Chinese are available, seek their advice for gift giving.**
- Autonomy is limited and is based on functioning for the good of the group. When a situation arises that requires independent decision making, many times the Chinese know what should be done but do not take action until the leader or superior gives permission.
- **Teach the U.S. expectations for assertiveness and autonomy to Chinese employees.**
- Language may be a barrier for some Chinese. The Chinese language does not have verbs that denote tense, as in Western languages. Intonation in Chinese is in the words themselves, rather than in the sentence. Chinese people who have taken English lessons can usually read and write English competently, but they may have difficulty in understanding and speaking it. **Ask for a demonstration of the activity to ensure understanding.**
- Most Chinese speak in a moderate to low voice tone and consider Americans to be loud. **Be cautious about tone of voice when interacting with Chinese employees.**
- When asked whether they understand what was just said, the Chinese invariably answer yes to avoid loss of face. **Collect information in a manner that cannot be answered with a "yes" or "no" answer.**
- Negative queries are difficult to understand. **Place instructions in a specific order such as "First, gather your supplies." Second, explain the procedure, etc. Do not use complex sentences with "ands" and "buts." Have employees demonstrate instructions to ensure that they are understood.**

### 10.1.4 Biocultural Ecology

- Skin color among Chinese is varied. Many have skin color with pink undertones; some have yellow tones, and others are very dark. Hair is generally black and straight, but some have naturally curly hair. Most men do not have much facial or chest hair.
- Mongolian spots—dark bluish spots over the lower back and buttocks—are present in about 80% of infants.
- Bilirubin levels are usually higher in Chinese newborns, with the highest levels occurring on the fifth or sixth day after birth.
- The ulna is longer than the radius. Hip measurements are significantly smaller than Westerners (Seidel et al. 1994). **Provide instruction on adequate calcium consumption to prevent osteoporosis and fractures. Ensure diet is low in lactose. Assess patients for genetic conditions before medications are prescribed.**
- The Rh-negative blood group is rare.
- Poor metabolism of mephenytoin occurs in 15–20% of Chinese. Sensitivity to beta blockers, such as propranolol, is evidenced by a decrease in overall blood levels accompanied by a more profound response. Atropine sensitivity is evidenced by an increased heart rate. Increased responses to antidepressants and neuroleptics occur at lower doses. Analgesics have been found to cause increased gastrointestinal side effects, despite a decreased sensitivity to them. Chinese people generally have an increased sensitivity to the effects of alcohol (Levy 1993). **Carefully monitor Chinese patients on medications such as propranolol, atropine, antidepressants, and neuroleptics.**
- Table 10.1 lists the commonly occurring health conditions for Chinese in three categories: lifestyle, environment, and genetics.

### 10.1.5 High-Risk Health Behaviors

- Smoking is a high-risk behavior for many men and teenagers. **Partner with schools to teach hazards of smoking and substance misuse.**
- Most Chinese women do not smoke, but recently the numbers Chinese women who smoke is increasing among males, as much as 68% smoke. **Screen newer immigrants for smoking-related health conditions.**

**Table 10.1** Health conditions common to people of Chinese heritage

| Health conditions | Causes |
|---|---|
| Lactose intolerance | Genetic, lifestyle |
| Thalassemia | Genetic |
| Hepatitis B | Environment, lifestyle |
| Tuberculosis | Environment, lifestyle |
| Pancreatic cancer | Environment, lifestyle |
| Diabetes | Genetic, environment, lifestyle |
| Glucose-6-phosphate dehydrogenase deficiency | Genetic |
| Nasopharyngeal cancer | Environment, lifestyle |
| Liver cancer | Environment, lifestyle |
| Stomach cancer | Environment, lifestyle, unknown? |
| Cardiovascular disease | Environment, lifestyle |
| Hepatitis B | Environment, lifestyle |

## *10.1.6 Nutrition*

- Food habits are important, and food is offered to guests at any time of the day or night. Foods served at meals have a specific order, with a focus on a balance for a healthy body.
- The typical diet is difficult to describe; each region in China has its own traditional diet. Traditional Chinese medicine frequently uses food and food derivatives to prevent and cure diseases and illnesses and to increase strength in weak and older people.
- Peanuts and soybeans are popular. Common grains include wheat, sorghum, and maize (a type of corn). Rice is usually steamed but can be fried with eggs, vegetables, and meats as well. Many Chinese eat beans or noodles instead of rice. Meat choices include pork (the most common), chicken, beef, duck, shrimp, fish, scallops, and mussels. Tofu, an excellent source of protein, is a staple of the Chinese diet and is fried, boiled, or served cold like ice cream. Bean products are another source of protein. Many desserts or sweets are prepared with red beans.
- Fruits and vegetables may be peeled and eaten raw. Vegetables are lightly stir-fried in oil with salt and spice. Salt, oil, and oil products are important parts of the Chinese diet. Their healthy selection of green vegetables limits the incidence of calcium deficiencies.
- Drinks with dinner include tea, soft drinks, juice, and beer. Foreign-born and older people may not like ice in their drinks. **Ask patients if they want ice in their drinks.**
- Foods that are considered ***yin*** and ***yang*** prevent sudden imbalances and indigestion. A balanced diet is considered essential for physical and emotional harmony. **Provide special instructions regarding risk factors associated with diets that are high in fats and salt.**

- Chopsticks should never be stuck in the food upright because that is considered bad luck.
- Heavy use of salted fish increases the risk for nasopharyngeal, esophageal, and stomach cancers. **Encourage low use of salty fish.**

### *10.1.7 Pregnancy and Childbearing Practices*

- Because of the previous "one couple, one child" law in China, abortions are common in China. Most Chinese families see pregnancy as positive and important in the immediate and extended family. Pregnancy is seen as women's business, although men are beginning to demonstrate an active interest in pregnancy and the welfare of the mother and baby. Women are usually very modest and may insist on a female midwife or obstetrician. Some agree to use a male physician when an emergency arises. **Respect different views on involving men in pregnancy issues.**
- Pregnant women usually increase meat in their diets because their blood needs to be stronger for the fetus. Pregnant women may avoid shellfish during the first trimester because it causes allergies. Some may be unwilling to take iron because they believe that it makes the delivery more difficult. **Explain factual necessity about taking iron preparations and dispel myths.**
- Traditional postpartum care includes 1 month of recovery, with the mother eating foods that decrease the yin (cold) energy. **Ask what foods the mother plans to eat postpartum and dispel myths.**
- Many mothers do not expose themselves to the cold air and do not go outside or bathe for the first month postpartum because cold air can enter the body and cause health problems. Some postpartum women wear many layers of clothes and are covered from head to toe, even in the summer, to keep the air away from their bodies.
- Drinking and touching cold water are taboo for women in the postpartum period.
- Raw fruits and vegetables are avoided because they are considered "cold" foods. They must be cooked and be warm. Mothers eat 5–6 meals a day with high-nutritional ingredients including rice, soups, and 7–8 eggs. Brown sugar is commonly used because it helps rebuild blood loss. Drinking rice wine is encouraged to increase the mother's breast-milk production. **Caution mothers that rice wine may prolong the postpartum bleeding time.**

### *10.1.8 Death Rituals*

- Death is viewed as a part of the natural cycle of life; some believe that something good happens to them after they die. Death and bereavement traditions are centered on ancestor worship, a form of paying respect. Many believe that their spirits will never rest unless living descendants provide care for the grave and worship the memory of the deceased.

- The dead are honored by placing food, money for the person's spirit, or articles made of paper around the coffin.
- The belief that the Chinese greet death with stoicism and fatalism is a myth.
- The number 4 is considered unlucky because it is pronounced like the Chinese word for death; this is similar to the bad luck associated with the number 13 in many Western societies. The color white is associated with death and is also considered bad luck. Black is a bad luck color. Red is the ultimate good-luck color. Mourners are recognized by black armbands on their left arm and white strips of cloth tied around their heads.
- The purchase of life insurance may be avoided because of a fear that it is inviting death.
- Traditional Chinese do not overtly express emotions to strangers. **Accept and be non-judgmental with a wide range of bereavement practices.**

### *10.1.9 Spirituality*

- The main religions in China are Buddhism, Catholicism, Protestantism, Taoism, and Islam. Prayer is generally a source of comfort. Some Chinese people do not acknowledge a religion such as Buddhism, but if they go to a shrine, they burn incense and offer prayers. As immigration increases, many who practice Christian religions have become more visible.
- "Life forces" are sources of strength. These forces come from within the individual, the environment, and the past and the future of the individual and society.
- The individual may use meditation, exercise, massage, and prayer. Drugs, herbs, food, good air, and artistic expression may also be used. Good-luck charms are cherished and traditional and nontraditional medicines are used.
- The family is usually a source of strength. Individuals draw on family resources and are expected to provide resources to strengthen the family. Resources may be defined as financial, emotional, physical, mental, or spiritual. Calling on ancestors to provide strength as a resource requires giving back to the ancestors when necessary.
- In the United States, churches play an important role in the local Chinese community by providing support to immigrants, students, scholars, and their families. **Network and partner with churches to provide additional support for Chinese patients.**

### 10.1.10 Health-Care Practices

- While many Chinese people have made the transition to Western medicine, others maintain their roots in traditional Chinese medicine, and still others practice both types of medicine. Younger people usually do not hesitate to seek allopathic health-care providers when necessary unless they believe that it does not work for them; then they use traditional Chinese medicine.
- Older people may try traditional Chinese medicine first and only seek Western medicine when traditional medicine does not seem to work. Even after seeking Western medical care, older people may continue to practice traditional Chinese medicine in some form. **Ask patients in a non-judgmental manner if they are using traditional Chinese medicine. Impress upon them the importance of disclosing all treatments because some may have antagonistic effects.**
- Traditional Chinese medical treatments are discussed in Table 10.2.
- The Chinese tend to describe their pain in terms of more diverse body symptoms, whereas Westerners tend to describe pain locally. The Western description includes words like "stabbing" and "localized," whereas the Chinese describe pain as "dull" and more "diffuse." They tend to use explanations of pain from the traditional Chinese influence of imbalances in the yin and yang combined with location and cause.

**Table 10.2** Traditional Chinese medical treatments

| |
|---|
| • Traditional Chinese medicine has many facets, including the five basic substances (*qi*, energy; *xue*, blood; *jing*, essence; *shen*, spirit; and *jing ye*, body fluids); the pulses and vessels for the flow of energetic forces (*mai*); the energy pathways (*jing*); the channels and collaterals, including the 14 meridians for acupuncture, moxibustion, and massage (*jing luo*); the organ systems (*zang fu*); and the tissues of the bones, tendons, flesh, blood vessels, and skin. The scope of traditional Chinese medicine is vast and should be studied carefully by professionals who provide health care to Chinese patients |
| • Acupuncture and moxibustion are used in many treatments. Acupuncture is the insertion of needles into precise points along the channel system of flow of the *qi* called the 14 meridians. The system has over 400 points. Many of the same points can be used in applying pressure (acupressure) and massage (acumassage) to achieve relief from imbalances in the system. The same systems approach is used to produce localized anesthesia |
| • Moxibustion is the application of heat from different sources to various points. For example, one source, such as garlic, is placed on the distal end of the needle after it is inserted through the skin, and the garlic is set on fire. Sometimes the substance is burned directly over the point without a needle insertion. Localized erythema occurs with the heat from the burning substance, and the medicine is absorbed through the skin |
| • Cupping is another common practice. A heated cup or glass jar is put on the skin creating a vacuum, which causes the skin to be drawn into the cup. The heat that is generated is used to treat joint pain |
| • Herbal therapy is integral to traditional Chinese medicine. Herbs fall into four categories of energy (cold, hot, warm, and cool), five categories of taste (sour, bitter, sweet, pungent, and salty), and a neutral category. Different methods are used to administer the herbs, including drinking and eating, applying topically, and wearing on the body |

- **Chinese cope with pain by applying oils and massage, using warmth, sleeping on the area of pain, relaxation, and aspirin.**
- The balance between yin and yang is used to explain mental as well as physical health. Because a stigma is associated with having a family member who is mentally ill, many families initially seek the help of a folk healer. Many Chinese still view mental and physical disabilities as a part of life that should be hidden. Traditionally, an ill person is viewed as passive and accepting of illness. Illness is expected as a part of the life cycle.
- Many individuals feel uncomfortable touching their own bodies, which may be problematic when they need to provide their own health care, for example, with breast self-examinations and scrotal exams. People of the same sex may use touch if they are close friends or family. Men and women do not touch each other, and even couples who have been married for a long time do not show physical affection in public. Most women feel uncomfortable being touched by male health-care providers and tend to seek female providers. **Always ask permission, and explain the necessity for touching patients during examinations and treatments.**
- Families may be reluctant to allow autopsies because of their fear of being "cut up." Most accept blood transfusions, organ donations, and organ transplants.

### *10.1.11 Health-Care Providers*

- Traditional Chinese medicine health-care providers are shown great respect by the Chinese. In many instances, they are shown equal, if not more, respect than Western health-care providers.
- Some distrust Western health-care providers because of the pain and invasiveness of their treatments.
- Older health-care providers receive more respect than younger providers, and men usually receive more respect than women. Physicians receive the highest respect, followed closely by nurses with a university education. Other nurses with limited education are next in the hierarchy.
- If Chinese patients disagree with the health-care provider, they may not follow instructions. Moreover, they may not verbally confront the health-care provider because they fear that either they or the provider will suffer a loss of face.

## 10.2 Reflective Exercises

1. What are the most common health conditions among Chinese in the United States?
2. What are the principle tenets of Traditional Chinese Medicine?

3. What are the primary Chinese languages? If an interpreter knows one of these languages and the patient speaks another Chinese language, what might the interpreter do to get the information?
4. Postpartum women might drink rice wine to increase lactation. What is the problem with this practice?
5. Your Chinese patient has been prescribed two medicines. One medicine is *tid* and the other is *qid*. Describe how you would teach your patient to take them?
6. What are the major religions practiced by Chinese in the United States?
7. What seating arrangement would you make to interview a Chinese patient?
8. What is meant by "saving face" among the Chinese? Give an example.
9. What are the principles of *yin* and *yang*? Give an example with dietary practices.
10. Identify some medications where metabolism is different for a patient with Chinese heritage.

## Bibliography

Bradford D (2013) Mental health news watch. Race, genetics, metabolism: drug therapy and clinical trials http://www.miwatch.org/2008/04/race_genetics_metabolism_drug_1.html

CIA (2024) World factbook: China. https://www.cia.gov/the-world-factbook/countries/china/

http://us2.mofcom.gov.cn/article/aboutchina/custom/200411/20041100004548.shtml

Lau Y (2012) Traditional Chinese pregnancy restrictions, health-related quality of life and perceived stress among pregnant women in Macao, China. Asian Nurs Res 6(1):27–34

Levy RA (1993) Ethnic and racial differences in response to medicines: preserving individualized therapy in managed pharmaceutical programmes. Pharm Med 7:139–165

Migration Policy Institute (2016) Chinese immigration in the United States. http://www.migrationpolicy.org/article/chinese-immigrants-united-states

National Center for Complementary and Integrative Health (2013) Traditional Chinese medicine: in depth. https://nccih.nih.gov/health/whatiscam/chinesemed.htm

Seidel H, Ball J, Dains J, Benedict W (1994) Quick reference to cultural assessment. Mosby, St. Louis

Shek DT (2010) Oxford handbook of Chinese psychology. Oxford handbooks, vol 1. Oxford University Press, Oxford. https://academic.oup.com/edited-volume/38605

Tsai HM (2013) People of Chinese heritage. In: Purnell L (ed) Transcultural health care: a culturally competent approach, 4th edn. F.A. Davis Company, Philadelphia

Wikipedia (2024) Demographics of China. https://en.wikipedia.org/wiki/Demographics_of_China

World Bank (2016) Smoking prevalence among males. http://data.worldbank.org/indicator/SH.PRV.SMOK.MA

Yasuda SU (2008) The role of ethnicity in variability in response to drugs: focus on clinical pharmacology studies. Clin Pharmacol Ther 84(3):417–423

Yasuda SU, Zhang L, Huang SS (2008) The role of ethnicity in variability in response to drugs: focus on clinical pharmacology studies. Clin Pharmacol Ther 84(3):417–423

Executive Yuan (2022) Same marriage legal. https://www.ey.gov.tw/Page/9277F759E41CCD91/9eaeb648-e4b3-4bde-b44e-e2274661b0dd

Zhong XH, Guo WQ (2018) Chinese super mothers in the new social risks: a qualitative study of child care among families in Guangzhou. J Chinese Women's Stud 2(146):67–101

# Chapter 11
# People of Cuban Heritage

**Jorge A. Valdés and Victor Delgado**

## 11.1 Overview and Heritage

The Republic of Cuba (*República de Cuba*), located 90 miles south of Key West, Florida, is a multiracial society with people of primarily Spanish and African origins. Other groups include Chinese, Haitians, and Eastern Europeans. Spain, the United States, and Russia (formerly the Soviet Union) significantly influenced Cuba's history and culture. Mistrust of government has reinforced a strong personalistic tradition and sense of national identity evolving from family and interpersonal relationships. The desire for personal freedom, hope of refuge, political exile, and promise of economic opportunities prompted migration to the United States from the 1960s to today. Cuban people in the United States self-identify as Cuban or Cuban American. Cubans in the United States take extraordinary pride in their heritage, possess a strong ethnic identity, speak Spanish, and adhere to traditional Cuban values and practices. Their highest concentration is in Florida (approximately 70%), although significant numbers live in New Jersey, New York, Illinois, and California.

### *11.1.1 Communications*

- Many live and transact business in Spanish-speaking enclaves. Cuban people speak amazingly fast, especially in Cuban Spanish, mainly by dropping the final consonant from words. Whereas the second generation speaks English, many converse with friends or peers in "Spanglish," a mixture of Spanish and English.

J. A. Valdés (✉) · V. Delgado
Nicole Wertheim College of Nursing and Health Sciences, Florida International University, Miami, FL, USA
e-mail: jvalde@fiu.edu; vadelgad@fiu.edu



E. A. Fenkl, L. D. Purnell (eds.), *Handbook for Culturally Competent Care*,
https://doi.org/10.1007/978-3-031-70492-5_11

The highly educated are more likely to speak English at home. **Assistance with required forms in the U.S. health-care system is needed by the less acculturated and those for whom English is a concern.**

- Many value *simpatía* and *personalismo* in their interactions with others. *Simpatía*, the need for smooth interpersonal relationships, is characterized by courtesy, respect, and the absence of criticism or confrontation. *Personalismo*, the importance of intimate interpersonal relationships, is valued over impersonal bureaucratic relationships.
- ***Choteo***, a light-hearted attitude with teasing, bantering, and exaggerating, is often observed in their communications with others. Conversations are characterized by animated facial expressions, direct eye contact, hand gestures, and gesticulations. Hand gestures can easily be considered exaggerated or animated; however, this should not be interpreted as disrespectful. Voices tend to be loud, and the rate of speech is fast. **Do not interpret these behaviors as family discord.**
- Handshakes, touching, and hugs are acceptable among family, friends, and acquaintances and may be used to express gratitude to the caregiver. Greetings among family and close friends usually involve kisses on the cheek. Touch is common between people of the same gender; older men and women rarely touch in public. **Explain the essential necessity and ask permission to touch private body areas during a physical examination.**
- Most tend to emphasize current issues and problems rather than future ones. *Hora Cubana* (Cuban time) is a flexible period that stretches 1–2 h beyond designated clock time. **When setting up appointments, assess the patient's level of acculturation with respect to time. Explain the necessity of punctuality.**
- Cubans use two surnames representing the mother's and father's family names. Married women may also add the husband's name. Thus, **ask which name is used for legal purposes. When addressing patients, especially older people, formal rather than familiar forms of speech should be used unless told otherwise (i.e., Señor [Mr.], Señora [Mrs.], or Señorita [Miss or Ms.]).**

## *11.1.2 Family Roles and Organization*

- Traditional family structure is patriarchal, characterized by a dominant and assertive male and a passive, dependent female. Traditionally, Cuban wives stay at home, manage the household, and care for children, whereas husbands are expected to work, provide financially, and make major decisions for the family. ***La casa* (the house), the province of the woman, and *la calle* (the street), the man's domain, demonstrate the distinction in gender roles and should be respected. With acculturation, more women are working outside the home; egalitarian decision making prevails in the** United States.
- Honor is attained by fulfilling family obligations and treating others with ***respeto*** (respect). ***Vergüenza***, a consciousness of public opinion and the judgment of the

entire community is considered more important for women than for men. **Machismo** dictates that men display physical strength, bravery, and virility and be the spokesperson, even though they might not make the decisions.

- *La familia* (the family, nuclear and extended, including grandparents, aunts, uncles, and godparents) is the most important source of emotional and physical support. Multigenerational (3–4 generations) households are common, including a high proportion of people 65 years and older who live with their relatives. **Family is considered the most important social unit and is usually involved in all decisions. Specifically, ask who makes which decisions for the family and who is the spokesperson. Verbal consent for invasive procedures from family members as well as from next of kin is often required. Extended family members are valuable resources for home care.**
- Compared with European American standards, Cubans tend to pamper and overprotect their children. Children are expected to study, respect their parents, and follow *el buen camino* (the straight and narrow path). Boys are expected to learn a trade or prepare for work and to stay away from vices. Girls are expected "to remain honorable while single," to prepare for marriage and to avoid the opposite sex. When a daughter reaches 15 years, a ***quinceañera***, or elaborate 15th birthday party, is typically held to celebrate this rite of passage. Adolescents may undergo an identity crisis and reject their heritage; parents may feel their authority is being challenged.
- **Explain American child abuse laws, taking into consideration the culturally complex definition of what may be considered child abuse.**
- Personal relationships, known as *compadrazgo* (relationships existing between a godparent or godparents and the godchild and its parents), are typical, with **heavier reliance for help on family and personal relationships than on government or organizations.**
- **Traditional families seldom discuss sex openly; initiating these discussions with health teaching may be welcome.**
- Little information is available on same-gender attraction. Same-sex behaviors among men may be regarded as a sign of virility and power rather than gay behavior. The gay lifestyle is contradictory to the machismo orientation of this culture. Same-sex couples may be alienated from their families.
- **Given the stigma associated with same-gender attraction, a matter of fact, a nonjudgmental approach must be used when questioning patients regarding sexual practices. Discussions should always be held in private. Do not disclose same-sex relationships to family members.**

### *11.1.3 Workforce Issues*

- A source of tension is the tendency of Cubans and Cuban Americans to speak Spanish or Spanglish with other Cuban or Hispanic co-workers. Speaking the same language allows them to form a common bond, relieve anxieties at work, and feel comfortable with one another. **Explain upon employment the rules for speaking a language other than English at work.**

- Traditional Cubans recognize supervisors as authority figures and treat them with respect and deference. Cubans value a structure characterized by *personalismo*, one that is oriented around people rather than around concepts or ideas. Personal relationships at work are considered an extension of family relationships. Because of the emphasis on the job or task in the American workplace, many Cubans view this workplace as being too individualistic, businesslike, and detached.
- **Explain the individualistic culture of the American workforce upon employment.**

### *11.1.4 Biocultural Ecology*

- Most Cubans are white, and only 5% are black, with physical features like those of African Americans. **Because of their European ancestry, most have skin, hair, and eye color that vary from light to dark. To assess for cyanosis and jaundice in individuals with dark skin, assess the color of the sclera and conjunctiva, palms of the hands, soles of the feet, and buccal mucosa rather than relying on skin tone.**
- Cuban Americans tend to have lower incidences of diabetes mellitus, obesity, and hypertension than other Hispanic groups or whites. Because of their diet, which is high in sugar, many exhibit a high prevalence of tooth loss, filled teeth, gingival inflammations, and periodontitis. Table 11.1 lists commonly occurring health conditions among Cubans in three categories: lifestyle, environment, and genetics.
- Specific information related to drug metabolism is limited; however, in general, many require lower doses of antidepressants and experience greater side effects than non-Hispanic white populations.

**Table 11.1** Health conditions common to Cubans

| Health conditions | Causes |
|---|---|
| Hypertension | Genetic, lifestyle |
| Coronary artery disease | Genetic, lifestyle |
| Obesity | Environment, lifestyle |
| Diabetes mellitus | Genetic, lifestyle |
| Lung cancer | Environment, lifestyle |

### 11.1.5 High-Risk Health Behaviors

- Cuban Americans tend to exhibit a higher incidence of smoking than other Hispanic or European groups. **Encourage smoking cessation starting in elementary school. In Cuba, it is estimated that over 40% of all men smoke tobacco daily. For Cubans living in the** United States **and Cuban American, these rates are lower.**
- Alcohol use is greater among males than females and among younger versus older groups. **When assessing alcohol consumption, determine the amount consumed during holidays and celebrations, not just daily.**
- Violent deaths account for high mortality rates among adolescents and young adults. Suicide rates also exceed those of the white non-Hispanic population.
- Use of preventive services depends on accessibility, which, in turn, is significantly influenced by education, annual income, and age.

### 11.1.6 Nutrition

1. Food allows families to reaffirm kinship ties, promotes a sense of community, and perpetuates customs and heritage. It is important to consider the significance of accepting food offered by others. **Rejecting food that is offered may be perceived as rejecting the person who is offering the food.**
2. Staple foods include root crops like yams, yuca, malanga, and boniato, plantains, and grains. **Many dishes are prepared with olive oil, garlic, tomato sauce, vinegar, wine, lime juice (*sofrito*), and spices.** Meat is usually marinated in lemon, lime, sour orange, or grapefruit juice before cooking.
3. **Determine food preparation practices as well as customary foods consumed. Fats or flavorings used for cooking can add significant calories to meals.**
4. **All major food groups are well represented in the Cuban diet; however, fiber content and leafy green vegetables may be lacking. Lactose intolerance is common among most Hispanic groups.**
5. A leisurely noon meal (*almuerzo*) and a late evening dinner (*comida*), sometimes as late as 10 or 11 p.m., are often customary.
6. **Determine mealtimes and adjust medication schedules accordingly.**
7. The strong and bittersweet coffee called *café cubano* is a standard drink after meals and throughout the day.
8. Being overweight, according to U.S. actuarial tables, is seen as positive, healthy, and sexually attractive.
9. **Patients should be coparticipants in deciding an acceptable weight. Counsel patients about the differences between being overweight and obese and the health hazards associated with obesity. Incorporate patients' traditional and preferred food choices into nutritional health planning.**

See Table 11.2 for Popular Cuban American foods.

**Table 11.2** Popular Cuban American foods

| | | | |
|---|---|---|---|
| ***Entrees and side dishes*** | | | |
| Roast pork | *lechon* | Fried pork chunks | *masas de puerco* |
| Sirloin steak | *Palomilla* | Shredded beef | *Ropa vieja* |
| Pot roast<br>Ground beef<br>Cuban sandwich<br>Fried green plantains<br>Fried ripe plantains | *Boliche*<br>*Picadillo*<br>*sándwich cubano*<br>*Tostones o mariquitas*<br>*Platanos maduros fritos* | Roasted chicken<br>Chicken and rice<br>Black beans and rice<br>Boiled cassava with garlic sauce | *Pollo asado*<br>Arroz con pollo<br>*Moros y cristianos*<br>*Yuca con mojo* |
| ***Desserts*** | | | |
| Caramel custard | *Flan* | Egg custard | *Natilla* |
| Rice pudding | *Arroz con leche* | Coconut pudding | *pudín de coco* |
| Bread pudding | *pudín de pan* | | |
| ***Beverages*** | | | |
| Sugar cane juice | *Guarapo* | Iced coconut milk | *Coco frio* |
| Cuban soft drinks<br>Cuban shakes | *Materva, ironbeer, jupiña,*<br>*Malta hatuey*<br>*Batidos* | Punch beverage | *Sangria* |
| Strong coffee | *café cubano* | | |

### *11.1.7 Pregnancy and Childbearing Practices*

- Cuban women's fertility rate is lower than that of other Hispanic American women. Cuba's current reproductive rate is among the lowest in the developing world. The mean age of Cuban women at first birth is 27 years. Even before the revolution, Cuba had the lowest birthrate in Latin America. This low fertility rate has been attributed to the fact that many women are in the workforce. Preterm births and neonatal and post-neonatal deaths are lower among Cuban American women than among other Hispanic American groups. Women are less likely to use oral contraceptives than other Hispanic American women.
- **Explain that condoms not only help prevent pregnancy but also decrease the incidence of HIV or *VIH* and other sexually transmitted infections. Determine on an individual basis the methods of acceptable birth control.**
- Prenatal care is higher than among other Hispanic and white non-Hispanics. **Stress that medical evaluations during pregnancy are necessary to help ensure the health of the mother and baby. When language barriers exist, use interpreters, videos, and literature in Spanish for teaching.**
- Mothers tend to use advice about child health given by their spouses, mothers, mothers-in-law, clerks, and pharmacists.
- Strangers are not permitted to touch the infant, child, or pregnant patient's abdomen-only family.
- Strangers are viewed as unclean, and their touch may harm the baby.

- Childbirth is a time for celebration, with family members and friends congregating in the hospital. Traditionally, men have not attended the births of their children, but younger, more acculturated fathers are frequently present. **Respect the cultural beliefs of families if fathers are not wanted or do not wish to be in the delivery room.**
- Table 11.3 lists Cuban dietary and folk beliefs and practices about pregnancy.
- During the postpartum period, ambulation, exposure to cold, and bare feet place the mother at risk for infection. Family members and relatives often care for the mother and baby for about 4-weeks postpartum. Most women consider breastfeeding better than bottle feeding; approximately half choose to breastfeed. **Mothers tend to wean infants from the breast and introduce solid foods at an early age, but bottle weaning often occurs at a median age of 4 months, fostering their beliefs that "a fat child is a healthy child" and that breastfeeding may contribute to a deformity or asymmetry of the breasts.**
- Cutting the infant's hair or nails in the first 3 months is believed to cause blindness and deafness. **Do not cut the baby's hair or nails without asking permission.**

### *11.1.8 Death Rituals*

- In death, as in life, the support of the extended family network is paramount. Bereavement is expressed openly as loud crying, with other physical manifestations of grief. Death is often seen as a part of life, and some, especially men, may approach death stoically. **Recognize that the lack of an open display of emotion may not imply indifference to the dying or deceased.**
- The dying person is typically attended by a large gathering of relatives and friends. In Catholic families, individual and group prayers are held for the dying to provide a peaceful passage to the hereafter. Religious artifacts such as religious

**Table 11.3** Cuban dietary and folk beliefs and practices about pregnancy

| |
|---|
| • Women eat for two during pregnancy |
| • Morning sickness is cured by eating coffee grounds |
| • Eating a lot of fruit ensures that the baby will be born with a smooth complexion |
| • Wearing necklaces during pregnancy causes the umbilical cord to be wrapped around the baby's neck |
| • Raising one's arms over the head while pregnant may cause the umbilical cord to wrap around the infant's neck |
| • Respect nonharmful cultural beliefs when prescribing care regarding nutrition<br>• Assess which foods are eaten or avoided during pregnancy to provide a culturally appropriate nutritional plan. For example:<br>– Some foods, such as beans and plantains, may be prohibited and rice may be limited during pregnancy<br>– Foods such as liver and salt cod are seen as beneficial during and after pregnancy |

statues, rosary beads, and crucifixes are placed in the dying person's room. For believers of *santería*, death rites may include animal sacrifices, chants, and ceremonial gestures.

- **Summon appropriate clergy to perform appropriate death rites. A place should be found for family members to gather during this time, preferably close to the dying person. Do not touch or remove little statues of saints without asking permission.**
- **Candles (electric candles are acceptable) are lighted after death to illuminate the path of the spirit to the afterlife.**
- A *velorio* (wake) lasts 2–3 days and is usually held at a funeral parlor or in the home where friends and relatives gather to support the bereaved family. Funeral rites may include handing out holy cards or *estampitas* to guests with the deceased's date of birth, date of death, and a religious prayer imprinted. Burial in a cemetery is customary practice, although some may choose cremation.
- The deceased are customarily remembered and honored on their birthdays or death anniversaries by lighting candles, offering prayers or masses, bringing flowers to the grave, or gathering with family members at the grave site.

### *11.1.9 Spirituality*

- Approximately, 85% of Cuban Americans are Roman Catholic; the remaining 15% are Protestants, Jews, and believers in African Cuban Santería. Roman Catholicism is personalistic and characterized by devotion and intimate, confiding relationships with the Virgin Mary, Jesus, and the saints. Cuban people are likely to have a shrine in the home for a saint, and patients or family may bring religious objects and herbs for religious healing to the health-care facility; saints reinforce identity as well as spirituality. **Religious beliefs and practices must be viewed in an open, sincere, and nonjudgmental manner. In the in-patient setting, privacy is important if patients and families need to perform certain rituals or prayers. A visit from a priest, rabbi, or *santero* may promote a sense of psychological support and spiritual well-being.**
- Significant religious holidays include *Noche Buena* (Christmas Eve), Christmas, *Los Tres Reyes Magos* (Three Kings Day), and the festivals of the *La Caridad del Cobre* and *Santa Barbara.*
- *Santería* or *Regla de Ocha*, a 300-year-old African Cuban religious system, combines Roman Catholic elements with ancient Yoruba tribal beliefs. Catholic saints often represent African deities in Ocha and Palo. Followers of *Santería* believe in the magical and medicinal properties of flowers, herbs, weeds, twigs, and leaves. Sweet herbs such as chamomile *(manzanilla),* verbena *(verbena)*, and marjoram *(mejorana)* are used to attract good luck, love, money, and prosperity. Bitter herbs such as epazote (*apasote)*, sarsaparilla (*zarzaparilla)*, and prodigiosa (*yerba bruja)* are used to banish evil and negative energies. *Santería*

is viewed as a link to the past and is used to cope with physical and emotional problems.

- **By law, if standards of safety and sanitation are maintained, allow space and privacy to be able to engage in specific religious ceremonies and rituals.**
- Physical complaints may be diagnosed and treated by a medical professional, whereas the *santero* may assist in balancing and neutralizing the various aspects of the illness.
- Belief in a higher power is evident in practices used to maintain health and well-being or cure illness, such as using magical herbs, special prayers or chants, ritual cleansing, and sacrificial offerings. Many tend to be fatalistic, believing that they lack control over circumstances influencing their lives. **Fatalistic beliefs do not mean that the patient is not willing to engage in preventive practices.**

### *11.1.10 Health-Care Practices*

- Cuban cultural treatments for ailments are listed in Table 11.4.
- African Cubans may seek biomedical care for organic diseases but consult a *santero* for spiritual or emotional crises. Conditions such as *decensos* (fainting spells) or *barrenillos* (obsessions) may be treated solely by a *santero* or simultaneously with a physician.
- Many tend to seek help only in response to crisis situations. **Encourage visits to health providers for wellness checks, explaining the long-term benefits.**
- Many Cuban Americans rely on the family as the primary source of health advice. Older women provide traditional home remedies such as herbal teas or mixtures to relieve mild or moderate symptoms or cure common ailments. **Include the entire family in health promotion and health teaching to increase compliance with health prescriptions. Because health often simply means the absence of pain, explain the importance of preventive measures and health promotion activities.**
- Many, especially older, Cuban Americans were socialized into a strong health ideology and successful primary-care system while still in Cuba.
- Many Cuban Americans use traditional medicinal plants in the form of teas, potions, salves, or poultices. In Cuban communities, stores called ***botanicas*** sell herbs, ointments, oils, powders, incense, and religious figurines to relieve maladies, bring luck, drive away evil spirits, or break curses. Santería necklaces and animals used for ritual sacrifices are often available at *botanicas.*
- **Determine which medicinal products are harmful and which will cause no harm.**
- **Incorporate nonharmful practices into the plan of care. Ascertain if patients are using over-the-counter medicine or sharing prescriptions of others. Explain the hazards of overuse of these medications as well as using medicines that were originally prescribed for other family members.** Cultural approaches used to treat common ailments are given in Table 11.4.

**Table 11.4** Cuban cultural treatment for ailments

| Herbal tea | Ailment |
|---|---|
| *Cosimiento de anis* (anise) | Stomachaches, flatulence, baby colic, anxiety |
| *Cosimiento de limón con miel de abeja* (lemon and honey) | Cough and respiratory congestion |
| *Cosimiento de apasote* (pumpkin seed) | Gastrointestinal worms |
| *Cosimiento de canela* (cinnamon) | Cough, respiratory congestion, menstrual cramps |
| *Cosimiento de manzanilla* (chamomile) | Stomachaches |
| *Cosimiento de naranja agria* (sour orange) | Cough and respiratory congestion |
| *Cosimiento de sábila* (aloe vera) | Stomachaches |
| *Cosimiento de tilo* (linden leaves) | Anxiety |
| *Cosimiento de yerba buena* (spearmint leaves) | Stomachaches, anxiety |
| Fruit or vegetable | Ailment |
| *Chayote* (vegetable) | Anxiety |
| *Zanaoría* (carrots) | Visual problems |
| *Toronja y ajo* (grapefruit and garlic) | Elevated blood pressure |
| *Papaya y toronja, y piña* (papaya, grapefruit, and pineapple) | Gastrointestinal parasites |
| *Remolacha* (beets) | Influenza, anemia, cough |
| *Cascara de mandarina* (tangerine peels) | Choleterol |
| Home remedies | Ailment |
| *Agua con sal* (saltwater) | Sore throat |
| *Agua de coco* (coconut water) | Kidney problems and infections |
| *Agua raja* (turpentine) | Sore muscle and joint pain |
| *Bicarbonato*, Limon y agua (baking soda, lemon, and water) | Upset stomach or heartburn |
| *Cebo de carnero* (fat of lamb) | Applied directly on skin for contusions and swelling |
| *Mantequilla* (butter) | Applied directly on burns to soothe pain |
| *Clara de huevos* (egg whites) | Applied directly to scalp to promote hair growth |
| Amulets | |
| *Azabache* is a black stone placed on infants and children as a bracelet or pin to protect them from the "evil eye" or "mal de ojo"<br>*La manito de coral*, symbolic of the hand of god protecting a person, may also be worn as a necklace or bracelet.<br>*Los ojitos de Santa Lucía*, or the eyes of Saint Lucy, may be hung on a bracelet or necklace for prevention of blindness and protection from the evil eye | |

- Language is the greatest barrier to health care. Additional barriers include bureaucratic paperwork and transportation problems, cost of services, and inconvenient hours. **Obtain interpreters when necessary and help them navigate the complex health-care system.**
- Many immigrants may suffer from loneliness, depression, anger, anxiety, insecurity, and health problems. **Determine physical, emotional, and financial**

**resources available for caring for sick patients either at home or in care facilities.**

- Dependency is a culturally acceptable sick role that allows the extended family network to assume the chores and tasks of the sick person. Help is sought in response to crisis situations, pain, and discomfort and is often a signal of a physical disturbance that warrants consultation with a traditional or biomedical healer. **Pain is expressed with verbal complaints, moaning, crying, and groaning that may or may not signify a need for pain medication. Individual assessments for patients with pain are needed to determine patients' perceptions of causes, remedies, and possible coping behaviors. Explain pain medication may promote healing.**
- Blood transfusions and organ donations are usually acceptable.
- **Dispel myths related to organ donation, blood transfusions, and organ transplantation. Assistance from a priest or other clergy may be helpful.**

### *11.1.11 Health-Care Providers*

- Cuban people believe that the mind, body, and spirit are one and are accepting of Western medical care. Both traditional and biomedical models of care are acceptable. **Cuban people may concurrently seek out alternative or spiritual care but may not admit to seeking such care.** Folk remedies may be used at home, but if the condition persists, folk health-care providers, such as ***santeros, practitioners of santeria,*** and biomedical health-care providers may be used either simultaneously or successively. *Santeros* may prescribe treatment or perform rituals to enable ill people to recover by invoking supernatural deities to intervene to help make them well. They may purchase teas, herbs, and other alternative or spiritual remedies from botanicas (folk pharmacy).
- Cubans may frequent small private Cuban health clinics, santeros (Santeria priest), or other faith healers, especially in urban areas. **Ask patients if they are using folk practitioners, their reasons for using them, and the treatments that have been prescribed. Collaborating with folk healers and allopathic providers may increase compliance with health prescriptions.**

## 11.2 Reflective Exercises

1. What are the primary principles and practice of *Santeria*?
2. What is meant by *simpatia*, *personalismo*, and *choteo*? How are they displayed by Cubans?
3. Explain the extensive name format used by Cubans.
4. Name at least three health conditions common among Cubans.
5. Identify at least three folk beliefs about pregnancy common to Cubans.

6. What are some common amulets used by Cubans to teat ailments?
7. A female Cuban new diabetic patient wants to use herbs to control her diabetes. Would you encourage her to try herbs rather than allopathic treatments? Why or why not?
8. What is a *botanica*? Is there one in your community?
9. Identify ways in which to integrate religious practices and preferences into the patient's health-care plan.
10. What is the most important source of emotional and physical support for most Cubans?
11. What is a *quinceañera?*

## Bibliography


About Santeria (2016). http://www.aboutsanteria.com/what-is-santeria.html

Bryant CA (1982) The impact of kin, friend, and neighbor on infant feeding practices: Cuban, Puerto Rican, and Anglo families in Cuba. Soc Sci Med 16(20):1757–1765

Burroughs Peña MS, Patel D, Rodríguez Leyva D, Khan BV, Sperling L (2012) Lifestyle risk factors and cardiovascular disease in Cubans and Cuban Americans. Cardiol Res Pract 2012:470705. https://doi.org/10.1155/2012/470705. Epub 2011 Nov 23. PMID: 22203917; PMCID: PMC3235660

Centers for Disease Control and Prevention (CDC), CDC Vitalsigns-Hispanic Health (2015). https://www.cdc.gov/vitalsigns/hispanic-health/index.html

Central Intelligence Agency (CIA) (2023) The World Factbook: Cuba. https://www.cia.gov/the-world-factbook/countries/cuba/

Countries and Their Culture (n.d.) Cuba. http://www.everyculture.com/Cr-Ga/Cuba.html

Factsheet: Santeria in Cuba (2021). https://www.uscirf.gov/sites/default/files/2021%20Factsheet%20-%20Santeria%20in%20Cuba.pdf

Global Affairs Canada (2023) Cultural information: Cuba. https://www.canada.ca/en/sr/srb.html?q=cultural+information+cuba&wb-srch-sub=

Kowalewska A, Makarevich E (2017) Cuba and US: food customs and health: FSHN17-8/FS303, 12/2017. EDIS 2017(6):1. https://doi.org/10.32473/edis-fs303-2017

Lippincott Advisor for Education (2023) Cultural perspectives: Cubans. https://advisor.lww.com/lna/document.do?bid=26&did=1280684

Migration Policy Institute (2015) Cuban immigrants in the United States. http://www.migrationpolicy.org/article/cuban-immigrants-united-states

Purnell L, Gil J (2013) People of Cuban heritage. In: Purnell L (ed) Transcultural health care: a culturally competent approach, 4th edn. F.A. Davis Company, Philadelphia

# Chapter 12
# People of European American Heritage

**Larry D. Purnell**

## 12.1 Overview and Heritage

The term European American culture in this chapter refers to the middle-class values of citizens of the mainland United States and is the world's oldest constitutional democracy. The European American culture evolved from those of early immigrants, primarily from Northern European countries. Today, the United States includes immigrants or descendants of immigrants from almost every nation and culture of the world.

Americans, and individuals from other groups who have assimilated, value individualism, free speech, freedom of choice, independence, self-reliance, confidence, "doing" rather than "being," egalitarian relationships, non-hierarchal status of individuals, individual responsibility, achievement status over ascribed status, volunteerism, friendliness, openness, futuristic temporality, technology, and the ability to control the environment. Many emphasize material possessions and physical comfort. The extent to which people conform to this dominant culture depends on the variant characteristics of culture discussed in Chap. 1. However, remember aggregate data are true for the group but not necessarily for the individual, so this profile may not be accurate for everyone who self-identifies with the European American culture.

L. D. Purnell (✉)
College of Health and Nursing Sciences, Excelsior University, Albany, NY, USA
e-mail: lpurnell@udel.edu



E. A. Fenkl, L. D. Purnell (eds.), *Handbook for Culturally Competent Care*,
https://doi.org/10.1007/978-3-031-70492-5_12

### 12.1.1 Communications

- The United States does not have an official language. However, the unofficial language in English and accents vary regionally but are usually understandable to all. Having well-developed verbal skills is considered important.
- Voice volume is loud compared with many other cultures.
- The rate of speech varies depending on location within the United States, with people from the South generally speak slower than people in the Northeast.
- Many individuals readily disclose personal information about themselves. Personal sharing is encouraged in a wide variety of topics, but not religion.
- People of the same sex (especially men) or opposite sex do not generally touch each other unless they are relatives or close friends. **The low-touch culture of the** United States **is reinforced by sexual harassment policies.**
- Regardless of class or social standing of the conversant, Americans are expected to maintain direct eye contact without staring. **Just because someone does not maintain direct eye contact does not mean they are not listening or not telling the truth.**
- Men and women extend the right hand when greeting someone for the first time.
- Most Americans are future-oriented but try to balance the past with the present. Punctuality is valued in both business and social settings.
- The usual name format is given name, middle name (optional), and family name. People known to each other use the given name, although it is common to have a nickname.
- **Greet patients using use Mr., Mrs., Miss, Ms., or appropriate title on first meeting someone and until told to do otherwise.**

### 12.1.2 Family Roles and Organization

- A value is placed on egalitarian relationships and decision making, although great variations exist within families. Women may have careers, and men often assist with child care, household chores, and cooking responsibilities.
- Great value is placed on children, and many laws help protect children.
- Autonomy is encouraged in children and teenagers. Children and teenagers are encouraged to have friends of the same and opposite sex.
- Teenagers are expected to refrain from premarital sex, smoking, recreational drug use, and alcohol use until they reach the age of 18 years or until they leave home. Most teenagers move out of their parents' home when their education is completed or when they turn age 18 years.
- Social attitudes toward homosexual activity vary widely and sometimes carry a stigma. In 2015, same-sex marriage became legal in the United States and same-sex relationships have become more acceptable among many.

- **Do not to disclose patients' same-sex relationships to family members or others. Do not assume that married individuals are necessarily in a male–female marriage. Provide an environment that is welcoming to all individuals regardless of their sexual orientation.**
- Terminology for the same sex relationship keep evolving.

Heterosexual, homophobia (fear or dislike someone based on prejudice or negative attitudes toward someone who is perceived to be bi, lesbian, or gay. Intersex: described as biological attributes of both sexes, and LGBTQIA: lesbian, gay, bisexual, transgendered, queer or questioning, intersex, and asexual.

### *12.1.3 Workforce Issues*

- Americans are expected to be punctual on their job, with formal meetings, and with appointments. If one is more than a few minutes late, an apology is expected.
- The American workforce stresses efficiency (time is money), operational procedures on how to get things done, task accomplishment, and proactive problem-solving.
- Intuitive abilities and common sense are not usually valued as much as technical abilities. The scientific method is valued and everything must be proved. Americans want to know *why*, not *what*. Many are obsessed with collecting facts and figures before they make decisions. Pragmatism is valued.
- American values expect that the needs of individuals are subservient to the needs of the organization. Most Americans place a high value on "fairness," and rely heavily on procedures and policies in the decision-making process, a necessity in individualistic cultures in order to reach consensus.
- All professionals are expected to be assertive without being aggressive, a concept that is difficult for some non-Americans. When others speak in their native language at work, it may become a source of contention for both patients and health-care personnel.
- **Explain rules for foreign language use upon employment.**
- Four generations are in the workforce in the United States: (a) Traditionalists, (b) Baby Boomers, (c) Generation X, and (d) the Millennials. Each group brings a different worldview, different perspectives, and varied strengths to the workforce.
- The Traditionalists were born before 1945 and are characterized as being loyal, patriotic, and hard working. Many from this group worked under "control and command" styles of management and may perceive changes in the health-care field as being too radical. They have a wealth of practical expertise to share. Most have also developed conflict resolution and negotiation skills.
- The Baby Boomers, born between 1945 and 1965, currently comprise almost half the workforce. Most are seen as idealistic, optimistic, competitive, and community-focused. They are also considered the "sandwich generation" because they are often caring for their children and their parents. This group also

has a wealth of practical expertise to share and has developed conflict resolution and negotiation skills.

- Generation Xers, born between 1966 and 1980, are characterized as being highly independent and skeptical and have a "free agent" mentality and change jobs easily to meet personal needs. Most are looking for experience rather than job security.
- Millennials, born between 1981 and 1996, like new ideas and change. Technology use is a given, and most are comfortable with diversity.
- Generation Z, born between 1997 and 2012, the first actual digital generation, they are characterized by their assertiveness in demanding flexibility, work-life balance, and advancement in their careers.
- **Capitalize upon the strengths of each generation. Generational descriptions may not be accurate with many newer immigrant and ethnic populations.**

### *12.1.4 Biocultural Ecology*

- The leading causes of death in Americans are heart disease, cancer, chronic obstructive lung disease, unintentional injury, diabetes, and HIV.
- Nearly one million people in the United States are HIV-positive or have AIDS, with the highest concentrations in large urban areas.

Table 12.1 lists commonly occurring diseases and disorders for European Americans in three categories: lifestyle, environment, and genetics.

**Table 12.1** Health conditions common to European Americans

| Health conditions | Causes |
|---|---|
| Appendicitis | Unknown |
| Diverticular disease | Lifestyle, genetic |
| Colon cancer | Genetic, lifestyle, environment |
| Hemorrhoids | Lifestyle, genetic |
| Osteoporosis | Lifestyle, genetic |
| Osteoarthritis | Lifestyle, genetic |
| Cystic fibrosis | Genetic |
| Phenylketonuria | Genetic |
| Skin cancer | Lifestyle, environment |
| Cardiovascular disease | Genetic |
| Varicose veins | Genetic, lifestyle |
| Obesity | Lifestyle, environment |
| Multiple sclerosis | Genetic, environment |

### 12.1.5 High-Risk Health Behaviors

- Cigarette-smoking has been declining in the United States. However, many people now vape rather than smoking cigarettes. **Assess for vaping habits**. Higher socioeconomic groups are more likely to drink but also more likely to drink without problems. **Encourage responsible drinking and smoking cessation beginning in elementary school.**
- High-risk sexual behavior continues with the incidence of STIs and HIV infections. **Encourage health screening and preventive behaviors such as condoms and safer-sex practices starting in elementary school.**
- Many traffic laws have been enacted, although they vary from state to state, requiring automobile seat belts, child restraints, and helmets for bicycle and motorcycle riders.
- **Encourage use of seat belts, child restraints, and helmets. Refer to state laws where they exist.**

### 12.1.6 Nutrition

- The typical diet is high in fats and cholesterol and low in fiber. **Encourage low-fat, low-cholesterol, and high-fiber diets using the U.S. Department of Agriculture food pyramids.**
- Given the varied topography of the United States, there are no food shortages for a healthy diet. However, socioeconomic resources prohibit some from purchasing healthy foods. **Teach patients about healthy foods and preparation practices.**
- Meal times and food choices vary according to ethnic background, the region of the country, urban versus rural residents, and weekdays versus weekends.
- Breakfast is usually consumed between 6:00 a.m. and 9:00 a.m., depending on the person's work schedule. The noontime meal, typically consumed by 1:00 p.m., is called lunch in urban areas and dinner in rural areas. The evening meal, dinner in urban areas or supper in rural areas, is served between 6:00 p.m. and 9:00 p.m. and is usually the largest meal of the day.
- **Determine meal times, and adjust medications accordingly.**

### 12.1.7 Pregnancy and Childbearing Practices

- Commonly used methods of fertility control include natural ovulation methods, birth control pills, foams, Norplant, the morning-after pill, intrauterine devices, condoms, abortion, and sterilization procedures such as vasectomy and tubal ligations. Sterilization in the United States is strictly voluntary. **Abortion remains a controversial issue.**

- Pregnant women are expected to seek preventive care, eat a well-balanced diet, and get adequate rest in order to have a healthy pregnancy and baby. Women are encouraged to breast-feed.
- **Discourage smoking, drinking alcohol, consuming large amounts of caffeine, and taking recreational drugs during pregnancy.**
- Prospective fathers are encouraged to take prenatal classes with expectant mothers and provide a supportive role in the delivery process. **Fathers who do not want to participate should not be made to feel guilty.**
- **Respect cultural beliefs associated with pregnancy and the birthing process.** Possible restrictive beliefs for pregnant women are listed in Table 12.2.
- **Integrate nonharmful cultural practices into preventive teaching and interventions.**

## *12.1.8 Death Rituals*

- Death and responses to death are difficult topics for many Americans to verbalize.
- The dying person should not be left alone. **Make accommodations for a family member to be with the dying person at all times. Health-care personnel are expected to care for the family as much as for the patient during this time.**
- Most people are buried or cremated within 3 days of death; extenuating circumstances may lengthen this period to accommodate family and friends who must travel a long distance to attend a funeral or memorial service. The family decides if the deceased will be cremated or will have an open or closed casket.
- More people are choosing to remain at home and participate in a hospice for end-of-life care to have their comfort needs met. **One requirement for hospice care**

**Table 12.2** Possible restrictive beliefs for pregnant women

| |
|---|
| • Pregnant women should refrain from being around loud noises for prolonged times |
| • Wearing an opal ring during pregnancy can harm the baby |
| • Eating strawberries or being frightened by a snake cause birthmarks |
| • Congenital anomalies can occur if the mother sees or experiences a tragedy during her pregnancy |
| • Nursing mothers should eat a bland diet to avoid upsetting the baby |
| • The infant should wear a band around the abdomen to prevent the umbilicus from protruding and becoming herniated |
| • A coin, key, or other metal object should be put on the umbilicus to flatten it |
| • Cutting a baby's hair before baptism can cause blindness |
| • If the pregnant woman raises her hands over her head while pregnant, it may cause the cord to wrap around the baby's neck |
| • Moving heavy items can cause a pregnant woman's "insides" to fall out |
| • Children born with physical or mental abnormalities are punishment from god for acts of the parents |

**is that the patient must sign documents indicating that he or she does not want extensive life-saving measures performed.**

- Although women may be more expressive than men, most Americans demonstrate their grief conservatively. Generally, tears are shed, but loud wailing and uncontrollable sobbing rarely occur. Bereavement support strategies are listed in Table 12.3.

## *12.1.9 Spirituality*

- Dominant religions in the United States are those of the Judeo-Christian faiths and include Catholicism, Protestantism, and Judaism. However, there are many other recognized religions that are consistent with religious freedom, a reason that many people immigrated to the United States. Although church and state are separate, many public events and ceremonies still open with a prayer.
- In times of illness, many people say prayers. **Ask patients if they need anything to say prayers. Inquire if the person wants to see a member of the clergy.**
- **Do not remove religious symbols or statues as they provide solace to the person. Removing them may increase or cause anxiety.**

## *12.1.10 Health-Care Practices*

- The United States continues to undergo a paradigm shift from curative and restorative medical practices and sophisticated technological care to health promotion and wellness; illness, disease, and injury prevention; and health restoration. Many view good health as a divine gift from God.
- Advance directives are an important part of medical care, which allow patients to specify their wishes concerning life and death decisions before entering an inpatient facility. These directives also designate the name of a family member or

**Table 12.3** Bereavement support strategies

| |
|---|
| • Being physically present |
| • Acknowledging the patient's pain |
| • Encouraging a reality orientation |
| • Openly acknowledging the family's right to grieve |
| • Recognizing varied behavioral responses to grief |
| • Assisting the patient and family to express their feelings |
| • Encouraging interpersonal relationships |
| • Promoting interest in a new life |
| • Making referrals to other resources such as a priest, minister, rabbi, or pastoral care person, with the patient's or family's permission. Allow a family member to remain with the patient |

significant other to speak for the patient and make decisions when or if the patient is unable to do so. Some patients have a living will that outlines their wishes in terms of life-sustaining procedures in the event of a terminal illness or a serious accident. **Each inpatient facility has these forms available and will ask the patient what his or her wishes are. Patients may sign these forms at the in-patient facility or elect to bring their own forms.**

- As an adjunct to biomedical treatments, many people use acupuncture, acupressure, acumassage, herbal therapies, massage therapy, aromatherapy, and other complementary and alternative treatments. An awareness of combined practices when treating or providing health education to individuals and families helps ensure that therapies do not conflict with each other, intensify the treatment regimen, or cause an overdose.
- **Nonjudgmentally inquire about all therapies being used, such as foods, teas, herbal remedies, non-food substances, over-the-counter medications, and medications prescribed or loaned by others.**
- **Often, traditional, folk, and magicoreligious non-harmful practices should be incorporated into the plan of care for patients.**
- **To help prevent contradictory or exacerbated effects of prescription medication and treatment regimens, ask patients about self-medicating practices.**
- Pain is the "fifth vital sign." Patients should be made comfortable and not have to tolerate high levels of pain. **Offer and encourage pain medication, and explain that it will help the healing process.**
- The Americans with Disabilities Act protects individuals with a handicap from discrimination. Rehabilitation and occupational health services focus on returning individuals with handicaps to productive lifestyles in society as soon as possible.
- **Traditional practice calls for fully disclosing health conditions to patients, who then disclose the information to the family.**
- Most Americans favor organ donation and transplantation and blood or blood products transfusions. **Some people refuse blood or blood products for fear of contracting HIV.**
- **Jehovah's Witnesses do not receive blood or blood products. Assist patients in obtaining a religious leader to support them in making decisions regarding organ donation or transplantation if requested.** Health-Care Providers.
- Most people combine biomedical health-care practices with traditional practices, folk healers, and magicoreligious healers.
- Standard practice is to assign staff to patients regardless of gender differences.
- **Recognize and respect differences in gender relationships when providing care. Respect patients' modesty by providing adequate privacy and assigning same-sex caregivers whenever the patient requests.**
- The advanced practice role of registered nurses is gaining respect as their numbers increase and research demonstrates and the public sees them as equal or preferable to physicians. The public holds nurses in high regard.

## 12.2 Reflective Exercises

1. What are the key attributes of the European American individualistic culture?
2. What are the five generations in the American workforce?
3. What is the official language of the United States?
4. A 40-year-old male, John David Armstrong arrives to the medical office. How should he be addressed?
5. What is your acceptance of gays, lesbians, bisexual, and transgendered people?
6. What are the leading causes of death among European Americans?
7. As a health-care provider, what might you do to help people decrease or eliminate smoking and substance misuse?
8. What types of birth control are acceptable to you?
9. How adherent are you to the tenets of your religion?
10. What are the bereavement practices of your family? Are they different from what is described in this chapter?

## Bibliography

Carr D (2012) Death and dying in the contemporary United States: what are the psychological implications of anticipated death? Soc Person Compass 6(2):184–195

CIA World factbook online (2023). https://www.cia.gov/the-world-factbook/

https://biztimes.com/communicating-with-the-five-generations-in-your-workplace/

McCool WF, Simeone SA (2002) Birth in the United States: an overview of trends past and present. Nurs Clin N Am 37(4):735–746

National Center of HIV/AIDS, STD, and TB prevention (2023). https://www.cdc.gov/nchhstp/index.html

Nimke A (2016) Communicating with 5 generations in the workforce

U.S. Department of Agriculture (2023) Guidelines for Americans. MyPlate|U.S. Department of Agriculture

# Chapter 13
# People of Filipino Heritage

**Nelson Tuazon**

## 13.1 Overview and Heritage

Filipinos, predominantly of Malayan ancestry, have been influenced by the neighboring Chinese, Japanese, East Indian, Indonesian, Malaysian, and Islamic cultures and by influences from Spanish and American colonization. The Philippine archipelago consists of 7107 around 7,640 islands. The tropical climate is suitable for year-round agriculture and fishing, but it is affected by the seasonal northeast and southwest monsoons.

Filipino Americans are a diverse group whose regional variations influence the spoken dialect, food preferences, and traditions as well as the other variant characteristics of culture discussed in Chap. 1. Aggregate data in this profile may not be accurate for some more acculturated Filipinos. Most Filipinos who live in the United States reside in California, Hawaii, Illinois, New Jersey, New York, Washington, and Texas. Since the passage of the 1965 Immigration Act, the Philippines have become the largest source of immigrants from Asia. Economic and educational opportunities and reunification with family members in the United States are the primary motivating factors for emigration. Filipinos view educational achievement as a pathway to economic success, status, and prestige for the individual and the family.

N. Tuazon (✉)
University Health, San Antonio, TX, USA
e-mail: Nelson.tuazon@uhtx.com



E. A. Fenkl, L. D. Purnell (eds.), *Handbook for Culturally Competent Care*,
https://doi.org/10.1007/978-3-031-70492-5_13

### 13.1.1 Communications

- Although more than 100 dialects are spoken in the Philippines, most Filipinos speak the national language, Filipino or **Tagalog**. English is used for business and legal transactions, and in school beyond the third grade. Business and social interactions commonly use a hybrid of Tagalog and English (Tag-Lish) in the same sentence.
- Filipinos have different levels of proficiency and may practice *translanguaging* where they speak more than one local dialect and use English simultaneously.
- Nouns are used in place of the generic and gender-neutral pronouns *siya* (singular "he/she") and *sila* (plural "they/them"). Hence, it is customary for Filipinos to use "he" and "she" interchangeably in reference to the same individual. *Sila* is also used as a sign of respect, usually when addressing the elderly or esteemed individuals.
- Communication is highly contextual. The individual is acculturated to attend to the context of the interaction and to adopt appropriate behaviors.
- Most Filipinos are observant, displaying an intuitive feeling about the other person and the contextual environment during interactions. Meanings are embedded in nonverbal communication.
- The emphasis on maintaining smooth interpersonal relationships brings a consequent ambiguity in communication to prevent the risk of offending others.
- **Filipinos may sacrifice clear communication to avoid stressful interpersonal conflicts and confrontations. Saying "no" to a superior may be considered disrespectful, which predisposes an ambiguous positive response.**
- Focusing on action-oriented strategies may be seen as coercive.
- Saving face, a characteristic pattern of behavior employed to protect the integrity of both conversants, is a consequence of the cultural value placed on maintaining smooth interpersonal relations.
- During a teaching session, a Filipino patient's nod may have several meanings, such as comprehension or agreement ("Yes, I hear you"; "Yes, we are interacting"; "Yes, I can see the instructions") or some other message that may be difficult for the patient to disclose. "Yes" does not always mean understanding or agreement.
- **Asking for a return demonstration of a procedure can help ensure understanding. Speak clearly and slowly. Allow time for a response to questions.**
- One may not disagree, talk loudly, or look directly at a person who is older or who occupies a higher position in the social hierarchy. Direct eye contact depends on the extent of acculturation, amount of time in the United States, age, and education. Some individuals may avoid prolonged eye contact with authority figures and older people as a form of respect. Older men may refrain from maintaining eye contact with young women because it may be interpreted as flirtation or a sexual advance. **Just because Filipino patients do not maintain direct eye contact does not mean they are not listening, are not being truthful, or do not care.**

- Most individuals are comfortable with silence and allow the other person to initiate verbal interaction as a sign of respect. Greater distance is observed when interacting with outsiders and people in positions of authority.
- Same-sex closeness and touching are normal behaviors. Young adults of the same sex may hold hands, put one arm over another's shoulder, or walk arm-in-arm.
- Most Filipinos have a relaxed temporal outlook. They have a healthy respect for the past, the ability to enjoy the present, and hope for the future. Past orientation is evident in their respect for elders, strong sense of gratitude, obligation to older generations, and honoring the memories of dead ancestors. Future orientation is manifested in the strong sense of family commitment to provide for the education of the young, parental participation in the care of their children and grandchildren, and a strong work ethic.
- Promptness for social events is determined situationally. "Filipino time" means arriving much later than the scheduled appointment, which can be from one to several hours.
- Children carry the surnames of their father and their mother. For example, Jose Romagos Lopez and Leticia Romagos Lopez are the children of Maria Romagos and Eduardo Lopez. The middle name or initial is the mother's maiden name. After marriage, Jose keeps the same name, whereas his sister's name becomes Leticia Lopez Lukban (her husband being Ernesto Lukban). The middle name is generally abbreviated as an initial. Nicknames symbolize affectionate regard for the person and are commonly used instead of the first name; hence, Nini, Baby, Bongbong.
- **Address Filipino adults by their title or professional affiliation such as Mr., Mrs., Miss, Ms., Dr., or attorney. Ask what name is used for legal purposes.**

### *13.1.2 Family Roles and Organization*

- In contemporary families, although the father is the acknowledged head of the household, authority is considered egalitarian. The mother plays an equal and often major role in decisions regarding health, children, and finances.
- Traditional female roles include caring for the sick and children, maintaining positive relationships with kin, and managing the home.
- Parents and older siblings are involved in the care and discipline of younger children.
- In extended family households, older relatives and grandparents share much authority and responsibility for the care and discipline of younger members. Institutionalization of aged parents is tantamount to abandonment of filial obligation.
- Parents traditionally expect their children to pursue a college education, have economically productive careers, and raise a family.

- Dating at an early age is discouraged for young daughters. Sex education and sex are not usually discussed openly within the family.
- As a country, although the Philippines is considered the most gender-equal in Asia, there remains discrimination against the LGBTQ+ community. Studies show that the negative attitudes toward the LGBTQ+ community may lead to high self-stigma among the members of the LGBTQ+ community. A survey conducted in 2023 shows growing support for gays and lesbians (Westerman, 2023). In the Philippines, a survey shows growing support for gays and lesbians, NPR.
- Current research suggests the complexity of the intersection of the Filipino identity and masculity; the experience of Filipino boys is likely to vary from their peers coming from different racial identities.
- Traditional families desire that their daughters remain chaste before marriage. Pregnancy out of wedlock brings shame to the whole family.
- Conditions such as mental illness, divorce, terminal illness, criminal offenses, unwanted pregnancy, and HIV/AIDS are not readily shared with outsiders until trust is established. In times of illness, the extended family provides support and assistance.
- A family visit to the hospital may take on the semblance of a family reunion.
- **Include nuclear and extended family for support in time of serious illnesses.**
- **Make arrangements for visitations by large numbers of family. Elicit help for visitor control from a respected family member.**
- Homosexuality may be recognized but is often considered an aberrant behavior. Thus, to save face and decrease stigmatization for the family, it may not be openly practiced or acknowledged.

### *13.1.3 Workforce Issues*

- Racism is a continuing theme voiced by Filipino nurses working with Americans. Filipino nurses have been recruited in large numbers where acute shortages of American-trained nurses exist. This has reinforced the cultural tendency toward collective solidarity by defining the context of interactions within the insider-outsider continuum. **Develop culturally specific staff development programs that include Filipinos and Americans.**
- The cultural concept of shared identity with other Filipinos creates a propensity among Filipino nurses to speak in their own dialect with each other to the exclusion of non-Filipino coworkers and patients. **Explain organizational rules about when and where a foreign language can be spoken in the American workforce.**
- Assertive communication is difficult for Filipinos who have been acculturated to avoid conflict. They may consider it impolite and disrespectful to confront or challenge the authority of a superior. When a problem with a manager occurs, a Filipino may communicate through a mediator, usually another Filipino, who is in the same level within the hierarchy as the manager.

- **Explain the cultural value of assertiveness in the American workforce.**
- Filipinos feel comfortable performing what they perceive as caring tasks for patients that American nurses expect patients to do for themselves. Initially they may not be inclined to teach and demonstrate procedures to patients because of their traditional belief in doing the caring tasks for patients. **Explain the American value of self-care with patients and families.**
- The Filipino values of shared perception and being one with others create a cooperative, rather than competitive, outlook. An individual produces for the group and puts the group above his or her own personal gain. The businesslike and competitive perspectives of Americans, in which behavior is internally motivated by individual gain, may be interpreted as selfish and uncaring. Self-proclamations of accomplishments may be seen as offensive. It is more meaningful recognition for a group to acknowledge and celebrate a member's achievement, which is assessed in terms of how the action benefited the group.
- **Bicultural development of Filipino and non-Filipino staff should be the goal of orientation and training. Biculturalism requires awareness of self and others and the ability to adapt behaviors that build positive relationships with others who may be different from oneself. Role model to help Filipinos assert themselves and feel confident in problem-solving and conflict resolution.**
- Filipinos place importance on maintaining self-esteem and dignity by saving face and avoiding shame. A Filipino may say "yes" to avoid hurting other people's feelings.
- Their sensitivity and attention to other people's feelings are often exhibited as indecisiveness, which many Americans interpret as lack of assertiveness. Filipinos are taught not to show open disagreement. Less acculturated Filipinos may not understand the directness of Americans and, thus, may find it insulting. **Explain the expectation of assertiveness by all professionals in the American workforce.**
- The concept of individual accountability and responsibility in a highly litigious society, such as the US, may initially be difficult for Filipino nurses to understand. **Explain the litigious nature of the American culture and the need for defensive charting.**

### *13.1.4 Biocultural Ecology*

- The typical Filipino may come from a multiracial genetic background. The youthful features of Filipinos may make it difficult to assess their age. Common Filipino physical features may include jet black to brunette or light-brown hair, dark- to light-brown pupils with eyes set in almond-shaped eyelids, deep brown to very light tan skin tones, mildly flared nostrils, and slightly low to flat nose bridges.

- The high melanin content of the skin and mucosa may pose problems when assessing signs of jaundice, cyanosis, and pallor. **Assess anemia, pallor, and jaundice in the conjunctiva and palms and soles of the feet.**
- Filipinos range in height from under 5 ft. to the height of average Americans.
- Approximately 40% have blood type B and a low incidence of the Rh-negative factor.
- Table 13.1 lists commonly occurring health conditions for Filipinos in three categories: lifestyle, environment, and genetics (see Chap. 2).
- Newborns may have Mongolian spots: bluish-green discolorations on their buttocks that are physiological and that eventually disappear.
- Liver cancer tends to be diagnosed in the late stages of the disease and appears to be associated with the presence of the hepatitis B virus. Silent carriers of the virus are common among Asians, and its presence is detected only when other problems are being evaluated. **Routinely screen for hepatitis B virus among recent immigrants.**
- Asians require lower doses of central nervous system depressants such as haloperidol, have a lower tolerance for alcohol, and are more sensitive to the adverse effects of alcohol.
- Because of availability of over-the-counter antibiotics and lack of adequate medical monitoring of these drugs in the Philippines, Filipino immigrants may be insensitive to the effects of some anti-infectives.
- A positive reaction to tuberculin or to the Mantoux test is observed because of the practice of giving bacille Guérin-Calmette vaccinations in childhood. **Chest X-rays and sputum cultures are recommended for screening and diagnosis of tuberculosis.**

**Table 13.1** Health conditions common to filipinos

| Health conditions | Causes |
|---|---|
| Thalassemias | Genetic |
| Glucose-6-phosphate dehydrogenase | Genetic |
| Lactose intolerance | Genetic, lifestyle |
| Coronary artery disease | Genetic, lifestyle |
| Diabetes mellitus | Genetic, lifestyle |
| Hypercholesterolemia | Genetic, lifestyle |
| Renal stones | Genetic |
| Hyperuricemia | Lifestyle |
| Gout | Lifestyle |
| Arthritis | Genetic?, lifestyle |
| Thyroid cancer | Lifestyle, environment |
| Breast cancer | Genetic, lifestyle, environment |
| Prostate cancer | Lifestyle, environment |
| Liver cancer | Lifestyle, environment |

### 13.1.5 High-Risk Health Behaviors

- Filipinos who are more acculturated have higher rates of alcohol misuse and increased smoking rates. **Explain the physical and psychological problems of tobacco and alcohol misuse.**
- Less educated Filipinos in the United States have increased rates of HIV/AIDS compared with other Asians/Pacific Islanders. **Take every opportunity to include HIV/AIDS education in elementary school and in health teaching.**
- The more recent immigrants in the United States frequently do not access health services until their illness is advanced because they are not aware of available services and may not have linguistic competence in the English language. **Educate the Filipino community about the availability of services. Provide interpreters as necessary.**

### 13.1.6 Nutrition

- Food and meal patterns emphasize generosity, hospitality, and thoughtfulness, which support group cohesiveness. Spanish, Chinese, and American influences are integrated into contemporary cuisine.
- Foods may be sautéed, fried, or served with a sauce. Rice is a staple food and is usually eaten at every meal, either steamed, fried, or as a dessert.
- *Balut* (egg that has been fertilized) is an iconic Filipino food but has received checkered reputation in the United States and the Western world.
- *Sawsawan* (condiment) is typical of Filipino cuisine and represents culinary significance in terms of esthetics, social relevance, and symbolic context.
- *Kakanin (local rice or root crop delicacies)* is part of the daily food consumption among Filipinos, typically as snacks (in between meals) or dessert (after meals).
- Except for babies and young children, milk is almost absent in the Filipino diet, partly due to lactose intolerance. Milk in desserts such as egg custard and ice cream seems to be tolerated.
- Cold drinks or foods such as orange juice or fresh tomatoes are not served for breakfast in order to prevent stomach upset.
- Dietary calcium is derived from green leafy vegetables and seafood.
- Households may keep potted medicinal plants that are used for common colds, stomach upsets, urinary tract infections, and other minor ailments.
- Daily consumption of garlic to combat hypertension is common. Ginger root is boiled and served as a beverage to relieve sore throats and promote digestion.
- Bitter melon is prepared and eaten as a vegetable and believed to prevent diabetes. Greens such as *malunggay* (moringa) and *ampalaya* leaves are used in stews to regain stamina for someone believed to be anemic or run-down. **Knowledge of indigenous food sources and meal patterns, nutritional content of Filipino**

**foods and American food substitutes, and accessibility of traditional ingredients are important aspects of nutritional assessment and counseling.**

### *13.1.7 Pregnancy and Childbearing Practices*

- The only acceptable methods of contraception are natural family planning methods. Abortion is considered a sin. Recent Filipino immigrants who come from urban areas and who are more educated may be open to alternative methods of birth control.
- Many Filipino women are embarrassed to have males perform vaginal examinations. **Obtain a female health-care provider whenever possible.**
- **The pregnant Filipino woman's network of family and community health advisers, whose opinions she respects, are important support for building trust and rapport in the patient-provider relationship.** Table 13.2 describes common beliefs about pregnancy.
- Some prefer to have their mothers rather than their husbands in the delivery room.
- Some women may prefer the squatting position for birthing.
- Common reasons for not breast-feeding are insufficient milk, mother working, nipple and breast problems, and mother's poor health.

**Table 13.2** Prenatal beliefs about pregnancy

| |
|---|
| • Some women refuse to take vitamins because they are afraid that vitamins could deform the fetus |
| • Some individuals believe that when pregnant women crave certain foods, the craving should be satisfied to avoid harm to the baby. They believe that if the mother craves dark-skinned fruit or dark-colored food, the infant's skin will be dark |
| • Sudden fright or stress may harm the developing fetus |
| • Eating blackberries will make the baby have black spots |
| • Eating black plums will give the baby dark skin |
| • Eating twin bananas will result in twins being born |
| • Eating apples will give the baby red lips |
| • When a woman's stomach is not round or the mother's face is blemished, the baby will be a boy |
| • Going outside during a lunar eclipse is harmful to the baby |
| • Going out in the morning dew is bad for the baby because evil spirits are present |
| • Funerals are avoided because the spirit of the dead person may affect the baby |
| • Wearing necklaces may cause the umbilical cord to wrap around the baby's neck |
| • Sitting by a doorway will make the delivery difficult |
| • Sitting by a window when it is dark may let evil spirits come to the pregnant woman |
| • Sweeping at night may sweep away the good spirits |
| • Knitting might tangle the baby's intestines at birth |
| • Naming the baby before it is born or after a dead person is bad luck |

- Supplementing breast-feeding with other liquids and foods occurs as early as 2 months. **Educate mothers and family members on the importance of breast-feeding.**
- Beliefs about the postpartum period are included in Table 13.3.

### *13.1.8 Death Rituals*

- Family members generally wish to provide intimate caregiving for the patient regardless of the setting. **Make arrangements for the family to participate in caregiving.**
- Illness and death may be attributed to supernatural and magic or religious causes such as punishment from God, angry spirits, or sorcery.
- Religiosity and fatalism contribute to stoicism in the face of pain or distress as a way of accepting one's fate.
- Before the decision is made to inform the patient about his or her terminal condition, a discussion among family members should occur. They may request the doctor to not divulge the truth to protect the patient.
- The ethical principles of beneficence and malfeasance take precedence over patient autonomy.
- Planning for one's death is taboo and may be considered tempting fate. Hence, many traditional Filipinos are averse to discussing advance directives or living wills.
- A priority for the family is to gather around the dying person and during the immediate period after death to pray for the soul of the departed. **When death is imminent, contact a priest if the family is Catholic. Do not remove religious**

**Table 13.3** Postpartum practices and beliefs

| |
|---|
| • Mothers should eat plenty of hot soups (chicken with papaya) to promote milk production |
| • Postpartum mothers should avoid exposure to cold and should use warm water to drink and bathe for a month |
| • Showers are prohibited because they may cause arthritis. However, the woman's mother may give her a sponge bath with aromatic oils and herbs, or a *hilot*, traditional healer, may give an aromatic herbal steam bath followed by full body massage, including the abdominal muscles, to stimulate a physiological reaction that has both physical and psychological benefits |
| • Eating sour or ice-cold foods may cause abdominal cramps for both the baby and mother |
| • The mother and baby should not go out for a month except to visit a doctor |
| • One should give money to charity or to the needy when a baby comes to one's house for the first time |
| • The baby's abdomen is wrapped with a cloth until the umbilical cord falls off to prevent an umbilical hernia |
| • Garlic, salt, or a rosary is placed near the baby's crib to ward off evil spirits |
| • Hanging the baby's placenta in a tree will make the baby a good climber |

**medallions, rosary beads, a scapular, or religious figures from the patient or from the bedside.**

- After death, a wake may last from 3 days to 1 week (to wait for the arrival of kin from other states or countries).
- Women generally show emotions openly by crying, fainting, or wailing. Men are expected to be more stoic and grieve silently.
- Cremation is acceptable to avoid the spread of disease.
- On the first-year anniversary of death, family and friends are reunited in prayer to commemorate this event.
- Many women wear black clothing for months or up to a year after the death. The 1-year anniversary ends the ritual mourning.

### *13.1.9 Spirituality*

- The dominant religious affiliations are Catholicism (the majority), Protestantism, Islam, and Buddhism. Most seek medical care; they believe that part of the efficacy of a cure is in God's hands or by some mystical power.
- *Penitensiya* (repentance) is a Lenten Filipino spiritual ritual that may be viewed as controversial because of the violent nature of this ritualistic act.
- Novenas and prayers are often said on behalf of the sick person. Performance of religious obligations and sacraments and daily prayers are some of the ways health and peaceful death are achieved. **Provide for spiritual needs of patients by making accommodations to their practicing beliefs.**
- Strength comes from an intimate relationship with God, family, friends, neighbors, and nature. The concept of self is formed from the relationship with a divine being.

### *13.1.10 Health-Care Practices*

- Some Filipinos are fatalistic, tending to accept fate easily, especially when they feel they cannot change a situation.
- Decisions about when, where, and from whom to seek help are largely influenced by the intimate circle of family.
- **Include nuclear and extended family in health-care decision making when the patient requests.**
- Many accept and adhere to medical recommendations while using alternative sources of care suggested by trusted friends and family members, which include indigenous medical practices. Often, major decisions are delegated to the physician rather than to the patient or family taking an active collaborative role in decision making.

- **Stress that medications need to be taken as prescribed, medications are ordered specifically for each ailment, unused drugs should be discarded.**
- A high value is placed on personal cleanliness. Keeping oneself clean and free of unpleasant body odors is viewed as good for one's health and face-saving. To be slovenly and disorderly is to be shamelessly irresponsible. Aromatic baths are taken both for pleasure and to restore balance.
- Illness in infancy and childhood may be attributed to the "evil eye." Healing rituals may involve religion (prayers and exorcism), sacrifices to appease the spirits, use of herbs, and massage.
- Balance and moderation are embedded in the hot and cold theory of healing. Change should be introduced gradually. Sudden changes from hot to cold, from activity to inactivity, from fasting to overeating, and so forth, introduce undue bodily stresses that can cause illness.
- Some believe that after strenuous physical activity, a rest should precede a shower; otherwise the person could develop arthritis. Exposure to sudden cold drafts may induce colds, fever, rheumatism, pneumonia, or other respiratory ailments. Some avoid handwashing with cold water after ironing or heavy labor. Exposure to cold such as showers is avoided during menstruation and during the postpartum period.
- Many accept minor ailments stoically and consider them natural imbalances that will run their normal course and disappear.
- Patients often do not complain of pain despite physiological indicators. Many view pain as part of living an honorable life and view this as an opportunity to reach a fuller spiritual life or to atone for past transgressions. **Offer and, in fact, encourage pain relief interventions.**
- Mental illness may be attributed to an external cause such as witchcraft, soul loss, or spirit intrusion.
- Minimal expressions of psychological and emotional discomfort may be observed. The discomfort in discussing negative emotions with outsiders may be manifested by somatic complaints or ritualistic behaviors, such as praying.
- **Explore the underlying meaning of somatization (loss of appetite, inability to sleep) and observe patient interactions with others for valuable information.**
- Many believe that mental illness is hereditary and that it carries a certain stigma.
- Beliefs and practices regarding depression vary from the different indigenous populations. These beliefs are influenced by magico-spiritual factors, relational problems, economic pressure, and emotional factors.
- Family members tend to take care of emotional problems to minimize exposing the problems to outsiders. **Involving a trusted family member or friends, initiating contact with a Filipino mental health worker, especially a Filipino physician, may increase the likelihood of getting the person into a treatment program.**
- The birth of a child with a developmental disability may be viewed as God's gift, an opportunity to become a better person or family, a curse from some unknown

"angry spirit," negligence while pregnant, or a family matter that should be kept private; institutionalization may not be readily accepted.
- Organ donation is not an option, except perhaps in cases in which a close family member is involved. Many Filipinos who follow Catholic traditions believe that keeping the body intact as much as possible until death is preparation for the afterlife. **Provide factual information on organ donation and transplantation.**

### *13.1.11 Health-Care Providers*

- Western medicine is familiar and acceptable to most. Some accept the efficacy of folk medicine and may consult both allopathic and indigenous healers.
- Folk healers are less common in the United States, with the exceptions of the West Coast and Hawaii. When available, they contribute by facilitating cultural rapport between health-care providers and patients.
- The ***hilot*** (traditional healer) is often willing to be included in the counseling session and provide support for the patient's compliance with medical treatment. The *hilot* may provide a special prayer to be incorporated into the medically prescribed treatment plan to increase the patient's sense that all available resources are being used.
- Linguistically and ethnically congruent health-care providers are preferred.
- A health-care provider of the same gender and the same culture may encourage more Filipinos to take advantage of disease prevention services.
- **Include a Filipino primary care provider whenever possible. A bilingual person is helpful to improve communication with older Filipinos.**

## 13.2 Reflective Exercises

1. What is the most common languages spoken by Filipinos? What is *Taglish*?
2. What is meant by "Filipino time?"
3. Describe the name format for Filipinos.
4. Describe Filipino communication practices.
5. What are the most common health conditions of Filipinos?
6. What are acceptable birth control practices for *traditional* Filipinos?
7. Describe at least three prenatal beliefs about pregnancy.
8. What is a *hilot*?
9. What healing rituals are used for a child with the "evil eye"?
10. To what might Filipinos attribute mental illness?

## Bibliography

Alibudbud R (2023) Gender in mental health: toward an LGBTQ+ inclusive and affirming psychiatry and mental healthcare in The Philippines. Front Psych 1–4:1189231. https://doi.org/10.3389/fpsyt.2023.1189231

Aster KGV, Castillo JRB, Cheng SR, Trillana JPL, Varlez YMB, Mercado JMT (2023) Sawsawan: explicating the culinary heritage significance of Philippine condiments using bibliometrics (1972–2018). J Ethnic Foods 10(1):1–15. https://doi.org/10.1186/s42779-023-00186-3

Benard S, Manago B, Acosta Russian A, Cha Y (2023) Mapping the content of Asian stereotypes in the United States: intersections with ethnicity, gender, income, and birthplace. Soc Psychol Q 86(4):432–456. https://doi.org/10.1177/01902725221126188

Cann V (2020) Negotiating identities: Being "boy," being "Filipino," being "other". Boyhood Stud 13(1):23–42. https://doi.org/10.3167/bhs.2020.130103

CIA (2024) The World Factbook: the Philippines. https://www.cia.gov/the-world-factbook/countries/philippines/

De Castro AB, Gee GC, Takeuchi DT (2009) Job-related stress and chronic health conditions among Filipino immigrants. J Immigr Minor Health 10(6):551–558

De Villa BDM, Domingo TMM, Ramirez RJC, Mercado JMT (2022) Explicating the culinary heritage significance of Filipino kakanin using bibliometrics (1934–2018). Int J Gastron Food Sci 28:100522. https://doi.org/10.1016/j.ijgfs.2022.100522

Galanti GA (2000) Filipino attitudes toward pain medication: a lesson in cross-cultural care. J Immigr Minor Health 10(6):551–558

Johnson-Kozlow M, Matt GE, Rock CL, de la Rosa R, Conway TL, Romera RA (2011) Assessment of dietary intakes of Filipino-Americans: implications for food frequency questionnaire design. J Nutr Edu Behav 43(6):505–510

Magat M (2020) Localizing the 'ethnic:' Balut-eating contests and identity performance among Filipino Americans. West Folk 79(1):105–126

Manielleas: 30 Superstitions Filipinos practice during funerals (2024). http://www.manillenials.com/filipino-superstitions-funerals-pamahiin-sa-patay/

Moreno-Lacalle RC, Bangsal MR, Bay-Ya MRT, Erese KAR, Gabol LAP, Geronimo MM, Legiralde SBB, Lomandas MMA, Menzi MMR, Real CDL, Solis JMBF, Ufina KU, Lacanaria MGC (2023) Beliefs and practices on depression among selected Filipino indigenous peoples: a focused ethnography. J Trans Nurs 34(5):321–329. https://doi.org/10.1177/10436596231183183

Munoz CC (2013) People of Filipino heritage. In: Purnell L (ed) Transcultural health care: a culturally competent approach, 4th edn. F.A. Davis Company, Philadelphia

Parba J (2021) Teaching critical vocabulary to Filipino heritage language learners. Educ Sci 11:260

Philippines guide: A look at Filipino language, culture, and etiquette (2016). http://www.commisceo-global.com/country-guides/phillippines-guide

Siojo R (n.d.) Filipino beliefs on pregnancy. https://philippines-events-culture.knoji.com/philippine-beliefs-on-pregnancy/

Wight AC, Victoria M (2022) ' Hayan na ang mga Hampas-dugo! (the penitents are coming!)': Penitensya as religious-dark tourism. Tour Stud 22(1):89–113. https://doi.org/10.1177/14687976211067261

# Chapter 14
# People of German Heritage

Eric A. Fenkl

## 14.1 Overview and Heritage

The Federal Republic of Germany is one of the most densely populated countries in Europe, with a population of over 80 million people. 60 million Americans are of German heritage. The first wave of German immigrants came to the United States for religious freedom and settled along the eastern seaboard. The second wave arrived between 1840 and 1860 and was fleeing political persecution, poverty, and starvation. Many worked as indentured servants. The 1930s and 1940s saw a third wave because of the rise of fascism in Germany.

Germans have a deep respect for education. Credibility, social status, and level of employment are based on educational achievement. Germans receive a stronger education than Americans, with the German undergraduate degree being equal to the American master's degree. Great diversity exists among German Americans. Based on individuality and the variant characteristics of culture as described in Chap. 1, the aggregate data in this profile might not fit some German Americans. Because many German immigrants were discriminated against in the 1940s, and after World War II, many Germans changed their names.

E. A. Fenkl (✉)
Nicole Wertheim College of Nursing and Health Sciences, Florida International University, Miami, FL, USA
e-mail: efenkl@fiu.edu



E. A. Fenkl, L. D. Purnell (eds.), *Handbook for Culturally Competent Care*,
https://doi.org/10.1007/978-3-031-70492-5_14

### 14.1.1 Communications

- German is the official language of Germany; however, many German dialects exist, making it difficult for some to understand each other. German is a low-contextual language, with a greater emphasis on verbal than nonverbal communication. A high degree of social approval is shown to people whose verbal skill in expressing ideas and feelings is precise, explicit, and straightforward.
- Feelings are considered private and are often difficult to share. Sharing one's feelings with others often creates a sense of vulnerability or is looked on as a weakness.
- Expressing fear, concern, happiness, or sorrow allows others a view of the personal and private self, creating a sense of discomfort and uneasiness. Therefore, philosophical discussions, hopes, and dreams are only shared with family members and close friends.
- Emotions are experienced intensely but are not always expressed. "Being in control" includes harnessing one's emotions and not revealing them to others. Newer generations are more demonstrative in sharing their thoughts, ideas, and feelings.
- **Do not assume that just because a person may not express or display emotions that the person does not care about his or her health or the situation.**
- In traditional families where the father plays a dominant role, little touching occurs between the father and children. This relationship may become more demonstrative as parents and children age. Affection between a mother and her children is more evident. In other families, there is an outward expression of love from parents, grandparents, and extended family members; hugs and kisses are a "reaffirmation of love."
- Whereas close friends are often extended warmth through handshakes, brief embraces, and sometimes kisses, strangers are kept at arm's length and greeted formally. Germans generally are careful not to touch people who are not family or close friends.
- Most individuals place a high value on privacy.
- Doors are used to protect privacy. A closed door requires a knock and an invitation to enter, regardless of whether the door is encountered in the home, business, or in-patient facility. Even looking into a room from the outside is considered a visual intrusion.
- **Do not enter a patient's room without knocking on the door, announcing yourself, and requesting permission to enter.**
- Eye contact is maintained during conversations. Staring at strangers is considered rude.
- To focus on the present is to ensure the future. The past, however, is equally important, and Germans often begin their discussions with background information.
- Most pride themselves on their punctuality. There are rarely good excuses for tardiness or delays that disturbs the "schedule" of events. **If the health-care**

**provider is late with an appointment or treatment, provide a thorough explanation.**

- Traditionally, Germans keep social relations on a formal basis. Those in authority, older people, and subordinates are always addressed formally. Only family members and close friends address each other by their first names. Younger generations or the more acculturated may be less formal in their interactions. Healthcare providers will find that many patients vary widely in their observance of these rules of etiquette.
- **Greet patients formally unless told to do otherwise. Ask patients how they would like to be addressed.**

### *14.1.2 Family Roles and Organization*

- Traditional families view the father as head of the household. In the United States, the husband and wife are more likely to make decisions mutually and share household duties.
- Older people are sought for their advice and counsel, although the advice may not always be followed. They are admired for maintaining their independence and their continued contributions to society. Many live alone or with aging spouses. Helping parents or grandparents to remain in their own home is important. For those who grow dependent, moving in with children or in a retirement nursing home is a viable choice.
- **Include the extended family in decision-making. Consider using family members to assist with personal care and treatments when the need arises.**
- Prescriptive behaviors for children include using good table manners, being polite, doing what they are told, respecting their elders, sharing, paying attention in school, and doing their chores.
- Prescriptive behaviors for adolescents include staying away from bad influences, obeying the rules of the home, sitting like a lady, and wearing a robe over pajamas.
- Restrictive and taboo behaviors for children include talking back to adults, talking to strangers, touching another person's possessions, and getting into trouble. Restrictive and taboo behaviors for adolescents include smoking, using drugs, chewing gum in public, having guests when parents are not at home, and having run-ins with the law.
- Family and lifelong friendships are highly valued. Concern for one's reputation is a strong value. One's family reputation is considered part of a person's identity and serves to preserve one's social position.
- Pregnancy outside marriage results in overt or subtle disapproval. Because families are concerned about their reputations in the community, the presence of an unwed mother taints their reputation and may result in the family being ostracized by others. If marriage follows the pregnancy, less sanctioning occurs, but

the fact that pregnancy existed before marriage creates a stigma for the woman, and sometimes for the child, that may last for the rest of their lives.
- Although same-sex marriage is not legal in Germany, registered partnerships for same-sex couples have been legal since 2001, which provides most of the same rights as opposite-sex married couples receive. However, many older gays and lesbians may fear exposure because of the extreme discrimination homosexuals experienced in Nazi Germany. **When encountering gays and lesbians who need support, a referral to one of the gay and lesbian religious or community groups may be helpful. Do not disclose same-sex relationships to others.**

### *14.1.3 Workforce Issues*

- For Germans being on time is important. Business communication should remain formal: shaking hands daily, using the person's title with the last name, and keeping niceties to a minimum.
- For the German manager, credentials and education confirm their credibility and lead to power. Germans, who have no translation for "fair play," are assertive by putting other people in their place. Nuances and jargon can frustrate the individual who does not understand idioms and colloquial expressions.
- **Help other employees understand the formality of German employees. Explain the culture of the American workforce and legalities of "fairness."**

### *14.1.4 Biocultural Ecology*

- Germans range from tall, blond, and blue-eyed to short, stocky, dark-haired, and brown-eyed. Because many Germans have fair complexions, skin color changes and disease manifestations can easily be observed. For those with fair skin, prolonged exposure to the sun increases the risk for skin cancer. **Counsel patients on exposure to sun and recommend protection.**
- Table 14.1 lists commonly occurring health conditions for German Americans in three categories: lifestyle, environment, and genetics.

### *14.1.5 High-Risk Health Behaviors*

- Smoking and excessive alcohol consumption remain high-risk behaviors for many Germans. **Encourage smoking cessation and limiting alcohol consumption.**
- Most individuals enjoy the outdoors, fresh air, and exercise. Sports are played for exercise and the pleasure of participating in group activities. Encourage water

**Table 14.1** Health conditions common to German

| Health conditions | Causes |
|---|---|
| Cardiovascular disease | Genetic, lifestyle, environment |
| Stomach cancer | Genetic, lifestyle, environment |
| Myotonic muscular dystrophy | Genetic |
| Hereditary hemochromatosis | Genetic |
| Sarcoidosis | Genetic |
| Dupuytren's disease | Genetic |
| Peyronie's disease | Genetic |
| Cystic fibrosis | Genetic |
| Hemophilia | Genetic |
| Cholelithiasis | Genetic |

sports and other activities among older people, people with disabilities, mothers, and small children. **Encourage health club memberships for exercise.**

### *14.1.6 Nutrition*

- Food is a symbol of celebration for Germans and is often equated with love. Children are rewarded for good behavior with food. Their infatuation with food can lead to overeating, resulting in obesity.
- Real cream and butter are used. Gravies, sauces, fried foods, rich pastries, and sausages are only a few of the culinary favorites that are high in fat content; meats are stewed, roasted, and marinated and are often served with gravies. Vegetables (fresh is preferred) are often served in a butter sauce. Foods are also fried in butter, bacon fat, lard, or margarine. Traditional food preparation methods use high-fat ingredients that add to nutritional risks. Garlic and onions may be eaten daily to prevent heart disease.
- One-pot meals such as string beans and potatoes, cabbage and potatoes, pork and sauerkraut, stews, and soups are served as family meals. Casseroles are also popular. Foods prepared with vinegar and sugar as flavorings are also favorites. Potato salad, cucumber salad, coleslaw, pickled eggs, pickled cucumbers, cauliflower, tongue, and herring are common examples of favored foods. **Encourage reducing portion size, overcoming harmful food rituals, and reducing fat intake.**
- Those who are ill receive egg custards, ginger ale, or tomato soup (without cream) to settle their stomach. Prune juice is given to relieve constipation. Soup from fresh tomato juice is used to treat a migraine headache.
- Ginger ale or 7-Up relieves indigestion and settles an upset stomach. After gastrointestinal illnesses, a recuperative diet is administered to the sick family member, beginning with sips of ginger ale over ice. If this is retained, hot tea and toast are offered. The last step is coddled eggs, a variation of scrambled eggs prepared

with margarine and a little milk. If these foods are tolerated, the sick person returns to the normal diet.

### 14.1.7 Pregnancy and Childbearing Practices

- All usual methods of contraception (*Verhütungsmittel*) are easily available in Germany. Heterologous artificial insemination, use of contraceptive pills, and unnatural contraception are forbidden among strict Catholic Germans. Therapeutic or direct abortion is forbidden as the unjust taking of innocent life. **Assist patients in identifying acceptable fertility practices within their religious practices.**
- Prescriptive practices during pregnancy include getting plenty of exercise and increasing the quantity of food to provide for the fetus. Restrictive practices during pregnancy include not stretching and not raising the arms above the head to minimize the risk of the cord wrapping around the baby's neck.
- A child born with the membrane (the amniotic sac, also knows as a "veil") over its head is believed to be a special child, a belief shared by many cultures.
- Prescriptive practices for the postpartum period include getting plenty of exercise and fresh air for the baby. If the mother is breast-feeding, she should eat foods that enhance the production of breast milk, such as malt beverages. **Assist mothers in identifying nonharmful foods and beverages that improve milk production.**

### 14.1.8 Death Rituals

- Death is a transition to life with God. Because illness is sometimes perceived as a punishment, the duration and intensity of the dying process may be seen as a result of the quality of the life led by the person. Death is considered part of the life cycle, a natural conclusion to life. Individuals who embrace a set of religious beliefs may look forward to a life after death, often a better life.
- Careful selection of the clothes to be worn by the deceased and the flowers that represent the immediate family is important. Traditionally a 3-day period of mourning is reserved after the death of a family member.
- The body of the deceased is prepared and "laid out" in the home, where support from family and friends is readily available. Neighbors come to do the chores and to sit with the family of the deceased until the burial. A short service is held in the home before the body is taken to the church, where family and friends can attend a funeral service. After the church services, the body is taken to the cemetery for burial. After a short graveside service, the minister invites everyone at the graveside service to go to the home of the deceased for food.

- Wearing black or dark clothing when attending a viewing or a funeral may be expected of both family and friends. Crying in public is permissible among some families, but in others the display of grief is private. **Accept varied expressions of grief.**

### *14.1.9 Spirituality*

- Major religions include Roman Catholicism, Methodism, and Lutheranism. Other religions, such as Judaism, Islam, and Buddhism, have substantial membership. **Recognize that individuals' decisions may vary from the formal position of their religious groups.**
- Prayer is used to ask for healing, for effectiveness of treatments, for strength to deal with the symptoms of the illness, and for acceptance of the outcome of the course of the illness. Prayers are often recited at the bedside with all who are present joining hands, bowing their heads, and receiving the blessing from the clergy.
- Most families have a family bible, which is passed down through the generations. It serves as spiritual comfort and as a reservoir of family historical data such as the dates of births, marriages, and deaths. Individual sources of strength are their beliefs in God and in nature, although they may not attend church on a regular basis.
- Family loyalty, duty, and honor to the family are strong values. **Assist patients with obtaining clergy as needed.**

### *14.1.10 Health-Care Practices*

- In traditional families, the mother usually ensures that children receive check-ups, get immunizations, and take vitamins. Women in the family often administer folk/home remedies and treatments. German Americans use a variety of over-the-counter drugs, believing that individuals are responsible for their own health.
- Common, natural folk medicines include roots, herbs, soups, poultices, and medicinal agents such as camphor, peppermint, and spirits of ammonia. Folk medicine includes "powwowing," use of special words, and wearing charms.
- **Ascertain if over-the-counter and folk remedies are being used so as to determine if there are contraindications with prescription medications.**
- Even when experiencing pain, many individuals continue to carry out their family and work roles. Many value being stoic when experiencing pain. **Health-care providers may not be able to identify verbal or nonverbal clues about pain among Germans. Careful interviewing and astute observation must be used**

**to accurately assess the level of pain they experience. Explain that pain medication will help the healing process.**

- Mental illness may be viewed as a flaw, resulting in this group being slow to seek help because of the lack of acceptance and the stigma attached to needing help. Physical disabilities caused by injury are more acceptable than those caused by genetic problems.
- People's discomfort with expressing personal feelings to strangers may impede the counseling process and influence the counseling methods used. Most need to discuss the past without expressing their feelings. **Recognize that patients may not readily discuss feelings, especially during the initial counseling session.**
- **Rehabilitation is consistent with Germans' health beliefs because it helps return them to productive lives.**
- Blood transfusions, organ donation, and organ transplants are acceptable medical interventions.

### *14.1.11 Health-Care Providers*

- Health-care providers hold a relatively high status among Germans. This admiration stems from the love of education and respect for authority.
- Most individuals accept care from either gender. Some younger and older, more traditional women prefer intimate care from a same-sex health-care provider.

## 14.2 Reflective Exercises

1. Distinguish German communication practices between family and outsiders.
2. What protocol does Germans expect with a "closed door"?
3. Describe taboo behaviors for German adolescents.
4. Name and describe at least three genetic diseases common among Germans.
5. What are the primary foods a German might use while recuperating from and illness?
6. Describe some prescriptive and restrictive German postpartum practices.
7. What might you do to encourage a stoic German to accept pain medication?
8. How is mental illness perceived by Germans?
9. Identify at least three common folk remedies commonly used by Germans.
10. Describe a traditional German funeral.

## Bibliography

CIA (2024) The World factbook: Germany. https://www.cia.gov/the-world-factbook/countries/germany/

FKulig JC, Fan H (2013) Understanding death and dying among the low-German-speaking Mennonites. https://www.researchgate.net/publication/271954003_Understanding_Death_and_Dying_Among_the_Low-German-Speaking_Mennonites

Jain A, Al Khalili Y (2023) Congenital myotonic dystrophy. In: StatPearls [internet]. StatPearls Publishing, Treasure Island, FL

Lehmann-Willenbrock N, Allen JA, Meinecke AL (2013) Observing culture: differences in U.S.-American and German team meeting behaviors. Psychology faculty. http://digitalcommons.unomaha.edu/cgi/viewcontent.cgi?article=1102&context=psychfacpub

Steckler J (2013) People of German heritage. In: Purnell L (ed) Transcultural health care: a culturally competent approach, 4th edn. F.A. Davis Company, Philadelphia

# Chapter 15
# People of Greek Heritage

Irena Papadopoulos

## 15.1 Overview and Heritage

This chapter presents two groups of people with Greek heritage. The first group refers to those or their ancestors who emigrated from Greece and the second group originated in Cyprus. Both groups have a common history, language, and religion. The Greek and Greek Cypriot diaspora is of considerable size and has spread to all continents and numerous countries. The largest Greek community outside Greece is in the United States, and the largest Greek Cypriot community outside Greece is in Britain.

Greece, a small country in southern Europe, enjoys a climate similar to that of southern California. The characteristics of members of the Greek and Greek Cypriot communities vary according to acculturation, individuality, and the variant characteristics of culture (see Chap. 1).

The core values of philotimo (honor and respect) and endropi (shame) are key when considering the experience of Greeks and Greek Cypriots. Although values of honor and shame are found in all societies, they are particularly important among Mediterranean groups. Philotimo is a Greek's sense of honor and worth derived from one's self-image, one's reflected image (respect), and one's sense of pride. Philotimo is enhanced through courage, strength, fulfilling family obligations, competition with other people, hospitality, and appropriate behavior. Shame results from any conduct that is considered deviant.

I. Papadopoulos (✉)
Middlesex University, London, UK
e-mail: R.Papadopoulos@mdx.ac.uk



E. A. Fenkl, L. D. Purnell (eds.), *Handbook for Culturally Competent Care*,
https://doi.org/10.1007/978-3-031-70492-5_15

### 15.1.1 Communications

All Greeks, whether in Greece, Cyprus, or the diaspora, use the same form of written language, although there are regional and country variations in spoken Greek. Greek communities regard the retention of the Greek language as an essential part of their identity; thus, much effort is expended encouraging second and subsequent generations to speak Greek. Many first-generation Greek Cypriots speak very little or no English.

Obtain an interpreter when necessary; do not rely on family members who may not fully disclose information because of honor and shame.

Greek and Greek Cypriots tend to be expressive in both speech and gesturing. They use their hands frequently while talking. They embrace family, friends, and others to indicate solidarity. Whereas innermost feelings, such as anxiety or depression, are often shielded from outsiders, anger is expressed freely, sometimes to the discomfort of those from less expressive groups. Eye contact is generally direct and speaking and sitting distance is closer than that of other European Americans. Do not take offense if patients stand closer to you than you are accustomed. Do not assume that prolonged eye contact is a sign of anger.

Patients often appear to be compliant in the presence of the healthcare worker, but this may be only superficial to ensure a smooth relationship. Stress the importance of letting the healthcare provider know the patient's intentions on complying with health prescriptions.

Greeks are oriented to the past as they are highly conscious of the glories of ancient Greece. They are present-oriented regarding philotimia, family life, and situations involving family members. However, they tend to be future-oriented regarding educational and occupational achievements. Greek Americans differentiate between "Greek time," which is used in family and social situations, and "American/clock time," which is used in business situations. Greek time emphasizes participating in activities until they reach a natural breaking point, whereas American/clock time emphasizes punctuality. Reinforce the necessity of punctuality if the situation requires it.

For Greeks having a Greek name is an important sign of their heritage. Honorific titles might be given to members of the community who are elders or otherwise respected. Terms such as Thia (aunt), Kyria (Mrs.), or giagia (grandma) may be used. First names come from either the Bible or from Greek mythology and history. Ideally, first daughters are named for the mother's mother and first sons for the father's father. Following tradition, middle names are the first name of the father.

In healthcare situations, it is not appropriate to call an older woman or man by the first name. The prefix Kyria or Kyrie should be used with the first name, for example, Kyria Maria or Kyrie Alexandre. Alternatively, the preferred mode of address is using the surname preceded by Mr., Mrs., Miss, or Ms.

### 15.1.2 Family Roles and Organization

Traditionally, the father was considered the head of the household in Greek and Greek Cypriot families. The complexity of household dynamics is noted in the well-known folk saying, "The man is the head but the wife is the neck that decides which way the head will turn." This acknowledges the primacy of fathers in the public sphere and the strong influence of women in the private sphere. In recent years, there has been increased recognition of a trend toward more equality in decision making.

The core values of philotimo and endropi set the pattern for the family and for the enactment of gender roles. The traditional roles of husband and wife are characterized by mutual respect (a partnership). However, their relationship is less significant than that of the family as a unit. Although in the past fathers were responsible for providing for the family, in modern times mothers are also providers as well as having the bigger share of responsibility for the management of the home and children. Protection of family members and maintenance of family solidarity tends to be foremost among their values. Consequently, they are often friendly but somewhat superficial and distant with those considered "outsiders."

Traditionally the cleanliness and order of the home reflect the moral character of the woman.

Children are included in most family social activities and tend not to be left with babysitters. The child is the recipient of intense affection, helpful interventions, and strong admiration. The child may be disciplined through teasing, which is thought to "toughen" children and make them highly conscious of public opinion. Accept alternative forms of childrearing as long as they are not psychologically or physically abusive.

Adolescents, particularly young women, tend to reside with their parents until they get married. Unlike previous generations, spouse selection is left to the adolescent, with parental approval. Girls have considerably less freedom in dating than their brothers and it is common for girls to be prohibited from dating until they are in the upper grades of high school. Adolescents in more traditional families may experience stress as the differences in family and peer values precipitate family conflict.

High stress areas include extreme dependence on the family, intense pressure for school achievement, and a lack of sexual education in the home although the young generation has access to a mobile phone, a tablet, and a computer which they use to educate themselves. School nurses and counselors are in a good position to provide sex education and to discuss issues related to personal freedom. Educate teachers on early warning signs of suicide ideology.

Prestige is connected to the idea that honor is collective and individual. Because a person loses honor if a family member acts improperly, the honor of each family member is a matter of concern for all family members. Older people hold positions of respect. Include the extended nuclear family in healthcare decision making.

Although in the past families were obliged to care for their parents in old age, this rule is now more flexible for several reasons. Many of the older people (70–80 years) are likely to have arrived in the United States as children or young adults and were able to become fluent in English, be educated to a high level, and find employment outside the Greek community. All these factors enabled them to be take care of themselves through finding and using relevant information or seeking help from the systems operating in the area where they live. However, families continue to provide support that will ensure the comfort and dignity of their parents. Failure to do so results in a sense of dishonor and guilt for the family. Treatment of the yiayia (grandmother) and the pappou (grandfather) reflects the themes of closeness and respect emphasized in the family. Grandparents tend to participate fully in family activities. Include the family when planning to or providing services to older persons.

If an older person is ill, living with the family is the first preference, followed by placement in a residential care facility. Although living alone is often the least preferred residential pattern, many older people are choosing to live alone with the support of family, friends, and healthcare providers. Help the family obtain needed support services for providing care to family members in the home setting.

The basis of social status and prestige is family philotimo and cohesiveness. However, social status is also received from attributes such as wealth, educational achievement, and achievements of its members. Honor is the social worth of the family as judged by the community.

Greeks tend to be relatively conservative. Consequently, alternative lifestyles encompassing premarital sex, divorce, and same-sex relationships are considered sources of concern for family members and the community. Do not disclose same-sex relationships and matters of personal privacy to family members or others.

### *15.1.3 Workforce Issues*

In the United States, the high achievement orientation and work ethic have resulted in Greeks serving as a "model" ethnic group. Greeks stress on self-reliance is sometimes seen as reluctance to be told what to do. Do not misinterpret self-reliance with not taking direction from superiors.

Eye contact is generally direct and speaking, and sitting distance is closer than that of other European Americans. Do not take offense if Greek employees stand closer to you than to which you are accustomed. Do not assume that prolonged eye contact is a sign of anger.

### 15.1.4 Biocultural Ecology

Greeks are most commonly of medium stature, shorter than northern Europeans but taller than other populations of southern Europe. Although some Greeks have blue eyes and blond hair, usually from the northern provinces of Greece, most Greeks have brown eyes, dark hair, and dark skin.

Assess recent immigrants for health conditions such as typhoid and hepatitis A and B.

Screen for glucose-6-phosphate dehydrogenase (G-6-PD) deficiency and do not administer oxidant drugs such as primaquine, quinidine, thiazolsulfone, dapsone, furazolidone, nitrofuran, naphthalene, toluidine blue, phenylhydrazine, chloramphenicol, and aspirin. Make patients aware that broad beans (fava beans) can induce hemolysis and an acute anemic crisis when ingested. Consider G-6-PD deficiency in Greek patients with unconjugated jaundice.

Most Greek and Greek Cypriot women choose to have an abortion if they are found to carry a fetus with thalassemia.

Screen Greek patients for thalassemia, and advise them of genetic risks.

Table 15.1 lists commonly occurring health conditions for Greek and Greek Cypriots in three categories: lifestyle, environment, and genetic.

**Table 15.1** Health conditions common to Greeks and Greek Cypriots

| Health Condition | Causes |
|---|---|
| Cardiovascular diseases | Genetic, lifestyle, environment<br>Major risk factors include smoking, diet, obesity, and lack of physical activity |
| Cancer | Genetic, lifestyle, environment<br>Cancer is one of the highest leading causes of death. Common cancers include colorectal, breast, prostate, and lung cancers |
| Mental health disorders | Genetic, lifestyle, environment<br>Mental health issues like depression, anxiety, and substance abuse are prevalent in the Greek population |
| Diabetes | Genetic, lifestyle<br>The diabetes rates have rapidly increased along with rising obesity. Genetics and changing diets are possible factors |
| Neurodegenerative diseases | Genetic, lifestyle, environment<br>Alzheimer's disease and Parkinson's disease have rates comparable to other ethnicities. As life expectancy increases, so will age-associated neurological conditions |
| Glucose-6-phosphate dehydrogenase (G6PD | Genetic<br>Deficiency in G6PD is more common in ethnic populations from regions like Africa, the Mediterranean, and Southeast Asia |
| Thalassemia | Genetic<br>Thalassemia is a blood disorder which affects Greek people. However, government initiatives in Greece and Cyprus have significantly reduced the number of people affected by it |

### 15.1.5 High-Risk Health Behaviors

- Alcohol is most often considered a food item and is consumed with meals. However, losing control by being "under the influence" engenders considerable gossip and social disgrace, focused not only on the individual but also on the family.
- Obesity among both sexes and smoking among men are high among Greeks. Although a high level of risk-taking appears to be part of survival, it is in fact perpetuated by the belief that "God will look after me," meaning that God will prevent anything untoward happening and that if anything should happen, God will heal and sustain the person.
- Network in elementary schools to educate about the hazards of smoking.
- Many Greeks tend to disregard standard health promotion behaviors. Safety measures such as seat belts and helmets are often viewed as infringements on personal freedom and are frequently ignored, particularly by the older generation. Explain the legal requirements for seat belt and helmet laws, and encourage their use.

### 15.1.6 Nutrition

- Greeks describe their culture as an "eating culture." Food is a centerpiece of everyday life as well as of social and ritual events.
- Fasting is an integral part of the Greek Orthodox religion. During fasts, it is forbidden to eat meat, fish, and animal products such as eggs, cheese, and milk. General fast days are Wednesdays and Fridays. At least 3 days of fasting must be observed by anyone wishing to take Holy Communion. Some first-generation Greeks observe the four major fasting periods which include the Great Fast, Lent, for 7 weeks before Easter; the Fast of the Apostles, from Monday 8 days after Pentecost to June 28; the eve of the Feast of the Saints Peter and Paul; the Assumption fast, from August 1 to August 14; and the Christmas fast, from November 15 to December 14.
- Accommodate for patients to fast if it is not contradictory to their health status. People with health conditions and small children are exempt from fasting.
- Traditionally, Greeks based their diet on cereals, pulses, lentils, vegetables, fruits, olive oil, cheese, and milk. They are also relatively high consumers of sweets and snacks. For adults, dairy products are consumed in the form of yogurt or cheeses such as feta, kopanisti, kefaloteri, kasseri, and halloumi.
- Acknowledge the presence of lactose intolerance and help patients choose alternative sources of calcium.
- Fats are consumed in the form of olive oil, butter, and olives. Meats include chicken and lamb or, in the United States and Britain, beef adaptations. Eggs,

lentils, and fish such as shrimp and other shellfish, whitefish, and anchovies are additional sources of protein.
- Vegetables such as potatoes, eggplant, zucchini, spinach, garlic, onions, peas, artichokes, cucumbers, asparagus, cabbage, and cauliflower are common.
- Bread choices include pita, crescent rolls, and egg breads. Other foods include rice, tabouli, macaroni, and cracked wheat (bourgouri). Fruit preferences include grapes and currants, figs, prunes, oranges, lemons, melons, watermelons, peaches, and apricots.
- Common seasonings used by Greeks are anise, basil, cumin, cinnamon, citron, cloves, coriander, dill, fennel, ginger, garlic, lemon, marjoram, mint, mustard, nutmeg, oregano, parsley, rose, sage, sesame, thyme, vinegar, bay leaf, and honey.
- Beverages such as coffee, tea, chocolate milk, and wine are common choices.
- Food choices and preparation practices should be assessed on intake.

### 15.1.7 Pregnancy and Childbearing Practices

- In North America, Greeks have deliberately limited family size so children can be adequately cared for and educated. A wide variety of birth control measures, such as intrauterine devices, birth control pills, and condoms are preferred. The strong pro-life Greek Orthodox Church condemns birth control while silently accepting the reality. Help patients identify acceptable birth control methods.
- Abortion is condemned as an act of murder except in circumstances that threaten the life of the mother or when a young woman becomes pregnant because of rape. In practice, a few women, particularly those who are unmarried, have legal abortions because of the negative consequences of having a baby out of wedlock. Inflicting endropi on the family is believed to be more severe than the consequences of abortion. A fetus with thalassemia may also be aborted. Do not disclose abortions to family or other members of the community.
- Pregnancy is a time of great respect for women and a time when women are given special considerations. Proscriptions include not attending funerals or viewing a corpse, refraining from sinful activity as a precaution against infant deformity, and praying to Saint Simeon.
- Pregnant women are encouraged to eat large quantities; foods high in iron and protein are particularly important. If a pregnant woman remarks that a food smells good or if she has a craving for a particular food, it should be offered to her; otherwise, the child may be "marked." This is the usual explanation for birthmarks.
- Factually explain the potentially harmful effects on the fetus and mother of excessive weight gain during pregnancy.
- After delivery, most traditional Greeks consider the mother ritually impure and particularly susceptible to illness for 40 days. During this time, she is admonished to stay at home and not attend church. At the end of the 40 days, the mother

and child attend church and receive a ritual blessing. Arrange for a visiting nurse if the mother is reluctant to see healthcare providers in their offices.
- For breastfeeding mothers, early showering is sometimes thought to result in the infant developing diarrhea and becoming allergic to milk. Newborns are generally breastfed and solids are not introduced early.
- When relatives visit an infant in the hospital, silver objects or coins may be placed in the crib for good luck. Do not remove good luck charms from the infant's crib.

### *15.1.8 Death Rituals*

- Anointing of the Sick is administered in the sacrament of Holy Communion given by a priest or occasionally by a deacon. Assist family members to contact a priest when needed.
- A klama (wake) is held in the family home or, more commonly in North America today, in a funeral home. All relatives and friends are expected to attend for at least a brief time. The wake ends when the priest arrives and offers prayers. Pictures and mirrors may be turned over. During the wake, women may sing dirges or chant. In some regions, people practice "screaming the dead," the miroloyi.
- In Greece and Cyprus, the kidia (funeral) is held the following day at the Orthodox Church, with internment in a cemetery. After internment, family and friends gather for a meal of fish, (symbolizing Christianity), wine, cheese, and olives at the family home or a restaurant.
- Based on the Orthodox belief in the physical resurrection of the body, Greeks reject cremation. The extent of adherence to this precept varies in North America, but Greeks in Britain do not practice cremation.
- Black is the color of mourning dress and is often worn by family members throughout the 40 days of mourning; for widows, it may be worn longer. Formerly, a widow was expected to wear black and no jewelry or makeup for the rest of her life. While black armbands are still worn in Greece and Cyprus, that custom is virtually non-existent in immigrant communities.
- After death, family and close relatives, who may stay at home, mourn for 40 days. Close male relatives do not shave as a mark of respect. A memorial service follows 40 days after burial and at 3 months, 6 months, and yearly thereafter. At the end of this service, koliva (boiled wheat, with almonds, pomegranate, sesame seeds and powdered sugar) is served to participants, and mourning is conducted with joyful reverence.
- Accept varied expressions of grief and rituals related to death and dying.

## 15.1.9 Spirituality

- Most Greeks in America are affiliated with the Greek Orthodox Church, whereas Greeks in Britain are affiliated with the Archdiocese of Thyateira and Great Britain. The central religious experience is the Sunday morning liturgy, which is a high-church service with icons, incense, and singing or chanting by the psaltes (main singers of hymns using Byzantine musical tradition) and a few background assistant singers.
- The Greek Orthodox religion emphasizes faith rather than specific tenets. The Greek faith does not emphasize Bible reading and study. Some parishioners attend church services weekly; others attend only a few times a year.
- Easter is considered the most important of holy days and nearly all Greeks in America and Britain attempt to honor the day.
- There is a strong belief in miracles, even among second and subsequent generations of Greeks in America and Britain. Daily prayers may be offered to the saints. Women often consider faith an important factor in regaining health. Family members may make "bargains" with saints, such as promises to fast, be faithful, or make church donations if the saint acts on behalf of the ill family member. They may call on an individual's namesake or a saint believed to have special affinity with healing.
- Particularly in immigrant communities, the Greek Orthodox Church serves as a base for spirituality, language, social, and political organization, and an ongoing identity with Greece and Cyprus. Assist with referrals to the local Greek community and the Greek Orthodox Church.
- A distinctive feature of the Greek Orthodox religion is the place it assigns to icons, such as paintings of saints, the Virgin Mary, and Christ. These icons have sacred significance as sources of connection to the spiritual world. Icons grace the walls and ceilings of churches and cathedrals and are found in personal altars in homes. In the homes, holy vigil candles are often kept burning. To ensure safety and health, many begin each day by kissing the blessed icons and making the sign of the Greek cross.
- When a person is ill, the icon of the family saint or the Virgin Mary may be placed above the bed. Do not remove religious art or icons from the patient's bedside.
- Many Greeks may sprinkle their homes with holy water from Epiphany Day church services to protect the members of their household from evil.

## 15.1.10 Healthcare Practices

- The amount of acceptance and use of biomedicine is highly related to one's level of education and generation of immigration. Although fourth-generation Greek Americans are highly traditional in many aspects of their lives such as religion, language retention, and food preferences, they have not retained many of the folk beliefs and practices concerning healthcare.

- Greek immigrants tended to be anxious about health, to lack trust in healthcare providers, and to rely on family and community for advice and remedies. Inquire about home treatments in a nonjudgmental manner.
- To be healthy means to feel strong, joyful, and content; to be able to take care of oneself; and to be free from pain. Greek immigrants believed that threats to health resulted from a lack of balance in life, departure from family, and failure to demonstrate right behaviors such as respect toward parents, sharing with family, upholding religious precepts, and staying out too late.
- Traditionally Greeks believed that problems originated outside the individual's control were attributed to God, the devil, spirits, and envy or malice of others.
- The family generally assumes responsibility and care for a sick member and works to control interactions with healthcare providers. Help the family find resources to care for their family member.
- Greeks are extremely reluctant to use welfare services or other forms of governmental assistance to meet their healthcare needs. Reliance on public assistance might indicate that the sick person and family are not self-reliant.
- Three traditional folk healing practices were particularly notable in the early immigrants: those related to matiasma (bad eye or evil eye), practika (herbal remedies), and vendousas (cupping). However, it is important to say that the current generations (particularly the fourth generation) of Greeks rarely use these methods and many do not believe in matiasma. Today, the acculturated Greeks believe in the use of modern medicine and may combine it with some of the traditional remedies.
- Early immigrants believed that matiasma results from the envy or admiration of others. While the eye is able to, harm a wide variety of things including inanimate objects, children are particularly susceptible to attack. Common symptoms include headache, chills, irritability, restlessness, and lethargy; in extreme cases, matiasma has resulted in death. Greeks employ a variety of preventive mechanisms to thwart the effects of envy or evil eye, including protective charms in the form of phylactos, amulets consisting of blessed wood or incense, or blue "eye" beads, which "reflect" the eye. Do not remove protective charms from the patient or from the bedside.
- When matiasma is suspected, the most common method of detection consists of placing olive oil in a glass of water. If the oil disperses, then the eye has been cast. Subsequent treatment consists of physical acts such as making the sign of the Greek cross over the glass of water or reciting ritual prayers passed on in families by members of the opposite sex. In particularly severe cases, the Orthodox priest may recite special prayers of exorcism and use incense to fumigate the afflicted person.
- Practika are herbal and humoral treatments used for initial self-treatment. Chamomile, the most popular herb, is used in teas for gastric distress or abdominal pain, including infant colic and menstrual cramps. It is also used as an expectorant to treat colds.

- Liquors, such as anisette, ouzo, and mestika, are used primarily for colds, sore throats, and coughs and are consumed alone or in combination with tea, lemon, honey, or sugar, either alone or in some combination. Occasionally, liquors are used for treatment of nevra (nerves). Raw garlic is used as prevention for colds and cooked garlic is used for blood pressure and heart disease.
- Encourage patients to fully disclose the use of complementary and alternative practices.
- Vendousas, a healing practice, is used as a treatment for colds, high blood pressure, and backache. It consists of lighting a swab of cotton held on a fork, then placing the swab in an inverted glass, thereby creating a vacuum in the glass, which is then placed on the back of the ill person. The skin on the back is drawn into the glass. This procedure is repeated 8–12 times. An alternative method, for particularly serious cases, is the kofte, "cut vendousa." Here, the same procedure is followed, except that a cut in the shape of a cross is made on the skin. When the glass is placed over this cut, blood is drawn into the glass. The therapeutic rationale for using vendousas surrounds its counterirritant effect; the technique increases and revitalizes the circulation, draws out poisons and "cold," and prevents coagulation of blood. Fourth-generation community members rarely use vendousas.
- Self-medicating behaviors are common. Herbal remedies and over-the-counter medications are used widely for specific symptoms.
- Encourage full disclosure of over-the-counter drug use and folk practices.
- Mental illness is accompanied by social stigma with negative consequences for the afflicted person as well as the family and relatives. Shame originates in the notion that mental illness is hereditary; afflicted people are viewed as having lifelong conditions that "pollute" the bloodline. As a result, families place a wide variety of behaviors within the range of "normal" to delay receiving the stigmatized label.
- Individuals with mental illness often present with somatic complaints such as dizziness and paraesthesia on initial visits to healthcare providers. Recent immigrants tend to have higher rates of mental disorders which perhaps result from the stress of culture and societal change.
- Do not disclose emotional or psychiatric disorders to people outside the immediate family.
- A folk model for nevra is a socially acceptable and culturally condoned medium for the expression of otherwise unacceptable emotions. Nevra is experienced most commonly by those in positions of least power, such as women and people living in poverty. It encompasses a wide variety of symptoms and provides a metaphor for social disorders such as conflict between close kin or intergenerational conflict. Ideally, nevra are treated through medications for the relief of symptoms instead of through talk therapy.
- Ponos (pain) is the cardinal symptom of ill health and an evil that needs eradication. The person in pain is not expected to suffer quietly or stoically in the presence of family. The family is relied on to find resources to relieve the pain or, failing that, to share in the experience of suffering. In the presence of outsiders,

the lack of restraint in pain expression suggests lack of self-control, and therefore it is considered endropi.

- Provide adequate pain relief to help the patient maintain control.
- The key aspect of the sick role is for the sick to fully rely on the family for sustenance. When an individual is ill, it is particularly important that he or she not be left alone.
- When hospitalization occurs, family members expect to stay with the individual, even during examinations and therapeutic procedures. Families are expected to ensure that sick family members are not harmed and are receiving the best care possible. Protection includes shielding the sick from a serious diagnosis such as cancer, until the family feels the individual is ready to learn about the diagnosis. Make arrangements to accommodate multiple visitors. Determine the spokesperson for the family and obtain his or her assistance in visitor control. Seek assistance from the family spokesperson or next of kin before informing the patient of a grave diagnosis.
- Based on the Christian Orthodox belief in the physical resurrection of the body, some Greeks may reject the concept of autopsy and do not readily accept organ donation. However, the Greek Orthodox Church is strongly pro-life and more recently has been encouraging organ donation as an act of love. Blood transfusions are wholly acceptable and are common for people with thalassemia. Explain the legalities of autopsy.

### *15.1.11 Healthcare Providers*

- A woman who cures, particularly one who cures the evil eye, is known as a magissa, which is usually translated as "witch" but means "magician"; she may also be called doctor. In the United States and Britain, "wise women" from one's own family conduct most lay healing. Occasionally, for particularly difficult cases of matiasma, a woman with particular gifts in diagnosis and healing may be called. The priest may also be called on for advice, blessings, exorcisms, and direct healing. Assist patients and family in obtaining a priest.
- Many Greeks display a general distrust of all healthcare providers and considerable shopping around for physicians and other professionals to obtain additional opinions is relatively common. The use of several physicians simultaneously may result in untoward drug interactions from conflicting interventions or overdoses. Encourage patients to disclose the use of all healthcare providers seen and treatments used.
- Hospitals are a particular source of mistrust for Greeks. When hospitalizations occur, the family may be perceived by staff as demanding or interfering as they enact their protective advocacy roles. Mothers may demand to sleep with their children and fear that the children may not receive appropriate care. There is also a fear that the sick person may be used as a subject for experimentation.

Encourage family members to participate in hospital procedures. Provide factual information and dispel myths.

## 15.2 Reflective Exercises

1. What do the terms philotimo and endropi mean and how are they displayed among Greeks?
2. What are typical nonverbal communication practices among traditional Greeks?
3. The core values of philotimo and endropi set the pattern for the family and for the enactment of gender roles. How are they displayed among men, women, and children in the family context?
4. List some genetic conditions common among Greeks. What dietary implications do they have?
5. Identify two high-risk health behaviors common among Greeks. What might you do to help decrease them?
6. Traditional Greeks fast for a number of religious holidays. What implications does fasting have for medication administration?
7. Under what conditions is an abortion acceptable for most Greeks?
8. The Greek Orthodox Church is important for Greeks. For what health problems might you use the church for health teaching?
9. Describe three folk illnesses. What is the treatment for each? What implications are there for the healthcare provider?
10. Hospitals are a particular source of mistrust for some Greeks. What might you do to gain their trust?

## Bibliography

Athanasopoulou M, Papadopoulos I (2011) Traditional and alternative therapies in Greece and Cyprus. In: Papadopoulos, Kalokerinou, Kouta (eds) Transcultural nursing and cultural competence for health professionals (in Greek). Paschalides, Athens

CIA (2024a) World factbook: Cyprus. https://www.cia.gov/the-world-factbook/countries/cyprus/

CIA (2024b) World factbook: Greece. https://www.cia.gov/the-world-factbook/countries/greece/

Jurgens J (n.d.) Greek Americans (n.d.). https://www.everyculture.com/multi/Du-Ha/Greek-Americans.html

Langer LA (2023). Beta-Thalassemia. https://www.ncbi.nlm.nih.gov/books/NBK1426/

Manoli I, Fryssira H (2015) Medical genetics and genomic medicine in Greece: achievements and challenges. Mol Genetics Genom Med 3(5):383–390

Osborn DK (2021) Greek medicine: herbal medicine. http://www.greekmedicine.net/therapies/Herbal_Medicine.html

Papadopoulos I (2021) People of Greek heritage. In: Purnell L, Fenkl EA (eds) Textbook for transcultural health care: a population approach, 5th edn. Springer, Cham, Switzerland

Scott-Sheldon LAJ, Carey KB, Carey MP (2008) Health behaviors and college students: does Greek affiliation matter? J Behav Med 3(1):61–70

# Chapter 16
# People of Guatemalan Heritage

**Cory Salcedo**

## 16.1 Overview and Heritage

Individuals of Guatemalan heritage are one of many cultures grouped under the umbrella of Hispanic/LatinX populations in the United States and elsewhere. Although similarities range among language paradigms for Hispanic/LatinX individuals, this unique cultural group has an abundance of linguistic differences in addition to religious, cultural, and economic structures that must be acknowledged to adequately care for this population. Healthcare providers must be knowledgeable regarding the specific cultural characteristics and behavioral practices for Guatemalan individuals in order to provide individualized care that is sensitive to cultural necessities. Although each methodology to practice may not adequately reflect the distinctiveness encompassed within each Guatemalan profile, understanding the basis of cultural sensitivity and competence and necessity for patient healing is a core component for successful nursing practice.

Guatemala, referred to as "eternal spring," is a country in Central America with an estimated population of over 17 million. Guatemala is inhabited by Mestizo (mixed Amerindian Spanish—in local Spanish called ***Ladino***), K'iche, Kaqchikel, Mam, Q'eqchi, and other Mayan, and indigenous non-Mayan sub-cultural groups. Following the civil war, during the 1960s–1980s, a widespread of mortality, loss of land, community right restrictions, economic instability, and cultural disruption arose, ultimately impacting the structural archetype of everyday life for most Guatemalans. As a result, as many as one million Guatemalan natives fled north toward Mexico and the United States seeking safety and political asylum. The increased migration to the United States created increased difficulties for Guatemalan

C. Salcedo (✉)
Nicole Wertheim College of Nursing and Health Sciences,
Florida International University, Miami, FL, USA
e-mail: csalc011@fiu.edu



E. A. Fenkl, L. D. Purnell (eds.), *Handbook for Culturally Competent Care*,
https://doi.org/10.1007/978-3-031-70492-5_16

individuals in regard to cultural differences, language restrictions, and governmental policies. Although much time has passed since the cultural disruption of Guatemala, residing individuals continue to be plagued with "abject" poverty with a majority of the population living below the universal poverty level.

### *16.1.1 Communications*

- The primary languages spoken in Guatemala include the official language, Spanish, which is spoken by 60% of the population, and Amerindian languages, spoken by the remaining 40%. Officially, 23 Amerindian languages are recognized and primarily spoken in rural areas throughout the country.
- Some Mayan men do not have a formal education but are able to speak Spanish because of frequent interactions with Spanish speakers. Due to the sporadicity of mayan cultural groups, these indivudials tend to adopt skills, customs, and traditions from one-another and through generational instruction. **Obtain an interpreter when needed. Note: Interpreters are not available for many of the Amerindian languages; however, providers must attempt to use established, accredited resources to adequately communicate with this population.**
- Although topics such as income and investments are taboo among many Guatemalans, they generally like to express their inner beliefs, feelings, and emotions once they get to know and trust a person. Meaningful conversations often become loud and seem disorganized from unfamiliar parties. To the outsider, the situation may seem stressful or hostile, but this intense emotion means the conversants are enjoying each other's company. **Do not assume that a higher voice amplitude is indicative of emotional anger.**
- **Within the context of respect (*respeto—goes on the glossary as well*), encourage open communication and sharing. Develop the patient's sense of trust by inquiring about family members and individual customs before proceeding with the usual business if the patient's condition and situation permits.**
- Guatemalans place great value on closeness, togetherness, and unity, including when they are in an in-patient facility and/or under stressful circumstances. This culture uses a higher degree of touch and embrace and support the involvement of relatives and significant others. Touch between men and women, between men, and between women is acceptable, including in the health-care environment. **Explain the necessity of and ask permission before touching during a physical examination to ensure patient consent and comfortability.**
- Traditional Guatemalans consider sustained eye contact when speaking directly to an older person to be rude. Direct eye contact with superiors may be interpreted as insolence. **Do not assume that lack of eye contact means the patient and/or family is being less than truthful or not listening. Avoiding direct eye contact with superiors is a sign of respect.**
- Guatemalans tend to value the past and live in the present, being more concerned with today than the future that is uncertain for many. **Inquiring about past**

**events in one's life may help build a rapport and establish comfortability with Guatemalan individuals.**

- Because work is a priority in the life of many Guatemalans, many seek healthcare only when their illness has progressed to the point of preventing them from working or carrying out their duties or roles within the family. Taking time to go to the doctor means time lost from work and loss of pay. **Educational initiatives, follow-up care, and take-home documents may be beneficial with continuation of care among this population.**
- Punctuality is difficult for many because of limited transportation and unexpected family needs. **Carefully explain the necessity of timeliness in keeping appointments in the health-care system.**
- Guatemalans who have a Hispanic heritage use the Spanish format for names. At birth, a child is given a first name (Ovidio) followed by the surname of the father (Garcia), and then the surname of the mother (Salvador), resulting in Ovidio Garcia Salvador. In referring to a male as Mr. (*Señor*) it would be appropriate to use Señor Garcia. Men's names remain the same through their lifetime. However, when a woman named Jovita Garcia Salvador marries Francisco Vasquez Gutierrez, she then becomes Jovita Garcia de Vasquez or simply Jovita Garcia Vasquez. She then becomes Mrs. (*Señora*) Vasquez. **Ask what the legal name is for medical record keeping and to ensure patient preferred name is used by interdisciplinary teams.**
- **To convey respect, address the Guatemalan in a formal manner unless otherwise requested by the patient.**
- Many Guatemalans shake hands softly. To give a firm handshake indicates aggressive behavior.

### *16.1.2 Family Roles and Organization*

- Traditionally, the man has been the head of household and is the primary decision maker, "breadwinner," and provider for the family.
- Women's roles have traditionally involved raising the children. **More Guatemalan women are entering the workforce outside of the home, resulting in more egalitarian male/female roles. Inquire about the familial structure and head-of-household before assuming.**
- Guatemalans place high value on familial unity. Most families are nuclear—comprised of a father, mother, and children. Extended family is important to Guatemalans and may include grandparents, aunts, uncles, and cousins. **Determine the spokesperson in the family.**
- Children are a gift from God and are highly valued in Guatemalan society. Sons are usually more valued than daughters. Children are taught to be obedient and demonstrate respect for older people.
- In Mayan communities, family members and other adults take an active part in raising a child. Values include being humble, content, and respectful of others,

working hard, avoiding arguments, and placing the needs of the family before one's own individual needs.

- Disobedience among children in Guatemala may be handled with physical punishment. **Explain child abuse laws in the** United States **and elsewhere.**
- Many Guatemalans do not believe in homosexuality, sexual activity among the unmarried, or infidelity. Persons involved in these activities must do so in secrecy. **Do not disclose same-sex relationships to the family.**
- Homosexuality in Guatemalan culture is considered taboo and advocates for discretion. **Provide information on sexual care practices when necessary and with discretion.**
- Larger cities in the offer organizations such as *Ellas*, a support group for Latina lesbians; El Hotline of Hola Gay, an organization with information and referrals in Spanish, and Dignity, a gay Catholic support organization. **Refer patients to gay support groups when requested.**

### *16.1.3 Workforce Issues*

- During the civil war, residents of Guatemala were permitted to migrate to the United States and apply for political asylum, allowing them to stay permanently in the United States. Political asylum allows national protection of an individual who has left their native country as a political refugee. The application of political asylum for this cultural group ensured Guatemalans were safe to migrate and protected from immigrative law upon entrance to the United States.
- Guatemalans who migrate today often find employment in the agricultural industry, including hand-harvesting crops and working on egg or chicken farms. Additional occupations of Guatemalan migrants include citrus processing plants, plant/flower nurseries, and construction services throughout the country.
- **Assess patients for diseases and deficits related to farming, construction and respiratory compromise, including pesticides and inhalation debris.**
- Migrants are able to find additional positions in housekeeping or maintenance in businesses or academic institutions. Because punctuality is not valued in Guatemala, Guatemalan employees may arrive late for work. They might not wear a timepiece, be able to tell time, or understand the importance of punctuality.
- **Stress the necessity of and importance placed on punctuality. Include teaching regarding employment related requirements and community services to assist with transportation.**
- Those of lower socioeconomic status or with formal education and English language skills usually acquire positions with responsibility but little authority. These individuals prefer to get along well with others and not criticize or voice complaints when treated poorly. Moreover, the Guatemalan is likely to remain in a position equal to his or her peers rather than seek a promotion. **Provide opportunities for advocacy and self-representation when caring for this population.**

### *16.1.4 Biocultural Ecology*

- Most Guatemalan natives are culturally constructed from Spanish and Mayan Indian heritage and have adopted lifestyle practices from generational growth.
- There is a small population of Guatemalans with ancestry from the Caribbean and Africa who may use cultural practices from these regions to maintain health adequacy. This accounts for variations in skin color, facial features, hair, body structure, and other biological variations. Guatemalans of Caribbean and African descent tend to have black hair, dark skin, dark eyes, and be of average (Male: 5′4; Female: 4′10) or taller based on a regional population with a medium to large build.
- Guatemalans who are predominantly Spanish have lighter features and produce similar characteristics to Caucasian-identifying individuals. This sub-cultural group may have blonde or brown hair, fair (white) complexion, blue eyes, and be of average or taller height based on regional population with a medium to large build.
- Guatemalans with predominantly Mayan Indian ancestry tend to have black hair, brown skin, dark eye color, and are of short height with a petite build.
- Jaundice, cyanosis, and petechiae are difficult to detect among dark-skinned Guatemalans. **Exam the sclera and buccal mucosa, tongue, lips, nailbeds, and palms of hands and soles of feet.**
- The leading causes of mortality in Guatemala are diabetes mellitus, ischemic heart disease, respiratory infections (pneumonia, influenza, COVID-19), nephritis/nephrosis, perinatal conditions, intestinal infectious diseases, and nutritional deficiencies. **Assess newer immigrants for health conditions common in their home country. Include areas of nutrition, familial history, activity level, and occupation.**
- Other common health conditions include malaria or dengue fever, gastritis, intestinal worms, pulmonary tuberculosis, and diarrhea resulting from shigella, amoebas, and giardia. Table 16.1 lists commonly occurring health conditions for Guatemalans in three categories: lifestyle, environment, and genetics. **Inquire about previous vaccinations and provide patient education regarding required/recommended vaccinations.**

**Table 16.1** Health conditions common to Guatemalans

| Health conditions | Causes |
|---|---|
| Diabetes mellitus | Genetic, lifestyle |
| Respiratory infections | Environment, lifestyle |
| Lactase deficiency | Genetic |
| Gastritis | Environment, lifestyle |
| Malaria | Environment, lifestyle |
| Tuberculosis | Environment, lifestyle |
| Eye disorders | Lifestyle |

- Eye disorders common among Guatemalan individuals include dacrocytosis and trachoma. Cataracts (even in children), congenital toxoplasmosis, toxic optic neuropathy (usually related to TB treatment), vitamin A deficiency, and actinic allergy are additional health concerns related to this population. **Assess Guatemalans for health conditions common in their home country. Provide opportunity for visual examination(s) and address dermatologic concerns.**

### *16.1.5 High-Risk Behaviors*

- Alcohol products are readily available and widely abused throughout Guatemala. These products are accessible to majority of the population and are highly consumed among early to middle-aged men. Men who immigrate from Guatemala may find themselves drinking alcohol excessively, even if they did not prior to migration. This may be due to such factors as (a) migrational stress of living in another country, (b) lack of communication and increased distance from family, friends, and support systems, (c) fear of inadequate work, financial struggle, and deportation, and (d) illness and (e) being victims of violence and injury.
- Guatemalans of low socioeconomic status have received little health education, reduced access to healthcare and provider coverage, and often believe illness is punishment from God or a religious entity.
- Guatemalan families readily participate in immunization programs for their children, yet do not participate themselves. Many parental figures strive for healthier lifestyles among children when compared to oneself. **Encourage immunizations as part of routine care and provide education for immunization rationale, timing, and accessibility.**
- Most women do not participate in routine screening for breast and cervical cancer. Few men participate in routine screening for testicular or colon cancer.
- **Educate Guatemalans in a family context for disease, illness, and injury prevention, assist with low-cost referrals, and provide milestone necessities for all individuals across the lifespan.**

### *16.1.6 Nutrition*

- Food to Guatemalans signifies physical, spiritual, and cultural wellness and is a catalyst for familial integration and support.
- Nutritional options vary among Guatemalans based on cultural traditions and accessibility. Gari'funa cuisine reflects the Caribbean coast and includes recipes from African ancestors. Common foods include sea bass, flounder, red snapper, tarpon, shrimp, pork, chicken, ceviche, coconut milk, tomato, onion, garlic, lime, and lemon.

- Corn is highly valued in the Mayan culture and is the chief crop and the basis for many food products and meals. The Mayan diet primarily consists of maize (corn grain), black beans, rice, chicken, squash, tomatoes, carrots, chilies, beets, cauliflower, lettuce, cabbage, chard, leek, onion, and garlic. These foods are used to make tortillas, *atole* (liquid corn drink), *pinol* (chicken-flavored corn gruel), *pepi'an* (chicken stew with squash seeds, hot chilies, tomatoes, and tomatillos [small green tomato]), and *caldos* (soups made of chicken stock and vegetables). Green leafy plants are often added to soups and stews. Preferred beverages are Hibiscus tea, limonadas/ponche de frutas (juices), and coffee.
- Ladino foods reflect their Spanish ancestry and include maize, rice, black or red beans, beef, chicken, pork, milk, cheese, plantains, carrots, peppers, tomatoes, squash, avocado, cilantro, chilies, onion, garlic, lemon, lime, and chocolate. Common dishes include *arroz con pollo* (chicken with rice), *chile rellenos* (peppers stuffed with pork or beef with carrot, onion, and tomato and fried in egg batter), and *tamales* (chicken or pork in a corn paste steamed in banana leaves or corn husks). Additional delicacies may stem from surrounding regions of Mexico, El Salvador, and Belize, and include *Paches* (Maize, potato) and Dobladas (masa dough, cheese, ham and/or chicharones). Again, green leafy plants are often added to soups and stews. Foods are seasoned with toasted squash seeds ground to a powder. Guatemalan food is not served spicy; however, traditional foods tend to be rich in sodium.
- **Complete a nutritional assessment as part of the intake interview. Include teaching related to health concerns and provide cultural alternatives to nutrition for at-risk individuals.**

### *16.1.7 Pregnancy and Childbearing Practices*

- For religious purposes, most Guatemalan individuals do not believe in contraception use and/or abortion practices. Women who desire contraception will not actually seek to obtain modalities for prevention due to lack of spousal, familial, and religious support. When the Guatemalan patient expresses a desire to learn more about family planning, women may request the session include her husband and supporting family members to assist with decision-making.
- **Arrangements should be made to accommodate the patient's request. Health-care providers should allow autonomy in patient decision-making and provide resources for reference.**
- Mayan women do not believe in traditional birthing processes which includes lying down to give birth and/or delivering in a hospital setting. The idea that a Guatemalan woman in the might need to deliver in the hospital may be perceived as foreign and frightening. Traditional practices of Mayan culture include ceremonial/religious environments and prefer the squatting technique when delivering offspring. **If possible, provide a tour of the labor and delivery unit well before her due date and provide options for religious and/or midwife**

**involvement for guidance. Allow women to guide their birthing process that is tailored in a safe and ethical structure.**

- If an individual eats in front of the pregnant woman without offering her food, she may believe she will have a miscarriage. Additional concerns include frequent touching of a mothers stomach during pregnancy. Instances as such may be indicative of fetal demise in Mayan culture.
- Following delivery of a newborn, many indigenous Mayan tribes require the placenta to be burned, not buried, because it is disrespectful to the earth to do so. In contrast to this, traditional Mayan culture may support the burial of a placenta under a chosen tree to symbolize protection over the newborn. **Inquire specific birthing practices and cultural necessities prior to, during, and following the birthing process.**
- Guatemalan women support breastfeeding and may continue breastfeeding until the child reaches the age of 5 years. They may also be breastfeed a new baby while continuing to breastfeed a toddler.
- A bag with garlic, lime, salt, and tobacco is hung around the baby's neck, and a red thread is used to tie the umbilical cord to protect the baby, provide strength, and to denote respect for the ancestors. Urbanized Guatemalan individuals seek to use amulets, gold, and munekitas (dolls) for newborn protection and to ward off evil spirits. **Do not remove amulets or other items that are seen as protective for the baby.**

### *16.1.8 Death Rituals*

- Some Guatemalans relate their illness or poor diagnostic outcomes to "punishment" or impending death. The notion that tragedy is a result of "God's will" is primary and some may refuse an intervention or heroic measures to align with spiritual and cultural practice.
- When a Guatemalan dies, the family may request repatriation of remains. It is important for the final resting place to be the home country. Often, immediate and extended family will pool their resources together in order to financially support the transportation of remains home.
- **Refer families to Social Services and funeral homes to assist with transporting the body to Guatemala. Additional resources may be available to support funding for repatriation.**
- Guatemalans believe in burial; they do not practice cremation. **Practices may differ in urbanized regions—include family, chaplain, and social services to support decision-making.**
- Yellow is the color of mourning. Yellow flowers are placed at the grave. Food is placed at the head for the spirit of the departed.
- **Ladinos mourn the dead by wearing black. Maya do not believe in this practice.**

### *16.1.9 Spirituality*

- Most Guatemalans are Roman Catholic. When the Spanish brought Roman Catholicism to Guatemala, some Maya converted to Christianity. Others continued to practice their Mayan religion. Currently, many Guatemalans use a combination of beliefs and practices adopted through generational influence. In some cases, Guatemalans integrate aspects of Catholicism into their lives while continuing to believe in the spirituality of their ancestors in private. **Utilize chaplain services to tailor religious practice regimines.**
- Two practices influenced by the Spanish are *guachibal* and *cofradia. Guachibal* involves the practice of keeping an image of a Christian saint in the home and celebrating on the particular saint's day. *Cofradia* refers to a "religious brotherhood" that serves to maintain the "cult" of a particular saint. *Cofradias* consist of dues-paying members and elected leaders. In addition to religious activities, this population often assists with care of the underserved in the community by visiting the ill, rehabilitation centers, community kitchens, and paying for funeral services.
- Family provides Guatemalans with meaning in their lives. Life revolves around the nuclear and extended family. Spirituality helps to explain life and the circumstances faced by Guatemalans. **Encourage the incorporation of familial involvement for supportive care.**
- Whether Catholic, Protestant, or traditional Mayan, many believe that life events happen for a reason. The reason may be attributed to favor from God(s) when positive experiences occur and to punishment or disfavor from God(s) when negative events occur. Some Guatemalans feel nothing can be done to change the outcome of these experiences. This belief is referred to as ***fatalism***.
- When illness occurs, many Guatemalans turn to their family and faith for strength, wisdom, and hope. **Do not remove statues or pictures of patron saints. Contact chaplaincy for spiritual care services for the patient and family through their church or hospital resources.**

### *16.1.10 Health-Care Practices*

- The preferred mode of treatment among Ladinos is medication administered by hypodermic injection. If an infant has a cold, Ladinos believe an injection is necessary to treat conditions effectively. If someone has the flu, an IV infusion is preferred. **Provide factual information about allopathic medical practice.**
- Healthcare and provider seeking among Guatemalans generally occurs by first seeking advice from a mother, grandmother, or other respected elder. If this approach is unsuccessful, then the family usually seeks healthcare from folk healers or spiritual guiders of preference.

- Several medications, including antibiotics, purchased in the United States only with a prescription are available without a prescription in Guatemala. **Ascertain if Guatemalan patients are receiving medicines from family or friends from their home country.**
- Modern medical care may be the last resort. Many are fearful of hospitals. In Guatemala, when hospital care is necessary, patients are often seriously ill, resulting in death, which perpetuates the belief that "hospitals are places where patients go to die."
- Most Guatemalans engage in folk medicinal practices and use a variety of prayers, herbal teas, and poultices to treat illness. Many of these practices are regional-specific and vary among families, sub-cultures, and generational influence. Many of these practices are harmless, however, some may contradict or potentiate therapeutic interventions. **Specifically ask patients whether they are using folk medicine and/or what modalities to care they are currently practicing.**
- A common belief among this population is that numerous disorders can be attributed to an imbalance in the thermoregulatory equilibrium of the human body (hot and cold). Therefore, the consumption of foods with contrasting properties has the potential to alleviate or preempt certain ailments and disorders associated with temperature imbalances.
- Physical or mental illness may be attributed to an imbalance or disconnect between a person and the environment. Influences include emotional, spiritual, and social state, as well as physical factors such as humoral imbalance expressed as either too much hot or cold.
- **Nonjudgmentally ask patients and families about the use of folk and traditional therapies and holistic modalities used.**
- Folk practitioners are consulted for several notable conditions. ***Mal de ojo*** (evil eye) is a folk illness that occurs when one person (usually older) looks at another (usually a child) in an admiring fashion. Another example of *mal de ojo* is if a person admires something about a baby or child, such as beautiful eyes or hair. Such eye contact can be either voluntary or involuntary. Symptoms are numerous, ranging from fever, anorexia, and vomiting to irritability. The spell can be broken if the person doing the admiring touches the person admired while it is happening. Children are more susceptible to this condition than women, and women are more susceptible than men.
- Guatemalans often delay seeking healthcare until they are incapacitated by illness, disease, or injury. Many times, they are unaware of the dangers associated with working in agriculture. These individuals may be exposed to pesticides and dangerous equipment without proper training. Although the government has specific laws in place to protect farm workers, enforcement is limited. Companies may not disclose to workers the dangers encompassed in a specific job role or the workers may not understand instruction due to language differences. **Be aware of the chemical exposures, laws, and resources available in order to adequately provide care for and advocate for the patient. Resources should be provided in the patient's native language and/or via a certified interpreter.**

- Barriers to healthcare include *availability, accessibility/reliability, affordability, appropriateness, accountability, adaptability, acceptability, awareness, attitudes, approachability, alternative practices and practitioners, and additional services* (see Chap. 3).
- Guatemalans tend to view health and illness in relation to their ability to perform duties associated with their roles. As long as women are functioning in their role of caring for the home and family and men are functioning in their job, then they feel "healthy."
- When an illness prevents normal functionability required for their specific role(s), then Guatemalans view this as a serious concern. The cause of debilitating illness or disease may be viewed as punishment from God rather than lack of prevention or early detection.
- When symptoms persist, fear may keep the Guatemalan patient from seeking medical attention.
- Family members would rather care for their loved one at home if at all possible. **Demonstrate respect for the cultural values by providing community resources that enable them to adequately care for a family member at home if they desire.**
- Blood transfusions and organ donation are not common in Guatemala, and with many living outside of the realm of modern medicine, residents may be uneducated regarding the indication, risks, and benefits of interventional practice. Questions related to organ donation may be puzzling and elicit fear and anxiety. **Ensure patient/family readiness to learn when providing complex educational objectives.**
- Some Guatemalans fear venipuncture from belief that taking blood leaves the body without enough blood to keep them strong and healthy. **Assess the level of education and understanding of the Guatemalan patient regarding blood and blood products. Use an interpreter who is competent in the language and culture of the patient to promote successful communication.**

### *16.1.11 Health-Care Providers*

- *Modern medicine* refers to healthcare provided by educated physicians and nurses, generally in urbanized regions. *Ladino folk medicine* is provided by Ladino pharmacists, spiritualists, and lay healers (***curanderos***).
- *Mayan Indians* seek medical care from Mayan *shaman*, *bonesetters*, and *herbalists*.
- When Ladinos and Mayan Indians have access to modern medicine, the utilization of these services increases. **Provide education on accessibility to patient preferred medicinal practices.**
- In the United States and elsewhere, Guatemalans use many Mexican-based traditional healers.

- Guatemalans have great respect and admiration for health-care providers who are viewed as authority figures with clinical expertise.
- Guatemalans expect their health-care provider to have the appearance and manners of a professional. When this is not the case, Guatemalans lose confidence in the provider.
- Many fear disclosure which may result in deportation or rejection. Patients also fear confidentiality will not be maintained in the health-care setting. **Educate patients on disclosure laws in the related to health-care information and the ability to receive care without fear of deportation.**
- Guatemalan women are usually very modest. They may refuse to discuss personal issues or receive an examination by a male health-care provider. Likewise, a male Guatemalan patient may refuse a female health-care provider. **Provide a same-sex provider when possible, especially for intimate care.**
- Because Guatemalans dislike conflict, they may not actually refuse care; yet they withhold personal information owing to discomfort with the health-care provider. **Ensure rapport building exercises are used to provide patient comfortability and potentiate the divulgence of the reasoning for seeking health-care advisory.**

## 16.2 Reflective Exercises

1. Explain the name format for Guatemalans at birth and at matrimony.
2. What languages besides Spanish are spoken by Guatemalans and Mayans?
3. Explain the concept of "evil eye". What could one do to break a spell of the "evil eye"?
4. Explain common infectious diseases among Guatemalan natives? What resources could be suggested to increase disease prevention?
5. What are common eye problems among Guatemalans? What are some reasons for these eye problems?
6. Identify common birthing practices that are used by women of Guatemalan heritage?
7. What folk illnesses are common among Guatemalans and Mayans?
8. What routes do Guatemalans prefer for medication administration?
9. If a Guatemalan dies in the United States (and probably elsewhere) and the family wants to have the body returned to Guatemala, what resources are available to assist the family?
10. How is *fatalism* displayed among Guatemalans?
11. How would you assess jaundice, cyanosis, and petechiae in dark-skinned Guatemalans?

## Bibliography

Callister LC, Vega R (1998) Giving birth: Guatemalan women's voices. J Obstet Gynecol Neonatal Nurs 27(3):289–295

Cerón A, Ruano AL, Sánchez S, Chew AS, Díaz D, Hernández A, Flores W (2016) Abuse and discrimination towards indigenous people in public health care facilities: experiences from rural Guatemala. Int J Equity Health 15:77. https://doi.org/10.1186/s12939-016-0367-z. PMID: 27177690; PMCID: PMC4866428

CIA world factbook: Guatemala (2023). https://www.cia.gov/the-world-factbook/countries/guatemala/

Ellis T, Purnell L (2013) People of Guatemalan heritage. In: Purnell L (ed) Transcultural health care: a culturally competent approach. F.A Davis, Philadelphia

Ellis TA, Purnell LD (2021) People of Guatemalan heritage. In: Textbook for transcultural health care: a population approach: cultural competence concepts in nursing care. Springer, Cham, pp 445–467

Johanson L (2016) Caring for the vulnerable: an exploration of Guatemalan healthcare issues. J Christ Nurs 33(1):44–49

Weller SC, Ruebush TK, Klein RR (1991) An epidemiological description of folk illnesses: a study of *empacho* in Guatemala. Med Anthropol 13(1–2):19–31

# Chapter 17
# People of Haitian Heritage

**Jessie M. Colin**

## 17.1 Overview and Heritage

Haiti, with a population of 11.4 million, is located in the Caribbean on the island of Hispaniola, which it shares with the Dominican Republic. The per capita annual income as of 2022 is $1743. Only 13% of the people have access to potable water. Haiti defines itself as a black, republic, consisting of 95% of the population, nation. In Haiti, the concept of color differs from the concept of race. The Haitian system has been described as one in which there are no tight racial categories but in which skin color and other phenotypic demarcations are significant variables. When Duvalier was elected president-for-life in 1964, many Haitians covertly immigrated to the United States in small sailboats, resulting in their being labeled "**boat people**," a term that is associated with extreme poverty.

Migration from Haiti to the United States since the mid-twentieth century has occurred in response to persistent political instability, persecution, frequent economic crises, and natural disasters including a devastating 2010 earthquake that killed over 300,000 people, injured another 300,000, and left over 1.5 million homeless.

The recent collapse of basic governmental functions and widespread gang violence followed the 2021 assassination of President Jovenel Moïse. As of 2022, nearly 731,000 Haitian immigrants resided in the United States, making them the country's 15th-largest foreign-born population.

The Haitian population in the United States is not well documented because many are undocumented immigrants. The top 10 metro areas with the most Haitians are within or close to the six states into which the majority of immigrants arrive—Florida, New York, New Jersey, Illinois, California, and Texas. The Haitian

J. M. Colin (✉)
College of Health and Wellness, Barry University, Miami, FL, USA
e-mail: jcolin@barry.edu





E. A. Fenkl, L. D. Purnell (eds.), *Handbook for Culturally Competent Care*,
https://doi.org/10.1007/978-3-031-70492-5_17

immigrant population has more than tripled in the United States, which is the top global destination.

The year 2023 offered more opportunities than ever before for Haitians to come into the United States legally. President Joe Biden's extensions of Temporary Protected Status (TPS) have allowed Haitian immigrants to stay and work legally. They participate in the labor force at a similar rate as all immigrants. About 69 percent of Haitian immigrants and 67% of all immigrants aged 16 and over were in the civilian labor force in 2022, compared to 62% of the native-born. Haitians are more likely than all immigrants in the United States to work in service occupations.

Even though Haitians value education, only 15% are privileged enough to obtain a formal education. The illiteracy rate of 61% continues to be a major concern in Haiti. Like other ethnic groups, Haitians are very diverse according to the variant characteristics of culture as described in Chap. 1.

Haitians are more likely to be naturalized US citizens, have obtained legal permanent residence (also known as getting a green card) through family-based categories, and work in service occupations. They are also less likely to have a college degree and tend to report lower household incomes.

### *17.1.1 Communications*

- Haiti has two official languages: French and Creole. French is the dominant language of the educated and the elite, whereas Creole is the language of the lower socioeconomic classes. Creole, a rich, expressive language, is spoken by everyone. **Develop video programs, audiocassettes, and picture brochures in Creole to provide health education.**
- Most are very expressive with their emotions, including loud animated speech. Pain and sorrow are very obvious in facial expressions.
- Most Haitians are affectionate, polite, and shy. The under-educated generally hide their lack of knowledge by keeping to themselves, avoiding conflict, and sometimes projecting a timid air or attitude. They smile frequently and often respond in this manner when they do not understand what is being said.
- Many individuals pretend to understand by nodding; this sign of approval is given to hide their limitations. **Because Haitians are very private, especially in health matters, it is inappropriate to share information with friends.**
- Many patients may prefer to use professional interpreters, who will give an accurate interpretation of their concerns. Most important, the interpreter should be someone with whom they have no relationship and will likely never see again. **Use simple and clear instructions. Ask family members to assist with interpretation only if an interpreter is not available.**
- When conversations are really animated, the conversants speak in close proximity and ignore territorial space. Sometimes the conversation is at such a high pitch and speed that to an outsider the conversation may appear disorganized or angry. **Do not interpret loud conversations as anger.**

- Traditionally, Haitians do not maintain eye contact when speaking with those in a position of authority. **Do not interpret lack of eye contact as not listening or not caring.**
- Haitians touch frequently when speaking with friends. They may touch you to make you aware that they are speaking to you. **Do not take offense from a casual touch on the arm or shoulder.**
- Haitians greet each other by kissing and embracing in informal situations. In formal encounters, they shake hands and appear composed and stern. Men usually do not kiss women unless they are old friends or relatives. Children greet everyone by kissing them on the cheek. Children refer to adult friends as uncle or auntie.
- Temporal orientation is a balance among the past, present, and future. The past is important because it lays the historical foundation from which one must learn. The present is cherished and savored. The future is predetermined and God is the only Supreme Being who can redirect it.
- Haitians have a fatalistic but serene view of life. Some believe that destiny or spiritual forces are in control of life events such as health and death.
- **To achieve acceptance, build trust, and ensure compliance, be clear, honest, and open when assessing individuals' perceptions and how they perceive the forces that have an influence over life, health, and illness.**
- Most do not respect clock time; flexibility with time is the norm, and punctuality is not valued. Arriving late, even for medical appointments, is not considered impolite. **Make reminder calls for appointments and encourage the patient about the importance of timeliness if it is important.**
- Most Haitians have a first, middle, and last name; sometimes the first two names are hyphenated as in Marie-Maude. The family name is very important in middle- and upper-class society, promoting and communicating tradition and prestige. Families usually have an affectionate name or nickname for individuals.
- When a woman marries, she takes her husband's full name. She loses her name except on paper. Most names are of French origin, although many Arabic names are also heard. Most are formal and respectful. **Address patients by their title: Mr., Mrs., Miss, Ms., Doctor, or other title until given permission to do otherwise.**

### *17.1.2 Family Roles and Organization*

- Traditionally, the head of the household has been the man, but today most families are matriarchal. The man is generally considered the primary income provider for the family, and daily decision-making is considered his province. Women are expected to be faithful, honest, and respectable.
- Children are valued among Haitians because they are key to the family's progeny, cultural beliefs, and values. Children are expected to be high achievers.

Children are expected to be obedient and respectful to parents and older people. They are not allowed to express anger to older people.

- Physical punishment, often used as a way of disciplining children, is sometimes considered child abuse by US standards. Fear of having their children taken away from them because of their methods of discipline can cause parents to withdraw or not follow through on health-care appointments if such abuse is evident (e.g., bruises or belt marks). **Educate parents about other methods of discipline and laws so they can learn new ways of disciplining their children without compromising their beliefs or violating US laws.**
- In the summer, parents engage their children in health promotion activities such as giving them *lok* (a laxative), a mixture of bitter tea leaves, juice, sugarcane syrup, and oil. Children are also given *lavman* (enemas) to ensure cleanliness. This rids the bowel of impurities and refreshes it, prevents acne, and rejuvenates the body.
- Boys are given more freedom than girls and are even expected to receive outside initiation in social and sexual life. Girls are educated toward marriage and respectability. Even when they are 16 or 17 years of age, girls cannot go out alone because any mishap can be a threat to her future and bring shame to her family.
- **Assist children and family members to work through cultural differences related to the proper age for dating while conveying respect for family and cultural beliefs.**
- The expression "blood is thicker than water" reflects family connectedness. An important unit for decision-making is the family council, which is composed of influential members of the family. Any action taken by one family member has repercussions for the entire family; consequently, all share prestige and shame. **Include the extended nuclear family in health-care decision-making.**
- The family system includes the nuclear, consanguine, and affinal relatives, some or all of whom may live under the same roof. **Include family members in the care of loved ones to achieve trusting relationships and compliance with treatment regimens.**
- When family members are ill and in the hospital, there is an obligation to be there for them; all family members try to visit. **Suggest alternative locations for family members to gather. Ask a family spokesperson to help regulate the number of visitors at any one time.**
- When grandparents are no longer able to function independently, they move in with their children. Older people are highly respected and are often addressed by an affectionate title such as aunt, uncle, grandma, or grandpa, even if they are not related. Their children are expected to care for and provide for them when self-care becomes a concern. Older people are family advisers, babysitters, historians, and consultants.
- Most Haitians are reluctant to place family members in nursing homes.
- Homosexuality is taboo; gay and lesbian individuals usually remain closeted. If a family member discloses homosexuality, everyone keeps it quiet; there is total

denial. Gay and lesbian relationships are not talked about; they remain buried. **Do not disclose lesbian and gay relationships to family members.**

- Although divorce is common, family members, friends, the church, and elders try to counsel the couple before the divorce becomes final.
- Single parenting, widespread in Haiti, is well accepted. It is a well-accepted practice for men to have both a wife and a mistress. Both women bear children. The mistress usually raises her children alone and with minimal support from the father. Haitian women in general know that their husbands are involved in extramarital relationships but pretend not to know. **Approach health education, birth control, and safe-sex issues with sensitivity and acceptance within cultural boundaries. Provide educational brochures in French, Creole, and English.**

### *17.1.3 Workforce Issues*

- Many Haitians in the United States work two jobs to provide for their family while sending money to Haiti for those left behind. Most readily conform to the rules and regulations of the workplace. Haitians tend to settle in clusters with other Haitians. When people live and work primarily in an ethnic enclave, the native culture becomes a barrier to assimilation and acculturation into the dominant society. **Encourage Haitian employees to socialize with other cultural groups.**
- Sometimes Haitians in the workplace greet each other in their native tongue because it is easier to articulate ideas and feelings and to express support in their native language. This may be irritating to non-Haitians who consider it rude. **Explain the rules for speaking a foreign language when at work.**
- Many individuals pretend to understand by nodding; this sign of approval is given to hide their limitations. **Because Haitians are very private, especially in health matters, it is inappropriate to share any information with friends.**
- When conversations are really animated, the conversants speak in close proximity and ignore territorial space. Sometimes the conversation is at such a high pitch and speed. To an outsider, the conversation may appear disorganized or angry. **Do not interpret loud conversations as anger or take offense at spatial distancing closer than what you are accustomed.**
- Traditionally, Haitians generally do not maintain eye contact when speaking with those in a position of authority. Maintaining direct eye contact can be considered rude and insolent, especially when speaking with superiors. **Do not interpret lack of eye contact as not listening or not caring.**
- Haitians touch frequently when speaking with friends. They may touch you to make you aware that they are speaking to you. **Do not take offense from a casual touch on the arm or shoulder.**

### 17.1.4 Biocultural Ecology

- Different assessment techniques are required when assessing dark-skinned people for anemia and jaundice. **Examine the sclera, oral mucosa, conjunctiva, lips, nail beds, palms of the hands, and soles of the feet when assessing for cyanosis and low blood hemoglobin levels. To assess for jaundice, examine the conjunctiva and oral mucosa for patches of bilirubin pigment because dark skin has natural underlying tones of red and yellow.**
- Table 17.1 lists commonly occurring health conditions for Haitians in three categories: lifestyle, environment, and genetics.
- **Assess newer immigrants for signs of malaria, hepatitis, and infectious diseases that are common in Haiti.**

### 17.1.5 High-Risk Health Behaviors

- Behaviors that may be considered high-risk in American society are generally viewed as recreational or unimportant among Haitians.
- Alcohol plays an important part in society and is culturally approved for men. Women drink socially and in moderation. Cigarette smoking is practiced by men, whereas Haitian women have a very low rate of tobacco use. **Encourage smoking cessation and start educational programs in elementary school.**
- Other high-risk behaviors include non-use of seat belts and helmets when driving or using a motorcycle or bicycle because these safety measures are not normally used in Haiti. **Educate patients about seat belt use, car seats for youngsters, and the need for helmets on bicycles.**
- **Use graphic videos or skits when instructing patients about these safety practices.**
- **Use Haitian radio stations for educational purposes when they are available.**
- **Promote behavioral changes using church and community group activities.**

**Table 17.1** Health conditions common to Haitians

| Health conditions | Causes |
|---|---|
| Cholera | Lifestyle, environment |
| Malaria | Lifestyle, environment |
| Tuberculosis | Lifestyle, environment |
| Hepatitis | Lifestyle, environment |
| Parasitosis | Lifestyle, environment |
| Diabetes mellitus | Genetic, lifestyle, environment |
| Cancer | Genetic, lifestyle, environment |
| Cardiovascular diseases | Genetic, lifestyle, environment |
| Cerebrovascular disease | Genetic, lifestyle, environment |

### 17.1.6 Nutrition

- For many, food means survival; however, food is also relished as a cultural treasure and most Haitians retain their food habits and practices after emigrating. They prefer eating at home, take pride in promoting their culture through their food choices to their children, and discourage fast food. When hospitalized, many individuals would rather fast than eat non-Haitian food. They do not eat yogurt, cottage cheese, or "runny" egg yolks. They drink a lot of water, homemade fruit juices, and cold fruity sodas.
- **Encourage family members to bring food from home for the hospitalized patient if necessary.**
- The typical breakfast consists of bread, butter, bananas, and coffee. Children are allowed to drink coffee that is not as strong as that consumed by adults.
- The largest meal is eaten at lunch and includes rice and beans, boiled plantains, a salad made of watercress and tomatoes, and stewed vegetables and beef or cornmeal cooked as polenta.
- "Hot and cold," "acid and non-acid," and "heavy and light" are the major categories of contrast when discussing food. Illness is caused when the body is exposed to an imbalance of ***fret*** (cold) or ***cho*** (hot) factors. For example, soursop, is a large green prickly fruit with a white pulp that is used in juice and ice cream, is considered a cold food and is avoided when a woman is menstruating; eating white beans after childbirth are believed to induce hemorrhage.
- Foods that are considered heavy, such as plantain, cornmeal mush, rice, and meat, are to be eaten during the day because they provide energy. Light foods, such as hot chocolate milk, bread, and soup, are eaten for dinner because they are more easily digested.
- To treat a person in the hot-and-cold system, a potent drink or herbal medicine of the class opposite to the disease is administered. For example, cough medicines are considered to be in the hot category, and laxatives are in the cold category.
- Teenagers are advised to avoid drinking citrus fruit juices such as lemonade to prevent the development of acne.
- After performing strenuous activities or any activity that causes the body to become hot, one should not eat cold food because that will create an imbalance causing a condition called ***chofret***. A woman who has just straightened her hair by using a hot comb and then opens a refrigerator may become a victim of *chofret*. This means she may catch a cold and/or possibly develop pneumonia.
- When ill, pumpkin soup, bouillon, or a special soup made with green vegetables, meat, plantains, dumplings, and yams is consumed.
- Eating right entails eating sufficient food to feel full and maintain a constant body weight, which is often higher than the weight standards medically recommended in the United States. Men like to see "plump" women. **Negotiate a desirable weight with patients.**
- Weight loss is considered one of the most important signs of illness. A healthy diet includes tonics to stimulate the appetite and the use of high-calorie

supplements such as *Akasan*, which is either prepared plain or made as a special drink with cream of cornmeal, evaporated milk, cinnamon, vanilla extract, sugar, and a pinch of salt. **Ask patients about food rituals when designing individualized dietary plans to facilitate compliance with dietary regimens that promote a healthier lifestyle. Include family in the counseling process.**

- Many women and children who come from rural areas have significant protein deficiencies. A cultural factor that contributes to this problem is the uneven distribution of protein among family members. The problem is not one of net protein deficiency in the community but rather the distribution of the available protein among family members. Whenever meat is served, the major portion goes to the men on the assumption that they must be well-fed to provide for the household. **Being aware that men receive the largest protein servings enables health-care providers to prepare nutritional plans that meet patients' dietary needs.**

### *17.1.7 Pregnancy and Childbearing Practices*

- Pregnancy and fertility practices are not readily used. Most Haitians are Catholic and are unwilling to overtly engage in conversation about birth control or abortion.
- Abortion is viewed as a woman's issue and is left to her and her significant other to decide.
- **Be cautious in assessing and gathering information related to fertility control.**
- Because pregnancy is not a disease or illness, many women do not seek prenatal care. Pregnancy does not relieve a woman from her work.
- Pregnant women are restricted from eating spices that may irritate the fetus. They eat vegetables and red fruits because these are believed to improve the fetus's blood. They are encouraged to eat large quantities of food because they are eating for two.
- Pregnant women who experience increased salivation may rid themselves of the excess at places that may seem inappropriate in the United States. They may even carry a "spit" cup in order to rid themselves of the excess saliva. They are not embarrassed by this behavior because they believe it is perfectly normal.
- During labor, the woman may walk, squat, pace, sit, or rub her belly. Generally, women practice natural childbirth and do not ask for analgesia. Some may scream or cry and become hysterical, whereas others are stoic, only moaning and grunting. **Women in labor need support and reassurance. Applying a cold compress on the woman's forehead demonstrates caring and sensitivity.**
- Cesarean birth is feared because it is abdominal surgery.
- Fathers do not generally participate in labor and delivery believing that this is a private event that is best handled by women. The woman is not coached; female

members of the family provide assistance as needed. **Do not make men feel guilty if they do not want to participate in the delivery.**

- Postpartum, the woman takes an active role in her own care. She dresses warmly after birth as a way to become healthy and clean.
- Many believe that the bones are "open" after birth and that a woman should stay in bed during the first 2–3 days postpartum to allow the bones to close. Wearing an abdominal binder is another way to facilitate closing the bones.
- The postpartum woman also engages in "the three baths." For the first 3 days, the mother bathes in hot water boiled with special leaves that are either bought or picked from the field. She also drinks tea boiled from these leaves. For the next 3 days, the mother bathes in water prepared with leaves that are warmed by the sun. At this point, the mother takes only water or tea warmed by the sun. Another practice is to take a vapor bath with boiled orange leaves to enhance cleanliness and tighten the internal muscles. At the end of the third to fourth week, she takes the third bath, which is the cold bath.
- A cathartic may be administered to cleanse her intestinal tract. When the process is completed, she may drink cold water again and resume her normal activities.
- In the postpartum period, women avoid white foods such as lima beans, okra, mushrooms, and tomatoes because they are believed to increase vaginal discharge. Strength foods include porridge, rice, and red beans, plantains boiled or grated with the skins and prepared as porridge (the skin is high in iron), carrot juice, and carrot juice mixed with red beet juice.
- Breast-feeding is encouraged for up to 9 months postpartum. Breast milk can become detrimental to both mother and child if it becomes too thick or too thin. If it is too thin, it is believed that the milk has "turned," which may cause diarrhea and headaches in the child and, possibly, postpartum depression in the mother. If milk is too "thick," it is believed to cause *bouton* (impetigo).
- **Provide factual information and dispel myths related to breast-feeding.**
- Breast-feeding and bottle feeding are accepted practices. If the child develops diarrhea, breast-feeding is immediately discontinued. *Lok*, a laxative similar to the one administered to the older children in the summer, is given to infants to hasten the expulsion of meconium. **Stress the risks associated with lok and the need to prevent dehydration.**

### *17.1.8 Death Rituals*

- Generally, Haitians prefer to die at home rather than in the hospital.
- When death is imminent, the family may pray and cry uncontrollably, sometimes even hysterically. **Meet the person's spiritual needs by encouraging family members to bring religious medallions, pictures of saints, or fetishes.**
- When the person dies, all family members try to be at the bedside and have a prayer service. **If possible and if it is not too disturbing to other patients, encourage a family member to assist with postmortem care.**

- The preburial activity is called ***veye***, a gathering of family and friends who come to the house of the deceased to cry, tell stories about the deceased's life, and laugh. Food, tea, coffee, and rum are in abundant supply. The intent is to show support and to join the family in sharing this painful loss.
- Another religious ritual is the *dernie priye*, a prayer service consisting of 7 consecutive days of prayer, which facilitates the passage of the soul from this world to the next.
- On the seventh day, there is a mass called *prise de deuil*, which officially begins the mourning process. After each of these prayers, there is a reception/celebration in memory of the deceased.
- Haitians have a very strong belief in resurrection and paradise; thus, cremation is not usually an option.
- Most Haitians are very cautious about autopsies. If foul play is suspected, they may request an autopsy to ensure the patient is really dead. This alleviates their fear that their loved one is being zombified. According to the belief, this can occur when the person appears to have died of natural causes but is still alive. **Carefully explain the legalities of autopsy.**

### *17.1.9 Spirituality*

- Catholic religious beliefs are combined with voodooism. Voodooism involves communication by trance between the believer and ancestors, saints, or animistic deities. Participants gather to worship the *loa*, deities, or spirits, who are believed to have received their powers from God and are capable of expressing themselves through the possession of a chosen believer.
- Believers in voodooism attribute their ailments or medical problems to the doings of evil spirits. In such cases, they prefer to confirm their suspicions through the *loa* before accepting natural causes as the problem, which would lead to seeking Western medical care. Belief in the power of the supernatural can have a great influence on the psychological and medical concerns of patients.
- **Recognizing and accepting patients' beliefs alleviates barriers and may make patients feel more at ease to discuss their beliefs and needs.**

### *17.1.10 Health-Care Practices*

- Good health is the ability to achieve internal equilibrium between *cho* and *fret*. To become balanced, one must eat well, give attention to personal hygiene, pray, and have good spiritual habits. To promote good health, one must be strong, have good color, be plump, and be free of pain. To maintain this state, one must eat right, sleep right, keep warm, exercise, and keep clean.

- Illness is perceived as punishment, considered an assault on the body, and may have two different causes: natural illnesses, known as *maladi Bondye* (disease of the Lord), and supernatural illnesses.
- Natural illnesses are of short duration and are caused by environmental factors such as food, air, cold, heat, and gas.
- Supernatural illnesses are caused by angry spirits. To placate these spirits, patients must offer feasts called *manger morts*. If individuals do not partake in these rituals, misfortunes are likely to befall them. Illnesses of supernatural origin are fundamentally a breach in the rapport between the individual and his or her protector. Health can be recovered if the patient takes the first step in determining the nature of the illness. This can be accomplished by eliciting the help of a voodoo priest and following the advice given by the spirit itself. **To accurately prescribe treatment options, differentiate between natural and supernatural causes of illness and disease.**
- Most individuals believe that gas may provoke pain and anemia. Gas can occur in the head, where it enters through the ears; in the stomach, where it enters through the mouth; and in the shoulders, back, legs, or appendix where it travels from the stomach. When gas is in the stomach, the patient is said to suffer *kolik* (stomach pain). Gas in the head is called *van nan tet* or *van nan zorey*, which literally means "gas in one's ears" and is believed to be a cause of headaches.
- Gas moving from one part of the body to another produces pain. Thus, the movement of gas from the stomach to the legs produces rheumatism, to the back causes back pain, and to the shoulder causes shoulder pain. **Ask patients what they think is causing their pain.**
- Foods that help dispel gas include tea made from garlic, cloves, and mint; plantain; and corn. To deter the entry of gas into the body, one must be careful about eating leftovers, especially beans.
- Most Haitians engage in self-treatment and consider these activities as a way of preventing disease or promoting health, trying home remedies as a first resort for treating illness. If they know someone who has had a particular illness, they may take the prescribed medicine from that person.
- Haitians tend to keep numerous topical and oral medicines on hand. In Haiti, many medications can be purchased without a prescription, a potentially dangerous practice.
- **Admonishing patients may cause them to withdraw and not listen to instructions. Be very discreet in assessing, teaching, and guiding the patient toward safer health practices. Inquire if the patient has been taking medication that was prescribed for someone else.**
- Many Haitians in America ask friends or relatives to send medications from Haiti. Such medications may consist of roots, leaves, and European-manufactured products that are more familiar to them. **Ascertain what the patient is taking at home to avoid serious complications.**
- A primary respiratory ailment is ***oppression***, which is a term used to describe asthma. However, the term really describes a state of anxiety and hyperventilation rather than the condition. *Oppression* is considered a "cold" state, as are

many respiratory conditions. A home remedy for oppression is to take a dry coconut and cut it open, fill it half with sugarcane syrup and half honey, grate one full nutmeg and add it to the syrup mix, reseal the coconut, and then bury it in the ground for a month. The coconut is reopened, the content is stirred and mixed, and one tablespoon is administered twice a day until it is finished. By the end of this treatment, the child is supposed to be cured of the respiratory problem.

- Many Haitians have low-paying jobs that do not provide health insurance and they cannot afford to purchase it themselves. Thus, economics acts as a barrier to health promotion. Those who do not speak English well have difficulty accessing the health-care system, explaining their needs fully, or understanding prescriptions and treatments. **Obtain an interpreter when necessary.**
- Diabetes is considered a natural illness; however, most do not seek immediate medical assistance when they detect the symptoms of polyuria, excessive thirst, and weight loss. Instead, they attempt symptom management by making dietary changes on their own by drinking potions or herbal remedies. When the person finally seeks medical attention, he or she may be very sick. **Dispel myths and explain the medical regimen, diet, and medications for diabetes. Reinforce at every opportunity.**
- Pain is commonly referred to as ***doule***. Many Haitians have a low pain threshold. They are verbal about the cause of their pain and sometimes moan. They are vague about the location of the pain because they believe that it is not important; they believe that the whole body is affected because disease travels, making it very difficult to assess pain accurately. Injections are the preferred method for medication administration, followed by elixirs, tablets, and capsules.
- Chest pain is *doule nan ke mwen*, abdominal pain is *doule nan vent*, and stomach pain is *doule nan ke mwen* or *doule nan lestomak mwen*.
- Nausea is expressed as *lestomak/mwen ap roule, M santi m anwi vomi, lestomak/mwen chaje*, or *ke mwen tounin*. Because of modesty, they may discard vomitus immediately so as not to upset others. **Specific instructions should be given regarding keeping the specimen until the health-care provider has had a chance to see it.**
- Oxygen should be offered only when absolutely necessary because the use of oxygen is perceived as an indicator of the seriousness of the illness.
- Fatigue and physical weakness are interpreted as signs of anemia or insufficient blood. Various external and internal environmental factors are believed to cause *sezisman* (*fright*), thereby disrupting normal blood flow. *Sezisman* may occur when someone receives bad news, is involved in a frightful situation, or suffers from indignation after being treated unjustly.
- The stigma attached to mental illness is strong; most do not readily admit to being depressed. Depression can be viewed as a hex placed by a jealous or an envious individual.
- **Ask patients what they think is causing their depression. Negotiate treatment accordingly.**
- In an unnatural illness, the person's poor health is attributed to magical causes such as a hex, a curse, or a spell, which has been cast by someone as a result of

family or interpersonal disagreement. Symptoms include burning skin, rashes, pruritus, nausea, vomiting, and headaches and often coincide with psychological problems.

- Most are extremely afraid of diseases associated with blood irregularities. Patients and their families become emotional about blood transfusions. Thus, they are received with much apprehension. Additionally, blood transfusions are feared because of the potential for HIV transmission.
- **Factually explain the need for a blood transfusion, and carefully explain the procedure along with the involved risks. Involve patients and their families in the care as much as possible. Explain precautionary measures that have been taken to prevent blood contamination.**
- The body must remain intact for burial. Thus, organ donation and transplantation are not generally discussed or practiced. **Assess patients' beliefs about organ donation and involve a religious leader to provide support and help facilitate a decision regarding organ donation or transplantation.**

### *17.1.11 Health-Care Providers*

- Most Haitians resort to symptom management with self-care first and then spiritual care. They commonly use traditional and Western health-care providers simultaneously.
- Physicians and nurses are well respected. Physicians are men, and nurses are women. Nurses are referred to as "Miss." **Explain and re-explain relevant points to compensate for patients' knowledge deficits or language limitations.**

## 17.2 Reflective Exercises

1. In what languages would you prepare educational brochures for general health education for Haitians?
2. What are the primary roles of traditional Haitian men and Women?
3. Explain the principles of *fret* and *cho*. How are these principles applied in dietary practices? In health conditions?
4. What are the primary health conditions common with Haitians?
5. Explaining breast-feeding practices common with Haitian mothers. Which practices are potentially dangerous and how would you address them with the mother?
6. Explain the tenets of *voodism*. How is *voodism* seen in health-care practices?
7. Distinguish natural versus supernatural illness among Haitians.
8. What is *oppression*? What education would you include in a patient who has *oppression?*

9. What infectious diseases would you look for in recent immigrants from Haiti?
10. What are traditional Haitian verbal and nonverbal communication practices?

## Bibliography


CIA (2024) World Factbook: Haiti. https://www.cia.gov/the-world-factbook/countries/haiti/

Cultural Competency and Haitian Immigrants: Death rituals (2003). https://nj.gov/humanservices/emergency/HaitiCultural_ClinicalCarePresentation.pdf

Dain B, Batalova (2023). Haitian Immigrants in the United States. https://www.migrationpolicy.org/article/haitian-immigrants-united-states#income-poverty

Death and Funeral Rituals in the Haitian Culture (2013) Haiti Observer. http://www.haitiobserver.com/blog/death-and-funeral-rituals-in-haitian-culture.html

Finnegan N (2023) Haitian Immigrants' information needs and behaviors: libraries, information professionals and Haitians in the United States. Student Res J 12(2):3. https://scholarworks.sjsu.edu/ischoolsrj/vol12/iss2/3/

Haiti: Global Health (2016) USAID. https://www.usaid.gov/haiti/global-health

Hasalhoef JO (2023) Haitian immigration to the Americas as of 2023, by the numbers. Haitian Times. https://haitiantimes.com/2023/12/19/haitian-immigration-to-america-as-of-2023-by-the-numbers/

Pardo C, Brutus N, Labatte D, Saint-Jean F, Tribié S, Joseph I, Clare CA, Su J, Sánchez JP (2021) An introduction to structural Competency for Haitian-identified patients: history, culture, and access to care. MedEdPORTAL 17:11207. https://doi.org/10.15766/mep_2374-8265.11207. PMID: 35018304; PMCID: PMC8710435. https://www.ncbi.nlm.nih.gov/pmc/articles/PMC8710435/

World Bank in Haiti. https://www.worldbank.org/en/country/haiti/overview

# Chapter 18
# People of Hindu and Sikh Heritage

**Arvinder Singh**

## 18.1 Overview and Heritage

More than 1.4 billion people inhabit India. Over 80% of the population are Hindus, followers of Hinduism. Other significant religious groups include Muslims (13.4%), Christians (2.3%), Sikhs (1.9%), and other (1.8%). Although different religious sectors share many common cultural beliefs and practices, they differ according to the variant characteristics of culture as presented in Chap. 1. India is divided into north and south, based primarily on Dravidian and Aryan cultural variations.

Immigrants to the United States come predominantly from urban areas, including all major Indian states, and they have come in two waves. The first wave began in the early twentieth century and continued to the mid-1920s. Conditions such as racial discrimination and lack of access to economic advancement made it difficult for the first wave of Asian Indians to sustain themselves or their culture in America. The second wave of immigration began after 1965 and still continues. Most individuals from this wave are highly educated. Most come to the United States to attain a higher standard of living, better working conditions, and job opportunities. Secondary reasons include opportunities for additional education and Indian perceptions of America as a country of opportunity and freedom. Trafficking, primarily internal, of men, women, and children for the purposes of forced or bonded labor and commercial sex exploitation continues to be a major problem for Indians.

A. Singh (✉)
Baptist Health South Florida, Miami, Florida, USA
e-mail: arvinder.singh@baptisthealth.net


E. A. Fenkl, L. D. Purnell (eds.), *Handbook for Culturally Competent Care*,
https://doi.org/10.1007/978-3-031-70492-5_18

### 18.1.1 Communications

- Many Indian state borders are reorganized in accordance with language limitations. Languages fall into two main groups: Indo-Aryan in the north and Dravidian in the south. Hindi with numerous dialects is spoken by 41% of the population and is the national language, along with English. In addition, there are 14 other official languages: Bengali, Telugu, Marathi, Tamil, Urdu, Gujarati, Malayalam, Kannada, Oriya, Punjabi, Assamese, Kashmiri, Sindhi, and Sanskrit. English is the language among the educated. **Provide an interpreter for those who may not speak English well.**
- Women often speak in a soft voice, making it difficult to understand or decipher what they say. Men may become intense and loud when they converse with other family members. To an onlooker, it might seem disruptive but, in general, this form of communication can be construed as meaningful when it is conducted with close friends. Although with changes in modern society, this trend is following west with women empowerment. In Sikhs, women have relatively more freedom and comparable status to men.
- Traditional women avoid direct eye contact with men. Direct eye contact with older people and authority figures may be considered a sign of disrespect. **Do not misconstrue a lack of eye contact as not caring or not listening.**
- Touching and embracing are not acceptable for displaying affection. Even between spouses, a public display of affection such as hugging or kissing is frowned upon, being considered strictly a private matter.
- Temporality is past-, present-, and future-oriented. Time is conceived in cycles of four ages, which start with the "age of perfection" and end with the "age of degeneration."
- Punctuality in keeping scheduled appointments may not be considered important. **Do not misconstrue being late for appointments as a sign of irresponsibility or not valuing health. Reinforce the necessity of punctuality when needed.**
- Older family members are usually not addressed by name but as elder brother, sister, aunt, or uncle. A woman never addresses a man by name because the woman is not considered an equal or superior. The trend has changed in past decade where women are now addressing their spouses by name.
- Strangers are greeted with folded hands and a head bow.

### 18.1.2 Family Roles and Organization

- No institution in India is more important than the family. The hierarchical structure of authority is the patriarchal joint family, based on the principle of superiority of men over women.

- The male head of the family is legitimized and considered sacred by caste and religion, which delineate relationships. Central relationships in this system are based on the continuation and expansion of the male lineage through inheritance and ancestor worship, related to the father–son and brother–brother relationships. Due to recent changes in law, women are considered equal shareholders of parental assets; however, socially, its still frowned upon.
- A matrilineal system exists in a few areas in the southwestern and northeastern regions of the country; however, power rests with the men in the woman's family.
- A submissive and acquiescent role is expected of women in the first few years of married life, with little or no participation in decision-making. Strict norms govern contact and communication with the men of the family, including a woman's husband. However, millennials are open, and women participate in decision-making.
- Parents want their children to be successful and strongly encourage and emphasize scholastic achievement in fields that promise good employment and high social status. Parents in America want their children to maintain ties with their families and the Indian community.
- Although many parents expect and accept the Westernization of their children, the question of marriage is still a concern for parents who have opinions about how their children should be married, whether "arranged" or partly arranged. Hindu parents or Indians from all religious traditions want their children to marry other Indians. However, with western influence, cross cultural marriages are seen among Indian and Indian Americans but frowned upon socially.
- Arranged marriages at a young age are considered most desirable for women. The practice of an arranged marriage continues in the United States in order to minimize the stress associated with differences in castes, lifestyles, and expectations between the male and female hierarchy.
- The two major types of transfer of material wealth accompanying marriage are bride price and a dowry. Bride price is payment in cash and other materials to the bride's father in exchange for authority over the woman, which passes from her kin group to the bridegroom's kin group.
- In the joint family structure, Hindu women are considered "outsiders" and are socialized and incorporated in such a way that "jointness" and residence are not broken up; therefore, a close relationship between the husband and wife is disapproved because it induces favoring the nuclear family and dissolving the joint family. Therefore, a marriage is regarded as indissoluble. Due to westernization, trends have changed where divorces are becoming common among Indians and Indian Americans. In Sikhs, historically and religiously, women are granted similar status as men where they participate and share religious responsibilities with men.
- Older family members are held in reverence and cared for by their children when self-care becomes a concern.
- **Understand the various types of families (joint, extended, or nuclear) and determine which individual has control within the hierarchy.**

- Single-parent, blended, and communal families are not well accepted by Hindus. With changing trends, nuclear families are more common among Indians and American Indians.
- Homosexuality may cause a social stigma. **Refer lesbian, gay, or bisexual Hindu Americans to national support groups such as TriKone, a nonprofit group for lesbian, gay, and bisexual South Asians located in the San Francisco Bay area; the National AIDS hotline; or Asians Together in Washington, DC. Do not disclose the same-gender relationships to other family members. Due to recent changes in law, gay or same-sex marriages have become legal in India, however, social stigma persists.**

### *18.1.3 Workforce Issues*

- At work, Hindus adopt American practices and cultural habits. Hierarchies of age, gender, and caste prescribe transactions among Hindus. At work, relationships are a reproduction of the authority-dependence characteristic of family and social relationships.
- In seeking to establish a personal and benevolent relationship, Hindus may be seen as too eager to please, ingratiating, or docile, all antithetical to the task of assertion and independence. **Explain the value of assertiveness expected of all health-care providers in the American workforce.**
- Punctuality may not be considered important. **Explain the necessity of timeliness in the American workforce.**
- **Similar trends may be noted among American Sikhs. Practicing Sikhs may be seen wearing turbans (head covers) and iron bracelets in dominant hand, even at work.**

### *18.1.4 Biocultural Ecology*

- Indian diversity of physical types can be divided into three general groups according to the color of their skin: white in the north and northwest, yellow in areas bordering Tibet and Assam, and black in the south.
- *Indids* (whites) have a light-brown skin color, wavy black hair, dark- or light-brown eyes; are tall or of medium height; and are either dolichocephalic (long-headed) or brachycephalic (short-headed).
- *Melanids*, often referred to as the Dravidians, are the population of southern India and have dark skin ranging from light brown to black, elongated heads, broad noses, thick lips, and black, wavy hair. They are usually shorter than 5 ft. 6 in. tall.
- **Pallor in brown-skinned patients presents as a yellowish-brown tinge to the skin. Pallor in dark-skinned individuals is characterized by the absence of the underlying red tones in the skin. Jaundice may be observed in the sclera.**

**Table 18.1** Health conditions common to Hindus

| Health conditions | Causes |
|---|---|
| Malaria | Environment, lifestyle |
| Filariasis | Environment, lifestyle |
| Tuberculosis | Environment, lifestyle |
| Pneumonia | Environment, lifestyle |
| Cardiovascular diseases | Genetic, environment, lifestyle |
| Diabetes | Genetic, environment, lifestyle |
| Rheumatic heart disease | Genetic, environment, lifestyle |
| Sickle cell anemia | Genetic, environment, lifestyle |
| Dental diseases | Environment, lifestyle |
| Lactose intolerance | Genetic, lifestyle |
| Cancer of the cheek, nose, and mouth | Lifestyle |
| Breast cancer | Genetic, environment, lifestyle |
| Ichthyosis vulgaris | Genetic |
| Beriberi | Lifestyle |
| Thiamine deficiency | Lifestyle |
| Lathyrism | Lifestyle |
| Goiter | Environment, lifestyle |
| Osteomalacia | Environment, lifestyle |
| Endemic dropsy | Lifestyle |
| Stomach cancer | Environment, lifestyle |
| Fluorosis | Environment, lifestyle |

**The oral mucosa of dark-skinned individuals may have a normal freckling or pigmentation. Inspection of the nail beds, lips, palpebral conjunctiva, and palms of the hands and soles of the feet shows evidence of cyanosis.**

- Table 18.1 lists commonly occurring health conditions for Hindus in three categories: lifestyle, environment, and genetics.
- **Be alert to possible signs and symptoms and risk factors associated with diseases linked to migration from India.**
- Many individuals require lower doses of lithium, antidepressants, and neuroleptics, and they may experience side effects even with the lower doses. They also are more sensitive to the adverse effects of alcohol, resulting in marked facial flushing, palpitations, and tachycardia. **Question therapeutic regimens that do not consider racial or ethnic differences.**

## *18.1.5 High-Risk Health Behaviors*

- Alcoholism and cigarette smoking among Hindu Americans, especially among men, cause significant health problems. Sikhs are more prone to drinking problems, but refrain from smoking due to religious reasons. **Refer to "stop smoking" clinics, and encourage low alcohol consumption. Partner with elementary school teachers to introduce anti-smoking at a young age.**

- Diabetes is becoming more common among Indian Americans due to genetic predisposition, dietary habits, and stressful lifestyle.
- Beriberi is found in people coming from rice-growing areas, pellagra in maize-millet areas, and lathyrism in khesari-growing areas of Central India. Thiamine deficiency is common among people who are mostly dependent on rice. Thorough milling of rice, washing rice before cooking, and allowing the cooked rice to remain overnight before consumption the following day results in the loss of thiamine.
- Lathyrism is a crippling disease-causing paralysis of leg muscles; it occurs mostly in adults who consume large quantities of seeds of the pulse khesari, *lathyrus sativus*, over a long period.
- Goiters are common along the sub-Himalayan tracts.
- Osteomalacia is prevalent in northwest India.
- Endemic dropsy is prevalent in west Bengal as a result of the use of mustard oil for cooking.
- Cancers of the mouth and lip are common because of chewing *pan* and tobacco.
- Fluorosis occurs in parts of Punjab, Haryana, Andhra Pradesh, and Karnataka, resulting from drinking water with large amounts of fluoride. Cancer is becoming increasingly common among North/Northwest Indians due to environmental changes, heavy use of pesticides for farming and food adulteration.

### *18.1.6 Nutrition*

- Dietary habits are complex and regionally varied. Most believe that food was created by the Supreme Being for the benefit of man. The influence of religion is pervasive in food selection, customs, and preparation methods.
- Classification of regional food habits can be twofold, based on the types of cereals and fresh foods consumed. In the first category are rice and bread eaters; in the second category are vegetarians and nonvegetarians. Vegetarianism is firmly rooted in culture; the term *nonvegetarian* is used to describe anyone who eats meat, eggs, poultry, fish, and sometimes cheese. Many Brahmins in North India consider eating meat to be religiously sanctioned. In some parts of India, eating fish is acceptable to Brahmins, whereas in other parts, eating meat is sacrilegious. Sikhs are primarily vegetarians, American Sikhs are more liberal and consume all variety of meats and animal products. Assess cultural food choices upon admission to the health-care facility.
- Dietary staples include rice, wheat, *jowar, bajra*, jute, oilseeds, peanuts, millet, maize, peas, sugarcane, coconut, and mustard. Cereals supply 70–90% of the total caloric requirements. A variety of pulses or lentils, cooked vegetables, meat, fish, eggs, and dairy products are also consumed. Heavily spiced (curry) dishes with vegetables, meat, fish, or eggs are favored, and hot pickles and condiments are common. Spice choices include garlic, ginger, turmeric, tamarind, cumin,

coriander, and mustard seed. Vegetable choices include onions, tomatoes, potatoes, green leaves, okra, green beans, and root vegetables.

- In North India wheat is the staple food. Other cereals are *jowar*, *bajra*, and *ragi*, consumed in porridges, gruels, and *rotis* (baked pancakes).
- *Bajra*, a staple food in Maratha families, is not considered favorably in Uttar Pradesh. People from Punjab do not favor fish, and people from the south generally dislike the idea of meat of any kind. In Saurashtra in the south, fish, fowl, flesh, and eggs are taboo practically everywhere.
- Women generally serve the food but may eat separately from men. Women are not allowed to cook during their menstrual periods or have contact with other members of the family. In Sikhs culture is very liberal and women may join men during meals, very less restriction is seen for women related to menstruation and cooking.
- **Assess food rituals practiced by Hindus and Sikhs in relation to meal times and food selections before attempting dietary counseling.**
- Foremost among the perceptions of Hindus and Sikhs is the belief that certain foods are “hot” and others are “cold” and therefore should only be eaten during certain seasons and not in combination. The geographic differences in the hot and cold perceptions are dramatic; many foods considered hot in the north are considered cold in the south. Such perceptions and distinctions are based on how specific foods are thought to affect body functions. The belief is that failure to observe rules related to the hot and cold theory of diseases results in illness. A more detailed description of the hot and cold theory of diseases is provided under Health-Care Practices. **Given the diversity of Hindus in America, individually assess dietary practices and nutritional deficiencies of patients according to their ethnic origins and area of residence.**

### *18.1.7 Pregnancy and Childbearing Practices*

- Birth control methods include intrauterine devices, condoms, and rhythm and withdrawal methods. **Women may desire education in family planning from a same-sex health-care provider as well as assistance with delivery from female physicians, midwives, or female nurse practitioners.**
- Grandmothers, mothers, and mothers-in-law are considered to have expert knowledge in the use of home remedies during pregnancy and the postpartum period. Many older women frequently travel to the United States to assist new mothers in antenatal and postnatal care that is consistent with traditional customs.
- The birth of a son is a blessing because the son carries the family name and takes care of the parents in their old age. The birth of a daughter is cause for worry and concern because of the traditions associated with dowry, a ritual that can impoverish the lives of those who are less affluent. With modernization, this thinking is changing and especially in urban areas, birth of daughter is welcomed.

**Table 18.2** Practices during pregnancy

| |
|---|
| • Based on the hot and cold theory of disease, certain "hot" foods like eggs, jiggery (a course brown sugar), coconut, groundnut, maize, mango, papaya, fruit, and meat are avoided during pregnancy because of a fear of abortion caused by heating the body or inducing uterine hemorrhage |
| • Pregnancy is a time of increased body heat; hence "cold" foods such as milk, yogurt, and fruits are considered good. Buttermilk and green leafy vegetables are avoided because of the belief that these foods cause joint pain, body aches, and flatulence. Minor swelling of the hands and feet is thought to result from increased heat and is not of much concern |
| • Morning sickness is caused by an increase in body heat. Burning sensations during urination, scanty urine, or a white vaginal discharge are considered serious signs of significant overheating |
| • Overeating and consumption of high-protein foods, including milk, are avoided because such foods result in an exaggerated growth of the baby that may lead to a difficult delivery |
| • Anemia caused by iron deficiency is one of the nutritional disorders affecting women of childbearing age. This condition may be aggravated because of the practice of reducing the consumption of leafy vegetables to avoid producing a dark-skinned baby |

- Table 18.2 describes various cultural practices related to pregnancy.
- No taboo against the father being in the delivery room exists, but men are usually not present during birthing. Instead, they tend to wait outside the delivery room and allow female relatives to support the pregnant mother during labor and delivery. **Report labor progress to fathers who prefer to stay in the waiting room.**
- Because self-control is valued, women suppress their feelings and emotions during labor and delivery. **Closely observe nonverbal communication, such as a change in body posture, restlessness, and facial expressions during labor and provide assistance as necessary.**
- Beliefs and practices used during the postpartum period are discussed in Table 18.3.

### *18.1.8 Death Rituals*

- A tenet of Hinduism is that the soul survives the death; death is a rebirth. Therefore, by performing a ritual bath, sprinkling holy river water over the body, covering the body with new clothes, daubing parts of the body with ghee, and chanting Vedic utterances, the deceased is considered purified and strengthened for the post-mortem journey.
- Hindus prefer to die at home. The eldest son is responsible for the funeral rites. In the absence of the oldest son, determine the spokesperson and decision maker for the family.
- The death rite is called ***antyesti,*** or last rites. The priest pours water into the mouth of the deceased and blesses the body by tying a thread around the neck or wrist. The priest may anoint with water from the holy Ganges River or put the sacred leaf from the *Tulsi* plant in the mouth. The eldest son completes prayers

**Table 18.3** Postpartum beliefs and practices

| |
|---|
| • During the postpartum period, the mother remains in a warm room and often keeps the windows closed to protect her against cold drafts. Exposure to air conditioners and fans, even in warm weather, may be considered dangerous. Provide warm clothing and additional blankets for the mother and baby to keep them warm |
| • After the birth, both the mother and the baby undergo purification rites on the 11th day. The postpartum mother is considered to be impure and is confined to a room. The pollution lasts for 10 days. This period of necessitated and mandatory confinement assists in bonding between the mother and the newborn and provides the mother with adequate rest and time to tend to the baby's needs. The baby is officially named on the 11th day during the "cradle ceremony," and several rituals are performed to protect the baby from evil spirits and to ensure longevity |
| • A sponge bath for the newborn is recommended until the umbilical cord falls off |
| • Soft massage to the extremities is recommended before bathing the infant |
| • Washing the infant's hair daily is believed to improve the quality of the hair |
| • During the postpartum period, hot foods such as brinjals, drumsticks, dried fish, dhal, and greens are good for lactation. Cold foods are thought to produce diarrhea and indigestion in the infant. Cold foods such as buttermilk and curds, gourds, squashes, tomatoes, and potatoes are restricted because they produce gas. Such abstentions are primarily practiced for the baby's health because harmful influences might be transmitted through the mother's breast milk |
| • Some believe that colostrum is unsuited for infants. Most women think that the milk does not "descend to the breast" until their ritual bath on the third day and, as a result, newborns are fed sugar water or milk expressed from a lactating woman |
| • Breast milk is commonly supplemented with cow's milk and diluted with sugar water. A child's stomach is considered weak as a result of diarrhea; therefore, the child is given diluted milk |
| • Sources of protein such as eggs, curds, and meat are avoided because they might adversely affect the baby |
| • The mother's diet the first few days is restricted to liquids, rice, gruel, and bread |
| • Boiled rice, eggplant, curry, and tamarind juice are added to the diet between 6 months and a year after the birth of the baby |
| • Obtain dietary preferences and practices from the family before planning nutritional counseling |

for ancestral souls, but all male descendants perform the rites; each offers balls of rice on behalf of the deceased.

- The body is usually cremated rather than interred. The ashes are immersed or sprinkled in the holy rivers. Such immersions are of great benefit to the souls of the dead. Hindus may save their family's ashes to later scatter them in holy rivers when they return to their homeland.
- Women may respond to the death of a loved one with loud wailing, moaning, and beating their chests in front of the corpse, attesting their inability to bear the thought of being left behind to handle situations by themselves. **Offer support and understanding with respect for death and grief behaviors.**
- Death rituals in Sikhs are like Hindus. Corpse is given final bath before cremating the body, followed by pouring ghee into mouth of the corpse. After cremation, ashes are immersed in holy river followed by last prayer at the deceased's home. Wake is usually observed for 3–5 days after death.

### 18.1.9 Spirituality

- Hinduism, the largest religion and oldest tradition practiced in India, represents a set of beliefs and a definite social organization. Hinduism denotes belief in the authority of Vedas and other sacred writings of the ancient sages, immortality of the soul and a future life, existence of a Supreme God, the theory of **karma** and rebirth, the theory of the four stages of life, and the theory of four *Purusarthas*, or ends of human endeavor. *Karma* stresses the individual's responsibility for one's actions and is interpreted in terms of past life. One's present condition is seen as a result of one's actions in a past life or lives.
- Sikhs believe in karma and reincarnation. However, they believe in one god and their holy book is Sri Guru Granth Sahib. Sikhs believe in hard work, sharing and meditation. Practicing Sikhs may also wear their five articles of faith which includes iron bracelet in dominant hand, small dagger for self-defense, long undergarment, small wooden comb and do not cut or shave hair from any part of their body.
- Orthodox Hindus view society as divinely ordained on the basis of the four castes: (a) *Brahmin*, the highest caste, priests and scholars, emerged from the head of God; (b) *Kshtriya*, warriors, from the arms; (c) *Vaisya*, merchants, from the waist; and (d) *Sudra*, menials, from the feet of God. The four-fold caste system is a theoretical division of society to which tribes, clans, and family groups are affiliated. Although religion does not bestow the caste system with a religious sanction, the great Hindu legal codes are based on the caste system. **Identify Hindu religious beliefs and practices, and incorporate these beliefs into patients' care. Respectfully ask Sikh patients if they would like to take off their articles of faith and get consent before cutting or shaving hair from any body part.**
- **Assess the extent to which religion is a part of the individual's life, how religious beliefs are related to the individual's perception of health and illness, and the individual's daily religious practices. Assessing spiritual life is essential for identifying resources and solutions for therapy.**
- Women often fast one day a week or for a lunar month to fulfill a vow made to a deity in supplication for a particular blessing. Sikh men/women do not believe in fasting.
- Shrines may be set up in the living room or the dining room but are most often located in a back room or in a closet. The shrine typically contains representations or symbols of one or more deities. Sikh shrines/temples usually have holy sculptures to read from, Sikhs do not worship idols or photos.

### 18.1.10 Health-Care Practices

- Physical examinations are especially traumatic to women who may not have experienced or heard about Pap tests and mammography examinations. **Provide female health-care providers for intimate care.**

- Most believe that illnesses attack an individual through the mind, body, and soul.
- Some believe that excessive consumption of sweets may cause roundworms and that too much sexual activity and worry are associated with tuberculosis. Others believe that diarrhea and cholera are caused by a variety of improper eating habits. **Explain factual information about parasitic and infectious diseases.**
- Suffering of any kind produces hope, which is essential to life. To maintain harmony between self and the supernatural world, the belief that one can do little to restore health by oneself provides a basis for ceremonies and rituals.
- Worshiping goddesses, pilgrimages to holy places, and pouring water at the roots of sacred trees are believed to have medicinal effects in healing the sick person and in appeasing the planets to help prevent illnesses and misfortunes. Sikhs may believe in consuming holy water or pouring holy water on affected organ/body part to heal.
- In *Ayurveda*, the traditional system of medicine in India, the primary emphasis is on the prevention of illnesses. Individuals have to be aware of their own health needs. One of the principles of Ayurveda includes the art of living and proper health care, advocating that one's health is a personal responsibility. The key to health is an orderly daily life in which personal hygiene, diet, work, and sleep and rest patterns are regulated. A daily routine has to be established and changed according to the season.
- A common health problem is self-medication. Those migrating to America are accustomed to self-medicating and may bring medications with them or obtain medications through relatives and friends. **Specifically ask patients about folk practices, use of over-the-counter medicines, and medicines brought from India.**
- The traditional healers, *nattuvaidhyars*, use Ayurvedic, Siddha, and Unani medical systems. These systems are all based on the Tridosha theory. The Ayurvedic system uses herbs and roots; the Siddha system, practiced mainly in the southern part of India, uses medicines; and the Unani system, similar to the Siddha, is practiced by Muslims.
- According to the Tridosha theory, the body is made up of modifications of the five elements: air, space, fire, water, and earth. These modifications are formed from food and must be maintained within proper proportions for health. A balance among three elements, or humors—phlegm or mucus, bile or gall, and wind—corresponds to three different types of food required by the body. Table 18.4 presents the three types of foods and the allopathic equivalents of diseases associated with them.
- Because of their religious beliefs of karma, Hindus may attempt to be stoic and may not exhibit symptoms of pain. Pain is attributed to God's will, the wrath of God, or a punishment from God and is to be borne with courage. **Because of stoicism, rely on nonverbal manifestations as well as verbal expressions when assessing pain.**
- Families tend to be protective of ill members. They may not want to disclose the gravity of an illness to the patient or discuss impending disability or death for fear of the patient's vulnerability and loss of hope, resulting in death. **The conflict between medical ethics and patients' values may pose a problem for**

**Table 18.4** Bodily manifestations of various foods

| |
|---|
| • Heat-producing foods: *Brinjals*, dried fish, green chilies, raw rice, and eggs. *Pittham* foods include cluster beans, cowgram, groundnuts (peanuts), almonds, millet, oil, and runner beans. Allopathic equivalents of *sudu* diseases include diarrhea, dysentery, abdominal pain, and scabies. Allopathic equivalents of *pittham* diseases include vomiting, jaundice, and anemia |
| • Cooling foods: Tomatoes, pumpkin, kul, gourds, greens, oranges, sweet limes, carrots, radishes, barley, and buttermilk. Cold, headache, chill, fever, malaria, and typhoid are allopathic equivalents of cool diseases |
| • Gas-producing foods: Root vegetables such as potato, sweet potato, and elephant yam; plantain; and drumstick. Joint pains, paralysis, stroke, and polio are disorders related to gas-producing foods |
| • Heating and cooling effects are produced in the body and hence are not related to the temperature or spiciness of foods. An imbalance leads to disease. If too much heat is in the body from consuming heat-producing foods, then cold foods need to be eaten to restore balance |

**health-care providers, who need to be cognizant of the importance of the family members' wishes and values regarding the care of their loved ones.**
- The sick role is assumed without any feeling of guilt or ineptness in doing one's tasks. The individual is cared for and relieved of responsibilities for that time.
- Psychological distress may be demonstrated through somatization, which is common, especially in women. The symptoms may be expressed as headaches, a burning sensation in the soles of the feet or in the forehead, and tingling pain in the lower extremities. Because of the stigma attached to seeking professional psychiatric help, many do not access the health-care system for mental health problems. Mental illness is considered to be God's will. **Do not disclose to outsiders that patients are receiving mental health counseling.**
- No Hindu policy exists that prevents receiving blood or blood products. Donating and receiving organs are acceptable.

## *18.1.11 Health-Care Providers*

- Although Hindus in general have a favorable attitude toward American physicians and the quality of medical care received in the United States, relatives and friends are usually consulted before health-care providers.
- Physicians are considered omnipotent because God grants cures through physicians. Patients tend to be subservient and may not openly question physicians' behavior or treatment. If they are not pleased with the treatment, they just change physicians. However, they tend to be appreciative of the information that physicians provide about their illness. **Provide factual information about diseases and illnesses. Explain the dangers of "shopping around" for health-care providers.**
- The physician is also viewed like an older person in the family; a protective, authoritative, and responsible relationship; a parent–child relationship; and a

*guru-chela* (teacher-disciple) relationship. **Patients expect physicians to teach them about the disease and how to get cured in a friend-to-friend relationship.**
- In mental health, traditional healers, such as *Vaids*, practice an empirical system of indigenous medicine; *mantarwadis* cure through astrology and charms; and *patris* act as mediums for spirits and demons. **Specifically ask patients if they are using folk health-care providers and what treatments have been prescribed.**
- Women are especially modest, generally seeking female health-care providers for gynecologic examinations. **Respect modesty by providing adequate privacy and assigning same-gender caregivers whenever possible.**
- **Explain the procedures, provide privacy, and assign a female health-care provider to decrease the stress and discomfort associated with a pelvic examination.**

## 18.2 Reflective Exercises

1. What are the primary religions practiced by Hindus?
2. What are the two national languages of India? What other languages are common?
3. How are older Hindu family members addressed?
4. Explain why arranged marriages are common and preferred by Hindus.
5. What is the difference between bride price and a dowry?
6. Distinguish between *Indids* and *Melanids.*
7. What health conditions are common to Hindus?
8. Identify at least three common beliefs related to pregnancy and at least three related to postpartum?
9. What are the principles of *Ayurveda?*
10. What are the principles of the Hindu religion?

## Bibliography

CIA (2024) World factbook: India. https://www.cia.gov/the-world-factbook/countries/india/

Heart of Hinduism: Family Structure (2014). https://iskconeducationalservices.org/HoH/lifestyle/lifestyle/

Hindu Beliefs Affecting Health Care (n.d.). https://www.health.qld.gov.au/__data/assets/pdf_file/0027/157608/hbook-hindu-s2.pdf

Hoffman A and Batalova J (2022) Indian immigrants in the United States. migrationpolicy.org. https://www.migrationpolicy.org/article/indian-immigrants-united-states

Importance of Marriage (2024). https://imperialmatrimonial.com/importance-of-marriage/

India. United States Census Bureau (2024). https://www.census.gov/popclock/world/in

Jambanathan J (2013) People of Hindu heritage. In: Purnell L (ed) Transcultural healthcare: a culturally competent approach, 4th edn. F.A. Davis Company, Philadelphia

National Center for Complementary and Integrative Health (2015) Ayurvedic medicine in depth. https://www.nccih.nih.gov/health/ayurvedic-medicine-in-depth

Sen CD (2009) Food culture in India. In: Food cultures around the world. Greenwood Publishing Group, London

# Chapter 19
# People of Hmong Heritage

**Jennifer Kue**

## 19.1 Overview and Heritage

Hmong (pronounced "*Mong*"—the "H" is silent) people lived primarily in the mountainous areas shared by China, Myanmar (Burma), Vietnam, Thailand, and Laos. Most Hmong who live in the United States (US) originate from Laos, a country with a population of more than 7.8 million, where only 9.2% identify as Hmong. The Hmong traditionally are an agrarian society that practiced "slash and burn" agriculture. Mountainous areas were cleared of underbrush, burned, and then used for crops. When the soil became depleted, they moved on, often relocating their village as well. They are believed to have originated in the Yellow River Valley of China.

Hmong people began immigrating to the US and other countries in 1975 following the Vietnam War. These refugees originated from the mountainous regions of Laos, where they had fought alongside the CIA during the "secret war" in Laos. They were targeted for genocide because they opposed the communist Pathet Lao, and as a result, they fled their country. They escaped to Thailand by traversing jungles and crossing the Mekong River. Many Hmong died due to war, genocide, or while attempting to flee. Hmong immigrants in the US bear the scars of war, including bullet and shrapnel wounds, as well as the lasting effects of exposure to biological warfare, which they referred to as "yellow rain." According to the US Census, as of 2023, there are approximately 336,000 Hmong people living in the US.

While the Central Valley of California was not initially settled by the Hmong, many now reside there. California has the largest Hmong population in the US, followed by Minnesota and Wisconsin. Small groups still tend to migrate to areas where they perceive economic opportunities exist. Hmong people have settled in

J. Kue (✉)
College of Nursing, University of South Florida, Tampa, FL, USA
e-mail: jkue3@usf.edu



E. A. Fenkl, L. D. Purnell (eds.), *Handbook for Culturally Competent Care*,
https://doi.org/10.1007/978-3-031-70492-5_19

North Carolina, Oregon, Colorado, Michigan, and other states. In 2004, the final refugee settlement in Thailand was shut down and 15,000 new refugees were relocated to the US.

The Hmong in Laos were preliterate farmers with generally limited or no access to education. They lived in very primitive and traditional circumstances and had no access to the modern world or modern medicine. When immigrating to the US, many Hmong experienced culture shock in a world that was completely unfamiliar to them. Depression, post-traumatic stress disorder, and suicide are prevalent in this population. More recent immigrants have received some education and have been exposed to the modern world. This aggregate profile may not fit well-acculturated Hmong individuals in the US based on length of time in the US and the diverse characteristics of culture (see Chap. 1).

***Paj ntaub* (pronounced *"pan dow"*)**, or "flower cloth," is an art form that includes batik designs, appliqué, reverse appliqué, and cross stitch created by Hmong women. Paj ntaub are used to decorate traditional clothing and create historical story cloths. Story cloths were used to pass down family history and culture from one generation to the next. Images and patterns carry representational and symbolic meanings. One such example is the triangle pattern, which symbolizes mountains, like the ones found in Laos where the Hmong people lived. Even today, remarkable story cloths are made depicting the Hmong fighting the communists in the jungles, Hmong being killed, yellow rain falling on villages, Hmong fleeing through jungles, and Hmong floating across the Mekong River to the refugee camps of Thailand.

### *19.1.1 Communications*

- **Ensure that a professional healthcare interpreter who can speak the patient's language is available.** Hmong in the US speak either a dialect of White Hmong (Hmoob dawb) or Green Hmong (Hmoob ntsuab), also known as Blue Hmong. These dialects are mutually intelligible, with variations in pronunciation and vocabulary.
- **Avoid providing written instructions in Hmong unless someone in the household can read them. Instructions in English may be preferable, as school-aged children may be able to read them.** The Hmong community has a rich tradition through which history, cultural knowledge, art, and ideas are passed from one generation to the next through oral communication. The Hmong did not have a written language until the 1950s, when Christian missionaries created a written form of their language. Due to the absence of a written Hmong language, many older Hmong individuals are unable to read or write in their native language.
- Several medical terms do not have equivalents in the Hmong language. For example, there are no equivalent words in the Hmong language for cancer,

cervix, or pancreas. Some words and phrases in Hmong may be challenging to translate into English.

- Be aware that Hmong immigrants born in Laos may not have their accurate age on their documents. For Hmong people born in Laos, their ages are often unknown; immigration officials assign an age to them before they immigrate to the US.
- **Avoid prolonged direct eye contact. Do not rush to ask questions; engage in small talk first if the situation is not urgent.**
- Many Hmong people believe that Americans are rude because they make direct eye contact when speaking and their questions are too straightforward. Proper communication when speaking to a Hmong person involves using fleeting glances without staring. Making small talk before inquiring about someone's health is polite and essential. Hmong people also use the word "yes" to indicate that they hear what you are saying. Saying "yes" does not necessarily indicate understanding or a commitment to follow through with your request.
- **To determine if treatments are understood, request a demonstration.** This is also known as the teach-back technique, a method used to assess and confirm whether the patient understands the provider's instructions.
- **When inquiring about Hmong individuals' adherence to treatment recommendations, it is most effective to request a demonstration of their actions.** For example, their method for conducting blood sugar testing. To determine the amount of medication taken, check the number of pills in the vial and the date the prescription was filled. In general, the Hmong community has a strong desire to be viewed positively by people in positions of authority. This social desirability factor may lead individuals to tell the healthcare provider what they believe the provider wants to hear, rather than what is happening; this is often seen as being respectful.
- **Obtaining informed consent is a legal requirement in the US, but it directly contradicts Hmong traditional decision-making.** In the Hmong culture, the male head of the family or clan is often responsible for making decisions for family members, including medical decisions.
- Due to the Hmong community's limited exposure to Western medicine and surgical procedures, they frequently experience significant fear in medical situations. They do not understand what is happening and they sometimes distrust medical personnel, especially if the person is a student. Rumors persist among the Hmong community that they are used for practice by medical students. If any treatment is labeled as experimental, this only reinforces their belief that they are being used for experiments, leading them to likely refuse the treatment. Obtaining consent from the head of a family may lead to treatment delays, as the head of the clan may live in another state.
- **To successfully obtain consent, it is crucial to respect the individuals' wishes, wait for family members to arrive, meet with the family members to explain the situation and accept their decision. If the patient's wishes differ from those of the head of the family, most patients will defer to the head of the family's decision.**

- The Hmong believe that it is inappropriate to say negative things in front of sick people. Words have great power, and the Hmong believe that speaking negative words in front of an ill person can make those things happen.
- **It is crucial to have a professionally trained healthcare interpreter who can work with the family to convey information in a culturally congruent manner.**
- Many Hmong individuals have attained higher education and degrees in fields such as nursing, medicine, psychology, and social work. In Western culture, attaining education and a high position commands respect and authority. In the Hmong culture, patients may not show the same deference toward Hmong healthcare providers and may treat them based on their position within the Hmong family or community hierarchy. This creates additional stressors for Hmong healthcare providers and interpreters, who may be required to prioritize the preferences of Hmong patients who have higher status in their community.
- **Be aware that Hmong healthcare providers and interpreters may be related to the patient, and they could be placed in an untenable situation because of a clan hierarchy that Western individuals may not fully comprehend.**

### *19.1.2 Family Roles and Organization*

- The Hmong people are organized into 18 clans, each with a surname that is shared by all men and children. Although wives and mothers typically retain the clan name from their birth, they are still regarded as members of their husband's clan. Hmong people have no single leader; instead, each clan has older males who serve as leaders. If a husband has died, the oldest son is expected to make decisions for the family. Daughters are expected to marry and live with their husbands and in-laws.
- Plural marriages were common in Laos, and they continue to exist today. Hmong men may marry as many women as they can afford to support. This is less common among Hmong people in the US and those who have moved away from traditional beliefs.
- Single Hmong individuals may find partners or spouses at the annual Hmong New Year celebrations, which are held between Christmas and New Year's Day. Marriages are forbidden between individuals who share the same clan name, regardless of how distant their relation. Young Hmong women are still victims of "capture bride," a practice in which a young girl is taken to her future husband's home overnight and then declared married. This practice is being discouraged because it has led to young women escaping and reporting their kidnappings to the police.
- Family is considered an important part of Hmong culture, so marriage is seen as a duty. There is significant social pressure for young Hmong girls to marry, often in their early teens. These marriages are not recognized as legal marriages by Western standards, but rather as traditional Hmong ceremonies; therefore, they do not violate state laws.

- Traditionally, Hmong men and women tend to marry young; however, this practice has changed, and it is now more acceptable to delay marriage for reasons related to socioeconomic status and shifts in cultural norms.
- **Do not reveal disgraces to family members.** The Hmong community in the US is well-connected through telephone, e-mail, and frequent family visits. Family "disgraces" are widely known even among Hmong people living in different parts of the US. "Disgraces" are considered taboo and can cause a family to "lose face." Examples may include birth defects, mental health issues, or a divorce in the family. Any of these "disgraces" are considered to reflect on every member of a family and, as a result, decrease their chances of forming a successful marriage.
- **Do not reveal same-sex relationships to the family members of others.** Same-sex relationships are not widely accepted in Hmong culture.

### *19.1.3 Workforce Issues*

- Hmong people have the highest unemployment rate among all Asians in the US. They have the lowest socioeconomic status among all Asian groups in the US. First-generation immigrants often face challenges in finding employment due to limited English language proficiency and a lack of workplace skills. Having a large family leads to lower employment rates for women.
- **Requesting demonstrations helps to assess knowledge so that better instructions can be given in the workplace.** Hmong individuals who acquire adequate English language skills may secure employment in low-skill occupations (e.g., factories or agriculture). Their desire for social acceptance may lead to them saying "yes" to questions about knowing how to do something, even when they do not understand.
- The Hmong community is rapidly attaining higher levels of education and socioeconomic status. This has led them to assist other family members in living better lives and attaining higher education. Hmong individuals are now occupying various professional roles and serving in government positions. They are hardworking and loyal to their employers.

### *19.1.4 Biocultural Ecology*

- Since resettling in the US, obesity has been rapidly increasing among children and adults. Hmong Americans are at significant risk of developing type 2 diabetes. One study found that Hmong patients were three times more likely to be diagnosed with type 2 diabetes compared to non-Hispanic White patients.
- Table 19.1 presents a list of frequently occurring health conditions for the Hmong in three categories: lifestyle, environment, and genetics. See Chap. 2.

**Table 19.1** Health conditions common to the Hmong

| Health conditions | Causes |
|---|---|
| Cervical cancer | Environment, lifestyle |
| Depression, anxiety, suicidal ideation, and post-traumatic stress syndrome | Environment, lifestyle |
| Tuberculosis | Environment, lifestyle |
| Hepatitis B | Environment, lifestyle |
| Parasitosis | Environment, lifestyle |
| Thalassemia | Genetic |
| Liver cancer | Environment, lifestyle |
| Trichinosis | Environment, lifestyle |
| Nasopharyngeal cancer | Environment, lifestyle |
| Lactase deficiency | Genetic |
| Diabetes | Environment, lifestyle, genetic |
| Hypertension | Environment, lifestyle, genetic |
| Asthma | Environment, lifestyle |

- **Encourage adherence to long-term mental health counseling and adherence to medication regimens.** Common health conditions among the Hmong community include depression, anxiety, suicidal thoughts, and post-traumatic stress disorder, stemming from their experiences during the war in Laos, their treacherous flight to safety, and cultural stressors. Many Hmong people continue to have nightmares and flashbacks to the terrors they experienced in Laos. Unemployment and evolving gender roles in families contribute to additional cultural stressors.
- Type 2 diabetes and hypertension are prevalent due to the rising obesity rates in both Hmong children and adults. Contributing factors to the increase in obesity rates include lifestyle and environmental changes, such as lack of physical activity and poor diet.
- **When evaluating nutrition, inquire about the patient's liquid intake;** this information is often not disclosed unless specifically asked and may reveal high consumption of sugary juices and drinks.
- **Avoid teaching about diet using measurements from the Western system.** Measuring foods in cups, tablespoons, etc., may not be understood by Hmong patients because baking is uncommon in traditional Hmong households, and they may not possess these utensils. A "cup" may refer to a drinking cup to the patient.
- **It is important to instruct patients in the proper use of inhalers and other medications.** Hmong of all ages in central California have a high prevalence of asthma. Exacerbations are frequent when air quality is poor.
- **Consider paragonimias in cases of pulmonary infections.** Paragonimiasis is a parasitic infection acquired in Southeast Asia. The parasites settle in the lungs, causing a widespread infection. Some elderly Hmong individuals may experience breathing difficulties due to paragonimiasis.

- **All new immigrants showing signs or symptoms of tuberculosis should undergo clinical evaluation for the disease.** Recent immigrants have been diagnosed with tuberculosis at high rates, and some have been infected with drug-resistant strains of the disease.
- Hepatitis B occurs at high rates and often affects all members of a family.
- **Assist the family in complying with public health monitoring. Screen all new immigrants for hepatitis B. If one family member has hepatitis B, all family members should be screened.**
- *Helicobacter pylori* is also commonly associated with peptic ulcers and adenocarcinoma of the gastrointestinal tract. *H. pylori* transmission may be due to sociocultural and environmental factors. The Hmong traditionally have high consumption of rice and preserved foods. Differences in health belief systems and use of Western medicine as a last resort may delay early diagnosis and treatment.
- When individuals are ill, traditional cupping or coin rubbing is practiced by pricking the skin to release blood, as it is believed to expel bad spirits. This practice involves using non-sterilized sewing needles, which may lead to the transmission of hepatitis B within the family.
- **Teach the family how to sterilize needles and other sharp instruments before performing coining.**
- The Hmong population has been found to have higher rates of cancers such as nasopharyngeal carcinoma, gastric cancer, hepatic cancer, and cervical cancer.
- Cervical cancer incidence and mortality rates among Hmong women are three to four times higher than those among non-Hispanic White women. Cervical cancer screening rates are low due to a lack of awareness, limited healthcare access, and cultural and linguistic barriers to screening.
- **Ensure that professionally trained healthcare interpreters are available at all times.** The greatest barrier to healthcare for the Hmong community in the US is the lack of accurate and culturally appropriate communication.

### *19.1.5 High-Risk Health Behaviors*

- **Encourage Pap smears as part of routine examinations and during other healthcare visits for women.** Hmong women have a high incidence rate of cervical cancer, yet they are the least likely among all Asian ethnic groups to undergo screening. Contributing factors to low screening rates include lack of awareness, discomfort or pain during the procedure, embarrassment, and modesty. Failure to undergo regular gynecological examinations with Pap smears increases the risk of advanced cervical cancer.
- **Explain the adverse effects of cigarette smoking, promote cessation, and collaborate with schools to educate about the dangers of smoking at a young age.** Many individuals are unaware that cigarette smoking can cause cancer. E-cigarette use is the most common form of tobacco use among Hmong adolescents.

### 19.1.6 Nutrition

- The Hmong people were primarily farmers, and many have developed small farms where they sell fruits and vegetables. Even Hmong living in apartments have small plots for growing vegetables, and their homes may have chicken pens in the backyards.
- Rice is the primary staple of the Hmong diet. Vegetables, fish, chicken, and pork are served with the rice. Spicy peppers are used to make into a condiment that accompanies the meal.
- **It is advisable to avoid consuming raw pork due to the potential for serious health risks. Encourage physical activity and maintaining a balanced, healthy diet.** Occasionally, a special dish called "*laub*" is made with raw pork, vegetables, and spices. This increases the risk for trichinosis, therefore, many young Hmong people have started to prefer American diets.

### 19.1.7 Pregnancy and Childbearing Practices

- Children are highly esteemed in Hmong families, and having large families is seen as an advantage. Hmong families tend to live in multigenerational households. Many older adults often assist in caring for their grandchildren.
- According to Hmong tradition, women are believed to be particularly susceptible to illnesses and misfortunes during the first 30 days after giving birth. Therefore, a woman must confine herself and follow strict postpartum practices, also known as *nyo dua hli.*
- **New mothers may not want to eat hospital food, so allowing families to bring traditional foods from home could be beneficial.** Women are prohibited from drinking cold beverages and spicy foods during pregnancy. Postpartum, white rice and chicken soup cooked with medicinal herbs are the traditional diet for 30 days. After giving birth, women must only consume "hot" food and drink, which is believed to help the body to expel postpartum blood.
- New mothers are expected to rest after giving birth. A woman should refrain from engaging in strenuous activities, such as lifting heavy objects or doing heavy cleaning. The mother-in-law and husband help the new mother during this period.
- Hmong women view regular menstruation as a sign of good health and may prefer not to use birth control methods that disrupt the regular menstrual cycle.

### 19.1.8 Death Rituals

- Hmong funerals are unique and can last for several days. The older or more respected the person, the longer the funeral may be. Family members are expected to attend the funeral and stay for the duration. Specific rituals are required, including animal sacrifice, to honor the deceased. These animals are used to provide food for people attending the funeral. A proper burial enables the deceased to transition to the spirit world in a positive manner, which is believed to lead to reincarnation. Most Hmong people believe in the existence of multiple souls that reincarnate. One soul stays in the vicinity of the body, while another remains in the present world, overseeing and caring for the family. Rituals are conducted to honor deceased ancestors.

### *19.1.9 Spirituality*

- Traditionally, the Hmong practiced animism and ancestor worship. They use traditional or spiritual healers (i.e., shamans) to prevent and treat illnesses.
- Most older and more recent immigrant Hmong adhere to traditional animist beliefs, where ancestors are revered and spirits are believed to be widely distributed in the world, residing in many inanimate objects or places such as trees, rivers, and houses. Many people believe that spirits have the power to cause harm, misfortune, illness, or death, as well as to provide protection or prevent negative events from happening. The Hmong believe that good spirits are ancestors who watch over and protect them. They also believe that a person's souls must live in harmony to maintain good health. The Hmong consider each person to have three souls. If a person's soul(s) wanders off or becomes frightened, it can lead to illness or death. This is called "soul loss." A shaman must perform a soul-calling ritual, known as *hu plig*, to guide the soul back to the person.
- There is also a large Christian Hmong community throughout the US. The Christian Hmong have beliefs and practices that are appropriate to their religion. Some people may denounce animist beliefs and traditional Hmong practices, such as the belief in soul loss and the performance of soul-calling ceremonies by shamans.

### *19.1.10 Healthcare Practices*

- Hmong in the US often seek Western medical care, and they may also consult traditional healers or shamans who perform rituals. They may consult herbalists and undergo multiple treatments for the same condition.
- **Do not mistake the regular patterns of coining or cupping for abuse. Encourage the sterilization of needles.** Some people practice home remedies such as coining or cupping. These traditional practices result in distinctive bruises, which are typically elongated or round, and are usually located over the affected area. For instance, a Hmong individual with a sore throat may exhibit bruises around the neck, while someone with chest symptoms may undergo cup-

ping on the front and back of the chest. They believe that pricking the center of the bruise releases bad spirits; and unsterilized sewing needles are typically used for this purpose.

- One cause of illness is believed to be the loss of one's soul. Some people, such as babies and children, are believed to have souls that struggle to remain their bodies. If the soul(s) leaves for too long, the baby can become ill or die. For this reason, parents may choose to tie a string around the baby's neck or wrist soon after birth. Older children and adults may have strings tied around their wrists, waists, or ankles. These must remain on until they fall off naturally, as removing them too soon can result in soul loss, illness, or even death. Since immigrating to the US, the traditional strings have been replaced with gold necklaces and bracelets.
- **Hmong people who practice shamanic ceremonies may wear amulets in small bags around their neck or waist. These contain objects that are believed to be protective against evil spirits. These should not be removed without the patient's or family's permission.**
- **Inquire about patients' herbal use upon admission.** Some Hmong people use herbs recommended by a Hmong herbalist. Studies have shown that these herbs may possess pharmaceutical properties.
- **Provide a detailed explanation of the benefits of blood transfusions when required.** More traditional Hmong people are opposed to blood transfusions and organ donation.
- **Caution patients about using unfamiliar herbs.** The traditional herbs found in Laos and Thailand are difficult or impossible to obtain in the US. Traditional herbalists sometimes use plants found in the US simply because the plant resembles a plant they had used in Laos. The efficacy of these herbs is unknown and may be harmful.

### *19.1.11 Healthcare Providers*

- Shamanic ceremonies are used for treating severe illnesses that have not improved with other forms of treatment. Shamans do not choose to become shamans; the occupation is a calling. Shamans undergo extensive training for many years and learn through apprenticeships. Shaman ceremonies are conducted within the home with the entire family present. During these ceremonies, the shamans enter the spirit world to discover the reasons behind the loss of the soul or its journey to the spirit world. During a second ritual, the shaman may call the person's spirit home, thereby restoring harmony. A shaman may sacrifice an animal to pay the spirit world and release the soul so it can return home. **Amulets from the ceremony may be placed on the patient to protect them, and they should not be removed.**

## 19.2 Reflective Exercises

1. When and under what conditions did the Hmong people begin immigrating to the US?
2. What languages do the Hmong speak? How can you obtain an interpreter for a Hmong individual who does not speak English?
3. Who makes healthcare decisions in Hmong society?
4. What are at least five diseases and health conditions that are common among the Hmong community?
5. What concerns might you have for a Hmong patient who practices animism?
6. What conditions do Hmong traditional healthcare providers treat?
7. Identify postpartum dietary practices among Hmong women.
8. What symptoms would you observe in a patient who practices "cupping" and "coining"?
9. What are some stressors that contribute to mental health conditions among the Hmong community?

## Bibliography

Ali AH, Kang MS, Kaur K, Al Adhami S, Yuvienco CR (2020) Review of Hmong-related health problems: a quick guide for healthcare providers. Cureus 12:e9808. https://doi.org/10.7759/cureus.9808

Budiman A (2021) Hmong in the U.S. fact sheet. Pew Research Center's Social & Demographic Trends Project. https://www.pewresearch.org/social-trends/fact-sheet/asian-americans-hmong-in-the-u-s/

Carroll W, Schaffer D (2020) Employment and wages of Hmong and other southeast Asian refugees in the United States. J Immigr Refug Stud 19(4):526–539. https://doi.org/10.1080/15562948.2020.1843099

Central Intelligence Agency (2024) Laos-East and Southeast Asia. Central Intelligence Agency. https://www.cia.gov/the-world-factbook/countries/laos/#people-and-society

Cobb TG (2010) Strategies for providing cultural competent health care for Hmong Americans. J Cult Divers 17(3):79–83

Corbett CA, Callister LC, Gettys JP, Hickman JR (2017) The meaning of giving birth: voices of Hmong women living in Vietnam. J Perinat Neonatal Nurs 31(3):207–215. https://doi.org/10.1097/JPN.0000000000000242. https://pubmed.ncbi.nlm.nih.gov/28244883/207

Duffy J, Harmon R, Ranard D, Thao B, Yang K (2004) An introduction to their history and culture. In: The Hmong culture profile no. 18. Center for Applied Linguistics, Washington, DC

Hamilton-Merritt J (1999) Tragic mountains: the Hmong, the Americans, and the secret wars for Laos, 1942–1992. Indiana University Press, Bloomington

Hmong in Minnesota (2016). Common medical issues and cultural concerns of Hmong patients. http://www.culturecareconnection.org/matters/diversity/hmong.html

Hmong Studies Resource Center (2016). www.hmongnet.org

Johnson SK (2002) Hmong health beliefs and experiences in the Western health care system. J Transcult Nurs 13(2):126–132. https://doi.org/10.1177/104365960201300205

Kue J, Zukoski A, Keon KL, Thorburn S (2013) Breast and cervical cancer screening: exploring perceptions and barriers with Hmong women and men in Oregon. Ethnicity Health 19(3):311–327. https://doi.org/10.1080/13557858.2013.776013

Lee-Yang M (2020) Flower cloth: A storytelling textile. American Craft Council. https://www.craftcouncil.org/magazine/article/flower-cloth-storytelling-textile#:~:text=For%20a%20people%20with%20no,Thai%20refugee%20camps%20after%201975

Lor M, Badenoch N, Yang MJ (2022) Technical meets traditional: language, culture, and the challenges faced by Hmong medical interpreters. J Transcult Nurs 33(1):96–104. https://doi.org/10.1177/10436596211039553

Owens CW (2007). Hmong cultural profile. Ethnomed. https://ethnomed.org/culture/hmong/hmong-cultural-profile

Pinzon-Perez H (2006) Health issues for the Hmong populations in the US: issues for health educators. Int Electronic J Health Educ 9:122–133

Purnell L, Fenkl EA (2019) People of Hmong heritage. In: Purnell L, Fenkl EA (eds) Handbook fore culturally competent care. Springer, Cham

Rice PL (2000) Nyo Dua Hli– 30 days confinement: traditions and changed childbearing beliefs and practices among Hmong women in Australia. Midwifery 16(1):22–34. https://doi.org/10.1054/midw.1999.0180

Stanford Medicine (2023). Tips for clinicians: Communication. Geriatrics. https://geriatrics.stanford.edu/ethnomed/hmong/understanding/tips_clinicians.html

Thao KK, Arndt B, Tandias A, Hanrahan L (2015) The prevalence of type 2 diabetes mellitus in a Wisconsin Hmong patient population. Wis Med J 114(5):190–195

United States Census Bureau (2021) Total population. Explore census data. https://data.census.gov/table?q=United+States&t=3766%3A3823&g=010XX00US

Warner ME, Mochel M (1998) The Hmong and health care in Merced. California Hmong Stud J 2(2):1–30

Wilhelm AK, Hammett P, Fu SS, Eisenberg ME, Pratt RJ, Allen ML (2023) Asian American adolescent e-cigarette use and associated protective factors: heterogeneity in a statewide sample. Addict Behav 145:107761. https://doi.org/10.1016/j.addbeh.2023.107761

Xiong N (2013) Households and marital status of Hmong in the United States, 2008-2010. In: Pfeifer ME, Thao BK (eds) State of the Hmong American community. Hmong National Development

Yang RC, Mills PK, Riordan DG (2004) Cervical cancer among Hmong women in California, 1988 to 2000. Am J Prev Med 27(2):132–138. https://doi.org/10.1016/j.amepre.2004.04.003

Yang RC, Mills PK, Riordan DG (2005) Gastric adenocarcinoma among Hmong in California, USA, 1988-2000. Gastric Cancer 8(2):117–123. https://doi.org/10.1007/s10120-005-0314-y

# Chapter 20
# People of Iranian Heritage

**Eric A. Fenkl**

## 20.1 Overview and Heritage

The number of Iranians in the United States is difficult to determine, with estimates ranging from 400,000 to over one million with half of them living in California. Of the 83 million people living in Iran, over 99% are Muslim. Iranians vary enormously, from highly traditional to highly acculturated, in their reasons for leaving Iran, as well as in the variant characteristics of culture (see Chap. 1) and individuality. The terms *Persian* and *Iranian* are used interchangeably in this chapter because some people call themselves Persian for historical and political reasons. In 1935, the country's name was changed from Persia to Iran to present an image of progress and in an attempt to unify into one nation the enormous diversity of urban dwellers and rural tribes, ethnic groups, and social classes.

Between 1930 and 1970, women were allowed to go unveiled and gained access to university education. After conservative Moslem protests that led to the 1978 violence, martial law was declared and an Islamic government was established. Since then, Iran resembles the more conservative Muslim countries.

Iranians are among the most highly educated immigrant group in the United States. Education is greatly valued, and advanced degrees are highly respected. Children are expected to do well in school and to attend college. Many middle-aged physicians, engineers, professors, army generals, and government officials who were unable to find comparable work in the United States have gone into business for themselves. Jobs that require manual labor are not respected, and some cannot accept taking such menial jobs often because of their limited skills in English.

E. A. Fenkl (✉)
Nicole Wertheim College of Nursing and Health Sciences, Florida International University, Miami, FL, USA
e-mail: efenkl@fiu.edu



E. A. Fenkl, L. D. Purnell (eds.), *Handbook for Culturally Competent Care*,
https://doi.org/10.1007/978-3-031-70492-5_20

### 20.1.1 Communications

- **Farsi** (Persian) is the national language of Iran. Nearly half the population speaks different languages: Turkish, Kurdish, Armenian, Baluchi, or other Iranian dialects. Well-educated and well-traveled immigrants may speak three or more languages, often using French as a cultural language or English in business settings.
- Many foreign invasions and strict control by each ensuing government have influenced Iranians' interaction with outsiders, making them suspicious of foreigners. The disclosure of personal thoughts to strangers is generally perceived to have detrimental consequences. Not verbalizing one's thoughts is considered a customary and useful defense; overt expressions of emotions to strangers may be culturally stigmatized.
- **Establish trust before asking questions of a highly personal nature.**
- Among women, patterns of speech may appear restrained or refined to avoid too much self-disclosure or to prevent loss of face. Most women give considerable detail rather than use blunt and succinct messages.
- Most Iranians are very concerned with respectability, a good appearance of the home, and a good reputation. They are embarrassed by financial troubles and even conceal such problems from relatives. *Zerangi* (cleverness) is valued but only among non-intimates; it means knowing how to manipulate bureaucratic structures.
- The influence of intimate versus public spheres and hierarchical social relationships is seen clearly in the practice of ***ta'arof*** (ritual expressed courtesy), which is not practiced with intimates. *Ta'arof* expresses the public face, with its respectful forms of speech and behavior, and is used when dealing with individuals whose status is unequal to one's own.
- Communication is structured by social hierarchy and varies according to ***baten*** and ***zaher***. *Baten* (inner self) is the true vulnerable self, a collection of freely expressed personal feelings. In contrast, *zaher* is proper and controlled behavior, a public face to protect and buffer the vulnerable world of the *baten* within. Silence is used to guard confidential matters and to manage impressions.
- Most Iranians refrain from showing anger or other strong emotions to outsiders because self-control is valued; showing anger can produce embarrassment, shame the family, or damage someone's reputation. To avoid embarrassment and to save face, they may agree or say that they understand an outsider, whether they do or not.
- **Because politeness and saving face prevail, do not assume that a positive response means a definite "yes."**
- **At the beginning of any health-care encounter, the provider should take time to "warm up" with social conversation if possible before "getting down to business." Any kind of bad news must be handled carefully by revealing it gently and gradually, in several meetings if possible, or only to the family spokesperson. A person should never be given bad news alone; for example, being informed of a death or serious diagnosis.**

- Among the more traditional, men and women do not hold hands or show affection toward each other in public. However, women often show affection for women and men for men by walking hand in hand or greeting each other with a kiss on each cheek.
- Strangers and health-care providers are greeted with both arms held at the sides or with a handshake. A slight bow or nod while shaking hands shows respect. **Offer something, e.g., a prescription, with both hands, shows respect.**
- Slouching in a chair or stretching one's legs toward another person is considered offensive. It is considered rude to show the sole of one's foot. Beckoning is done by waving the fingers with the palm down. **Do not beckon patients; call them formally by name.**
- Tilting the head up quickly means "no." Tilting the head to the side means "what?" and tilting it down means "yes." Extending the thumb (as in "thumbs up") is considered a vulgar sign. **Refrain from using nonverbal behaviors that may be offensive.**
- Personal distance is generally closer than that of European Americans. Respect for a provider's role and education might be demonstrated by keeping a wide distance.
- **Do not take offense at patients who stand closer to you than you wish.**
- Most individuals maintain intense eye contact between intimates and equals of the same gender, but the traditional tend to avoid eye contact. Younger people and those of lower status do not sustain eye contact with those they perceive as being older or of higher status.
- Temporal relationships are a combination of present and future orientation. **The future orientation enhances the effectiveness of health education.**
- Iranians are not clock-watchers; rather, they are mood- and feeling-oriented. While social time is flexible, Iranians meet the expectations for timeliness in work and appointments. **Stress the importance of timeliness for appointments in the United States.**
- In highly traditional families, privacy demands that husbands do not mention their wife's first name to other men.
- Close friends and children may be called by their first name by family members, but others should not do so. Shaking hands with a child shows respect to the parents.
- **A man should wait for a woman to extend her hand first. One is expected to greet every member of the family with a handshake. Heed nonverbal cues when greeting Iranians.**

### *20.1.2 Family Roles and Organization*

- Iranian culture is patriarchal. The father rules the family and expects obedience and submission from family members. In the father's absence, the oldest son has authority.

- Families were traditionally large in Iran, with male children being highly desirable. In more traditional families, older male siblings have the authority to make decisions about younger siblings, even in the father's presence.
- **Never underestimate the power of this hierarchy or of the family in general in decision-making. Traditional individuals usually cannot or will not make decisions.**
- Men see their role as protecting and providing for the family, managing the finances, and dealing with matters outside the home. Women maintain the home; even working women may place their priorities on the family and home.
- In traditional families, both daughters and sons stay at home until married, and in rural areas marriages are often arranged. In the United States, young people are free to select their own marriage partners, but families usually want to approve because of the importance of marriage alliances between families.
- Children are expected to be loyal to their families and behave respectfully toward older people. Manners are considered important even outside the home, where children are expected to be clean, well behaved, and refrain from rowdiness.
- Girls are expected to behave and dress more modestly than boys, especially as they approach adolescence. Children and teens are usually included in adult gatherings. Young children are rarely left with babysitters.
- Taboo behaviors for teens include smoking, drugs, alcohol, and sex, in that order.
- Young women are expected to remain virgins until they marry, but sexual activity by men outside marriage is tolerated.
- Family members often live close to each other so they can visit whenever possible. Most maintain strong intergenerational involvement with grandmothers, mothers, and children in the domestic sphere and with male family members. **Take advantage of this living situation when health care becomes a concern for Iranians.**
- Iranians often remove their shoes at the door and wear slippers. Outside the home, they tend to dress conservatively, but more traditional or religious women avoid bright colors, cover their arms and legs, and conceal their heads with head covers or scarves (***hijab***). In Iran, these coverings are mandatory.
- Age is a sign of experience, worldliness, and knowledge. Regardless of kinship or relationship, an older person is treated with respect. When grandparents reach an age when they can no longer care for themselves, they live with and are cared for by the family. Caring for older people is an obligation.
- Older and more traditional Iranians may be uncomfortable with unrelated members of the opposite sex. Although divorce is viewed negatively, its rate has been increasing among Iranians in the United States. Cultural mores also advocate ignoring or denying minor marital discord and accepting suffering for the sake of maintaining family stability.
- In Iran, out-of-wedlock teen pregnancy is neither talked about nor prevalent, and it can have a devastating outcome. If it happens in the United States, it may be taken care of quietly to preserve the face of the family. **Do not disclose pregnancy outside of marriage to family members.**

- Homosexuality is highly stigmatized and not discussed; gays and lesbians remain "in the closet." Since 1979, when the legal and religious systems became synonymous, homosexuality, which is considered unnatural and sacrilegious, has been a capital offense punishable by death. **Do not expose lesbian and gay relationships to family members or others.**

### 20.1.3 Workforce Issues

- Iranians may perceive and actually experience a degree of bias at work. More acculturated immigrant professionals respond flexibly in the workplace. Most newcomers may not be familiar with American vernacular or slang. **Do not use idiomatic expressions and slang with Iranian employees.**
- To avoid embarrassment and to save face, they may agree or say that they understand an outsider, whether they do or not. **Because politeness and saving face prevail, do not assume that a positive response means a definite "yes."**
- Personal distance is generally closer than that of Americans or northern Europeans.
- **Do not take offense if spatial distancing is closer than what you are accustomed to**.
- Most individuals maintain intense eye contact between intimates and equals of the same gender, but the traditional tend to avoid eye contact. Younger people and those of lower status do not sustain eye contact with those they perceive as being older or of higher status.
- **Do not assume that intense eye contact is anger or that avoiding eye contact means that the person is not listening, does not care, or is being less than truthful.**

### 20.1.4 Biocultural Ecology

- As white Indo-Europeans, coloring ranges from blue or green eyes, light-brown hair, and fair skin to nearly black eyes, black hair, and brown skin. **Because of the variations in skin color, assess jaundice and anemia in Iranians by examining the sclera and oral mucosa rather than by relying solely on skin assessments.**
- Table 20.1 lists commonly occurring health conditions for Iranians in three categories: lifestyle, environment, and genetics (see Chap. 2).
- **Screen newer immigrants for cholera, malaria, and parasitic disease. Screen for glucose-6-phosphate dehydrogenase before administering primaquine. Avoid fava beans, which can cause a hemolytic crisis.**

**Table 20.1** Health conditions common to Iranians

| Health conditions | Causes |
|---|---|
| Malaria | Environment, lifestyle |
| Bacterial meningitis | Environment, lifestyle |
| Hookworm/parasites | Environment, lifestyle |
| Cholera | Environment, lifestyle |
| Ischemic heart disease | Genetic, environment, lifestyle |
| Birth defects | Genetic, lifestyle |
| Hemophilia | Genetic |
| Thalassemia | Genetic |
| Glucose-6-phosphate dehydrogenase | Genetic |
| Dubin-Johnson syndrome | Genetic |

### 20.1.5 High-Risk Health Behaviors

- Smoking is more prevalent in Iran than in the United States among both men and women. **Encourage decreasing or stopping smoking and network with elementary school teachers to teach smoking hazards at a young age.**
- Some alcohol and street-drug misuse occur in the immigrant population, but the rate is no higher than that of the population at large.
- Alcohol is prohibited by the Qur'an (Holy Book), although many are not religious and drink socially, a few to excess. In some, years of opium use in Iran can create both psychological and physical addiction. Family responses to drug use range from complete support of the family member to disownment.
- Most comply with seat-belt and child-restraint laws, valuing the safety of their children.

### 20.1.6 Nutrition

- Food is a symbol of hospitality. Iranian food is flavorful and takes hours to prepare. Presentation is important.
- Food is classified into two categories, ***garm*** (hot) and ***sard*** (cold), which sometimes corresponds to high-calorie and low-calorie foods. The key is balance. At any given table, there is usually a pleasing mixture of foods of different colors and ingredients, composed of a balance of *garm* and *sard*. Too much of one category can cause symptoms of being "overheated" or "chilled." Symptoms are treated by eating food from the opposite group. Becoming overheated, sweating, itching, and rashes may result from eating too many hot foods, such as walnuts, onions, garlic, spices, honey, or candy. The stomach may become chilled, causing dizziness, weakness, and vomiting, after eating too many cold foods, such as grapes, rhubarb, plums, or yogurt. **Seek assistance from family members in selecting foods while in an inpatient facility.**

- Iranians prefer to use only the freshest foods, although in some cases cost is a factor in their using some dried herbs.
- Hot tea is the most popular drink among Iranians. The most common starchy foods are rice and wheat bread. Bread is usually baked flat like pita. Corn and potatoes are used but less favored. Beans and legumes, for example, lentils, pinto, mung, kidney, lima and green beans, and split and black-eyed peas make up a fairly high proportion of the dietary intake and are commonly used in rice mixtures.
- Dairy products are dietary staples, particularly eggs, milk, yogurt, and feta cheese as well as dairy by-products, such as *doog* (yogurt soda) and *kashk* (milk by-product).
- Favorite meats are beef, chicken, fish, and lamb. Shellfish is sometimes eaten. Fresh fruit is always found in Iranian houses.
- Green leafy vegetables are used in cooking, and herbs such as parsley, cilantro (coriander), dill, fenugreek, tarragon, mint, savory, and green onions are served fresh at a meal.
- Islam has a strict set of dietary prescriptions *(halal)* and proscriptions *(haram)*. Slaughter of poultry, beef, and lamb must be done ritually to make the meat *halal*. **Strict Muslims avoid pork and alcoholic beverages; a few avoid shellfish.**
- Food is eaten with the right hand, and foods or objects are passed with the right hand alone or with both hands. **Make adjustments to accommodate traditional food practices; make provisions for the family to bring food from home if the patient prefers. Incorporate Iranian foods and dietary practices into health teaching to improve compliance with special dietary restrictions.**

## *20.1.7 Pregnancy and Childbearing Practices*

- Menstrual blood is believed to be ritually unclean and physically polluting to the body. Menstruating women are not allowed to touch holy objects or to have intercourse. Menstruation is also considered a time of great fragility when a woman should not exercise or shower excessively, because these activities might cause a hemorrhage. At the end of the menses, the woman must wash and purify herself thoroughly before partaking in any religious rituals.
- Birth control methods used by Iranians include the pill, intrauterine devices, and natural methods. Vasectomies are beginning to gain acceptance.
- A woman's prestige is at its height when she delivers her first child, particularly if the child is a boy. Delivering the first child gives the young wife a more respected and cherished position with her in-laws. **Providing factual information regarding family planning is one area in which health-care providers can improve family care.**
- Food cravings must be satisfied lest a miscarriage occurs from not meeting the fetus's needs.

- Pregnant women avoid fried foods and foods that cause gas; fruits and vegetables are recommended, with special attention to the balance of hot *(garm)* and cold *(sard)* foods.
- In more traditional families, the father is usually not present at the birth. **Do not make the father feel guilty if he chooses not to be at the delivery.**
- To strengthen the postpartum woman, a *ghorse kamar* (a brown, flat disk of dried herbs) is mixed with eggs and placed on her lower back a few hours before bathing.
- New mothers avoid cold water for bathing, ablutions, or cleaning, although they now bathe sooner than the traditional first 30 days.
- Baby boys are considered "hotter" than baby girls, and mothers of sons are considered to have hotter bodies, hotter milk, and hotter temperaments than mothers of girls.
- Mothers of girls are given a mixture of honey and other nutrients, an herbal extract called *taranjebin*, to raise their bodily heat to ensure that the next child is a boy.
- Some families keep an infant home for the first 40 days, at which time the baby is strong enough to fight off environmental pathogens.
- The baby is given a ritual bath between the tenth and 40th days.
- **Ask the childbearing family about prescriptive, restrictive, and taboo cultural practices that they customarily follow, and incorporate nonharmful practices.**

### *20.1.8 Death Rituals*

- Muslims may need only a gentle reminder that death is not a termination of life but rather the beginning of a new and better life. **To discuss termination of life support with a practicing Muslim, begin the conversation by noting God's will for human beings and His power over our destiny, followed by a dialog about life and death as necessary steps toward immortal life in heaven.**
- Some individuals may oppose stopping life support, viewing it as "playing God," even though they may have no objection to beginning life support, viewing it as the "gift" of medical technology. Some families may demand that strenuous efforts be made to prolong life. **Assess each patient and family individually about life-prolonging medical treatments.**
- Among religious Muslims, the deathbed should be turned to face Mecca so family members can read prayers from the Qur'an to ensure that the dying person hears this at the time of death. **Determine the direction of Mecca so that the dying person can face Mecca.**
- After death, another Muslim should wash the body in a ritual manner, using soap and water and proceeding from head to toe and front to back. All body orifices must be closed and slightly packed with cotton to prevent leakage of bodily fluids (considered unclean). The final rinse is performed with water. The body is

then wrapped in a special white cotton shroud. Prayers and verses from the Qur'an are read during the procedure.
- **A non-Muslim health-care provider should wear gloves when touching the body.**
- Cremation is not practiced in Iran; it is unlikely to practice in the United States.
- No specific religious rules exist against autopsy. However, the body of the dead is to be respected. **The reason for the autopsy must be made clear, and some may still refuse. Explain legal requirement of autopsy in the** United States.
- Iranians may express their grief over the death of a loved one by crying, wailing loudly, or even striking themselves or an object.

## *20.1.9 Spirituality*

- Specific Muslim practices include prayers, read in the name of one of the 12 imams, to provide peace of mind or to plead for a miracle. Devout Muslims pray five times daily and need privacy and water for ritual washing. **Assist the patient to make arrangements for praying.**
- Most Muslims fast from sunrise to sundown during Ramadan, although pregnant women and the ill are exempt from fasting. **Arrange meals and medications to accommodate Ramadan. See Chap. 8 for Ramadan celebration dates through 2021.**
- Family relationships and friendships are sources of strength and meaning in life for many individuals. **Adjust visiting policies to accommodate family and friends.**
- Death is not a finalization; rather, it is a graduation to a higher level of being.

## *20.1.10 Health-Care Practices*

- Traditional health beliefs and therapeutic processes are a combination of Galenic (humoral), Islamic (sacred), and modern biomedicine. In classic humoral theory, illness arises from an imbalance (excess or deficiency) in the basic qualities such as hot/cold or wet/dry. The purpose of treatment is to restore balance.
- In Galenic-Islamic thought, every individual has a distinctive balance of four humors, resulting in a unique temperament.
- Sacred medicine is from the Qur'an and *hadith*, and holy men are considered able to heal. The sacred tradition includes beliefs in the "evil eye" and *jinns* as disease agents as well as healing by means of manipulating impurity.
- ***Narahati*** is a general term used to express a wide range of undifferentiated, unpleasant emotional or physical feelings, such as feeling depressed, uneasy, nervous, disappointed, or not fully well. Many individuals somaticize in a

subconscious effort to communicate *narahati* or distress that cannot be otherwise expressed verbally.

- **By somaticizing, they construct an illness that is culturally sanctioned and socially understood.**
- When *narahati* is caused by fright, the evil eye, or *jinns*, it may be treated with religious cures. If it is considered caused by problems of blood, nerves, or humoral imbalance, it is treated with herbs or biomedicine.
- *Ghalbam gerefteh (narahatiye qalb)* or distress of the heart may be expressed as a feeling that the heart is being squeezed and can range in severity from mild excitation of the heart or palpitations to fainting and heart attack.
- A sudden ailment or symptom of puzzling origin may be attributed to the evil eye, or ***cheshm-i-bad***, which is the belief that the eyes of another person can cause illness. *Cheshm-i-bad* can be unintentional (enthusiastically complimenting someone without saying "In the name of God") or intentional (cast out due to jealousy or enmity).
- Most individuals expect immediate relief or cure from the health-care system and may shop around until they find a provider they like. At the same time, they may seek advice from those who can suggest herbal remedies. **Explain the dangers of "shopping around" and not disclosing all treatments and home therapies used.**
- Most Iranians practice self-medication and use prescription and over-the-counter medications as well as homemade herbal remedies. Antibiotics, codeine-based analgesics, mood-altering drugs in the benzodiazepine family, and intramuscular vitamins are available over-the-counter in Iran, and immigrants often bring these medications with them to the United States.
- **Ask patients in a nonjudgmental manner about full disclosure of prescription and nonprescription medicines and herbal treatments.**
- Self-adjusting dosage of prescribed medications is common. **Explain the dangers of self-adjusting medications.**
- When ill, most individuals are inclined to be passive and cared for by health-care providers and family members.
- When an individual is hospitalized, family members are expected to visit frequently or to be present continuously until the patient is discharged from the hospital or is fully functioning again. **Self-care should be implemented by encouraging family members to assist with care of the sick person.**
- Common herbal remedies include dried flowers, seeds, leaves, and berries, steeped in hot or cold water and drunk for a variety of purposes, such as digestive problems, "cleaning the blood or kidneys," coughs, aches and pains, fevers, nerves, or fear. Some common herbal medications include *gol-i-gov zabon* (dried foxglove flowers) for an imbalance in the digestive system or nervous upsets, which is sometimes taken with *nabat*, a concentrated sugar. *Khakshir* (flat, brown rocket seed) is used for stomach problems and "dirty blood"; *razianeh* is used for halitosis; quince seeds are sucked for sore throats; and *sedr* prevents and treats dandruff. *Neshasteh* (wheat starch) is combined with boiling water and drunk for

sore throats or coughs and is also used to stop diarrhea. Mint extracts are used to relieve excess stomach gas. *Shatareh* is thought to cure fever.
- Most Iranians are expressive about their pain. Some justify suffering in light of later rewards.
- Mental illness is highly stigmatized and is thought to be genetic. If a family member has a mental illness, it is likely to be called a "neurological disorder" to avoid stigmatizing the family. Psychotherapeutic help may be avoided either because of stigma or because it is perceived as irrelevant.
- Rehabilitation, such as physical therapy and music therapy, is embraced as the way to bring the disabled into the mainstream.
- Blood transfusions, organ donations, and organ transplants are widely practiced. In Iran, donation of organs is often a business transaction—if a kidney is needed, it is purchased.

### *20.1.11 Health-Care Providers*

- Many immigrants think that Iranian doctors make more authoritative and quicker diagnoses, using minimal technology, even if they may be uncertain or wrong. American physicians, who are tentative, ask the patient to describe the problem, or order too many tests may be viewed as incompetent.
- Most Iranians prefer to be cared for by providers of the same sex. Iranian women are modest in front of men. Some very traditional families may consider taking a female patient elsewhere or avoid care until the situation is acute, if only male providers are available. **Male health-care providers should not ask women to undress fully for an examination or procedure.**
- The most respected health-care provider is an experienced, middle-aged to elderly male physician, with several degrees, and preferably with a high position in the hospital or university. He is considered the authority and is expected to act like one, making diagnoses quickly and prescribing remedies that cure the patient.
- Male nurses or women who have gray hair and positions of authority are accorded more respect than young, single, female nurses. Nurses who encourage self-care may be perceived as uncaring or even incompetent. **Provide factual information about the benefits of self-care as a method of rehabilitation.**

## 20.2 Reflective Exercises

1. What is the national language of Iranians? What other languages spoken by Iranians?
2. What is *zerangi* and how is it displayed?
3. How should traditional Iranian men and women be addressed?

4. Describe Iranian family hierarchy and decision-making practices.
5. How are Iranian gays and lesbians seen in Iran? What ramifications might this have in the US?
6. What are the primary health conditions among Iranians?
7. Explain *garm* and *sard*? What might you do if you are unfamiliar with this category of foods?
8. An Iranian patient is near death. How do you discuss this situation with the Iranian family?
9. How would you vary meals and medication administration during Ramadan?
10. What is *narahati,* what is its significance, and what are its symptoms?

## Bibliography

Bahrami H, Sadatsafavi M, Pourshams A, Kamangar F, Nouraei M, Semnani S, Brennan P, Boffetta P, Malekzadeh R (2006) Obesity and hypertension in an Iranian cohort study; Iranian women experience higher rates of obesity and hypertension than American women. BMC Public Health 6:158. https://doi.org/10.1186/1471-2458-6-158

CIA (2024) World factbook: Iran. https://www.cia.gov/the-world-factbook/countries/iran/

Hafizi H, Hafizi H (2013) People of Iranian heritage. In: Purnell L (ed) Transcultural health care: a culturally competent approach, 4th edn. F.A. Davis Company, Philadelphia

Heisey DR (2011) Iranian perspectives on communication is the age of globalization. http://web.uri.edu/iaics/files/03D.RayHeisey.pdf

https://paaia.org/wp-content/uploads/2017/11/iranian-americans-immigration-and-assimilation.pdf

Intimate Spaces: Coming Out in Iran (2015) The guardian. https://www.theguardian.com/world/iran-blog/2015/jun/11/iran-gay-coming-out-intimate-spaces

Iran guide: a look at Iranian language, culture, customs, and etiquette (2014). http://www.commisceo-global.com/country-guides/iran-guide

Iranian Americans: Immigration and Assimilation (2017) Public Affairs Alliance of Iranian Americans

Mahdi AA (1988) Ethnic identity among second-generation Iranians in the United States. Int Soc Iran Stud 31(1):77–95

# Chapter 21
# People of Irish Heritage

**Francisco Brenes and Rosa Roche**

## 21.1 Overview and Heritage

The Republic of Ireland, also known as Eire and the Emerald Isle, covers most of the island bearing its name. The remainder of the island, Northern Ireland, is part of Great Britain. During the potato famine between 1846 and 1848, thousands of Irish died from malnutrition, typhus, dysentery, and scurvy, and millions immigrated to America. Over 38 million people of Irish descent live in the United States. The Irish in America are a diverse group and vary in their beliefs according to acculturation, individuality, and the variant characteristics of culture as described in Chap. 1.

Most Irish immigrants initially settled in industrial areas in northeastern United States along the Atlantic Coast. The Irish attained success in America because they spoke the same language, had the same physical appearance as other European Americans, and mastered the political system. They have made significant contributions in politics, the labor movement, the Catholic Church, the arts, and service to their country.

### *21.1.1 Communications*

- The major languages spoken in Ireland are English and Irish (Gaelic); the latter is the official language and is spoken primarily in West Ireland. The Irish enjoy puns, riddles, limericks, and storytelling.

---

F. Brenes (✉) · R. Roche
Nicole Wertheim College of Nursing and Health Sciences, Florida International University, Miami, FL, USA
e-mail: fbrenes@fiu.edu; rroche@fiu.edu



E. A. Fenkl, L. D. Purnell (eds.), *Handbook for Culturally Competent Care*,
https://doi.org/10.1007/978-3-031-70492-5_21

- The Irish accent has a nasal quality and is spoken with a strong inflection on the first syllable of a word, resulting in a loss of weak syllables. When one becomes accustomed to hearing the Irish-accented English used by newer immigrants, there is little difficulty in understanding the speaker. The Irish language is low-context, using many words to express a thought.
- Some common Gaelic words and their meanings are *shamrock* for "emblem," *limer* for "folklore character," *colleen* or *lassie* for "girl," *sonsie* or *sonsy* for "handsome," *cess* for "luck," *brogue* for "shoe," *dudeen* for "pipe tobacco," and *paddy* for "Irishman."
- Even though most Irish delight in telling long stories, when discussing personal matters, they are much less expressive unless they are talking with close friends and family. Even then, many are still reluctant to express their innermost thoughts and feelings. **The collaboration of care with a mental health specialist of the same ethnocultural group or of European descendant may promote the expression of feelings in the Irish with psychiatric morbidity. Timely accessible mental health care increases positive patient outcomes; however, in parts of Ireland there are delays in mental health services due to lack of referrals between general practitioners and other health-care providers.**
- Humility and emotional reserve are considered virtues. Displays of emotion and affection in public are avoided and are often difficult in private. To many, caring actions are more important than verbal expressions.
- The Irish use direct eye contact when speaking with each other. Not maintaining eye contact may be interpreted as a sign of disrespect, guilt, or evidence that the other person cannot be trusted.
- Personal space is important to the Irish, who may require greater distance in spatial relationships than other ethnocultural groups.
- Although the Irish may be less physically expressive with hand and body gesturing, facial expressions are readily displayed with frequent smiling even during times of adversity. **Many Irish underreport symptoms of pain and use more nonpharmacological interventions before seeking medical care. Primary care providers should assess cues of physical pain in this cultural group such as grimacing, facial flushing, guarding, or blood pressure elevation.**
- The Irish in America, with their strong sense of tradition, are typically past-oriented. They have an allegiance to the past, their ancestors, and their history. While respecting the past, they balance "being" with "doing," and they plan for the future by investing in education and saving money. Many Irish see time as being elastic and flexible. **Altering cultural perceptions of punctuality requires patience and understanding. The goal is not to force punctuality but to initiate a respectful dialog and facilitate an open exchange of ideas that may lead to positive shifts in attitudes toward punctuality.**
- *Mac* before a family name means "son of," whereas the letter *O* in front of a name means "descended from."

### 21.1.2 Family Roles and Organization

- The traditional Irish family is nuclear, with parents and children living in the same household.
- Irish families emphasize independence and self-reliance in children. Boys are allowed and expected to be more aggressive than girls, who are raised to be respectable, responsible, and resilient. Children are expected to have self-restraint and self-discipline and to be respectful and obedient to their parents, elders, and church and community figures.
- Adolescent years are a time for experiencing emotional autonomy, independence, and attachment outside the family while remaining loyal to the family and maintaining the traditional Irish belief in the importance of family. Peer-group pressures at school may have a significant influence and are often incongruent with the belief systems of Irish Americans. **Cultural differences between Irish immigrant parents and their acculturated children may cause distress in families and place the latter group at risk for conduct or behavioral disorders. Because it may be difficult for some Irish Americans to express their feelings, encourage openness between parents and teenagers.**
- Provisions are made in Irish homes for care of older family members, a task that becomes increasingly difficult when both parents work outside the home. Irish respect the experience of older people and seek their counsel for decision making.
- **Actively including the older population into decision making regarding health-care activities will help bridge the gap between age groups and foster an understanding of each other's perspectives.**
- The Irish value physical strength, endurance, work, the ability to perform work, children, and the ability to provide their children with the needed education to attain respectable socioeconomic status and professional accomplishments. Same-sex relationships continue to carry a stigma for some; however, they are becoming more acceptable in the United States and in Ireland. **Overall, the disclosure of same-sex relationships in the Irish culture has become more accepted and supported in recent years. However, there is ongoing work needed to create an inclusive society where all individuals, regardless of their sexual orientation, feel comfortable and safe in expressing their identities and relationships.**

### 21.1.3 Workforce Issues

- Because cultural differences between Ireland and the United States are minimal, Irish assimilate into the workforce easily.
- When change is necessary to improve the status quo, the Irish readily relinquish traditional beliefs and adjust to the workforce. Even though the Irish are typical of past-oriented groups in other ways, they tend to question the status quo of the

U.S. workforce. The Irish in America have taken many leadership positions and helped with unionization and other activities to express their collective autonomy.
- The low contextual use of language, in which most of the message is in an explicit mode rather than an implicit mode, enhances pragmatic communications in the workforce.
- Personal space is important and may require greater distance in spatial relationships than other ethnocultural groups. **In the Irish culture personal space is generally respected but the concept and norms of personal space can vary depending on the context and individual preferences. Do not take offense if the Irish stand farther from you than expected.**

### *21.1.4 Biocultural Ecology*

- Most Irish have dark or red hair, ruddy cheeks, and fair skin; however, other variations exist in hair and skin color. The fair complexion of the Irish places them at risk for skin cancer. **Explain the harmful effects of the sun and encourage the use of sun block and other types of protection from the sun.**
- Because mining is an important economic activity in Ireland, miners are at increased risk for respiratory diseases. In addition, the cool maritime climate of Ireland increases susceptibility to respiratory diseases. **Assess newer immigrants who worked in mining industries for respiratory illnesses.**
- Table 21.1 lists commonly occurring health conditions for Irish Americans in three categories: lifestyle, environment, and genetics.
- **Provide education and counseling regarding lifestyle and dietary changes to reduce risks associated with cardiovascular diseases.**
- Screen for osteoporosis in women in their late 60s or even younger if assessment data indicate it.
- The major cause of infant mortality in Ireland is congenital abnormalities. Other conditions with a high incidence among Irish newborns are phenylketonuria (PKU), neural tube defects, and fetal alcohol syndrome.
- **Encourage women who give birth at home to seek PKU screening for their infants.**
- **Explain the hazards of alcohol intake during pregnancy.**

**Table 21.1** Health conditions common to Irish Americans

| Health condition | Cause |
|---|---|
| Coronary heart disease | Genetic, environment, lifestyle |
| Phenylketonuria | Genetic |
| Osteoporosis | Genetic, environment, lifestyle |
| Alcoholism | Genetic, environment, lifestyle |
| Skin cancer | Genetic, environment, lifestyle |

- **Counseling or cognitive behavioral therapy (CBT) could be offered as a first-line treatment option for Irish pregnant women who misuse alcohol. Provide effective coping strategies such as deep breathing, imagery, journaling, prayer, and exercise. Explain the importance of avoiding places that promote alcohol use.**

### *21.1.5 High-Risk Health Behaviors*

- Smoking has been identified as a major risk factor causing premature mortality from cancer in Ireland and among the Irish in America. **Primary efforts include partnering with teachers beginning in elementary school to teach the negative effects of smoking and vaping on health. Encourage patients to stop smoking, and assist them in finding smoking cessation programs with group and one-to-one counseling. Cognitive behavioral therapy combined with pharmacological interventions can assist chronic smokers with smoking cessation.**
- The use of alcohol and IV drugs are major health problems among Irish Americans. Alcohol problems in Ireland are among the highest internationally. Alcoholism researchers generally agree that Irish ancestry puts individuals at risk for developing drinking problems. Irish pubs are popular establishments that have become synonymous with alcohol intake, lively music, and a vivacious time. **Collaboration between government agencies, schools, communities, families, and health-care providers is essential in creating a comprehensive approach to primary prevention of alcohol and drug use in Ireland.**
- **This image of the Irish pub perpetuates the stereotype of Irish as heavy drinkers; do not ascribe this label to all Irish Americans. Because drinking may be a way of coping with problems, assist Irish patients in exploring more effective coping strategies, and caution them against the dangers of mixing alcohol with medications.**

### *21.1.6 Nutrition*

- Irish food is unpretentious and wholesome if eaten in recommended proportions. Food is an important part of health maintenance and celebrations. Vitamins are commonly used as a dietary supplement.
- Meat, potatoes, and vegetables are dietary staples. Lamb, mutton, pork, and poultry are common meats. Seafood includes salmon, mussels, mackerel, oysters, and scallops. Popular Irish dishes include Irish stew made with lamb, potatoes, and onions.
- Potatoes are used in a variety of ways. *Colcannon* is made with hot potatoes, mashed with cabbage, butter, and milk, and seasoned with nutmeg. This dish

may be served at Halloween. *Champ* is a popular dish made with mashed potatoes and scallions. The scallions are cut in small pieces, including the green tops, boiled in milk until tender, and then added to the mashed potatoes and served with butter. Potato cakes, made with mashed potatoes, flour, salt, and butter are shaped into patties and fried in bacon grease. Potato cakes are served hot or cold with butter and sometimes molasses or maple syrup. Another popular dish is Dublin coddle, made with bacon, pork sausage, potatoes, and onions. Oatmeal is popular in Ireland. Soda bread, another popular food, is made with flour, baking soda, salt, sugar, cream of tartar, and sour milk.
- Mealtimes are important occasions for Irish families to socialize and discuss family concerns. Meals are eaten three times a day, with a large breakfast in rural areas, lunch around noon, and a late dinner. Some Irish Americans continue the afternoon tradition of "tea," a light sandwich or biscuit with hot tea.
- **Because the Irish diet has the potential for being high in fats, cholesterol, and carbohydrates, assist patients with balanced food selections and preparation practices that reduce their risk of cardiovascular disease.**

### *21.1.7 Pregnancy and Childbearing Practices*

- The Irish believe that pregnant women who do not eat a well-balanced diet or do not eat the right kinds of food may cause the baby to be deformed.
- The Irish share the belief, common to many other ethnic groups that the mother should not reach over her head during pregnancy because the baby's cord may wrap around its neck. A taboo behavior in the past, which some women still respect, is that if the pregnant woman sees or experiences a tragedy, a congenital anomaly may occur. **Providing factual information and dispelling myths regarding pregnancy will help ensure the health and well-being of expectant mothers and their babies.**
- Eating a well-balanced diet after delivery continues to be a prescriptive practice for the Irish to ensure a healthy baby and maintain the mother's health. Plenty of rest, fresh air, and sunshine are also important for maintaining the mother's health. Going to bed with wet hair or wet feet causes illness in the mother.

### *21.1.8 Death Rituals*

- Many Irish are fatalists and acknowledge the inevitability of death. The American emphasis on technology and dying in the hospital may be incongruent with the Irish belief that family members should stay with the dying person. **Making arrangements for families to stay with patients in the hospital and in long-term care facilities may provide comfort and disclosure to patients and their families.**

- Whereas men are expected to be more stoic in their bereavement, women are usually more expressive. **Expect and accept a wide variety of grief expression among the Irish.**
- A wake continues as an important phenomenon in contemporary Irish families and is a time of melancholy, rejoicing, pain, and hopefulness. The occasion is a celebration of the person's life.
- Cremation is an individual choice, and there are no proscriptions against autopsy if required.

### *21.1.9 Spirituality*

- The predominant religion of most Irish is Catholicism: the church is a source of strength and solace. Other religions common among Irish in America include various Protestant denominations, such as the Church of Ireland, Presbyterian, Quaker, and Episcopalian.
- In times of illness, Irish Catholics receive the Sacrament of the Sick, which includes anointing, communion, and a blessing by the priest. The Eucharist, a small wafer made from flour and water, is given to the sick as the food of healing and health. Family members can participate if they wish.
- **When dealing with sick individuals, it is important to prioritize and respect their preferences as it contributes to patient centered care and overall well-being.**
- Attending mass daily is a common practice among many traditional and devout Irish Catholic families. Prayer is an individual and private matter. In times of illness, the clergy may offer prayers with the sick as well as with the family. **Give patients privacy for prayer whether or not a clergy member is present.**
- For those who practice their religion regularly, holy day worship begins at 4 p.m. the evening preceding the holy day. All Sundays are considered holy days. The obligation to fast and abstain from meat on specified days is relinquished during times of illness.
- Some wear religious medals to maintain health. **Religious emblems provide solace and should not be removed.**

### *21.1.10 Health-Care Practices*

- Many Irish use denial as a way of coping with physical and psychological problems. The Irish view of life is illustrated in the belief that life is black with long suffering and the less said about it, the better.
- Many Irish ignore symptoms and delay seeking medical attention until symptoms interfere with the ability to carry out activities of daily living. Irish limit and

understate problems and handle problems by using denial. Because Irish may not be very descriptive about their symptoms, treatment may be more difficult.

- **Encourage early intervention and seeking health professionals for illnesses. Factually explain the importance of early intervention and primary care.**
- Illness or injury may be linked to guilt and is considered to be the result of having done something morally wrong. Restraint is a *modus operandi* in the Irish culture; temptation is ever-present and must be guarded against.
- Most Irish believe one is obligated to use ordinary means to preserve life. Extraordinary means may be withheld to allow the person to die a natural death. The sick person and family define extraordinary means. Finances, quality of life, and effects on the family usually influence the decision. **Arrange for a family conference to address issues and concerns in hospice and palliative care.**
- In most Irish families, nuclear family members are consulted first about health problems. Mothers and older women are usually sought for their knowledge of folk practices to alleviate common problems such as colds. The Irish believe that having a strong religious faith, keeping one's feet warm and dry, dressing warmly, eating a balanced diet, getting enough sleep, and exercising are important for staying healthy. **Optimize compliance for reducing health risks by emphasizing these important cultural values.**
- Irish practices include wearing religious medals to prevent illness, using cough syrup made from honey and whiskey, taking honey and lemon for a sore throat, drinking hot tea with whiskey for a cold, drinking hot tea for nausea, drinking tea and eating toast for a cold, and putting a damp cloth on the forehead for a headache. Some folk practices may be harmful, such as the use of senna to cleanse the bowels every 8 days, eating a lot of oily foods, and avoiding seeing a physician. **Encourage patients to reveal all traditional and home remedies being used to treat symptoms. Provide factual information about potentially harmful folk and traditional practices.**
- The behavioral response of the Irish to pain is stoicism, usually ignoring or minimizing it. **Providing effective coping techniques including deep breathing, prayer, guided imagery and encouraging pain medication when necessary help with the healing process.**
- One explanation for high rates of mental illness **and suicide** may be associated with the Irish having difficulty describing emotions and expressing feelings. **Encouraging the expression of emotions and feelings can lead to healthier emotional processing and stronger social connections, thus leading to the promotion of mental health awareness and the reduction of suicide.**
- Blood transfusions are acceptable to most Irish Americans. Many participate in organ donation and indicate their willingness to do so on their driver's licenses.
- **Obtain organ-donor status on an individual basis, be sensitive to patients and family concerns, explain procedures involved with organ donation and procurement, answer questions factually, and explain the risks involved.**

### 21.1.11 Health-Care Providers

- The Irish respect all health-care professionals.
- Although the Irish are not noted for being overly modest, some may prefer to receive intimate care from someone of the same gender. In general, men and women may care for each other in health-care settings as long as privacy and sensitivity are maintained.

## 21.2 Reflective Exercises

1. Identify four common health problems among Irish Americans. Provide a recommendation to address each problem.
2. What are some childrearing practices common among the Irish?
3. How are emotions displayed among the traditional Irish?
    - What does *Mac* before a family name mean? What does the letter *O* in front of a name mean?
4. Identify two high-risk behaviors common among the Irish. What health education would you provide to decrease these behaviors?
5. The family of a deceased patient invites you to a wake. What might you expect from a traditional wake?
6. What is the dominant religion among the Irish? How might you partner with the church to improve the health of the Irish?
7. Most Irish speak low-contexted English. What does this mean and how might it be displayed on an intake assessment?
8. Identify at least three common traditional health-care practices.
9. List at least three common traditional beliefs practiced by pregnant women.
10. Identify at least common health conditions among newborns. What health teaching would you do to decrease these conditions?
11. **What are some cultural differences and challenges experienced by acculturated children and their Irish immigrant parents?**
12. **What treatment modalities are available for pregnant women at risk for alcoholism?**

## Bibliography

CIA (2023) World factbook: Ireland. https://www.cia.gov/the-world-factbook/countries/ireland/#introduction

Dejone K, Olyaei A, Lo J (2020) Alcohol use in pregnancy. Clin Obstet Gynecol 62(1):142–155

Durand H, Monahan K, McGuire B (2021) Prevelance and impact of dysmenorrhea among university students in Ireland. Pain Med 22(12):2835–2845

Fitzpatrick M, Newton J (2005) Profiling mental health needs: what about your Irish patients? Br J Gen Pract 55(519):739–740

Ireland guide: a look at Irish language, culture, customs, and etiquette (2016) Commisceo Global. http://www.commisceo-global.com/country-guides/ireland-guide

Irish potato famine: gone to America (2010) The history place. http://www.historyplace.com/worldhistory/famine/america.htm

Linton J, Green A, Chilton L, Duffee J, Dilley K, Gutierrez R, Keane V, Krugman S, McKelvey C, Nelson J (2019) Providing care for children in immigrant families. Pediatrics 144(3):e20192077

Mauger A (2021) A great race of drinkers? Irish interpretations of alcoholism and drinking stereotypes, 1945–1975. Med Hist 65(1):70–89

Myers Shim S (2013) People of Irish heritage. In: Purnell L (ed) Transcultural health care: a culturally competent approach, 4th edn. F. A. Davis, Philadelphia

Noonan M, Galvin R, Jomeen J, Doody O (2019) Public health nurses' perinatal mental health training needs: a cross sectional survey. J Adv Nurs 75(11):2535–2547

Parents' perspective on parenting styles and disciplining children (2010) National children's strategy research series. https://www.tcd.ie/childrensresearchcentre/assets/pdf/Publications/Parents'_Perspectives_on_parenting_styles.pdf

Purnell L, Foster J (2003a) Cultural aspect of alcohol use: part I. Drug Alcohol Prof 3(3):1723

Purnell L, Foster J (2003b) Cultural aspect of alcohol use: part II. Drug Alcohol Prof 2(3):3–8

Rapple BA (2003) Irish Americans. Countries and their culture. http://www.everyculture.com/multi/Ha-La/Irish-Americans.html

Reynolds C, Egan B, O'Malley E, McMahon L, Sheehan S, Turner M (2019) Fetal growth and maternal alcohol consumption during early pregnancy. Eur J Obstet Gynecol Reprod Biol 236:148–153

# Chapter 22
# People of Italian Heritage

**Alessandro Stievano, Rosario Caruso, and Gennaro Rocco**

## 22.1 Overview and Heritage

Italy is a parliamentarian Republic located in Southern Western Europe. Positioned in the middle of the Mediterranean Sea, it consists of a peninsula delimited by the Alps and surrounded by islands like Sicilia and Sardinia. Its population consists of nearly 60 million inhabitants; 5.3% of them come from different countries with a million of Romanian people. However, this chapter mainly focuses on the beliefs and practices of the Italians from mainland Italy. The large populations of Italians in the United States are not a homogeneous group and vary according to acculturation, individuality, and the variant characteristics of culture as described in Chap. 1. Italian Americans are concentrated in New York, New Jersey, Massachusetts, Pennsylvania, and California. Italian enclaves, or "Little Italy" areas, can be found in major cities throughout the United States.

### *22.1.1 Communications*

The official language of Italy is Italian; however, there are more than 20 dialects. Many second-generation Italian Americans do not speak Italian well or at all or speak in a jargon that sometimes is not understood by native Italians.

---

A. Stievano (✉) · G. Rocco
Centre of Excellence for Nursing Scholarship, Opi Rome, Rome, Italy
e-mail: alessandro.stievano@unicamillus.org

R. Caruso
Department of Biomedical Sciences for Health, University of Milan, Milan, Italy
e-mail: Rosario.caruso@grupposandonato.it



E. A. Fenkl, L. D. Purnell (eds.), *Handbook for Culturally Competent Care*,
https://doi.org/10.1007/978-3-031-70492-5_22

- **When a dialect-specific interpreter is not available, the interpreter should select words that have pure meanings from Tuscan Italian, which is the official language in the country (the language prompted in the Middle-Age by Dante Alighieri who is considered the Father of the Italian language).**
- Italian discussions can become quite passionate, with voice volume raised and many people speaking at once. **High voice volume does not mean anger, but rather passionate communication and this is true especially for inhabitants of the Southern parts of the country who immigrated in the United States.**
- Their willingness to share thoughts and feelings among family members is a major distinguishing characteristic. Positive and negative emotions and sentiments are permissible and encouraged with family members. Expressions of affection might erupt, and frequent kissing reaffirms the emotional bond among Italians.
- The typical kiss is a kiss on each cheek. Gestures demonstrate a range of feelings, from eloquence to intense anger, and are conveyed in an economical, subtle, flowing, and almost imperceptible manner. For example, a slowly raised chin means “I don’t know.”
- **Observe patients for nonverbal cues to obtain the full meaning of verbal communication**.
- Touching and embracing family and friends are common. Touching between men and women, between men, and between women can be observed during verbal communication.
- Past orientation is evidenced by the pride Italians take in their home country’s rich Roman heritage. A future-oriented temporality is evidenced by the importance given to planning and saving for the future. First-generation and newer immigrants view time as an approximation rather than categorically imperative; second- and third-generation Italians adhere to clock time at least in the work situation and for appointments.
- The order of first and last names is frequently reversed in Italy. Pietropaolo Vincenzo is often used when Vincenzo is the person’s first name and Pietropaolo is the last name. Nicknames are common and used among family and friends.
- A person is called by the first or given name in social situations and by a title such as Miss, Ms., Mr., Mrs., or Dr. and his or her last or family surname in the health-care environment. Ask recent immigrants the order of the name to ensure an accurate health-care record.

### *22.1.2 Family Roles and Organization*

Traditional families recognize the father’s authority. The father’s decision may be accepted as a strong principle, even among his married children. To criticize one’s father is considered disrespect toward the entire family.

- Women usually dominate decision-making on child-rearing issues and family social events. Even though traditional roles remain strong in second- and third-generation families, a trend toward more egalitarian relationships is evolving.
- Children are taught to have good manners and respect for older people. Boys and girls are encouraged to be independent and expected to contribute to the family's support as soon as they are old enough to work. **Adolescent girls are expected to remain virgins until they marry even; this habit is rapidly evolving.**
- The home is the focus of family gatherings. Sons and daughters visit frequently during the week and after church on Sundays to share a large meal at the parents' house.
- To many older people, the ideal living situation is to maintain one's own home near one's children. Older people receive respect, gratitude, and love in return for their many sacrifices.
- Most Italian Americans have an actively functioning kinship and extended family system that is the primary focus of solidarity for the nuclear family. **Because the extended family is close and frequent visits are important, Italians make special arrangements for visitation when patients are in acute or long-term care facilities.**
- Social status for most families comes from family lineage. Titles are more important than names. It is important to i**dentify Italians with a title such as Mr., Mrs., Miss, Ms., or Dr. to increase compliance with health treatments and regimens.**
- Most do not reject another family member because of an infraction or alternative lifestyle such as divorce, living together before marriage, or being involved in a same-sex relationship. The individual is accepted without consequences.
- When the need arises, Italians foster social support referrals for gays or lesbians and assist patients in contacting dedicated services or religious groups such as Dignity USA.

### *22.1.3 Workforce Issues*

- Italians living in the United States are punctual, and rarely miss work commitments due to a cold, headache, or minor illnesses. If completing their work requires staying later, they do so.
- Although the family is of utmost importance to Italians, work takes priority over family unless serious family situations arise. This cultural predisposition parallels the European American work ethic.
- Because of language barriers, first-generation Italians and newer immigrants are more likely to accept assigned tasks without expecting any decision-making authority.
- Italian discussions can become quite passionate, with voice volume raised and many people speaking at once. **High voice volume does not necessarily mean anger but rather passionate communication.**

### 22.1.4 Biocultural Ecology

- Italians as a group have varied physical characteristics. Those from a predominantly northern background have lighter skin, lighter hair, and blue eyes, whereas those from the south of Rome, particularly from Sicily, have dark, often curly hair; dark eyes; and olive-colored skin.
- In dark-skinned patients, the skin turns ashen instead of blue in the presence of cyanosis and decreased hemoglobin levels. Examine the sclera, conjunctiva, buccal mucosa, tongue, lips, nailbeds, and palms of the hands and soles of the feet. To assess for jaundice, look at the conjunctiva and in the buccal mucosa for patches of bilirubin pigment.
- Table 22.1 lists commonly occurring health conditions for Italians in three categories: lifestyle, environment, and genetics.
- Mediterranean-type G-6-phosphate dehydrogenase (G-6-PD) deficiency and β-thalassemia has a profound effect on drug metabolism. Because of conditions such as hypoxemia and acidosis, the ingestion of fava beans and the administration of sulfonamides, antimalarial agents, salicylates, and naphtha quinolones can exacerbate these conditions; take extra precaution when prescribing these drug therapies especially from Italians who originated from Sardinia.

### 22.1.5 High-Risk Health Behaviors

- Many immigrants continue to smoke. Alcoholism also presents a risk in this group.
- **Assess for illnesses related to smoking and alcohol misuse, especially among newer immigrants. Encourage patients to cease or decrease smoking. Incorporate anti-smoking campaigns in elementary school. Promote responsible drinking.**

**Table 22.1** Health conditions common to Italian Americans

| Health condition | Cause |
|---|---|
| Familial Mediterranean fever | Genetic |
| G-6-phosphate dehydrogenase deficiency | Genetic |
| Thalassemia | Genetic |
| Coronary artery disease | Genetic, environment, lifestyle |
| Vogt-Koyanagi-Harada syndrome | Genetic |
| Nasopharyngeal cancer | Lifestyle |
| Stomach cancer | Lifestyle |
| Liver cancer | Lifestyle |

### 22.1.6 *Nutrition*

- Food is symbolic of a connection between a child and the parents; food represents the product of the father's labor, prepared with care by the mother. A mother may demonstrate her affection by feeding her family and anyone else she likes. To the average "mom," love is a four-letter word: food.
- The diet, rich in vegetables, pasta, fruit, fish, and cheese, varies according to the region of Italy from which the individual originated. Northern Italian foods are rich in cream and cheese, resulting in a potentially high intake of fat. Southern Italian foods are prepared in red sauces, spices, and added salt and constitute the famous Mediterranean diet.
- **Because of regional variations in food selections and preparation practices, specifically inquire about the diet, preferably during intake assessments.**
- Dietary staples are spaghetti, lasagna, ravioli, fettuccine, special spaghetti called linguine or trenette with pesto, tortellini, bucatini, and rigatoni. Vegetables, fresh fruit, and beans are common. Other popular Italian foods include lentils; sausage; eggplant parmigiana; salami; olive oil; espresso and cappuccino coffee; wine; ice cream (gelato); pastries such as cannoli and biscotti; and cheeses such as provolone, gorgonzola, ricotta, pecorino cheese, and parmesan cheese. Other common dishes include escarole, Caesar salad, calzone, and pizza. Table 22.2 lists the Italian names of popular foods with their descriptions and ingredients.

**Table 22.2** Italian foods

| Common name | Description | Ingredients |
|---|---|---|
| Calamari | Squid, fried or on pasta | Floured squid fried in olive oil or in red sauce |
| Frittata | Italian omelet | Eggs, peppers, and onions cooked in olive oil |
| Minestrone | Soup with greens | Escarole and beans with garlic and other herbs |
| Parmigiana di melanzana | Eggplant parmesan | Eggplant, tomato sauce, bread crumbs, Parmesan cheese, and mozzarella cheese |
| Pasta con pesto | Pesto sauce served over linguine or trenette | Sauce of basil, nuts, olive oil, and garlic |
| Pasta e fagioli | Macaroni and beans | Shell-shaped pasta, kidney beans, and tomato sauce |
| Pasta marinara | Pasta in tomato sauce | Tomato sauce |
| Pizza fritta | Fried dough with sugar and cinnamon | Bread dough fried in oil and sprinkled with sugar and cinnamon |
| Prosciutto | Thinly sliced ham | Delicate, thin ham served with melon |
| Spaghetti aglio e olio | Spaghetti with olive oil and garlic | Spaghetti, olive oil, garlic, and red pepper |
| Tortellini | Little rounds of pasta in white or red sauce | Pasta dough stuffed with meat and/or cheese |
| Veal scaloppine | Medallions of veal | Veal sautéed with wine, butter, and lemon |

Wine is taken at almost every meal, and red wine, mixed with water, is given to children with meals to maintain healthy blood.

- A common practice in Italian alternative medicine for health promotion is eating a clove of garlic every night before going to bed to prevent upper respiratory infections. Garlic may also be worn around the neck when there is an epidemic of influenza or other upper respiratory ailments to prevent the wearer from getting the infection. Eating a fresh raw egg every morning keeps the person strong. Fresh dandelions are used to make a salad or are boiled to make soup to give the person strength.

### *22.1.7 Pregnancy and Childbearing Practices*

- Many third-generation Italian Americans use birth control from the beginning of marriage; sex is commonly discussed in the family.
- Sprinkling salt under and around the bed of a newly married couple makes them fertile.
- Common beliefs and customs about pregnancy are shown in Box 22.1.
- Some women still prefer having their children delivered at home by a family physician or a midwife. If labor does not progress rapidly enough, a neighbor has to spit out the window. This ritualistic spitting has the power to break any magic spell that might have brought the ill fortune of slow labor. Nowadays, f**ew Italian Americans believe in many of these traditional practices.**
- A traditional postpartum woman is not allowed to wash her hair, take a shower, or resume her domestic chores for 2 or 3 weeks after birth so she can rest. The woman's mother and other female family members tend to the chores and assist with the care of the new baby.
- Mothers are expected to breastfeed to restore the health of the reproductive organs and keep the mother and baby free of infections. The size of the baby is perceived as an index of the successful maintenance of maternal and wifely responsibilities. **However, People of Italian origins tend to comprehend that to be born oversized does not always mean to be in a good health.**

**Box 22.1 Beliefs About Pregnancy**

Traditional beliefs related to pregnancy include the following:

- Coffee spills may result in the baby being born with a birthmark where the coffee was spilled.
- Women must abstain from sexual relationships while pregnant.
- If the expectant mother craves for a particular food that is not satisfied, a congenital anomaly may occur or the baby will be marked.
- If a pregnant woman is not given the food she smells, the fetus moves and a miscarriage results.
- A pregnant woman should not reach over her head because harm may come to the baby.

### 22.1.8 Death Rituals

- Death is a great social loss and brings an immediate response from the community. It means sending food and flowers (chrysanthemums), giving money, and congregating at the home of the deceased. Friends and distant relatives bring food. The funeral procession to the cemetery is a symbol of family status, which is determined by the number of cars in the procession.
- Within the context of fatalism in Catholicism, many Italians view death as God's will; thus, a fatal diagnosis may not be discussed with the ill family member.
- Emotional outpourings can be profuse. Women, especially form the South, may mourn dramatically, even histrionically, for the whole family. They do not merely weep; they may rage against death for the harm it has done to the family. Family members may moan and scream for the deceased throughout the church service. Screaming is an effort to ensure that Jesus, Mary, and the saints hear what the bereaved are thinking and feeling.
- The real time of sorrow comes at the end of the ceremony when the priest and nonfamily congregation say good-bye. At this time, the family is left on its own for a time with the loved one. Older women may throw themselves onto the casket to prevent it from leaving the church.
- While men mourn, they do so in the fashion of *pazienza* (patience). Their constant, silent, and expressionless presence may be their only act of public mourning. When the loss is a child or spouse, expressions of grief continue for years.
- **In a more globalized and interconnected society, these death rituals tend not to be so striking anymore; however, for very traditional families they still have a strong significance.**
- More traditional families hold anniversary masses for the deceased and wear black for months or years.

### 22.1.9 Spirituality

- The predominant religion is traditional Roman Catholicism, the center of which is the celebration of Mass, the Eucharist, which is the commemoration of Christ's sacrificial death and of his resurrection. Other sacraments are baptism, confirmation, confession, matrimony, ordination, and anointing of the sick. When a loved one becomes ill, prayers are said at home and in church for the person's health.
- **To help patients to obtain the basic rites of the Sacraments of the Sick, which includes anointing, communion, and, if possible a blessing by the priest is crucial for many Italians.**
- Most pray to the Virgin Mary, the Madonna, and a number of saints.
- Even though Catholics are obligated to fast or abstain from meat and meat products on certain days of the year, the sick are not bound by this practice.

- Despite the church's exception for the sick, many first-generation immigrants choose to fast. Respect fasting if it is not harmful to the patient's health condition is seen as a sign of respect.
- Prayer and having faith in God and the saints help Italian Americans through illnesses.
- Niceties of Catholic orthodoxy and enlightened learning are employed together with pagan practices such as beliefs in witches, ghosts, and supernatural entities. The lessons of common sense and science are freely mingled with those of magico-religious beliefs. Many Italian American families display shrines to the Blessed Virgin in their backyards.
- **Respect and acknowledge religious beliefs to gain the trust of patients**.

### *22.1.10 Health-Care Practices*

- In traditional terms, illnesses are attributable to (a) wind currents that carry disease, (b) contamination, (c) heredity, (d) supernatural (God's will) or human causes, and (e) psychosomatic interactions.
- Within the context of fatalism, diseases largely run their own course; thus, it is better to leave the investigation into health problems until a condition becomes so obvious that it cannot be neglected. **Health-care providers should provide factual information and dispel myths.**
- Nervousness, hysteria, and many other mental illnesses are attributed to an evil spirit entering the body and remaining in the body until it is cast out by making its abiding place so unpleasant that it is forced to leave.
- The extended family may be the frontline resource for intensive advice on emotional problems. Mental health specialists are frequently perceived as inappropriate agents for meeting problems that are beyond the expertise of the family and local community.
- ***Il mal occhio*** (the evil eye), which is also called *occhio cattivo* (bad eye), *occhio morto* (eye of death), and *occhio triste* (wicked eye), has its roots in ancient Greece. Individuals can protect themselves from the evil eye by using magical symbols and by learning the rituals of the ***maghi*** (wizards). Amulets, miniature representations of natural or man-made weapons that fight off the evil eye include teeth, claws, and replicas of animal horns that are worn on necklaces or bracelets, held in a pocket, or sewn into clothing. ***Cornicelli*** (little red horns) are very common and can be purchased in Italian neighborhoods as good-luck charms. Many believe in obtaining as much protection from the evil eye as possible. An array of pictures and statues of the Madonna and saints are liberally distributed throughout their homes and supplement these red horns, which are often hung over a door. **Do not remove amulets; they provide great solace to the patients.**
- Common plant derivatives used in folk healing are olive oil, lemon juice, wine, vinegar, garlic, onion, lettuce, and tobacco. A crown of lemon leaves is believed

to cure a headache, as are wild fennel, deadly nightshade, and sorrel. The leaves and flowers of the wild mallow herb, *malva*, are used to make tea, providing cool energy and positive effects on the lungs and stomach. When suffering from a fever, a person is given hot rather than cold drinks. For indigestion, a mixture of coffee grounds and sugar is taken. Grain sprouts, especially those grown in the dark, in consecrated ground, in the churchyard, or in a crypt, are believed to protect against Satan and the forces of chaos.

- Sulfur and lemon juice are mixed as an ointment for scabies, and potato or lemon slices are bound to the wrists to reduce fever.
- Many individuals do not use available resources because they have little faith in health-care providers. In addition, the high cost of institutional care may be a deterrent for many. **Provide a cultural agent for newer immigrants who have difficulty with the English language and who are unfamiliar with the health-care system.**
- Women are more likely than men to report pain experiences and express symptoms fully; they expect immediate treatment. Most are expressive with chronic pain.
- The sick role is not entered without personal feelings of guilt; thus, individuals may keep sickness a secret from their family and friends and are not inclined to describe the details because they blame themselves for the health problem. Suppressing emotions and stress from fear, guilt, and anxiety can cause illness. Most believe that people who have disabilities should be cared for by the family at home; very few individuals are placed in long-term care facilities.
- A person with a physical or mental disability is not stigmatized because the condition is believed to be God's will.
- Judicious use of medications and blood transfusions are permissible and morally acceptable as long as the benefits outweigh the risks to the individual; thus, there is little objection to accepting a blood transfusion. Organ donation is morally permissible when the benefits to the recipient are proportionate to the loss of the organ to the donor and when the organ does not deprive the donor of life or the functional integrity of the body. **Organ transplant is an individual decision.**

### *22.1.11 Health-Care Providers*

- Mystical powers are not limited to saints. For the traditional, certain humans are believed to have immediate and potent access to magical powers. These are the *maghi*, "wizards" and the *streghe*, "female witches," who are granted various amounts of black magic power at birth. A powerful sorcerer is called *lupo mannaro* (werewolf). These extraordinary people are thought to possess or influence the evil eye. They have the power to cast spells, cause or cure ailments, and change events by using their own force—even their gaze is thought to be potent. These traditional beliefs may be practiced by first-generation Italian immigrants but hold little value for second- and third-generation Italian Americans.

- **Accept and incorporate folk practices into treatment plans, along with providing written instructions for biomedical treatments.**
- Success in persuading children to take medicine depends on the trust the mother has in the health-care provider. If the health-care provider is Italian or makes an effort to understand the Italian culture, the mother is more compliant. **Assigning providers of the same culture is advantageous for the patient.**

## 22.2 Reflective Exercises

1. Describe traditional Italian American gender roles. What are their implications for health-care decision-making?
2. How do you determine cyanosis and jaundice in dark-skinned Italians?
3. What are common health conditions for Italian Americans?
4. What are the dangers associated with the ingestion of fava beans and the administration of sulfonamides, antimalarial agents, salicylates, and naphthaquinolones in Italian Americans?
5. Identify several foods used for health promotion. What illnesses will they help prevent?
6. Identify some common beliefs and customs about pregnancy? Are any of them harmful?
7. In traditional terms, to what are illnesses attributed?
8. Many traditional Italians believe in *Il mal occhio*. What is this and what is done to protect oneself?
9. What are *maghi* and *streghe*? What are their functions?
10. Explain the basic rites of the Sacrament of the Sick. What is the purpose?

## Bibliography

CIA (2024) World factbook: Italy. https://www.cia.gov/the-world-factbook/countries/italy/

Dignity USA (2008). https://www.dignityusa.org/

Giordano J, McGoldrick M, Kages JG (2008) Italian families. http://www.dgapractice.com/documents/ItalianFamilies.pdf

Italy (2016) World Health Organization. http://www.who.int/countries/ita/en/

Italy: Countries and their Culture (2016). http://www.everyculture.com/Ge-It/Italy.html

Italy versus the United States: Family and Social Structures (2019). https://lifeinitaly.com/italy-vs-united-states-family-and-social-structures/

Stievano A, Rocco G, Shaffer FA, Caruso R (2021) People of Italian heritage. In: Purnell L, Fenkl EA (eds) Textbook for transcultural health care: a population approach, 5th edn. Springer, Cham

Taruscio D, Kodra Y, Ferrari G, Vittozzi L (2014) The Italian National Rare Disease Registry. Blood Transfus 12(Suppl 3):s606–s613

# Chapter 23
# People of Jamaican Heritage

**Audrey P. Miller, Chanadra Young-Whiting, Yasmine Campbell, and Valerie J. Diaz**

## 23.1 Overview and Heritage

Jamaica is the third largest island (next to Puerto Rico, Hispaniola, and Cuba) within a chain of islands that form the Greater Antilles in the northern area of the Caribbean Sea. Sitting between latitudes 17° and 19°N and longitudes 76° and 79°W, Jamaica is about 146 miles east to west; 22–51 miles north to south; and a total area of about 4400 miles$^2$. Its highest peak (Blue Mountain) is 7404 ft (about 2.26 km) and harvests some of the richest coffee. In some places, white and yellow limestone (bauxite) runs about 8200 ft (about 2.5 km) deep. Bauxite, the ore from which aluminum is made, is found in enormous quantities and Jamaica was the largest exporter in the world. Other exports include sugarcane, banana, coffee, and tobacco.

The capital of Jamaica is Kingston, located at the southeast end of the island. The currency is the Jamaican Dollar (JMD). Due to economic failure, the worth of the JMD is about \$155 to \$1 United States (US) dollar. The population in Jamaica is about 2.8 million people, divided amongst three counties and 14 parishes. Please see below approximate demographic and division of Jamaica in Table 23.1 and 23.2.

The island has a tropical climate with temperatures ranging from 60 °F (inland/mountains) to 90 °F (coastal). Unfortunately, due to location, Jamaica is vulnerable to Hurricanes and tropical storms. Fifteen hurricanes have directly hit Jamaica, with the most devastating being Hurricane Charlie (1951) and Hurricane Gilbert (1988). Earthquakes are another hazard of the island as it is located at the northern edge of the Caribbean Plate. The earthquake of 1692 in Port Royal measured a 7.5 on the Richter Scale. It killed about 5000 people and sank 2/3 of the city below sea level.

---

A. P. Miller (✉) · C. Young-Whiting · Y. Campbell · V. J. Diaz
Nicole Wertheim College of Nursing and Health Sciences, Florida International University, Miami, FL, USA
e-mail: apemille@fiu.edu; youngcd@fiu.edu; ycampbel@fiu.edu; vdiaz@fiu.edu



E. A. Fenkl, L. D. Purnell (eds.), *Handbook for Culturally Competent Care*,
https://doi.org/10.1007/978-3-031-70492-5_23

**Table 23.1** Demographic population of Jamaica

| Ethnic group | % population |
|---|---|
| Negro/Black | 90 |
| Negro/Mixed | 7 |
| East Indian | 1 |
| Chinese, white, other | 2 |
| Total | 100 |

**Table 23.2** Division of counties and Parish

| Cornwall | Middlesex | Surrey |
|---|---|---|
| St. Elizabeth | St. Catherine | Kingston and St. Andrew |
| Trelawny | Clarendon | St. Thomas |
| St. James | St. Ann | Portland |
| Hanover | St. Mary | |
| Westmoreland | Manchester | |

In 1907, a 6.2 magnitude earthquake with subsequent fires ravished Kingston for about 3–4 h and was followed by an 8-ft tsunami. About 1K people were killed.

The original people of Jamaica (Amerindians) settled for several centuries before Christopher Columbus landed on the island in 1494. In later years, they named themselves Taino (Arawaks) and the island was known as Xaymaca by their neighboring islands. Spain took formal possession of Jamaican in 1509; English captured Jamaica in 1655; the maroons (freed slaves) waged war until 1660. The abolition of slave trade to British colonies ended in 1808; the abolition of slavery dated 1834; and full Emancipation of Slavery ended in 1838. Jamaica remained under the British rule until 1848. However, to govern the country, the Jamaica Labor Party formed in 1943 and the People's National Party formed in 1938. Jamaica claimed their independence from British rule in 1962.

### *23.1.1 Communication*

- Due to British rule, the people of Jamaica speak, read, and scribe English.
- Jamaican Creole (Patois), developed in the seventieth century, is known for its lyrical styles, phrases, and unique terms when slaves from West and Central Africa were exposed to and learned the dialect forms during their enslavement.
- By the end of the eighteenth century, their culturally native language, Patois was the accepted language developed from a combination of English, Spanish, West African, and French.
- Jamaica continues to follow the British Education System with English, Math, and Sciences at the forefront.
- There are four levels of education in Jamaica: Basic, Primary, Secondary, and University.
- Basic education (ages 1–6) is provided by the government. However, primary education (while compulsory) is financed by the family. Secondary and univer-

sity education is for those who attain exemplary scores on the equivalency exams and have the financial support to attend.

- Literacy is at an all-time high of about 96% with almost 98% of those being female. Verbal and non-verbal communication is vital to the friendly population of Jamaica.
- Use greetings and salutations for the time of day: "good morning," "good afternoon," "good evening," or "good night."
- Status is respected in Jamaica, so it is important that you show reverence and respect to those in positions of authority and to the elderly. Address people by their title (Mr., Mrs., or Miss) and their surname until a personal relationship has developed. Always wait until you have permission to call someone by their first name or their nickname.
- The most common greeting for a Jamaican is the handshake, with direct eye contact and a warm smile. Jamaicans are very friendly and stand very close when conversing.
- Once a friendship has been established, women may hug and kiss on each cheek and men pat each other's shoulder or arm during the greeting process or while conversing.
- While Jamaicans are outwardly warm and friendly, they often appear standoffish at the initial introduction because they are reserved and observing behavior until they get to know someone. Do not appear overly familiar at the initial greeting. This might be seen as rude.
- Jamaicans can be direct communicators and are not afraid to say what they think expect others to be the same. However, they value tact, sensitivity, and dislike overt aggression.
- Most importantly, out of respect to adults and elders that are speaking, children are to remain quiet, seen, and not heard.

### *23.1.2 Family Roles and Organization*

- Family roles among people of Jamaican heritage are influenced by traditional values, historical factors, and contemporary dynamics. The roles often vary based on socioeconomic status, urban verses rural residence, and generational differences.
- Jamaican families emphasize collectivism and interconnectedness. Families are seen as the central support system, and individuals are expected to contribute to the well-being of the family unit. This can involve financial support, caregiving for elders, and helping younger family members.
- Jamaican families tend to be extended and include not only parents and children but also aunts, uncles, cousins, and sometimes close family friends. These extended networks provide a powerful sense of community, emotional and financial support which have proven vital especially in times of economic hardship.

- Respect for elders is a fundamental value in Jamaican culture. Older family members are given high regard and often hold positions of authority within the family. Children are taught to show deference and care for their elders.
- Women play a central role as matriarchs, managing household concerns and serve as the emotional anchors for the family.
- Traditionally, Jamaican men were expected to provide the financial support and protect the family from harm. However, recent years demonstrate gender role evolution where women assume professional careers and contribute to the household income.
- Parenting in Jamaican families is viewed as stern, emphasizing discipline and respect for authority. While corporal punishment was common many years ago, attitudes have evolved, and many families now advocate for non-violent forms of discipline.
- Education is highly valued. Jamaican parents encourage their children to excel academically to secure a better future for the family.

### 23.1.3 Workforce Issues

- Jamaica is equipped with a hospital in each parish except for Kingston, St. Catherine, and Clarendon where there are several hospitals.
- According to World Health Organization (2018), Jamaica has 0.8 nurses and midwives per 1000 population. This is compared to the United States which has an average of 17.6 nurses and midwives per 1000 population.
- When looking at physicians, Jamaica has 1.3 per 1000 population and 2.6 per 1000 in the US.
- Healthcare providers, nurses, and physician upon receiving specialized education and training migrate to other countries for a better lifestyle and economic stability.
- The migration of healthcare personnel contributes to continued shortages in Jamaica.
- Increased bureaucracy and the inability to compensate with foreign healthcare workers worsen the nursing workforce shortage.
- Additional workforce issues include shortages in medical supplies, equipment, and infrastructure.
- The island has a relaxed pace of life that extends to the workplace. There is a slow response to service and punctuality is non-existent. "Soon come" is a phrase commonly used and translates to acknowledgment only, not urgency or immediacy.

### 23.1.4 Biocultural Ecology

- Given the influx of varied races and ethnicities into Jamaica over time, its biocultural ecology is rather complex, and in some ways unique.
- The people exhibit influences from various races including Black, White, East Indian, Chinese, and Indigenous populations not only in appearance but also in cultural practices.
- Exploring various aspects of Jamaica's biocultural ecology, it has to be appreciated that variations among different subsets of people do exist.
- Jamaicans tend to exhibit a mixture of races even though specific individuals categorize themselves as belonging to a specific ethnic or racial group.
- Influences involving the Spanish, English, and African peoples have resulted in a population that have varied skin colors and biological features.
- Jamaicans acknowledge their multicultural background and verbalize beliefs that racism is not a major force in class structures.
- In surveys of Jamaicans, a perceived ethno-racial homogeneity is often described among respondents. This explains why 92% of those individuals cite their identity as either Black or Black mixed. This reflects the process of creolization, a term specific to the Caribbean region that reflects the intermixture of indigenous, European, and African races over time.
- In terms of biological features, the vast majority of Jamaicans fall into racial categories of African descent.
- Asian, indigenous, and European racial influences exist within the population and within genetic histories, however the larger influence came from slaves brought over from Central and West Africa.
- This is important from a biological perspective in that many health risks and tendencies common to African descendants are also common to Jamaicans.
- The number one cause of death in Jamaica is stroke, which combined with other cardiovascular diseases accounts for 37% of all-cause mortality in the country. These figures have risen over the past few decades in part due to rising overweight and obesity issues.
- A high incidence of hypertension, heart disease, and diabetes are prevalent.
- In addition to cardiovascular conditions, other noncommunicable diseases are also prevalent in Jamaica.
- Prostate cancer and lung cancer account for neoplastic conditions, which, along with other cancers, account for 14% of deaths.
- Most concerning communicable condition involves that of human immunodeficiency virus (HIV) infections, which remains a top 10 causes of death in the country.
- Poor health education and social practices account for its continued prevalence among the Jamaican population.
- Biological differences among individuals living in Jamaican or of Jamaican descent, variations can exist in drug response and metabolism.

- Given the high degree of cultural, racial, and ethnic influence on the population from its African-related heritage, pharmacologic variations related to this racial segment is likely relevant.
- Particularly in conditions like diabetes, hypertension, and other chronic diseases, adherence is critical to optimize quality care.
- Due to cultural beliefs and practices as well as health education limitations, medication adherence is often low among Jamaican patients.

### 23.1.5 High-Risk Health Behaviors

- In addition to biological aspects of Jamaican culture, inherent cultural beliefs, traditions and practices are also highly relevant to health outcomes and optimal care.
- In order to appreciate these influences, it is worth noting that Jamaicans currently experience two healthcare systems. One involves westernized biomedical care strategies while the other relates to traditional medicinal practices. Both systems are important in understanding health behaviors among the population as Jamaicans tend to weave these together in their healthcare strategies.
- Based on prior population surveys, most Jamaicans initially seek the advice of a primary care doctor when symptoms develop in order to understand a diagnosis.
- Jamaicans routinely use a variety of plants and herbs to prevent and treat a variety of conditions.
- Each of these have been handed down via oral tradition from either African or indigenous heritage, and many have now been studied through more scientific means to determine efficacy.
- Several such plants and herbs are commonly used today for various ailments.
- Limited literacy and health knowledge often fail to guide sexual and other behaviors to reduce disease.
- Combined with beliefs in God's power to protect and heal as well as punish, challenges exist in getting Jamaicans to adopt best health practices in some instances.

### 23.1.6 Nutrition

- Jamaicans are experiencing a state of flux when it comes to dietary practices. Traditional foods and rituals continue to play a significant role in dietary behaviors but increasingly these practices are being combined with westernized foods and diets.
- Over the last 30 years, there has been a process of remote acculturation taking place, which describes the adoption of other cultures' dietary practices despite being immersed in that culture.

- Jamaicans perceive food as a source of nutrition and sustenance.
- Diets have always consisted of a variety of foodstuffs that offer a complex array of vitamins and minerals as well as caloric support.
- Desserts and sweets play a role in celebrating occasions, but at the same time foods chosen represent Jamaica's historical struggles to a great degree.
- In this regard, meanings of food serve to remind Jamaicans of their background and heritage.
- In the traditional diet of island Jamaica, there are several foods that are commonly consumed on a regular basis.
- Rice, peas, and beans are included in most meals, as are endemic fruits and vegetables around the region.
- Guava and papaya are tropical fruits common in addition to other berries and citrus.
- Meat is also regularly consumed but differs from other cultures in terms of its type.
- Food rituals are a centerpiece for various festivals and celebrations.
- In the past, access to food was limited; therefore high-density carbohydrates offered the chance to gain the calories needed to survive the day.
- While dietary changes have occurred through the influence of other countries and through remote acculturation, some rituals and tendencies persist in terms of foods.
- Foods do have a role in promoting health in addition to its role in preserving cultural identity for Jamaicans.
- Though most Jamaicans today do not avoid salt religiously, salt is still considered to prevent spiritual connections.
- Though jerked meats no longer are essential to prevent spoilage and secondary illnesses, they remain a staple within the Jamaican diet.
- Jamaica struggles with some nutritional deficiencies.
- One of the most notable nutrient deficiencies involves iron, as anemia is common among children and women.
- Young preschool children have been shown to have an anemia rate of over 30% while iron deficient anemia averages about 10% in older children.
- In pregnant Jamaican women, low hemoglobin and ferritin levels indicative of iron deficiency approaching nearly 40%.
- Educational programs and social support are needed to overcome barriers to healthy eating and to improve existing nutritional limitations among the population.

### *23.1.7 Pregnancy and Childbearing Practices*

- As of 2022, the population is about 2.8 million; the life expectancy is 75-years-old, and there were 15 births per 1000 people and 10 deaths per 1000 live births.

- There are several things that Jamaicans say that are signs that a woman is pregnant. The most popular saying is that if a lizard jumps on a woman, she is pregnant. Another widespread belief is that if you have a random dream about fish, it is a sure sign that someone close to you has a child on the way.
- When you are pregnant, be careful. According to Jamaican folklore, using needle and thread while pregnant can cause the baby to become tied up in the umbilical cord.
- Do not hold your hands above your head, the baby's umbilical cord will somehow wrap around its neck causing inevitable death.
- An expectant mother is believed to "mark" her child with the things she craves during pregnancy. So, if the mother craves mangoes, the baby is likely to be born with a birthmark resembling a mango.
- In Jamaica, as with many cultures, a midwife, or nana, was the main overseer in birth and other transitional rituals. No matter the place, class or race of the family, all midwives shared a deep connection to and respect for the spirit world. Delivering at home was most common.
- Pregnant people were encouraged to drink okra to facilitate a healthy pregnancy and birth.
- Prior to the birth, the room would have to have an open bible on display, often open to a Psalm. The midwife would anoint the mother's belly with castor oil, which would later be given to the baby and mother.
- The navel string (umbilical cord) of a newborn may be buried and a tree planted on the spot. This tree is called the birth tree or navel string tree. It is usually a source of immense pride to children as they are growing up. It also indicates an unbreakable connection with the land on which the tree is planted.
- An open Bible is often placed in a newborn's crib or bed to protect the baby and ward off evil. The Bible is usually left open at a Psalm. A red string may be tied around a newborn's wrist or in the hair. This is to "run duppy" (chase ghosts away) and ward off evil.
- Many parents will not cut an infant's hair before the child has started to speak. It is believed that cutting their hair will delay the start of speech. The fingernails of an infant should not be cut with scissors as this might make the child grow up to be a thief.
- New mothers are encouraged to stay indoors for a few days so as not to catch a "baby cold." It used to be customary for both mother and newborn to be isolated up to 8 days after birth.
- On or about the tenth day after the baby's birth, the mother and child would be taken out into the community to receive greetings and gifts from family and friends. The child would also be named on this day. This tradition has roots in West African birth beliefs and traditions surrounding birth.
- Belly banding postpartum was also a customary practice, done for 2–3 months, using a type of bark that was heating.

### 23.1.8 Death Rituals

- A plethora of Jamaican customs associated with death were exercised, many originating in rural parts of the country. Though modern values and ways of thinking have shifted the culture, several customs have survived.
- When someone dies, family, friends, and community members gather at night to sing Christian hymns and read Bible passages for 9 days. This ritual, known as set-up, dates back to slavery and is still practiced today. On the ninth day ("dead-yard" or nine-night"), a ceremony is performed to usher the spirit of the dead, or duppy, out of the dwelling to forever rest in peace.
- Depending on how the deceased lived his life, Jamaicans believe that in death, the surviving ghost perpetuates the same traits.
- To restrict the movement of the ghost of the deceased and prevent his return to haunt the living, several measures are taken, including cutting out the pockets of the burial suit and inserting pins in the soles of both feet while chanting to prevent the ghost from walking about.
- The widow of a nagging husband would cut a handkerchief in halves, placing one half into the hand of the departed while burning the remaining half. It is believed that the spirit of the departed spends all his time searching for the other half of the handkerchief, allowing no time to plague his widow. The widow would also wear red panties for an extended time as an additional ghost repellant.
- Unusual crying of animals, birds or insects and the crowing of a cock inside the house represent omens of death in the Jamaican culture.
- Superstition surrounds graves and grave digging. It is a customary practice to attend a burial in odd numbers to prevent odd occurrences in the cemetery.

### 23.1.9 Spirituality

- Obeah is a form of folk magic or folk religion that is practiced in some Caribbean countries, including Jamaica. Originating from African spiritual traditions, obeah represents a belief system that encompasses spiritual and mystical practices, witchcraft, and sorcery. Obeah is often associated with healing, divination, and protection.
- Obeah is not an organized religion with a central doctrine. It is a set of practices that vary among those who follow its beliefs. Obeah receives mixed reviews as some herald it as a respected aspect of Jamaican heritage while others regard it as a superstition or even an illegal practice. Obeah.
- Obeah is often performed secretively by someone paid for services including shadow-catching, ointment and poison preparations, and talismans.
- The term Obeah is use throughout the Caribbean, but the phrase Myal is specifically a Jamaican term to describe the good obeah.

- Many Jamaicans believe that the purpose of Myal is to counteract the evil spirits of Obeah.

### *23.1.10 Healthcare Practices*

- Infection control is standard on the island due to fear of disease. Natural remedies are used as preventative care.
- Jamaicans generally mistrust mainstream medicine and often use herbal medicine instead of prescribed medicine by health professionals to treat illnesses.
- Physicians can prescribe medication in the country, or patients may go to the drug store, speak to the Pharmacist, and obtain medicine based on symptoms.
- Jamaicans in the lower socioeconomic group do not access healthcare due to possible out-of-pocket fees and the location of services.
- Especially in rural areas, there is minimal access to transportation and the cost, which decreases the incidence of patients accessing healthcare services.
- Private healthcare is available to those with health insurance or who can afford to pay cash.
- Public healthcare is available at no cost; however, access is still poor due to long wait times and a lack of facilities and healthcare workers available to provide care.
- Immunization is mandatory for children to attend school.
- There are currently five pathologists on the island. A police officer must order an autopsy. Most families are reluctant to come and identify the body; hence, there are many delays and a lack of storage due to no public facilities provided by the government.
- The Jamaican culture believes in bodily integrity at death; hence, this is a barrier to obtaining deceased donors. The access to living donors is also diminished due to Jamaica's high immigration rate to North America, Canada, Europe, and other Caribbean islands.
- Jamaicans have an increased fear of needles, which precludes them from donating blood. As a result, the island has an extreme shortage of banked blood due to a lack of volunteers.
- Mental illness is underfunded, and this diagnosis has a considerable amount of stigma, leading to avoidance, rejection, and victimization.
- Marijuana is viewed as an acceptable way to treat various illnesses, especially pain.
- Jamaicans have a poor adherence to pharmacotherapy, and using pills daily is not a usual practice and is usually guided by religious and cultural traditions. Different teas and purification methods are common antidotes that are acceptable.

### *23.1.11 Healthcare Providers*

- Most Jamaicans refer to the physician as "di docta man or bush doctor."
- Ensure the patient is comfortable, which will increase open communication.
- Jamaicans may use herbal medicine in conjunction with mainstream medicine.
- Most Jamaicans use herbal medicine as a primary source of healthcare despite scientific advancement in medicine.
- The practice of herbal medicine is independent of socioeconomic status.
- Traditional medicine is an afterthought due to access to care, lack of insurance, location, and socioeconomic status.
- Jamaicans' perception of quality versus efficiency also influences the choice between public and private healthcare facilities.
- There are two primary sectors of care (public and private), and not all providers work in the public sector, further decreasing access to care.
- Although unregulated, the private sector of the health system is more technologically advanced and provides outpatient services and cutting-edge procedures.
- Due to the lack of physicians, junior physicians are trained to bridge healthcare facilities' gaps.
- The 2030 Vision of the Ministry of Health in Jamaica is to provide Universal Healthcare to the entire population of Jamaica.

## 23.2 Reflective Exercises

1. Describe the language and/or dialect spoken by native Jamaicans.
2. Compare and contrast communication styles between men and women of Jamaican Heritage.
3. Name two folklore that will identify if a woman is pregnant.
4. What rituals are displayed and adhered to in the post-partum period of pregnancy?
5. What scale was used to describe varied skin colors?
6. In terms of biological features, the vast majority of Jamaicans fall into what racial category?
7. Fill in the blank: Currently Jamaicans are experiencing a ______epidemic fueled by changes in diet and behaviors?
8. What is the most concerning communicable condition which remains a top 10 cause of death in Jamaica?
9. Jamaicans perceive the meaning of food in ways that extend beyond its nutritional aspect. Many foods are used to treat health conditions and are therefore viewed as medicinal as well. TRUE OR FALSE
10. List five foods commonly consumed regularly.
11. Jamaicans have a strong focus on the use of fruits for healing and for wellness. TRUE OR FALSE

12. Vitamin D levels are deficient in as many as half of children under 5 years of age and in roughly a fifth of older children. TRUE OR FALSE
13. Due to cultural beliefs and practices as well as health education limitations, medication adherence is often low among Jamaican patients? TRUE OR FALSE
14. Statistics show that 73% of all Jamaicans self-medicate with plants and herbs with many being home sourced. TRUE OR FALSE
15. Jamaicans currently experience how many healthcare systems?
16. Describe three family values in the Jamaican culture.
17. Discuss several death rituals and their significance in Jamaican culture.
18. How is spirituality displayed in the Jamaican culture?
19. Discuss the use of Patois in the workplace.
20. Identify social determinants of healthcare in the rural sector.
21. Discuss the reason for herbal medicine being the first choice for healthcare in the Jamaican culture.

## Bibliography


Ayre S (2002) Bush doctor. LMH Publishing Ltd.

Barned C, O'Doherty K (2019) Understanding fatness: Jamaican women's constructions of health. Fat Stud 8(1):25–43. https://doi.org/10.1080/21604851.2019.1532231

Becca (2023a) Taboos in Jamaican culture. Wah Deh Gwaan. https://wahdehgwaan.com/taboos-in-jamaican-culture/

Becca (2023b) Jamaican Patois. Wah Deh Gwaan. https://wahdehgwaan.com/jamaican-patois/

Brown R, Bateman CJ, Gossell-Williams M (2022) Influence of Jamaican cultural and religious beliefs on adherence to pharmacotherapy for non-communicable diseases: a pharmacovigilance perspective. Front Pharmacol 817. https://doi.org/10.3389/fphar.2022.858947

Campbell A (2018) The Jamaican public healthcare system from the 17th-21st centuries: a policy and structural perspective. University of Technology, Jamaica Press

Cassells ED, de Carvalho G, Frisina Doetter L (2022) The colonial legacy and the Jamaican healthcare system. In: International impacts on social policy: short histories in global perspective. Springer, Cham, pp 89–103

Chevannes B (2001) Learning to be a man: culture, socialization, and gender identity in five Caribbean communities. University of the West Indies Press

Davis N (2011) Jamaica—the essential guide to customs and culture: culture smart. Kuperard, London

Ferguson JA, Bryan P, Black CV, Buisseret DJ (2023) Jamaica. Encyclopedia Britannica. https://www.britannica.com/place/Jamaica

Kelly MD (2020) Examining race in Jamaica: how racial category and skin color structure social inequality. Race Soc Probl 12(4):300–312. https://doi.org/10.1007/s12552-020-09287-z

Madu E, Edwards P (2022) Manpower shortages in Jamaica's health care. The Jamaica Observer

Ministry of Health and Wellness (MOH). Vision for health 2030. https://www.moh.gov.jm/wp-content/uploads/2019/05/MOHW-Vision-for-Health-2030-Final.pdf

Ministry of Health and Wellness (MOH). Location of hospitals and health centres. https://www.moh.gov.jm/location-of-hospitals-and-health-centres/

Mordecai M, Mordecai P (2001) Culture and customs of Jamaica. Greenwood Press, Westport, CT

Nelson MR, Ahn RJ, Ferguson GM, Anderson A (2020) Consumer exposure to food and beverage advertising out of home: an exploratory case study in Jamaica. Int J Consum Stud 44(3):272–284. https://doi.org/10.1111/ijcs.12563

Pan American Health Organization (PAHO) (1998) Micronutrient deficiencies in the English-speaking Caribbean and Guyana. Pan Am J Public Health 4(5):367–370. https://iris.paho.org/handle/10665.2/8550

Picking D, Delgoda R, Vandebroek I (2019) Traditional knowledge systems and the role of traditional medicine in Jamaica. CABI Rev 17:1–13. https://doi.org/10.1079/PAVSNNR201914045

Rodriques J (2022) "It's enough to survive through this hell to make ourselves immortal in the eyes of our descendants:" Myal, death and mourning in die the long day. Cult Dyn 34(1–2):28–44. https://doi.org/10.1177/09213740211011193

Rohman M (2023) Genetic variations influence drug metabolism in patients of African descent. Northwestern University Feinberg School of Medicine Website. https://news.feinberg.northwestern.edu/2023/02/07/genetic-variations-influence-drug-metabolism-in-patients-of-african-descent/

Smart CA (2019) African oral tradition, cultural retentions and the transmission of knowledge in the West Indies. IFLA J 45(1):16–25. https://doi.org/10.1177/0340035218823219

Smith CA (2003) The Jamaican family: strengths, realities, and reforms. University of the West Indies Press

Snider (2020) Rastafari culture. Black History Month 2023. https://www.blackhistorymonth.org.uk/article/section/real-stories/rastafari-culture/. Accessed 27 Sept 27 2023

Stone JH (2004) Culture and disability: providing culturally competent services. Sage Publications

Vandebroek I, Picking D (2020) Popular medicinal plants in Portland and Kingston, Jamaica. Springer

Whelan M, Ulrich E, Ginty J, Walsh D (2018) Journeys to Jamaica: a healthy dose of culture, competence, and compassion. J Christ Nurs 35(2):E21–E27. https://doi.org/10.1097/CNJ.0000000000000495

# Chapter 24
# People of Japanese Heritage

**Ivette M. Hidalgo and Carmen Rizzetto Duque**

## 24.1 Overview and Heritage

**Nihon**, or **Nippon**, as Japan is called in the Japanese language, is a 1200-mile chain of islands in the northwestern Pacific Ocean. The land is mountainous with steep ravines and prone to various geological phenomena, mostly earthquakes and active volcanos. The main islands are Hokkaido, Honshu, Shikoku, Kyushu, and Okinawa. Japan borders Russia, Korea, and China, and its modern history has, until recently, been shaped by conflict with these countries. Japanese citizens residing in North America tend to settle in large commercial and educational centers.

Education is highly valued; the literacy rate in Japan is nearly 100%. About 56% of all young people go on to higher education. The alumni network primarily provides job placements; the school one attends determines, to a great extent, where one is employed after graduation. *Issei* (first-generation Japanese immigrants) vary widely in their English-language ability. ***Nisei*** (second-generation immigrants) and ***sansei*** (third-generation immigrants) were educated under the American educational system to the extent that they were permitted; for example, educational access was limited or segregated during the World War II internment of American citizens of Japanese ancestry. Acculturation, individuality, and the variant characteristics of culture determine the degree of traditionality of Japanese in the United States and elsewhere. Therefore, this profile will not fit all Japanese.

I. M. Hidalgo (✉) · C. Rizzetto Duque
College of Nursing and Health Sciences, Florida International University, Miami, FL, USA
e-mail: ihidalgo@fiu.edu; caduque@fiu.edu



E. A. Fenkl, L. D. Purnell (eds.), *Handbook for Culturally Competent Care*,
https://doi.org/10.1007/978-3-031-70492-5_24

### 24.1.1 Communications

- Japanese is the language of Japan with the exception of the indigenous Ainu people.
- Japanese communicate in high-context style. Meaning communication is sophisticated, subtle, and layered. Messages are both spoken and read "between the lines." Messages may be often implied but not openly expressed.
- Speaking Japanese without an accent is considered a criterion for being Japanese. If other non-Japanese languages are spoken, the person is considered half-Japanese or non-Japanese.
- Because high-school graduates in Japan complete 6 years of English, even newer Japanese immigrants and sojourners can speak, understand, read, and write English to some extent.
- **While the language barrier may be an obstacle to verbal instructions or explanations in English-speaking health-care settings, Japanese patients are likely to use written materials effectively. Simple communication tools printed in Japanese and English can aid in communication.**
- Men tend to speak more coarsely and women with more gentility or refinement.
- Light social banter and gentle joking are a mainstay of group cohesiveness.
- Polite discussion unrelated to business, often over *o-cha* (green tea), precedes business negotiations. Relationship-building and respect for personal privacy are important aspects of working relationships in all sectors.
- Open communication is discouraged, making it difficult to learn what people think. In particular, saying "no" is considered extremely impolite; rather, one should let the matter drop.
- A high value is placed on "face" and "saving face." Asking someone to do something he or she cannot do induces loss of face or shame. For people to be shown wrong is deeply humiliating. People feel shame for themselves and their group, but they are respected when they bear shame in stoic silence. Because of ethnic homogeneity, an ingrained sensitivity to the feelings of others, and close contact with one's family, classmates, and work group, vague and intuitive communication is well understood by fellow group members.
- **Giving a yes answer does not necessarily mean yes as it does in the Western context.**
- Traditional Japanese exhibit considerable control over body language. Anger or dismay may be quite difficult for Westerners to detect.
- Avoiding confrontation and disruption is considered an appropriate behavior for maintaining a good relationship. Expression of feelings and opinions is not often done to maintain harmony.
- Smiling and laughter are common shields for embarrassment or distress.
- Prolonged eye contact is not polite even within families.
- Social touching occurs among group members but not among people who are less closely acquainted. In general, body space is respected.

- Intimate behavior in the presence of others is taboo, although this is changing among the younger generations and in the United States.
- When people greet one another, whether for the first time or for the first time on a given day, the traditional bow is performed. The depth of the bow, its duration, and the number of repetitions reflect the relative status of the parties involved and the formality of the occasion.
- Overall orientation is toward the future. Punctuality is highly valued.
- Family names are stated first, followed by given names. Seki Noriko would be the name of a woman, Noriko of the Seki family. The family names of both men and women, married or single, are designated by the suffix *-san*, but one does not use that designation when referring to oneself.
- Women generally assume their husband's family name upon marriage.
- Elders are referred to respectfully. The designation *sensei* (master) is a term of respect used with the names of physicians, teachers, bosses, or others in positions of authority.
- **Greet Japanese patients with a handshake and address them formally by their last name until invited to do otherwise.** An offer to shake hands by a Westerner is reciprocated graciously.

### *24.1.2 Family Roles and Organization*

- The predominant family structure is nuclear. The role of wife and mother is dominant.
- Traditionally, the father is believed to be the authority figure and therefore decides what is right for the family.
- Children are socialized to study hard, make their best effort, and be good group members. They are taught to take care of each other, and girls are taught to take care of boys. Self-expression is not valued. They are expected to have good manners, be respectful, obedient, and do well in school.
- Japanese parents do not commonly display public affection as Western parents do.
- **Be aware of differences in spousal relationships when assessing the quality of family dynamics and communication, sexual health, and sensitivity to risk for sexually transmitted infections.**
- The primary relationship within a family is the mother–child relationship, particularly that of mothers and sons. It is customary for a mother to sleep with the youngest child until that child is 10 years old or older, and when a new baby is born, the older sibling may sleep with the father or a grandparent. **Be aware of Japanese family sleeping practices and refrain from judgmental evaluation.**
- Babies are not allowed to cry; they are picked up instantly. Women constantly hold their babies in carriers on their chests and sleep with them.
- Corporal punishment is acceptable in Japan. **Explain US child abuse laws to patients.**

- Traditional teens and college students generally do not date. They typically join clubs, membership in which is taken seriously; most social activities, such as ski trips, are club activities. Japan has the lowest incidence of teen birth in the world. **Do not assume that dating holds the amount of concern for Japanese young people that it does for American teens. Do not assume that Japanese youth are well informed about sexuality and sexual health risks.**
- Older people are respected and cared for by the family in the home, if at all possible, with the eldest son being the responsible family member. **Be sensitive to Japanese patients' sense of obligation and commitment to older people. Help families network within the Japanese American community for both social support and for resources or for good long-term care facilities.**
- In Japan, a small segment of women have long lived outside the usual constraints for their gender. Women of "the floating world," or the entertainment industry, enjoy a fair amount of autonomy. The most traditional of these, the *geisha*, live in all-female communal arrangements. Geisha are not prostitutes, and they are now recognized as a cultural treasure. **Geisha entertainment industries exist in North America.**
- Tolerance for marriage of a Japanese person to a foreigner is less than in the United States.
- A gay and lesbian social network is evident, although they are not generally talked about. **Do not disclose same-sex relationships to family members.**

### *24.1.3 Workforce Issues*

- American practices designed to avoid liability, such as informed consent, are not routinely implemented in Japanese health-care settings. **Explain American practices such as informed consent, the legal and professional requirements of patient autonomy, and the legalities of defensive charting on orientation.**
- To be loyal and attached to a group or organization is an embedded characteristic of Japanese society. The company's prosperity may become more important than individual prosperity making the well-being of the group more important than their self-interests.
- **Explain to new employees about the direct communication practices of Amculture.**
- Japanese workers are sensitive to colleagues and superiors. Saying "no" or delivering bad news is extremely difficult; they may avoid issues or indicate that everything is fine rather than state the negative. **Ask less acculturated Japanese to demonstrate procedures instead of asking directly if they understand the procedure.**
- The orientation toward family ties and intimacy, combined with beliefs based on Buddhist and Confucian traditions, leads to long-term respect for people and the development of processes.

- A high value is placed on "face" and "saving face." Asking someone to do something he or she cannot do induces loss of face or shame. For people to be shown wrong is deeply humiliating. **Use tactful communication and approach sensitive and controversial topics carefully to allow the person to "save face."**
- Prolonged eye contact is not polite even within families and among friends. **Do not assume that lack of eye contact means the person is not listening, is not caring, or is being less than truthful.**

### *24.1.4 Biocultural Ecology*

- Racial features include epicanthal skin folds that create the distinctive appearance of Asian eyes (dark in color), a broad and flat nose, "yellow" skin that varies markedly in tone, and a physique that is not too heavy. Hair is straight and naturally black, with differences in shade. **Rely on color changes in the mucous membranes and sclerae to assess oxygenation and liver function.**
- Negative blood types account for less than 1% of the population.
- Table 24.1 lists commonly occurring health conditions for Japanese in three categories: lifestyle, environment, and genetics. The leading causes of death are cancer, heart disease, and cerebrovascular disease.
- Asthma, related to dust mites in the *tatami* (straw mats that cover floors in Japanese homes), is one of the few endemic diseases. The most common seasonal allergen in Japan is cedar pollen and it causes allergic rhinitis in the spring. Cedar pollinosis is termed as "a folk disease" in Japan. Asthma death rates are lower in Japan, but the prevalence of severe asthma disease is the same as that in Western countries. **The same condition can affect patients in the United States.**
- Drug dosages may need to be adjusted for the physical stature of Japanese adults. Many Asians are poor metabolizers of mephenytoin and related medications, potentially leading to increased intensity and duration of the drugs' effects. Dosages for patients with heart failure and reduced ejection fraction are on the lower end. It has been hypothesized that this is due to their small body size and increased drug sensitivity. The patient's geographic region of residence influences the prescribed dosage as well.
- **Be aware that most Asians tend to be more sensitive to the effects of some beta blockers, many psychotropic drugs, and alcohol.**
- Most individuals require lower doses of some benzodiazepines and neuroleptics.
- Opiates may be less effective analgesics, but gastrointestinal side effects may be greater than among white people. **Take patients' body mass into consideration in dosing; even with that precaution, patients' responses to drugs need to be monitored carefully.**

**Table 24.1** Health conditions common to Japanese Americans

| Health conditions | Causes |
|---|---|
| Heart disease | Genetic, environment, lifestyle |
| Tuberculosis | Environment, lifestyle |
| Renal disease | Lifestyle |
| Asthma | Environment, lifestyle |
| Vogt–Koyanagi–Harada syndrome | Genetic |
| Takayasu disease | Genetic |
| Acatalasemia | Genetic |
| Cleft lip/palate | Genetic, lifestyle |
| Oguchi disease | Genetic |
| Lactase deficiency | Genetic |
| Stomach cancer | Genetic, environment, lifestyle |

### 24.1.5 High-Risk Health Behaviors

- Smoking rates are high among Japanese and Japanese Americans. **Children exposed to secondhand smoke may demonstrate emotional and behavioral problems. Begin teaching smoking cessation in elementary school.**
- Alcohol (rice wine) is part of many social rituals, such as picnics, to celebrate cherry blossoms, autumn leaves, or moon viewing. Adults commonly drink beer and sake in the home, and children may be seen purchasing alcohol.
- Nomikai "after-work drinks" is an accepted event meant to help spark conversation and build relationships among colleagues. Once alcohol is consumed, people can relax and speak freely; they are forgiven for what they say because of the alcohol. **Be aware of the prevalence of smoking and heavy alcohol consumption, particularly among men. Give specific medical reasons why they must abstain, thus providing a socially acceptable excuse.**
- Mothers' time-honored strategy of rewarding academic diligence with candy and other treats contributes to the issue of the fitness of youth. **Encourage mothers to offer other rewards instead of candy and non-nutritious treats for academic excellence.**
- Public safety consciousness is high; Japanese readily use seat belts and other safety measures such as child safety seats and helmets.

### 24.1.6 Nutrition

- Dietary staples include rice, beef, poultry, pork, seafood, root vegetables, cabbage, persimmons, apples, and tangerines.
- Rice is the mainstay of the traditional diet and is included in all three meals as well as snacks. Rice consumption has greatly decreased. At breakfast, more bread is eaten compared to rice. A traditional breakfast includes fish; pickles, *nori* (various seaweeds used to flavor or garnish meals); a raw egg stirred into the hot rice; *miso* (soybean-based) soup; and tea. Some people prefer a Western breakfast of toast or cold cereal and coffee.
- Rice has a symbolic meaning related to the **Shinto** religion, analogous to the concept of the "bread of life" among Christians. A staple of schoolchildren's *o-bento* (lunch box) is a white bed of rice garnished with a red plum pickle, reminiscent of the Japanese flag. Meals combine elements of land and sea.
- Because of women's societal role change and children's attendance in after-school programs, more children are eating alone, which has led to an increased consumption of instant foods.
- It is believed that people who eat with others consume a well-balanced meal versus those who eat alone. Even though most people believe that eating with their families is important, those that are in their 20s and 50s do not eat as frequently with their families.

- Eateries for children (kodomo shokudo) provide meals for free or at low cost to children and other people, such as the elderly living alone, disabled people, and people who want to talk with others about their child-rearing problems.
- Because more people are active at night, they are eating later in the evening which has led to more individuals skipping breakfast.
- Malnutrition used to be the primary problem and now obesity is prevalent. Despite the latter, Japan is one of the countries with the highest longevity. The average lifespan for males is 81.09 years and for females is 87.26 years.
- Schoolchildren eat lunch or their *o-bento*, packed with rice, pickles, and meat or fish. A popular lunch among working people is a steaming bowl of ramen (noodles) in broth or cold noodles on a hot summer day. Instant broth, although high in sodium, is another popular quick lunch.
- A traditional dinner is a pot of boiled potatoes, carrots, and pork seasoned with sweet sake, garlic, and soy sauce or a stir-fried meat and vegetable dish.
- The daily intake of sweets can be high and often includes European-style desserts, pastries, and cookies, sweet bean cakes, soft drinks, and heavily sweetened coffee, which may contribute to the high incidence of tooth decay.
- Increasingly, westernized food tastes, resulting in higher fat and carbohydrate intake, have contributed to the rise in obesity and associated increases in diabetes, heart disease, and premature death.
- Obesity and depression are increasing in Japan, and they are positively associated with high processed food intake.
- There is growing public awareness that the sodium content of traditional soups and sauces contributes to the high rate of cerebrovascular accidents. **Undertake dietary assessment on admission. Incorporate traditional food practices and preparation practices in dietary counseling sessions.**
- Green tea, although high in caffeine, is a good source of vitamin C. Garlic and various herbs are used widely for their medicinal properties.
- Many individuals have difficulty digesting milk products because of lactose intolerance. **Encourage use of reduced-lactose milk, tofu, and unboned fish to meet calcium needs.**
- **Iron deficiency anemia is a concern among young women and can be alleviated with dietary counseling or dietary supplements. Nori is a traditional food source of iron.**

### *24.1.7 Pregnancy and Childbearing Practices*

- Oral contraceptives became legal in Japan in 1999. Condoms remain the most common contraceptive method. Most women have several abortions during their married lives. It is difficult for women to obtain emergency contraceptive pills in Japan. They require a medical prescription, are expensive, and are not covered by national health insurance.

- Japan has low infant and perinatal mortality rates. It is one of the safest countries in the world in which to give birth.
- Pregnancy is highly valued within traditional culture as a woman's fulfillment of her destiny. Women may enjoy the attention and pampering that they get at no other time. They may prepare themselves for the possibility of pregnancy when they become engaged and eliminate alcohol, caffeine, soft drinks, and tobacco.
- Complementary and alternative medicine (CAM) are favored and implemented during pregnancy by Japanese women.
- Women often return to their mother's home for the last 2 months of their pregnancy and through the first 2 months postpartum. **Explore a woman's expectations during pregnancy and the possibility that she might return to Japan. Finding another Japanese woman who has experienced childbearing in the United States and who can share her experiences would support the pregnant patient.**
- Loud noises, such as a train or a sewing machine, are thought to be bad for the baby.
- Shinto shrines sell amulets for conception and easy delivery. Husbands do not commonly attend the births of their children. **Do not make the husband feel guilty if he does not want to be in the delivery room.**
- Japanese women have poor access to obstetric care and analgesia medication during labor. There is a shortage of anesthesiologists. Vaginal deliveries are usually performed without medication. To give in to pain dishonors the husband's family. Many of the older generations of Japanese women believe that the bond between the mother and child is strengthened through the labor process and the pain that is experienced. Japan enjoys one of the world's lowest infant mortality rates, at 1.9 per 1000 live births and neonatal mortality rate at 0.9 per 1000 live births.
- Traditionally, postpartum women do not bathe, shower, or wash their hair for at least the first week.
- Breast-feeding is taken seriously. According to a national survey, more than 90% of Japanese women expressed a desire to breastfeed. Maternal rest and relaxation are deemed essential for success. If the mother is asleep, the grandmother feeds the baby formula. Women who give birth in the United States may resent the American expectation that they will resume self-care and child-care activities quickly. **Explain the expectations for postpartum care; exercise sensitivity and help plan for assistance upon early discharge.**

### *24.1.8 Death Rituals*

- The taboo against open discussion of serious illness and death is evident. The frequency of end-of-life discussions by nurses, physicians, and care staff is minimal. Hospice care in Japan is scarce but needed due to the increased aging population and decline of three-generational households. Hospice patients may not

want to be told their prognosis in order to allow a peaceful death and to spare both the patient and the family the difficulty of having to discuss the situation. They generally worry about becoming a burden and not being able to carry out their responsibilities within their social or family circle.
- **Hospice and palliative care professionals may have to rely on implicit behaviors to admit patients for care.**
- When a person is dying, the family should be notified of impending death so they can be at the dying person's bedside. Traditionally, the eldest son has particular responsibility during this time.
- The Japanese wake and funeral are now typically held in the funeral parlor or ceremonial hall instead of in the home. The average number of attendants has markedly declined from 180 to 46.
- Prior tradition is that when death occurs an altar is constructed in the home. Currently, most families do not have altars in their home and monthly memorial services are also not being done. The Japanese seldom cry at funerals, and the majority have a limited belief in the afterlife. The Buddhist belief is that if you shed your tears for the dead, the dead river will be flooded and the deceased will not be able to reach the other world. This can cause the spirit to wander and cause harm to the living. Photographs of the deceased are displayed, and floral arrangements are placed within and outside the home. A bag of money is hung around the neck of the deceased to pay the toll to cross the river to the hereafter. Visitors bring gifts of money and food for the bereaved family.
- Infants in Japan may have their own cemeteries and family members may or may not be expected to visit the grave thereafter.
- The mourning period is 49 days, the end of which is marked by a family prayer service and the serving of special rice dishes. Beliefs are common that the dead need to be remembered and that failure to do so can lead the dead to rob the living of rest; therefore, special prayer services can be conducted for the 1st, 3rd, 7th, and 13th anniversaries of the death.

### *24.1.9 Spirituality*

- Many people in Japan adhere to more than one religion (Shintoism 70.5%, Buddhism 67.2%, Christianity 1.5%, and others 6.2%).
- Shinto, the indigenous religion, is the locus of joyful events such as marriage and birth.
- Many festivals are marked by offerings, parades, and a carnival on the grounds of the shrine.
- Buddhism has permeated artistic and intellectual life. Very few people regularly attend services, but most are registered temple members, if only to ensure a family burial plot.

- One-half (1.5) percent of Japanese people are Catholic or Protestant. Most do not identify themselves solely with one religion or another and even a baptized Christian might have a Shinto wedding and a Buddhist funeral.
- Many accept the Buddhist belief in reincarnation; the eternal life of the soul is also recognized in the Shinto faith.
- *Kampo* (healers) often set up shop in the vicinity of the temple or shrine, and a person might be seen scooping incense smoke onto an ailing body part. Prayer boards might bear requests for special healing. Newborns are taken to a shrine for a blessing. Additional blessings take place on November 15 when a child is 3, 5, or 7 years old. Visits to shrines and temples are social, recreational, and spiritual outings. Souvenirs and refreshments are usually available, and the hike into the prayer area provides exercise.

### 24.1.10 Healthcare Practices

- There is great tolerance for self-indulgence even during minor illnesses. Because Japanese people are less likely to express feelings verbally, this indulgence may be a way for people to affirm caring for one another nonverbally. Termination of pregnancy when the health of the fetus is in doubt is common, and most parents want medically compromised neonates not to be treated aggressively when prognoses are not favorable.
- **Help patients with legal obligations who face value conflicts with termination of pregnancy and not treating medically compromised newborns.**
- The concept of *ki*, the life force or energy and how it flows through the body is integral to traditional Chinese healing modalities, including acupuncture. Good health requires the unobstructed flow of *ki* throughout the body.
- The concepts of *yin* and *yang* are reflected in modern attitudes, as is the need to balance five energy sources: water, wood, fire, earth, and metal. Strategies that help to restore balance include the use of herbal medicines, bed rest, bathing, and having a massage. One traditional form of massage, *shiatsu* (acupressure), involves redirection of energy along the Chinese meridians by application of light pressure to what we might recognize as acupuncture points.
- Leisure activities that promote health include hobbies and amusements, travel, sports, self-education, and volunteer activities.
- While Chinese tradition calls for a restoration of balance when one is ill, Shinto calls for purging and purification. Preoccupation with germs and dirt is not likely to interfere with daily life.
- Traditionally, in Japan, people use temples and shrines for their power to heal illness, promote health, and fulfill special wishes. **Health-care providers who visit Japanese homes should note or even ask whether the family removes their shoes upon entering.**
- Many pharmacies stock traditional herbal *kampo* preparations.

- Most individuals make liberal use of both modern medical and traditional providers of health care. Residents in the United States have Internet and mail-order access to traditional medications, if they are not available locally. **A complete health assessment includes inquiring about home therapies.** Table 24.2 lists common Japanese therapies.
- **Be sensitive to family issues that may underlie illnesses among patients. If psychotherapy is indicated, the therapist must be someone familiar with the Japanese culture. Resources may be obtained through Japanese churches or other religious organizations where Japanese people gather.**
- Japanese high regard for the status of physicians decreases the likelihood of their asking questions or making suggestions about their care. The idea that patients should be given care options may be alien. Japanese are cautious about adopting surgery. **Provide ample opportunity for dialogue and explain the choices that are offered. Use Japanese and Japanese American health-care providers to bridge gaps in understanding.**
- Some individuals may need assistance in seeking care. Their verbal English skills may be an impediment to making their needs known and understanding the care they are offered, even though their ability to understand written information may be very good.
- ***Itami*** (pain) may not be expressed, and bearing pain is considered a virtue and a matter of family honor. Addiction is a strong taboo in Japanese society, making patients reluctant to accept pain medication.
- **Use a schedule of analgesic administration rather than an as-requested or patient-controlled approach to ensure adequate pain management. Explain that physiological status and healing are actually enhanced by pain control.**
- Mental illness carries a significant stigma. Because emotional problems cannot be discussed freely, somatic manifestations are common and acceptable. Be aware that this may limit access to mental health care. The government is trying to increase knowledge about mental illness and decrease social stigma.
- Handicapped people may bring shame or heartache to the family, although they are treated kindly.
- Assumption of the sick role is highly tolerated by families and colleagues, and a long recuperation period is encouraged.
- Donating blood is encouraged in Japanese society and blood transfusions are accepted.
- Critical care and organ transplantation and donation issues need to be approached sensitively. People rely more heavily on the physician's opinion, and the family

**Table 24.2** Common Japanese therapies

| |
|---|
| • ***Morita*** therapy is an indigenous strategy for addressing *shinkei shitsu*, excess sensitivity to the social and natural environment. Introspection is seen as harmful. *Morita* therapy focuses on constructive physical activity to help patients accept reality as it is |
| • ***Naikan*** therapy is one of reflection on how much goodness and love are received from others |

may have difficulty negotiating the cessation of treatment. **The family is considered a unit and health care providers must communicate with both patients and families.**
- Many Japanese are uncomfortable talking about end-of-life. **The concept of advance directives is unknown to many. Carefully explain advance directives to patients.**

### 24.1.11 Health-Care Providers

- Physicians, referred to as *sensei*, are highly esteemed.
- Japan has one of the largest elderly populations in the world. Individuals aged 65 years and older accounted for 28.1% of the total population in 2018, and that percentage is estimated to reach 33.3% in 2036 and 38.4% in 2065. Japan is focused on increasing the number of home care medicine practicing physicians to help meet this population's medical needs.
- Due to the COVID-19 pandemic, health care workers endured increased stress which has led them to suffer from conditions such as depression, insomnia, and anxiety.
- Self-care as a philosophy is not evident among most. Being told what to do by the physician or ***kampo*** health-care provider is expected, and his (or, occasionally, her) authority is not questioned.
- Nurses and nursing are highly regarded, reflecting traditional taboos against illness and impurity as well as the status of women. Currently, in Japan, nurses are well respected, even though women in general are not. In the past, nurses were not highly respected because "good women" did not touch people with an illness unless they were immediate family members. If she did touch "sick bodies," the woman would become tainted and less pure.
- Home care and the orchestration of many community-based providers may be resisted by residents who expect long recuperations in the hospital. **Explain how the health-care delivery system works in the United States and the functions of the different health-care providers.**

## 24.2 Reflective Exercises

1. Japanese use the term *sensei* when referring to older people. What does this term mean?
2. Distinguish between male and female communication styles.
3. What does "saving face" mean? How is "saving face" used when communicating with health-care providers?
4. List and explain at least three genetic health conditions common among the Japanese.

5. What is the word for Japanese traditional healers? When might one purchase Japanese herbs?
6. What terms are used to distinguish three generations of Japanese?
7. What is the traditional religion of the Japanese? What are some of its principal tenets?
8. Explain the concepts of *yin* and *yang* when referred to in health care.
9. Explain *morita* and *naikan* therapies. For what conditions are they used?
10. What is *itami* and how it is expressed?

## Bibliography


Adachi N (2019) Dietary transition and food and nutrition policies in Japan. Nagoya J Nutr Sci 5:1–14

Becker CB, Taniyama Y, Kondo-Arita M, Yamada S, Yamamoto K (2020) How grief, funerals, and poverty affect bereaved health, productivity, and medical dependence in Japan. OMEGA J Death Dying 85(3):669–689. https://doi.org/10.1177/0030222820947573

Benedict TO (2023) Spiritual ends: religion and the heart of dying in Japan. DOAB Home. https://directory.doabooks.org/handle/20.500.12854/95731

CIA (2023) World Factbook: Japan. https://www.cia.gov/the-world-factbook/countries/japan/

Duranto PA, Nakayama S (2005) Japanese communication avoidance, anxiety, and uncertainty during initial encounters. https://www.semanticscholar.org/paper/Japanese-Communication-Avoidance%2C-anxiety%2C-and-Duronto-Tsukuba/76b89f2206f1f91e8359d75b58a2ed5de7340b27

Galanti GA (2001) Japanese Americans and self-care: a lesson in cross-cultural care. West J Med 174(3):208–209

Hamano J, Hanari K, Tamiya N (2019) Attitudes and other factors influencing end-of-life discussion by physicians, nurses, and care staff: a nationwide survey in Japan. Am J Hosp Palliat Med 37(4):258–265. https://doi.org/10.1177/1049909119876568

Hidalgo I, Brooten D, Youngblut JM, Roche R, Li J, Hinds AM (2020) Practices following the death of a loved one reported by adults from 14 countries or cultural/ethnic group. Nurs Open 8(1):453–462. https://doi.org/10.1002/nop2.646

Hikoyeda N, Tanabe MKG (2010) Health and health care of Japanese American elders. https://geriatrics.stanford.edu/wp-content/uploads/downloads/ethnomed/japanese/downloads/japanese_american.pdf

Hirata K, Matsumoto K, Hatakeyama Y, Onishi R, Seto K, Hasegawa T (2023) Social burden of three major diseases in Japan: a time trend and future projections using the comprehensive cost of illness method. PLoS One 18(1):e0280311. https://doi.org/10.1371/journal.pone.0280311

Imamura A (2003) The Japanese family. Asia Society. https://www.asianstudies.org/wp-content/uploads/the-japanese-family-faces-twenty-first-century-challenges.pdf

Ito M, Hattori K (2013) People of Japanese heritage. In: Purnell L (ed) Transcultural health care: a culturally competent approach, 4th edn. F.A. Davis Company, Philadelphia

Iwata H, Matsushima M, Watanabe T, Sugiyama Y, Yokobayashi K, Son D, Satoi Y, Yoshida E, Satake S, Hinata Y, Fujinuma Y (2020) The need for home care physicians in Japan—2020 to 2060. BMC Health Serv Res 20(1):752. https://doi.org/10.1186/s12913-020-05635-2

Kavanaugh CM, Jong J (2020) Is Japan religious? J Stud Religion Nat Cult 14(1):152–180

Lai J, Ma S, Wang Y, Cai Z, Hu J, Wei N, Wu J, Du H, Chen T, Li R, Tan H, Kang L, Yao L, Huang M, Wang H, Wang G, Liu Z, Hu S (2020) Factors associated with mental health outcomes among health care workers exposed to coronavirus disease 2019. JAMA Netw Open 3(3):e203976. https://doi.org/10.1001/jamanetworkopen.2020.3976

Lem WG, Kawata KH, Kobayashi T, Oyama H (2023) Public stigma related to people with mental health conditions among Japanese company employees. J Clin Med Res 15(3):139–147. https://doi.org/10.14740/jocmr4868

Maeda Y, Takahashi K, Yamamoto K, Tanimoto T, Kami M, Crump A (2019) Factors affecting the provision of analgesia during childbirth, Japan. Bull World Health Organ 97(9):631–636. https://doi.org/10.2471/blt.19.230128

Murakami K, Livingstone MBE, Sasaki S (2018) Thirteen-year trends in dietary patterns among Japanese adults in the National Health and Nutrition Survey 2003—2015: continuous westernization of the Japanese diet. Nutrients 10(994):1–15

Nagaishi M (2020) Diagnostic and dialogic organization development assumptions: their fit with Japanese culture. Organ Dev J 38(3):33–44

Nagase H (2019) Severe asthma in Japan. Allergol Int 68(2):167–171. https://doi.org/10.1016/j.alit.2019.02.004

OECD (2021) Tackling the mental health impact of the COVID-19 crisis: an integrated, whole-of-society response. https://oecd.org/coronavirus/policy-responses/tackling-the-mental-health-impact-of-the-covid-19-crisis-an-integrated-whole-of-society-response-0ccafa0b/

Pagliai G, Dinu M, Madarena MP, Bonaccio M, Iacoviello L, Sofi F (2021) Consumption of ultra-processed foods and health status: a systematic review and meta-analysis. Br J Nutr 125(3):308–318

Pappa S, Ntella V, Giannakas T, Giannakoulis VG, Papoutsi E, Katsaounou P (2020) Prevalence of depression, anxiety, and insomnia among healthcare workers during the COVID-19 pandemic: a systematic review and meta-analysis. Brain Behav Immun 88:901–907. https://doi.org/10.1016/j.bbi.2020.05.026

Parkes CM, Laungani P, Young B (1997) Death and bereavement across cultures. Routledge, London

Peng-Ng L, Chye-Goh PS (2014) Incidence of discontinuation of angiotensin-converting enzyme inhibitors due to cough, in a primary healthcare centre in Singapore. Pub Med 55(3):146–149. https://doi.org/10.11622/smedj.2014034

Sato Y (2021) 'Others' among 'us': exploring racial misidentification of Japanese youth. Jpn Stud 41(3):303–321. https://doi.org/10.1080/10371397.2021.1982645

Shinozaki N, Murakami K, Asakura K, Masayasu S, Sasaki S (2023) Consumption of highly processed foods in relation to overall diet quality among Japanese adults: a nationwide study. Public Health Nutr 26(9):1784–1797. https://doi.org/10.1017/s1368980023000721

Shiraishi M, Matsuzaki M, Kurihara S, Iwamoto M, Shimada M (2020) Post-breastfeeding stress response and breastfeeding self-efficacy as modifiable predictors of exclusive breastfeeding at 3 months postpartum: a prospective cohort study. BMC Pregnancy Childbirth 20(1):730. https://doi.org/10.1186/s12884-020-03431-8

Smith O (2017) Mapped: the world's most (and least) religious countries. The Telegraph. https://www.telegraph.co.uk/travel/maps-and-graphics/most-religious-countries-in-the-world/

Sonoda K, Takedai T, Salter C (2022) Communication sheet eases barriers for Japanese patients and health professionals. BMC Health Serv Res 22(1):1–6. https://doi.org/10.1186/s12913-022-08371-x

Sorano S, Emmi S, Smith C (2021) Why is it so difficult to access emergency contraceptive pills in Japan? Lancet Reg Health West Pac 7:100095. https://doi.org/10.1016/j.lanwpc.2021.100095

Statistics Bureau Ministry of Internal Affairs and Communications. (2022) Statistical handbook of Japan 2021—統計局ホームページ. Statistics Bureau of Japan. https://www.stat.go.jp/english/data/handbook/pdf/2021all.pdf

Tahara M, Mashizume Y, Takahashi K (2020) Coping mechanisms: exploring strategies utilized by Japanese healthcare workers to reduce stress and improve mental health during the COVID-19 pandemic. Int J Environ Res Public Health 18(1):131. https://doi.org/10.3390/ijerph18010131

Taneri P, Wehrli F, Roa-Diaz ZI, Oche AI, Dante S, Hamidreza R, Bally L, Minder B, Kiefter-de Jong J, Laine JE, Bano A, Glisic M, Muka T (2022) Association between ultra-processed food intake and all-cause mortality: a systematic review and meta-analysis. Am J Epidemiol 191(7):1323–1335. https://doi.org/10.1093/aje/kwac039

Teng TK, Tromp J, Tay WT, Anand I, Ouwerkerk W, Chopra V, Wander GS, Yap JJ, MacDonald MR, Xu CF, Chia YM, Shimizu W, Richards AM, Voors A, Lam CS (2018) Prescribing patterns of evidence-based heart failure pharmacotherapy and outcomes in the Asian-HF registry: a cohort study. Lancet Glob Health 6:1008–1018

Teng TK, Cooper L, Tay WT, Luo N, Sharma A, Whellan D, Anand I, Pina IL, Tromp J, Richards AM, Kraus WE, Connor CM, Mentz RJ, Lam CSP (2019) Association between body surface area and prescribed doses of guideline-directed medications among international patients with heart failure and reduced ejection fraction. Eur J Heart Fail 22(4):754–758. https://doi.org/10.1002/ejhf.1648

Wada K, Ueyama J, Konishi K, Goto Y, Koda S, Mizuta F, Tamura T, Watanabe K, Ando K, Kondo T, Nagata C (2020) Associations between exposure to tobacco smoke and behavioral problems in preschool Japanese children. J Environ Public Health 2020:7591263. https://doi.org/10.1155/2020/7591263

Watanabe A, Inoue S, Okatsu A, Eto H, Oguro M, Kataoka Y (2023) Complementary and alternative medicine use by pregnant women in Japan: a pilot survey. BMC Complement Med Ther 23(1):298. https://doi.org/10.1186/s12906-023-04126-1

World Bank Open Data (2023) Health. https://data.worldbank.org/topic/health

# Chapter 25
# People of Jewish Heritage

**Teri Aronowitz and Polly Zavadivker**

## 25.1 Overview and Heritage

*Jewish* people, or Jews, are a group who share a common cultural, ethnic, and religious background. Individual Jewish communities are diasporic, or globally dispersed, and vary widely in their distinct practices and customs. This chapter focuses on the needs of religiously observant individuals and their families, or Orthodox Jews, but also provides an overview of widely shared forms of Jewish heritage.

The central sacred scripture in Judaism is the Torah or the Five Books of Moses (Genesis, Exodus, Leviticus, Numbers, and Deuteronomy). In the Torah, Jews are referred to variously as *Hebrews*, *Israelites*, or the *children of Israel*. Jews regard Abraham and Sarah as their first ancestors. Abraham's grandson, Jacob, was renamed by God as Israel (in Hebrew, "struggles with God"), and his 12 sons and their descendants became known as the children of Israel. The later terms *Jew* and Jewish were derived from the name Judah, one of Jacob's sons.

Distinct Jewish communities practice the religion along a wide continuum in the observance of Jewish law, or **halakhah**. There are three main branches of Judaism in the United States: liberal Reform, Conservative (a mix of tradition and reform), and Orthodoxy (observance of all Jewish law). Some individuals in Orthodox or

The original version of the chapter has been revised. A correction to this chapter can be found at https://doi.org/10.1007/978-3-031-70492-5_36

T. Aronowitz (✉)
Tan Chingfen Graduate School of Nursing, University of Massachusetts, Boston, MA, USA
e-mail: terese.aronowitz@umassmed.edu

P. Zavadivker
Department of History and Program in Jewish Studies, University of Delaware, Newark, DE, USA
e-mail: pollyz@udel.edu



E. A. Fenkl, L. D. Purnell (eds.), *Handbook for Culturally Competent Care*,
https://doi.org/10.1007/978-3-031-70492-5_25

ultra-Orthodox communities may choose to consult their rabbi before making important decisions regarding their lives and health, such as fertility treatment, abortion, and organ transplants. Religious observance and beliefs can be determining factors for Jews of various subgroups. For example, according to Jewish law, a child born to a Jewish mother is identified as being Jewish. However, a child born from the union of a Jewish father and a non-Jewish mother would be recognized as a Jew only by the Reform movement but not by the Conservative or Orthodox movements. Reform Judaism approves of intermarriage between Jews and Judaism, whereas the Conservative and Orthodox movements tend to oppose it.

The population of Jews in the United States as of 2020 was estimated at 7.5 million people. The size of the American Jewish population grew rapidly between the mid-nineteenth and early twentieth centuries as a result of mass migration from Central and Eastern Europe. The greatest influx occurred between 1880 and 1920. Many Jews came from Russia and Eastern Europe, primarily to escape poverty, and because of legal inequality and *pogroms* (anti-Jewish riots or mob violence). Most families in America today are descendants of those Eastern European and Russian immigrants and are referred to as **Ashkenazi** Jews. **Sephardic** Jews, however, trace their origins to Spain, Portugal, the Mediterranean, and North Africa. *Israelis* are citizens of the state of Israel, which has a Jewish majority and Muslim and Christian Arab minority. Jews who trace their family roots to Middle Eastern countries, including Iraq, Yemen, and Iran are sometimes called **Mizrahi**, or "Eastern" Jews. *Falasha* are black Jews who originated in Ethiopia, most of whom emigrated to Israel starting in the mid-1980s.

Throughout their history, Jews have placed a major emphasis on education and social justice through social action. American Jews support a wide range of political causes from liberal to conservative. They often have strong ties to the state of Israel and worry about the rise of anti-Jewish hostility (antisemitism). In general, this population is well educated, and a high percentage obtain higher education and pursue professional vocations.

### *25.1.1 Communications*

- English is the primary language of Jewish Americans. Although Hebrew is used for prayers and Torah study, it is generally not used for conversation. Many elderly Ashkenazi Jews who immigrated early in the twentieth century or who are first-generation Americans might speak **Yiddish**, a hybrid language with Hebrew and Germanic elements written in Hebrew characters. In many ultra-Orthodox communities, Yiddish is a living spoken and written language.
- Many Yiddish terms have worked their way into English, such as *kvetch* (someone who complains a lot); *chutzpah* (audacity or nerve); *bagel* (round rolls with a hole in the middle); *tush*, *tushie*, or *tuchus* (buttocks); *ghetto* (a densely settled and often poor urban neighborhood); *klutz* (a clumsy person); *shlep* (drag or carry); *kosher* (legal or okay); and *oy vey* (oh my). For Orthodox Jews, other important terms in Yiddish include *davenn* (pray); *Shabbos* (Sabbath); *kashrus* (kosher food); and *rebbe* (rabbi).

- Common expressions include *l'chaim* (to life), which is said on joyful occasions and/or as a toast; *shalom aleichem* (peace be with you), a traditional salutation; *mazel tov* (congratulations); and *shabbat shalom* (a good and peaceful Sabbath), which is said from Friday evening at sunset until Saturday at sunset. Yiddish pronunciation differs slightly from Hebrew: for example, not Shabbat but Shabbos.
- Hebrew is read from right to left, and books are opened accordingly, with the spine on the right.
- Jews frequently use humor to communicate and as a coping mechanism. However, jokes that use mainstream stereotypes about Jews, such as implying that Jews are cheap, powerful, or pampered (e.g., Jewish American princess), are culturally offensive. **Any jokes that refer to the Holocaust or concentration camps are also inappropriate.** However, self-criticism through humor is acceptable.
- Modesty, which involves humility, is a primary value, especially among Orthodox communities. It is seen in the Orthodox style of dress and culture.
- Some Orthodox Jews observe the custom known as "guarding touch" (*shomer negiyeh*), by restricting touch of persons of the opposite sex to their children, spouses, and other close family members. Some Orthodox men and women may refuse to shake hands or have other physical contact with adults of the opposite sex, **and their failure to do so when one's hand is extended should not be interpreted as a sign of rudeness.**
- Some Orthodox men and women may have limited social contact with persons of the opposite sex, including conversation and eye contact. **In such cases, Orthodox men and women should only be touched when providing direct care. "Therapeutic touch" is not appropriate.**
- While most Jews live for today and plan for and worry about tomorrow, history is very important in Judaism, especially of the Holocaust. They are warned to "never forget," lest history be repeated. Therefore, their time orientation is simultaneously to the past, the present, and the future.
- The Jewish format for names follows the Western tradition. The given name comes first and is followed by the family surname. Only the given name is used with friends and in informal situations. **In more formal situations, the surname is preceded by the appropriate title of Mr., Miss, Ms., Mrs., Rabbi, or Dr. In Orthodox circles, children are not referred to by their names until after the *bris* or *brit milah* (circumcision) (bris and brit milah are also glossary terms).**

### *25.1.2 Family Roles and Organization*

- The family is the core of society, and the needs of all family members are respected.

- Conservative and Reform Jews typically send school-age children to Hebrew school at least two afternoons a week after public school throughout the school year. Most Orthodox families send their children to private Jewish schools.
- Children play an active role in holiday celebrations and services. Respecting and honoring one's parents is one of the Ten Commandments. Children are expected to show gratitude to their parents for giving them the gift of life.
- In Judaism, the age of majority is 13 years for a boy and 12 for a girl, at which age children are deemed capable of differentiating right from wrong and of committing themselves to performing the commandments. Recognition of adulthood occurs during a religious ceremony called a *bar* or *bat mitzvah* (son or daughter of the commandment). This rite of passage is usually accompanied by a celebration.
- The goal of the Orthodox family is to observe and raise children to observe *halakhah* (Code of Jewish Law), which emphasizes maintaining health, promoting education, and helping others.
- Women who follow the Jewish laws of family purity must physically separate themselves from all men during their menstrual periods and for 7 days after, at which time they visit *mikveh*, a ritual bath. No man may touch a woman or sit where she sat until she has been to the *mikveh*.
- Older people receive respect, especially for the wisdom they must share. Honoring one's parents is a lifelong endeavor; one is obligated to maintain their dignity by feeding, clothing, and sheltering them, even if they experience senility.
- The Bible, as interpreted by Orthodox Jews, prohibits intercourse between men but says nothing about sex between women. Some of the objections to gay and lesbian lifestyles include the inability of these unions to fulfill the commandment of procreation and the possibility that acting on the recognition of one's homosexuality could ruin a marriage. The liberal movements within Judaism support legal and social equality for lesbians and gays, and welcome them as members of the community. **Do not disclose same-sex relationships to others.**

### *25.1.3 Workforce Issues*

- Jews who observe the Sabbath will not work from Friday evening to Saturday evening. The definition of "work" in Jewish law includes driving and using phones and many other forms of technology.
- **They may work on the Sabbath in critical jobs such as health care when needed. They may work on Sundays.**
- Jewish staff should be allowed to request off the major Jewish holidays. **Be sensitive to the needs of Jewish staff and recognize the holiness of the Sabbath, taking into account current US Equal Employment Opportunity Commission laws.**
- Judaism's beliefs are consistent with the values that American society places on the individual and family. For some newer Jewish immigrants, especially the

elderly, English may pose a challenge. **Speak clearly and ask for clarification of understanding from those for whom English is a second language.**

- Most Jews believe in autonomy and willingly take leadership positions in the workforce when the need arises.

### 25.1.4 *Biocultural Ecology*

- Skin coloring for Ashkenazi Jews ranges from fair skin and blonde or red hair to darker skin and brunette hair. Sephardic Jews have slightly darker skin tones and hair coloring, similar to people from the Mediterranean area. There also are Jewish groups throughout Africa who are black, most notably the *Falasha* from Ethiopia.
- Genetic risk factors vary based on whether the family emigrated from Ashkenazi or Sephardic areas. There is a greater incidence of some genetic disorders among Ashkenazi individuals. Most of these disorders are autosomal recessive, meaning that both parents carry the affected gene.
- Table 25.1 lists commonly occurring health conditions for Jews in three categories: lifestyle, environment, and genetics.
- Ashkenazi Jews have a higher rate of side effects with clozapine; 20% develop agranulocytosis.
- **Test for agranulocytosis when Jewish patients are prescribed clozapine.**

### 25.1.5 *High-Risk Health Behaviors*

- Any substance or act that harms the body is not allowed. This includes smoking, suicide, illegal medications, and permanent tattooing.
- Most Jews are health-conscious and practice preventive health care, with routine physical, dental, and vision screening. This is also a well-immunized population. **Encourage these positive health-promotion and disease-prevention practices.**

### 25.1.6 *Nutrition*

- For those who follow the dietary laws, much attention is given to the slaughter, preparation, and consumption of food.
- Perhaps the food identified as culturally "Jewish" that receives the most attention is chicken soup, which has frequently been referred to as "Jewish penicillin" and is often served with *knaidle* balls in it (dumplings made of *matzoh* meal). Although it has no intrinsic meaning or religious value, it is a staple in religious

**Table 25.1** Health conditions common to Jews

| Health conditions | Causes |
|---|---|
| Tay-Sachs disease | Genetic |
| Gaucher's disease | Genetic |
| Canavan's disease | Genetic |
| Familial dysautonomia | Genetic |
| Torsion dystonia | Genetic |
| Niemann–Pick disease | Genetic |
| Bloom syndrome | Genetic |
| Fanconi's anemia | Genetic |
| Mucolipidosis IV | Genetic |
| Lactase deficiency | Genetic |
| Werdnig–Hoffmann disease | Genetic |
| Kaposi sarcoma | Genetic |
| Phenylketonuria | Genetic |
| Ataxia-telangiectasia | Genetic |
| Metachromatic leukodystrophy | Genetic |
| Myopia | Genetic |
| Polycythemia vera | Genetic |
| Cardiovascular diseases | Genetic, environment, lifestyle |
| Diabetes mellitus | Genetic, environment, lifestyle |
| Breast cancer | Genetic, environment, lifestyle |
| Ovarian cancer | Genetic, environment, lifestyle |
| Inflammatory bowel disease | Genetic, environment, lifestyle |
| Colorectal cancer | Genetic, environment, lifestyle |

homes, especially on Friday evenings to usher in the Sabbath and during times of illness. It is frequently associated with a mother's warmth and love.

- Common Ashkenazi foods include *gefilte* fish (ground freshwater fish molded into oblong balls and served cold with horseradish); *challah* (braided white bread); *kugel* (noodle pudding); blintzes (crepes filled with sweet cottage cheese); chopped liver (served cold); *hamentashen* (a triangular pastry with different types of filling); and lox and bagel sandwiches. Lox is cold smoked salmon, served with cream cheese and salad vegetables, on a bagel.
- Religious laws regarding permissible foods are referred to as ***kashrut***. The term ***kosher*** means "fit to eat." Foods are divided into those that are considered *kosher* (permitted or clean) and those considered ***treyf*** (forbidden or unclean). A permitted animal may be rendered *treyf* if it is not slaughtered, cooked, or served properly. All blood is drained from the animal before eating it.
- Milk and meat may not be mixed together in cooking, serving, or eating. To avoid mixing foods, utensils used to prepare foods and the plates used to serve them are separated, requiring two sets of dishes, pots, and utensils: one set for milk products and the other for meat. Because glass is not absorbent, it can be used for either meat or milk products, although religious households still usually

have two sets. Therefore, cheeseburgers, lasagna made with meat, and grated cheese on meatballs and spaghetti are unacceptable. Milk cannot be used in coffee if served with a meat meal. Non-dairy creamers can be used instead, as long as they do not contain sodium caseinate, which is derived from milk.

- Some foods are ***pareve*** (neutral) and may be used with either dairy or meat dishes. These include fish, eggs, anything grown in the soil (vegetables, fruits, coffee, sugar, and spices), and some chemically produced goods.
- Mammals are considered kosher if they meet the other requirements for their slaughter and consumption and have split (cloven) hooves and chew their cud. These animals include buffalo, cattle, goats, deer, and sheep. The pig is an example of an animal that does not meet these criteria. Although liberal Jews decide for themselves which dietary laws they will follow, many still avoid pork and pork products out of a sense of tradition and symbolism.
- **Serving pork products to a Jewish patient, unless specifically requested, is insensitive.**
- Poultry is acceptable as well as fish if it has both fins and scales. Nothing that crawls on its belly is allowed, including shellfish, tortoises, and frogs.
- People who keep kosher will buy kosher-certified products, including meats and cheeses.
- A U with a circle around it is the seal of the Union of Orthodox Jewish Congregations of America and is used on food products to indicate that they are kosher. A circled K and other symbols may also be found on the packaging to indicate that a product is kosher. **Do not bring food into a patient's home without knowing whether they adhere to kosher standards. If the patient keeps a kosher home, do not use any cooking items, dishes, or silverware without knowing which are used for meat and which are used for dairy products.**
- **Health-care providers must fully understand the dietary laws so they do not offend the patient and can plan medication times accordingly.**
- Care must be taken in serving cheese to ensure that no animal substances are served at the same time.
- Breads and cakes made with lard are *treyf*, and breads made with milk or milk by-products (for example, casein) cannot be served with meat meals.
- Eggs from non-kosher birds, milk from non-kosher animals, and oil from non-kosher fish are not permitted.
- Butter substitutes are used with meat meals. Honey is allowed because it is produced from the nectar of flowers.
- Kosher meals are available in most hospitals. They arrive on paper plates and with sealed plastic utensils. For in-patient facilities that do not have readily available kosher capabilities, frozen meals can be ordered online from a number of places.
- **Do not unwrap the utensils or change the foodstuffs to another serving dish. If health-care providers have difficulty locating a supplier of kosher foods, they should contact a local rabbi. Determining a patient's dietary preferences and practices regarding dietary laws should be done during the**

**admission assessment. Frozen kosher meals can be ordered online if the health-care facility does not have a kosher kitchen.**

- Religious Jews wash their hands and say a blessing before eating bread. **Make facilities for patients to wash their hands before eating.**
- During the week of Passover, no bread or product with yeast may be eaten. *Matzoh* (unleavened bread) is eaten instead. Any product that is fermented or that can cause fermentation (souring) may not be eaten. Rather than attend synagogue, the Jewish house of worship, the family conducts the service, ***seder***, around the dinner table during the first two nights and incorporates dinner into a service that includes all participants and retells the story of Moses and the exodus from Egypt.
- The Jewish calendar has several fast days. The most widely observed fast falls on the holiest day of the year, *Yom Kippur* (Day of Atonement). Jews abstain from food and drink as they pray to God for forgiveness for the sins they have committed during the past year. They eat an early dinner on the evening the holiday begins and then fast until after sunset the following day.
- Ill people, the elderly, the young, pregnant, and nursing women, and the physically incapacitated are absolved from fasting and may need to be reminded of this exception to Jewish law.
- **If concerns arise about fasting, a consultation with the patient's rabbi may be necessary.**
- **Schedule appointments and treatment to avoid religious days.**

### *25.1.7 Pregnancy and Childbearing Practices*

- Couples who are unable to conceive should try all possible means to have children, including infertility counseling and interventions and egg and sperm donation. Orthodox opinion is virtually unanimous in prohibiting artificial insemination when the semen donor is not the woman's husband. When all-natural attempts have been made, adoption may be pursued.
- Unless pregnancy jeopardizes the life or health of the mother, contraception is not looked on favorably among the ultra-Orthodox. Condom use is supported, especially when unprotected sexual intercourse poses a medical risk to either spouse.
- To the Orthodox, barrier techniques are not acceptable because they interfere with the full mobility of the sperm in its natural course. The birth control pill does not result in any permanent sterilization nor does it prevent semen from traveling its normal route. Therefore, the use of this method is the least objectionable to most branches of Judaism. Sterilization implies permanence, and Orthodox Jews probably oppose this practice, unless the life of the mother is in danger. **Reform Judaism allows free choice in fertility control.**
- The fetus is not considered a living soul or person until it has been born. Birth is determined when the head or "greater part" is born. If the physical or mental

health of a pregnant woman is endangered by the fetus, all branches of Judaism consider the fetus an aggressor and require an abortion. Abortion in cases where the mother's health is not compromised is not permitted by the Orthodox branch because the fetus is regarded as part of the mother's body according to Jewish law and one must not do harm to one's body. The liberal branches of Judaism hold that a woman maintains control over her own body and that it is up to her whether to abort a fetus.

- A Hasidic husband may not touch his wife during labor and may choose not to attend the delivery because he is not permitted to view his wife's genitals. These behaviors should never be interpreted as insensitivity.
- **During the delivery of a baby in an ultra-Orthodox family, initiate the following interventions: give the mother a hospital gown that covers her in the front and back to the greatest extent possible. Provide a surgical cap so that her head remains covered (because the hair is considered a private part of her body). Give the father the opportunity to leave during procedures and during the birth or, if he chooses to stay, drape the mother so that the husband may sit by his wife without viewing her perineum, including by way of mirrors or other means.**
- Pain medication during delivery is acceptable.
- For male infants, circumcision (*bris*), which is both a medical procedure and a religious rite, is performed on the eighth day of life by a ***mohel***, an individual trained in the circumcision procedure, asepsis, and the religious ceremony. Although a rabbi is not necessary, it is also possible to have the procedure completed by a physician with a rabbi present to say the blessings.
- Attending a *bris* is the only mitzvah for which religious Jews may violate the Sabbath so that it can be completed at the proper time.
- **Provide a space for a family celebration if the circumcision is done in the hospital.**

### *25.1.8 Death Rituals*

- Traditional Judaism believes in an afterlife where the soul continues to flourish, although many dispute this interpretation. A dying person is considered a living person in all respects.
- Active euthanasia is forbidden for religious Jews. Passive euthanasia may be allowed depending on its interpretation. Nothing may be used or initiated that prevents a person from dying naturally or that prolongs the dying process. **However, anything that artificially prevents death (cardiopulmonary resuscitation, ventilators, and so forth) may possibly be withheld, depending on the wishes of the patient and his or her religious views. The dying person should not be left alone.**
- Taking one's own life is prohibited. To the ultra-religious, suicide removes all possibility of repentance.

- Any Jew may ask God's forgiveness for his or her sins; no confessor is needed. Some Jews feel solace in saying the ***Sh'ma*** in Hebrew or English. This prayer confirms one's belief in one God. At the time of death, the nearest relative can gently close the eyes and mouth, and the face is covered with a sheet.
- The body is treated with respect and revered for the function it once filled. **Ask the closest relative of the deceased specifically about the family's preferred practices to follow after death.**
- Autopsy is usually not permitted among religious Jews because it results in desecration of the body, and it is important that the body be interred whole. Allowing an autopsy might also delay the burial, something that is not recommended. Conversely, autopsy is allowed if its results would save the life of another patient who is immediately at hand. Many branches of Judaism currently allow an autopsy if (a) it is required by law; (b) the deceased person has willed it; or (c) it saves the life of another, especially an offspring. **Explain legal requirements for an autopsy.**
- Cremation is prohibited because it unnaturally speeds the disposal of the dead body. Embalming is prohibited because it preserves the dead. However, in circumstances when the funeral must be delayed, some embalming may be approved. Cosmetic restoration for the funeral is discouraged.
- Funerals and burials usually occur within 24–48 h after the death. The funeral service is directed at honoring the departed by only speaking well of him or her. It is not customary to have flowers either at the funeral or at the cemetery. The casket should be made of wood with no ornamentation. The body may be wrapped only in a shroud to ensure that the body and casket decay at the same rate. There is no wake or viewing. The prayer said for the dead, ***kaddish***, is usually not said alone.
- In cases where the body cannot be removed to a funeral home, many Jewish communities have volunteers that care for the dead (Kaevra Kodusha) who are contacted and practice the custom of watching the body around the clock and prepare the body for burial by washing and wrapping in linen. This work is considered the holiest of all mitzvahs (good deeds).
- After the funeral, mourners are welcomed at the home of the closest relative. Outside the front door is water to wash one's hands before entering, which is symbolic of cleansing the impurities associated with contact with the dead. The water is not passed from person to person, just as it is hoped that the tragedy is not passed. At the home, a meal is served to all the guests. This "meal of condolence" or "meal of consolation" is traditionally provided by the neighbors and friends.
- *Shiva* (Hebrew for "seven") is the 7-day period that begins with the burial. *Shiva* helps the surviving individuals face the actuality of the death of the loved one. During this period when the mourners are "sitting *shiva*," they do not work. When health-care providers are the ones experiencing the loss, it is important for supervisors to understand the mourning customs. In some homes, mirrors are covered to decrease the focus on one's appearance; no activity is permitted to divert attention from thinking about the deceased. Evening and morning services

may be conducted in the closest relative's home. Condolence calls and the giving of consolation are appropriate.

- Crying, anger, and talking about the deceased person's life are acceptable.
- A common sign of grief is the tearing of the garment that one is wearing before the funeral service.
- In liberal congregations, a black ribbon with a tear in it is a symbolic representation of mourning. During *shiva*, the mourner sets the tone and initiates the conversation. Because there are such discrete periods of mourning and an emphasis on the needs of the living, Jewish tradition holds that it is wrong to mourn more than 30 days for a relative and for more than 1 year for parents.
- Mourning is not required for a fetus that is miscarried or stillborn. This is also true of any premature infant who dies within 30 days of birth. However, parents are required to mourn for full-term infants who die at birth or shortly thereafter.
- Within Orthodoxy, when a limb is amputated before death, the amputated limb and blood-soaked clothing are buried in the person's future gravesite because the blood and limb were part of the person. No mourning rites are required. **In the case of an amputation, assist with arrangement for the burial of the body part as needed.**

### *25.1.9 Spirituality*

- Jewish scripture, or the Hebrew Bible, includes 24 books in 3 sections (Torah, Prophets, and Writings). This body of scripture is what Christians refer to as the Old Testament. Judaism is a monotheistic faith that believes in one God as the creator of the universe. No physical qualities are attributed to God; making and praying to statues or graven images are forbidden.
- The spiritual leader is the rabbi (teacher). He (or she, in liberal branches) is the interpreter of Jewish law. All Jews pray directly to God. They do not need the rabbi to intercede, to hear confession, or to grant atonement. However, ultra-Orthodox groups venerate their most important rabbis as spiritual intercessors to God.
- Some of the major principles that guide Judaic bioethics are shown in Table 25.2.
- The practice of Judaism spans a wide spectrum. Although there is only one religion, there are three main branches or denominations of Judaism in the United States. The Orthodox are the most traditional. They adhere most strictly to the *halakhah* and try to follow as many of the laws as possible while also taking part in American society. They observe the Sabbath by abstaining from work, spending money, and driving, and many attend synagogue on Friday evening and Saturday morning. Orthodox Jews observe the Jewish dietary laws at home and when eating out; men always wear a yarmulke or kippah (head covering) in reverence to God. Women wear long sleeves and modest dress and cover their hair with scarves, hats, or wigs. In many Orthodox synagogues, the services are conducted primarily in Hebrew, and men and women sit separately.

**Table 25.2** Principles of Judaic bioethics

| |
|---|
| • Humanity's purpose on earth is to live according to certain God-given guidelines, the highest of which are to sanctify and repair all aspects of life on earth. In addition, Jews must follow God's laws as part of a covenant |
| • Life possesses infinite value. The obligation to preserve and save human life (known as *pikuakh nefesh*) is the guiding principle for Jewish law and bioethics. If a patient's life is in danger, all medical treatment that can potentially avert the threat to life must be carried out, overriding all other religious considerations |
| • All human beings are created equal and in God's image |
| • Our lives are not our own exclusive private possessions but "on loan" from the Creator and must be treated with dignity and care for that reason. Accordingly, Judaism accords great importance to medicine and medical treatment |
| • The first five books of the Hebrew Bible, also known as the Five Books of Moses, make up the Torah. They are handwritten in Hebrew on parchment scrolls, kept in the "Holy Ark" within each synagogue under an "eternal light," and read aloud in the community on the Sabbath, holidays, and other occasions. The Torah's commandments instruct Jews how they should live their lives in relation to other human beings and to God |
| • The commandments stated in the Torah and the oral law that was derived from the biblical commands are the basis of all codes of Jewish law. These commandments demand certain actions and ethical commitments. Observant Jews believe that the commandments reflect the will of God, and it is their duty to carry them out to fulfill their covenant with God |

- Ultra-Orthodox men wear a special garment with strings under their shirts every day (*tzitzit*).
- A *mezuzah* is a small container with a small piece of scripture inside. Many Jews will affix a mezuzah to all the doorways of their home. Some Jews wear a mezuzah as a necklace. Other religious symbols include the Star of David, a six-pointed star that has been a symbol of the Jewish community, and the *menorah* (candelabrum).
- While some Conservative Jews observe most of the *halakhah*, they do make concessions to modern society. Many drive to the synagogue on the Sabbath, and men and women sit together during services. Many keep a kosher home, but they may or may not follow all of the dietary laws outside the home. Women can be ordained as rabbis and unlike in Orthodox Judaism, they are counted in a *minyan*, the minimum number of 10 people that is required for a prayer quorum. While head coverings are required in the Conservative synagogue, they are optional outside of that environment.
- The liberal or progressive movement is called Reform. Reform Jews claim that only the moral laws of the **Torah** are binding. They observe fewer rituals and laws, although they frequently have a mezuzah on the doorpost of their homes, celebrate the holidays, have their children circumcised, and have an ethnic and cultural identity.
- **Reform Jews may or may not follow the Jewish dietary laws, but they may have specific unacceptable foods (for example, pork or shellfish), which they abstain from eating.**

- Of the many small groups of Orthodox Jews, **Hasidic** (sometimes spelled **Chasidic**) Jews are perhaps the most recognizable. They usually live, work, and study within close-knit urban communities. Men typically have full beards, uncut hair around the ears (*payos*), black hats or fur ***streimels***, dark clothing, and no exposed extremities.
- Like other Orthodox women, Hasidic women dress modestly; in addition, married women will cover their hair completely with scarves, wigs, and sometimes a hat.
- A relatively new denomination, **Reconstructionism**, is a mosaic of the three main branches. It views Judaism as an evolving religion of the Jewish people and seeks to adapt Jewish beliefs and practices to the needs of the contemporary world.
- The Jewish house of prayer is called a synagogue, temple, or *shul*. Jews may pray alone, in small groups, or as part of a minyan (ten Jews over the age of 13 who have had their bar/bat mitzvah). Many Jews will wear a large prayer shawl (*tallit*) during prayer. Orthodox Jews pray three times a day: morning, late afternoon, and evening. They wash their hands and say a prayer on awakening in the morning and before meals. Men will wrap *tefillin* (phylacteries) around their arms during morning prayers. Religious patients in hospitals may want their prayer items (*yarmulke* or *kippah*, *tallit*, *tefillin*) and may request a minyan. **Disregard hospital policies regarding the number of visitors in the sick person's room in such instances.**
- The Sabbath begins 18 min before sunset on Friday. During this time, religious Jews do no manner of work, including answering the telephone, operating any electrical appliance, driving, or operating a call bell from a hospital bed. If an Orthodox patient's condition is not life-threatening, medical and surgical procedures should not be performed on the Sabbath or holy days. A gravely ill person and the work of those who need to save him or her are exempted from following the commandments regarding the Sabbath. **Because Orthodox Jews cannot summon an elevator, patient care facilities with a Jewish population should consider installing a "Sabbath elevator" that automatically stops on every floor.**
- Table 25.3 provides a list of Jewish holidays for the years from 2024 to 2028.

### *25.1.10 Health-Care Practices*

- All denominations recognize that religious requirements may be laid aside if a life is at stake or if an individual has a life-threatening illness. **Hospice and palliative care are fully consonant with Jewish beliefs.**
- In ultra-Orthodox denominations of Judaism, taking medication on the Sabbath that is not necessary to preserve life may be viewed as "work" and is unacceptable. This belief may result in some people with conditions such as asthma not recognizing the severity of their condition; they may also be unaware of the laws

**Table 25.3** Jewish holidays: 2024–2028

| Jewish holidays begin at sundown the evening before the date recorded on this type of calendar; holidays end at sundown on the date shown | | | | | |
|---|---|---|---|---|---|
| Holiday | 2024 | 2025 | 2026 | 2027 | 2028 |
| Rosh Hashanah | 10/2–4 | 9/22–24 | 9/11–13 | 10/1–3 | 9/20–22 |
| Yom Kippur | 10/11–12 | 10/1–2 | 9/20–21 | 10/10–11 | 9/29–30 |
| Sukkot | 10/16–18 | 10/6–8 | 9/24–26 | 10/15–17 | 10/4–6 |
| Chanukah | 12/25–1/2 | 12/14–22 | 12/4–12 | 12/24–31 | 12/12–19 |
| Passover | 4/22–30 | 4/12–20 | 4/1–9 | 4/21–29 | 4/10–18 |
| Shavuot | 6/11–13 | 6/1–3 | 5/21–23 | 6/10–12 | 5/30–6/1 |

A calendar of all Jewish holidays with start and end times can be found at https://www.hebcal.com/

that allow them to take their necessary medications. **Teach patients about the potential life-threatening sequelae of their condition as well as the exceptions to Jewish law that permits them to take their medications.**

- The verbalization of pain is acceptable and common. Individuals want to know the reason for the pain, which they consider just as important as obtaining relief from pain.
- The sick role for Jews is highly individualized and may vary among individuals according to the severity of symptoms.
- Judaism opposes discrimination against people with physical, mental, and developmental conditions. The maintenance of one's mental health is considered just as important as the maintenance of one's physical health. Mental incapacity has always been recognized as grounds for exemption from all obligations under Jewish law.
- Jewish law considers organ transplants from four perspectives: those of the recipient, the living donor, the cadaver donor, and the dying donor. Because life is sacred, if the recipient's life can be prolonged without considerable risk, then transplant is ordained. For a living donor to be approved, the risk to the life of the donor must be considered. One is not obligated to donate a bodily part unless the risk is small. Examples include kidney and bone marrow donations. Organ donation at the time of death is acceptable if it saves a person's life. **Conservative** and **Reform** Judaism approve using the flat EEG as the determination of death so that organs, such as the heart, can be viable for transplant. Burial may be delayed if organ harvesting is the cause of the delay. **Assist patients in finding a rabbi if they desire to consult one when making a decision regarding organ donation or transplant.**
- The use of a cadaver for transplant is usually approved if it is to save a life. No one may derive economic benefit from the corpse. Use of skin for burns is also acceptable, although no agreement has been reached on the use of cadaver corneas.

### 25.1.11 Health-Care Providers

- Physicians are held in high regard. Although physicians must do everything in their power to prolong life, they are prohibited from initiating measures that prolong the act of dying.
- Only supportive care is required while the patient is dying and includes such care as food and water, good nursing care, and maximal psychosocial support.
- **More traditional Orthodox patients prefer that care be delivered by a same-sex health-care provider.**

## 25.2 Reflective Exercises

1. What is meant by ***halakhah*** and what does it emphasize?
2. Name and describe at least three genetic conditions common among Jews.
3. What is meant by ***kashrut***? Distinguish between kosher and *treyf.*
4. What are the tenets of the four broad branches of Judaism? What are the primary differences among them?
5. When does Sabbath begin and end?
6. Under what conditions are observant Jews able to work?
7. What are some of the major principles that guide Judaic bioethics?
8. What is meant by *shiva*? Describe its practices.
9. An orthodox Jewish couple is expecting their first child. The expectant father wants to be in the delivery room with his wife. What preparations do you need to make for this to occur?
10. The hospital where you work does not have a kosher kitchen. Where might you purchase kosher meals? Once you have a kosher meal, how do you handle the food?

## Bibliography

Ben Ben-Tzion Halevi HY, Lavine JB (2008) The thirteen principles of Jewish medical ethics. https://www.science.co.il/jewish-studies/articles/Medical-ethics.php

Death and mourning: Jewish customs, traditions, and practices (2016) http://www.shiva.com/learning-center/death-and-mourning/

Guidelines for health care providers interacting with Jewish patients and their families (2002). https://www.advocatehealth.com/assets/documents/faith/cgjewish.pdf

Jewish perspectives on the birthing experience (2016) https://www.mikvah.org/article/jewish_perspectives_on_the_birthing_experience

National Foundation for Jewish Genetic Diseases (2012) http://www.mazornet.com/genetics/

Pew Research Center (2021) Jewish Americans in 2021. https://www.pewresearch.org/religion/2021/05/11/jewish-americans-in-2020/

Satlow M (2006) Creating Judaism: history, tradition, practice. Columbia University Press, New York

Selekman J, Zavadivker P (2020) People of Jewish heritage. In: Fenkel E, Purnell L (eds) Textbook for transcultural health care: a population approach. Cultural competence in nursing care, 5th edn. Springer, Cham

U.S. Equal Opportunity Commission (2016) www.eeoc.gov/

# Chapter 26
# People of Korean Heritage

**Eun-Ok Im and Yaelim Lee**

## 26.1 Overview and Heritage

### *26.1.1 Communications*

- The dominant Korean language, *han'gul*, was the first phonetic alphabet in East Asia.
- Most Koreans in America can speak, read, write, and understand English to some extent. However, some Americans may have difficulty understanding their English, especially those who learned English from Koreans who spoke with their native intonations and pronunciations.
- A high value is placed on harmony and the maintenance of a peaceful environment. Most are comfortable with silence. Small talk may appear senseless and insincere.
- Touch in the realm of health care is readily accepted. Touching among friends and social equals is common and does not carry a sexual connotation, as it might in Western societies. Hugging and kissing are uncommon among parents and children as well as among children and older aunts or uncles.
- Age, gender, and social status determine the use of eye contact. Respect for those in senior positions is shown by not looking them directly in the eye.
- Feelings are infrequently communicated in facial expressions.

E.-O. Im (✉)
The University of Texas at Austin, School of Nursing, Austin, TX, USA
e-mail: eunok.im@austin.utexas.edu

Y. Lee
College of Nursing, The Catholic University of Korea, Seoul, South Korea



E. A. Fenkl, L. D. Purnell (eds.), *Handbook for Culturally Competent Care*,
https://doi.org/10.1007/978-3-031-70492-5_26

- Koreans prefer indirect communication because they perceive direct communication as an indication of intention or opinions as rude. Koreans may agree with the healthcare provider in order to avoid conflict. **A response of "yes" may not have the same connotation as it does in European American cultures.**
- More traditional Koreans are past-oriented. Much attention is paid to the ancestry of a family. Yearly, during the Harvest Moon in Korea, *chusok* (respect) is paid to ancestors by bringing fresh fruits from the autumn harvest, dry fish, and rice wine to gravesites.
- The younger and more educated generation is more futuristic and achievement-oriented.
- **Punctuality is the norm for keeping important appointments, making transportation connections, and reporting to work.**
- The number of surnames in Korea is limited, with the most common ones being Kim, Lee, Park, Choi, and Jung. Korean names contain two Chinese characters, one of which describes the generation and the other the person's given name. However, some younger generations do not use Chinese characters in their names.
- The surname comes first; however, because this may be confusing to many Americans, some Koreans in the United States follow the Western tradition of using the given name first, followed by the surname. **Address Koreans by their surname, with the title Mr., Mrs., Miss, Ms., Dr., or Minister, unless otherwise indicated. Determine Korean clients' language ability, comfort level with silence, use of eye contact, and spatial distancing practices when completing health assessments.**
- While speaking to someone directly, you use the appellation "Hoching," and while referring to someone to a third party, you use the designation "Jiching." They are collectively known as "Chingho (title)."
- In South Korea, using just one hand to receive something can be seen as rude. Therefore, use both hands to receive something, or you can accept it with your right hand while holding your right wrist with your left.

Younger individuals do not typically initiate a handshake with older people, whereas older people usually do.

### 26.1.2 Family Roles and Organization

- Men used to be the primary financial providers. Women used to stay home and care for the children and house choirs unless they were professionals (e.g., medical doctors, lawyers, teachers) or needed to work for financial purposes. Today, however, most men and women are both economically active, and the rate of economically active women population is increasing. Women have long been degraded in Korean society and seen as appendages of males. In earlier times, a woman's identity was determined by her role as someone's daughter, wife, or mother. Although many still practice these gender relationships, more educated women and men no longer adhere to these Confucian values.

- Parenting in Korea is authoritative, although class differences play a more influential role in determining parenting styles and family roles.
- Children are expected to be well-behaved because the whole family is disgraced if a child behaves in an embarrassing manner.
- Discussing domestic violence violates Korean cultural norms. **When family violence is suspected, approach the topic in an indirect manner.**
- Dating is common among high schoolers. More than half of high schoolers have dating experience.
- Once young adults have entered a university, they receive their freedom and are permitted to make their own decisions about personal and study time. Group outings are common for meeting the opposite sex.
- With rapid acculturation, children often take on the values of the dominant society or culture, challenging parents who support traditional values and ideals. **Assist parents in exploring alternative parenting practices.**
- Parents expect their children to care for them in old age. *Hyo* (filial piety) is the obligation to respect and obey parents, care for them in old age, give them a good funeral, and worship them after death. The obligation to care for one's parents is written into the civil code in Korea.
- Older people are frequently consulted on important family matters as a sign of respect for their life experiences.
- Old age begins when one reaches the age of 65 years.
- Women who divorce may suffer social stigma. Living together before marriage is not customary in Korea. If pregnancy occurs outside marriage, it may be taken care of quietly and without family and friends being aware of the situation.
- In the past, status was customarily assigned, but presently achieved status is more common.
- Lesbian and gay relationships are frowned upon. Personal disclosure to friends and family jeopardizes the family name and may lead to ostracism. The negative perception of homosexuality is now slightly improving.

### *26.1.3 Workforce Issues*

- The skills and work experiences Koreans bring from their home country are often not accepted in the American workforce, forcing them to take jobs in which they may be over-skilled.
- **Orientation programs need to address language skills, practice differences, and communication and interpersonal relationships to help Koreans adjust to the workforce.**
- A supervisor is treated with much respect in work and social settings. Informalities and small talk may be difficult for Korean immigrants.

- For an employee to refuse a request of an employer is unacceptable, even if the employee does not want or feel qualified to complete the request. **Asking Korean employees to demonstrate procedures is better than asking them if they know how to perform them.**
- Slang and colloquial language are difficult for Koreans to understand. **Be clear in your communication style and refrain from using slang and idiomatic expressions.**
- Most stand close when conversing. **Do not take offense if a Korean stands closer than how you are accustomed.**
- Age, gender, and social status determine the use of eye contact. Respect for those in senior positions is shown by not looking them directly in the eye. **Do not assume that lack of eye contact means the person is not listening, does not care, or is being less than truthful.**
- The foundation of the Korean leadership style is *jeong* and *woori*. *Jeong* is a fostered inner condition of filial compassion, intimacy, and empathy in relationships that break down interpersonal barriers. *Woori* is a powerful sense of belonging to a group in which incomplete, interdependent people come together to form a whole.
- South Korea is a highly hierarchical society with a large power gap.

### *26.1.4 Biocultural Ecology*

- Common physical characteristics include dark hair and dark eyes, with variations in skin color and hair darkness. Skin color ranges from fair to light brown, with those residing in the southern part of South Korea being darker.
- Epicanthal skin folds create the distinctive appearance of Asian eyes.
- Table 26.1 presents commonly occurring health conditions for Koreans in three categories: lifestyle, environment, and genetics.
- **Assess for stomach and liver cancer, tuberculosis, hepatitis, renal impairment, hypertension, parasites, and asbestos-related health problems.**
- **Be aware that many Koreans require lower dosages of psychotropic drugs.**
- In newborn Korean children, a salmon patch is the most prevalent vascular birthmark and a Mongolian spot is the most prevalent pigmentary birthmark.

### *26.1.5 High-Risk Health Behaviors*

- Smoking by women in public is taboo, but some women smoke at home.
- Men have a high incidence of alcohol consumption.
- Seat belts are required for all car seats in Korea. **Encourage moderate alcohol intake and smoking cessation. Explain the legal mandates of seat belt and child restraint laws in the United States.**

**Table 26.1** Health conditions common to Koreans

| Health conditions | Causes |
|---|---|
| Schistosomiasis | Environment, lifestyle |
| Renal failure | Genetic, environment, lifestyle |
| Asbestosis | Environment, lifestyle |
| Hypertension | Genetic, environment, lifestyle |
| Tuberculosis | Environment, lifestyle |
| Hepatitis | Environment, lifestyle |
| Stomach cancer | Environment, lifestyle |
| Liver cancer | Environment, lifestyle |
| Lactase deficiency | Genetic |
| Osteoporosis | Genetic, lifestyle |
| Peptic ulcer disease | Environment, lifestyle |
| Insulin autoimmune deficiency disease | Genetic |

- Teenagers in South Korea sleep less than those in other Organisation for Economic Co-operation and Development (OECD) (on average). The rate of women's smoking was reported to be low in the past because women's smoking was taboo. However, as the negative perception of female smokers has decreased, the smoking rate among females has increased (higher than before).

### *26.1.6 Nutrition*

- The traditional Korean diet includes steamed rice; hot soup; *kimchee*; and side dishes of fish, meat, or vegetables served in some variations for breakfast, lunch, and dinner. Breakfast is traditionally considered the most important meal.
- Rice is served with 5–20 small side dishes of mostly vegetables and some fish and meats.
- Food is flavorful and spicy. Cooking includes a variety of seasonings such as red and black pepper, garlic, green onion, ginger, soy sauce, and sesame seed oil.
- Most Korean Americans are at high risk for calcium deficiencies due to lactose intolerance. **Determine individual cultural food choices that are high in calcium.**
- A cultural treatment for the common cold is a soup made from bean sprouts, anchovies, garlic, and other hot spices.
- The saying, "foods are the best medicine," gave rise to the long-held belief that foods and medications have the same effect.
- Fermentation, a biological process that helps food mature so it can be preserved longer, is one of the important concepts for understanding traditional Korean cuisine. Kimchi, soy sauce, red pepper paste, salted fish, and soybean paste are the foods that most closely exemplify the Korean fermented culinary culture.
- Common Korean American dishes are listed in Table 26.2.

**Table 26.2** Common Korean American dishes

| |
|---|
| • *Kimchi*: A spicy fermented cabbage made from a variety of vegetables but made primarily from a Chinese, or Napa, cabbage. Spices and herbs are added to the previously salted cabbage, which is allowed to ferment over time, and it is served with every meal in a variety of forms |
| • *Bibimbap*: Rice, finely chopped mixed vegetables, and a fried egg served in a hot pottery bowl. Hot pepper paste is usually added |
| • *Bulgolgi*: Thinly sliced pieces of beef marinated in soy sauce, sesame oil, green onions, garlic, and sugar and served after being barbecued |
| • *Japchae*: Clear noodles mixed with lightly stir-fried vegetables and meats |
| • Adapt health teaching to Korean American food choices and practices |

### *26.1.7 Pregnancy and Childbearing Practices*

- Pregnancy is a highly protected time for women. Both pregnancy and the postpartum period are ritualized. See Table 26.3 for some possible rituals.
- Once a woman is pregnant, she starts practicing ***Tae-Kyo***, which literally means "fetus education." The objective of *Tae-Kyo* is to promote the health and well-being of the fetus and mother by having the mother focus on art and beautiful objects.
- Advice on prenatal care from the husband's family may seem like an invasion of privacy and can be a stressor to pregnant women.
- Pregnant women who practice *Tae-Kyo* well before giving birth are believed to have better maternal–fetal attachment and lower the risk of postpartum depression.

### *26.1.8 Death Rituals*

- Death and dying are fairly well accepted in Korean culture.
- Prolonging life may not be highly regarded in the face of modern technology. Families are expected to stay with family members and assist in feeding and personal care around the clock. **Include family in caring for their hospitalized family member.**
- Many believe that patients should not be told they have a terminal illness. **Take cues from the patient and, if necessary, disclose terminal illnesses to the oldest son or designated family spokesperson, who will inform the patient when the family deems it necessary.**
- Crying and open displays of grief are common and signify respect for the dead.
- Relatives and friends pay respect by viewing photographs of the deceased instead of viewing the body.

**Table 26.3** Pregnancy beliefs

| |
|---|
| • Pregnant women who handle unclean objects or kill a living creature may experience a difficult birth |
| • Women wear tight abdominal binders beginning at 20 weeks gestation or work physically hard toward the end of the pregnancy to increase the chance of having a small baby |
| • Expectant mothers should avoid chicken, fish with scales, squid, or crab because eating these foods may affect the child's appearance. For example, eating duck may cause the baby to be born with webbed feet |
| • Women attribute a variety of complaints to *naeng* (chill), a cold imbalance of the womb that brings on a heavy vaginal discharge and can cause women who experience it to become sterile |
| • Eating blemished fruit can cause a skin disease in the infant or result in an unpleasant face |
| • Women commonly labor and deliver in the supine position. After the delivery, women are traditionally served seaweed soup, a rich source of iron, which is believed to facilitate lactation and promote the healing of the mother |
| • Bed rest is encouraged after delivery for 7–90 days |
| • Women are encouraged to keep warm by avoiding showers, baths, and cold fluids or foods in order to prevent chronic illnesses such as arthritis |
| • The baby should be wrapped in warm blankets to prevent harm from cold winds |
| • Postpartum women should care for their bodies by augmenting heat and avoiding cold, resting without working, eating well, protecting the body from harmful strains, and keeping clean. To improve postpartum care, develop bilingual pamphlets that include medical terms used in the US health-care system |

- An ancestral burial ceremony follows death, with the body being placed in the ground facing south or north. Rice wine is sprinkled around the gravesite.
- The eldest son or male family member sits by the deceased, sometimes holds a cane, and makes a moaning noise to display his grief. The cane is a symbol of needing support.
- **Provide a private room so the family may grieve in culturally congruent ways.**
- The deceased wears special clothing, and the deceased's family prepares mourning garments for the funeral. In general, clothing for the dead is prepared while they are still alive. This clothing might be different depending on the location, including hemp clothing in either black or white.
- Guests at *Munsang*, which is a visit to the family of the deceased to express condolences, are advised to dress in black or white rather than bright colors. It is important to show respect and pray with the family at *Munsang*. As an additional expression of consolation, money is given to the family.

## 26.1.9 Spirituality

- Organized religions include Christianity, Buddhism, and ***Chondokyo***. The church is a powerful social support group for Korean immigrants.

- Christians believe the spirit goes to heaven; Buddhists believe the spirit starts a new life as a person or an animal. **Individually assess each patient's spiritual practices.**
- Family and education are central themes that give meaning to life.
- In South Korea, Christianity is the most prevalent religion.
- Different religions, philosophical systems, and lifestyles have left their marks on Korean history, and these varied religions are harmonious with people's daily lives.

### *26.1.10 Health-Care Practices*

- Many prescription drugs in the United States, such as anti-inflammatory and cardiac medications, as well as certain pain-control medications, can be purchased over-the-counter in Korea.
- Herbal medicine may be used in conjunction with Western biomedicine. Herbal remedies include ginseng, seaweed soup, and *haigefen* (clamshell powder), which contains high levels of lead, causing abdominal colic, muscle pain, and fatigue.
- **Ask about the use of traditional medicine, and include nonharmful practices in prescriptions.**
- ***Hanbang*** is the traditional Korean medical care system and works on the principle of restored energy. Complementary health practices include acupuncture, acumassage, acupressure, and moxibustion therapy.
- Some Korean Americans are stoic and are slow to express emotional distress from pain. Others are expressive and discuss their smallest discomforts.
- **Monitor nonverbal cues and facial expressions for pain.**
- Mental illness may be stigmatized. ***Hwa-Byung***, a traditional Korean illness, occurs from the suppression of anger or other emotions. These emotions are expressed as physical complaints, ranging from headaches and poor appetite to insomnia and lack of energy.
- Disability may carry a stigma, especially for older, more traditional Koreans.
- **Assist families to access resources to care for parents and children with disabilities.**
- Whereas blood donation and transfusions are acceptable, organ donation and organ transplantation are rare, reflecting traditional attitudes toward integrity and purity.
- Korean men smoke a lot. In addition, Korean men and young women use a lot of alcohol at least once a month, and "binge drinking" is a common and growing health problem.
- The inability to express their feelings openly could lead to a condition known as *Hwa-byung* (literally, "fire illness"), which could be attributed to illnesses.

### *26.1.11 Health-Care Practitioners*

- More traditional individuals frequently prefer health-care providers who speak Korean and are older.
- Because of modesty, women prefer women for performing Pap smears, mammography, and breast examinations.
- **Provide a same-gender health-care provider for intimate care whenever possible.**
- The appearance of symptoms can be interpreted as bad luck, misfortune, or "karma," which means retribution for wrongdoing in the past.
- When asked about unpleasant things, Koreans try to keep their emotions under control.

## 26.2 Reflective Exercises

1. For Koreans, a response of "yes" may not have the same connotation as it does in the European American culture. What does it mean in the Korean culture?
2. Explain the name format for Koreans and how you would address them.
3. When family violence is suspected, how might you approach the topic?
4. What is *hyo* and how is it displayed?
5. What needs to be included in an orientation program for Koreans to be successful in the workforce?
6. Name at least three common health conditions among Koreans.
7. What is *Tae-Kyo* and what is its purpose?
8. What is the term for traditional Korean medicine and what are its principles?
9. What is *Hwa-Byung* and what are its causes and symptoms?
10. How do more traditional Koreans see disability?

## Bibliography

An ancient medical tradition (2012). http://www.korea.net/NewsFocus/Society/view?articleId=99249

Bang KS, Lee I, Kim S, Yi Y, Huh I, Jang SY et al (2020) Relation between mother's taekyo, prenatal and postpartum depression, and infant's temperament and colic: a longitudinal prospective approach. Int J Environ Res Public Health 17(20):7691

CIA (2024) World Factbook: Korea. https://www.cia.gov/the-world-factbook/countries/korea-south/

Connor P (2014) 6 Facts about South Korea's growing Christian population. Pew Research Center. http://www.pewresearch.org/fact-tank/2014/08/12/6-facts-about-christianity-in-south-korea

Embassy of the Republic of Korea to Norway (2020) https://overseas.mofa.go.kr/no-en/brd/m_21237/view.do?seq=88

Health problems of Korean Americans: access to health care (2016) Stanford School of Medicine. https://geriatrics.stanford.edu/ethnomed/korean/patterns/access.html

Hwang JJ (2020) S. Korean teens suffer from lack of sleep, survey finds. https://en.yna.co.kr/view/AEN20200803009800315

Im E-O (2013) People of Korean heritage. In: Purnell L (ed) Transcultural health care: a culturally competent approach, 4th edn. F.A. Davis Company, Philadelphia

Kim SS (2017) A culturally adapted smoking cessation intervention for Korean American: preliminary findings. J Transcult Nurs 28(1):24–32

Kim YW (2021) The smoking rate for men...Why do women bloom more in 2030. https://www.bizhankook.com/bk/article/22011

Kim HK, Baik K, Kim N (2019) How Korean leadership style cultivates employees' creativity and voice in hierarchical organizations. SAGE Open 9(3):2158244019876281

Kim H, Baek S, Hong SM, Lee J, Jung SM, Lee J et al (2020) 1, 25-dihydroxy vitamin D3 and interleukin-6 blockade synergistically regulate rheumatoid arthritis by suppressing interleukin-17 production and osteoclastogenesis. J Korean Med Sci 35(6):e40

Kim-Godwin YS (2003) Postpartum beliefs and practices among non-western cultures. MCN Am J Matern Child Nurs 28(2):74–78

Korea.net (2023) https://www.korea.net/AboutKorea/Korean-Life/Food

Merkin R (2009) Cross-cultural communication patterns: Korean and American communication. Cross-cultural communication patterns-Korean and American communication. J Intercult Commun 20. https://doi.org/10.36923/jicc.v9i2.481. https://www.researchgate.net/publication/261363991_Cross-cultural_communication_patterns-Korean_and_American_Communication

Ministry of Employment and Labor (2023) https://www.moel.go.kr/english/resources/statistics.do

Nam SW, Ko SY, Lee YK, Sin SM, Kim IH (2014) The incidence of birthmarks in Korean newborn infants. Neonatal Med 21(3):151–157

National Agency for Management of Life-sustaining Treatment (2023) https://www.lst.go.kr/eng/index.do

OECD (2020) https://www.oecd.org/health/health-systems/OECD-Reviews-of-Public-Health-Korea-Assessment-and-recommendations.pdf

Pew Research Center (2023) https://www.pewresearch.org/global/2013/06/04/the-global-divide-on-homosexuality/

South Korea guide: a look at South Korean language, culture, customs, and etiquette (2016) http://www.commisceo-global.com/country-guides/south-korea-guide

Stanford medicine (2023) https://geriatrics.stanford.edu/ethnomed/korean/fund.html

The guidebook for living in Korea (2022a) https://krguide.kr/ENG/1160/contents.do

The guidebook for living in Korea (2022b) https://krguide.kr/ENG/1168/contents.do

The Long-Term Care Insurance Act for the Elderly (2023) https://www.law.go.kr/LSW/lsInfoP.do?lsId=010436&ancYnChk=0#0000

The Traffic Laws (2023) https://www.law.go.kr/%EB%B2%95%EB%A0%B9/%EB%8F%84%EB%A1%9C%EA%B5%90%ED%86%B5%EB%B2%95

University of Washington Medical Center (2016) https://www.lacrosseconsortium.org/uploads/content_files/files/Korean%20Culture.pdf

# Chapter 27
# People of Mexican Heritage

**Rick Zoucha and Anelise Zamarripa-Zoucha**

## 27.1 Overview and Heritage

Great variations exist among people of Mexican heritage, depending on acculturation, individuality, and the variant characteristics of culture as described in Chap. 1. However, a core of commonly held beliefs and practices is shared by those who self-identify as Mexican, Mexican American, **Latino**, Latinx, Latin@, Chicano, or **Hispanic**. Hispanic and Latino are historically used interchangeably; however, Latino includes Brazilians who are not Hispanic. Contemporary verbiage to include gender variant and gender non-conforming people of Latin American heritage is becoming more common. This includes the use of Latinx and Latin@ to promote gender inclusivity. In addition, some who identify as Mexican have no Hispanic heritage because they are from indigenous native groups who have never mixed with/married people from Spanish backgrounds; however, they are still considered Mexican.

Mexican Americans include newer immigrants and people who have been in the United States for six or seven generations. Some originate from high mountains and plateaus, some from low-lying swamp areas, and others from sea- and ocean-side communities. The topography of the region of origin in Mexico may be the key to accurate health assessments for dietary preferences, illnesses, and disease. Many Mexicans come to the United States because of the high poverty rate of over 20.2% in Mexico.

R. Zoucha (✉)
Duquesne University School of Nursing, Pittsburgh, PA, USA
e-mail: zoucha@duq.edu

A. Zamarripa-Zoucha
Central Outreach Wellness Center, Pittsburgh, PA, USA



E. A. Fenkl, L. D. Purnell (eds.), *Handbook for Culturally Competent Care*,
https://doi.org/10.1007/978-3-031-70492-5_27

### *27.1.1 Communications*

- Great diversity exists in the Spanish language; many dialects are spoken in Mexico. Although most Mexicans use Spanish as their primary language, some speak only English, some speak both English and Spanish, and some speak only an indigenous native language.
- Significant importance is placed on verbal communication. Conversants may stand very close, even with people who are not well known to them. Loud voice volume in formal settings may connote anger.
- **Whereas more traditional and older individuals do not maintain eye contact, acculturated and more educated people usually do maintain eye contact.**
- Many Mexicans tend to be present-oriented. Time is relative, not categorically imperative; thus, many arrive late for appointments and delay seeking health care until the condition is more serious. Offer flexibility in scheduling health care appointments if possible in the United States.
- Touch is common between people of the same sex. **Explain the necessity and ask permission before touching during a physical examination.**
- Formal names may be extensive and include a middle name and the mother's and father's surnames. A married woman may take her husband's surname, thereby having three surnames. **Greet adults formally with Señor, Señora, or Señorita unless told to do otherwise. Ask which name is preferred as well as which name is used for legal purposes.**

### *27.1.2 Family Roles and Organization*

- In traditional families, men are expected to provide financial support for the family. However, this is changing due to economic challenges in the United States.
- The stereotype of "**machismo**," in which men are the primary decision-makers in the household, is not accurate for all people of Mexican heritage. **Do not stereotype machismo.**
- Women care for children and maintain the home. In most homes where the man and woman both work outside the home, they tend to share household chores and child-rearing responsibilities.
- Some households are patriarchal, some are matriarchal, and some are egalitarian. Regardless of who makes the decision, the male traditionally is expected to be the spokesperson for the family. **Ask specifically who makes which decisions for the family.**
- Priorities for children are getting an education, having good manners, and respecting older people and people in high-status positions.
- Children born out of wedlock are loved regardless of the parents' marital status. Men usually remain involved with their children who are born out of wedlock.

- When possible, extended family members prefer living close to each other. The elderly frequently live with their children when self-care becomes a concern; in addition, the elderly are always respected for their wisdom.
- **Extended family members are a good resource for home care.**
- Social status is gained through formal academic education, having a respected position, and having children and family members with good manners.
- As of December 2022, all states in Mexico have sanctioned same-sex marriages and Mexico's Supreme Court has ruled that all 31 states must recognize same-sex marriages. Many continue to be stigmatized. **Do not disclose sexual orientation status to family members.**

### *27.1.3 Workforce Issues*

- People educated in Mexico are likely to have been exposed to pedagogical approaches that include rote memorization and an emphasis on theory with little practical application. American educational systems usually emphasize an analytical approach, practical applications, and a narrow, in-depth specialization.
- **Teach practical application of knowledge during orientation and ongoing.**
- Because family is a first priority for most Mexicans, activities that involve family members usually take priority over work issues. **Be flexible with work hours when feasible.**
- Most Mexican Americans tend to shun confrontation for fear of losing face. Many are very sensitive to differences of opinion, which are perceived as disrupting harmony in the workplace.
- Truth is tempered by diplomacy and tact. When a service is promised for tomorrow, even when they know the service will not be completed tomorrow, it is promised to please, not to deceive. Truth is seen as a relative concept, not an absolute value, and people are expected to give direct yes and no answers.
- **Explain the value of "absolute" truth in the American workforce.**
- For most Mexicans, titles and positions may be more important than money.
- Many Mexicans believe that time is relative and elastic, with flexible deadlines, rather than stressing punctuality and timeliness. **Explain the expectation of punctuality in the American workforce and repercussions for chronic tardiness.**
- Many are used to having traditional autocratic managers who assign tasks but not authority. A Mexican worker who is not accustomed to responsibility may have difficulty assuming accountability for decisions.
- More traditional and some older individuals do not maintain eye contact, whereas acculturated and more educated people usually do maintain eye contact. **Do not assume that the lack of eye contact means the person is not listening, does not care, or is being less than truthful.**

### *27.1.4 Biocultural Ecology*

- Most Mexicans have dark hair and dark eyes, but some may have blond hair and blue eyes. Many have a diverse gene pool that includes indigenous Native groups.
- **Assess for cyanosis and jaundice in those with dark skin by observing the sclera and conjunctiva, palms of the hands, soles of the feet, and buccal mucosa and tongue rather than relying on skin tone.**
- Table 27.1 lists commonly occurring health conditions for Mexicans in three categories: lifestyle, environment, and genetics.
- **Assessments need to include exposure to pesticides, parasitic diseases, and illnesses that are common in their home country.**
- Mexicans may require lower doses of antidepressants and experience more intense side effects than non-Hispanic white populations. Many are also poor metabolizers of debrisoquine. **Observe for side effects of antidepressants.**

### *27.1.5 High-Risk Health Behaviors*

- Smoking is more common among the more acculturated, but they smoke fewer cigarettes than European Americans when they do smoke. **Encourage smoking cessation beginning in elementary schools.**
- Some Mexicans use alcohol to make them emotionally and socially extroverted and are more likely to engage in binge drinking than other groups. **Assess alcohol consumption by determining the amount of alcohol used both daily and during celebrations.**

**Table 27.1** Health conditions common to Mexicans

| Health conditions | Causes |
|---|---|
| Lactase deficiency | Genetic |
| Diabetes mellitus | Genetic, environment, lifestyle |
| Parasitic diseases | Environment, lifestyle |
| Malaria | Environment, lifestyle |
| Dengue fever | Environment, lifestyle |
| Cleft lip/palate | Lifestyle |
| Dental caries | Lifestyle |
| Cardiovascular disease | Genetic, environment, lifestyle |
| Hypertension | Genetic, environment, lifestyle |
| Tuberculosis | Environment, lifestyle |
| Sickle cell anemia | Genetic |
| HIV/AIDS | Lifestyle |
| Pesticide/herbicide poisoning | Environment, lifestyle |

- Many newer immigrants may be reluctant to use seat belts and helmets because they are not used in their home country. **Explain the legal requirements of using seat belts and helmets.**
- Men may object to condom use because it decreases sensitivity. Women may object to the use of condoms because of the suggestion that the woman is "dirty."
- **Explain that condoms help prevent pregnancy and decrease the incidence of HIV/AIDS (SIDA) and other sexually transmitted infections.**

### *27.1.6 Nutrition*

- Food choices vary widely, depending on the region of Mexico from which the person comes. Rice, beans, and tortillas are food staples.
- **Determine preparation practices because some people use coconut oil and animal fat for flavoring. Determine flavorings and ingredients as well as the major food choices because these can add significant calories to the diet. Encourage use of corn tortillas instead of flour tortillas, leafy green vegetables, and soups and stews that use bones in them to increase calcium content in the diet and avoid side effects with lactose intolerance.**
- Being overweight according to the US height/weight tables is seen as positive. **Patients should be co-participants in deciding an acceptable weight.**
- Sweetened fruit drinks are popular, and adding sugar to fresh fruit juice is common.
- **Discouraging the use of sweetened fruit drinks and encouraging the use of natural juices without added sugar is one means of reducing calorie and carbohydrate consumption.**
- Mealtimes may not coincide with the dominant American schedule. For many, the noontime meal is the largest meal of the day, and the evening meal may be served very late in the evening. **Determine the individual's and family's mealtimes, and adjust medication schedules accordingly.**
- Many individuals balance food choices according to the "**hot" and "cold**" theory. Variations occur in the assignment of different foods for hot and cold properties. Herbal teas are commonly used to maintain health and treat illnesses.
- **Determine the patient's use of the hot and cold theory of food choices, which herbs and teas are used, how frequently they are used, and the amount used each time. Nonharmful practices should be incorporated into the plan of care.**

### *27.1.7 Pregnancy and Childbearing Practices*

- Fertility practices are primarily connected with religious beliefs usually associated with the Catholic Church. The church discourages the use of condoms, believing they promote promiscuity.

- The rhythm method or Norplant is usually deemed acceptable because they are more natural means of helping to prevent an unwanted pregnancy. Foams, creams, and intrauterine devices may not be acceptable because women are not supposed to touch their genitals except for purposes of bathing.
- Abortion is decriminalized in Mexico, but the social and cultural stigma remains. A few individuals may see sterilization and other methods of birth control as acceptable.
- **Determine on an individual basis acceptable methods of birth control. Refer to planned parenthood if appropriate.**
- Multiple births are more common among Mexicans. Pregnancy is considered a natural condition, with most advice for health care coming from family members; this may deter prenatal care. **Stress that medical evaluation during pregnancy is necessary to help ensure the health of both the mother and the baby. Encouraging female relatives to accompany the pregnant woman for prenatal checkups may be helpful.**
- For many, the delivery room is not a place for men, and many believe that allowing the father to see the woman or baby during the delivery may harm the mother or baby. **However, with acculturation this cultural view may be changing. Ask the mother who they would prefer in the delivery room.**
- Various cultural customs and beliefs are included in Table 27.2.

**Table 27.2** Common cultural beliefs and customs

| |
|---|
| • Walking in the moonlight while pregnant may cause birth defects |
| • Safety pins, metal keys, or other metal amulets are worn to prevent birth deformities |
| • Raising one's arms over the head while pregnant will cause the cord to wrap around the infant's neck |
| • Postpartum women are discouraged from taking baths, sitting in a bathtub, taking sitz baths, or washing their hair for 6 weeks |
| • Bathing with a cloth, including the hair, is acceptable. Encourage use of warm compresses instead of sitz baths |
| • Cutting the baby's hair or nails in the first 3 months is believed to cause blindness and deafness in the baby. Ask permission before cutting an infant's nails or hair |
| • Placing a key, coin, or other metal object on the infant's umbilicus is believed to promote healing. Teach how to clean the object to prevent infection |
| • Wearing an abdominal binder can prevent air from entering the uterus in postpartum women |
| • Covering their ears, head, shoulders, and feet can prevent blindness, mastitis, frigidity, or sterility in women |
| • Recognize which practices are harmless, and respect the woman's beliefs as appropriate |
| • Using videos and literature in Spanish and with pictures of Hispanics may help compliance with health interventions |

### 27.1.8 Death Rituals

- Death is an extended family affair, and members are obligated to visit in the hospital or long-term care facility. Expect many visitors in the inpatient setting. Among more traditional Mexicans, a family member must remain with the dying person.
- **Find a place where family members can be together, preferably in the unit with the dying person. In alternative settings, find another quiet place where the family can gather. Elicit a family spokesperson to assist with crowd control.**
- Some families may want to have lighted candles in the room of the dying patient. Although inpatient facilities will not permit open candle flames because of fire safety, electric candles are acceptable to most.
- Death is considered part of life, and some, especially men, may approach death stoically.
- **Just because someone does not openly display emotions about death does not mean the person does not care about the deceased.**
- Some, especially women, may have an *ataque de nervios* on hearing about the death of a loved one. In this culture-bound syndrome, the person exhibits hyperkinetic and seizure-like activity that releases strong emotions. **This is a normal reaction, and treatment is usually not necessary except to remain with the person and provide emotional support. Use family members for assistance when possible.**
- When a person dies, family and friends gather for a ***velorio***, a festive occasion to celebrate the person's life.
- Burial practices vary, but the more traditional do not practice cremation. Autopsy is not welcomed. **Explaining the legalities of autopsy is conducive to having the family accept an autopsy.**

### 27.1.9 Spirituality

- Although many other religions are practiced, the predominant religion is Roman Catholicism. However, do not assume that the practice of the Catholic religion is the same for all people. Each person's religious practices must be assessed on an individual basis. Many families have altars in their homes.
- The family is foremost. The individual's source of strength comes from being with family members. Priests and other religious leaders are major sources of emotional strength and support for most Mexicans. **On admission to the healthcare facility, inform the patient and family of the availability of referral for pastoral care and support.**

### *27.1.10 Health-Care Practices*

- The family is considered the most credible source of health-care information among Mexicans. This practice can impede health-seeking behaviors. For many, good health means being free of pain and is largely due to the "will of God."
- Many migrant workers are not aware of the necessity of protecting themselves from pesticides and herbicide poisoning because they are not common in Mexico.
- **Health teaching on herbicides and pesticides should be a family affair.**
- Many Mexicans use herbal medicines and teas. The specific herbs and teas vary among families. Most of these teas and herbs are beneficial or at least not harmful. A few can be harmful by themselves, whereas others may be harmful when included with prescription and over-the-counter medications. **Ascertain if patients are using over-the-counter medicines and explain the hazards of excessive use of over-the-counter medications as well as using medicines that were originally intended for use by another family member.**
- **Assess for use of herbs and teas. Two herbs commonly used by Mexican Americans are *azarcon* and *greta*, which are used for colic and stomach conditions in children. Both of these herbs contain lead and can be toxic, especially in children.**
- Some Mexicans practice the hot and cold theory, according to which many diseases and illnesses are caused by a disruption in the hot and cold balance of the body. Thus, if too many cold forces in the body cause an illness, treatment is aimed at balancing the condition by introducing hot treatments and foods that are considered hot. *Hot and cold do not always coincide with temperature* but perceived metabolic properties.
- Hot conditions include infection, diarrhea, and sore throats and, therefore, are treated with such cold foods as fruits, vegetables, and dairy products. Cold conditions, such as cancer, malaria, and earaches, are treated with such hot foods as liquor, beef, pork, and spicy foods. Tremendous variations exist between and among hot and cold conditions, depending on the family.
- **Include the entire family in health promotion and health teaching to increase compliance with health prescriptions and interaction.**
- Frequently seen cultural illnesses or conditions are shown in Table 27.3.
- Perform an individual pain assessment, determine usual treatment modalities used for pain, ask what the patient thinks caused the pain, and determine what the patient usually does to relieve pain. Have pain scales in Spanish as well as visual scales for those who do not read Spanish or English. **Explaining that pain medication will promote healing may encourage the patient to accept pain medication.**
- The sick role is easy to enter without personal feelings of inadequacy or blame. Family members readily take on the sick person's responsibilities. Family members usually care for the ill family member at home, if at all possible. **Query the family to determine if they have the resources to care for a member at home.**

**Table 27.3** Common cultural illnesses among Mexicans

| |
|---|
| • ***Empacho*** (blocked intestines) may result from an incorrect balance of hot and cold foods causing a lump of food to stick in the gastrointestinal tract. Treatment includes massaging the stomach and back to dislodge the food bolus |
| • ***Mal de ojo*** (evil eye) occurs when an older person looks at a younger person in an admiring fashion. Such eye contact can be voluntary or involuntary. Symptoms are numerous, including fever, anorexia and vomiting, or irritability. The spell can be broken if the person doing the admiring touches the person while admiring him or her. Allopathic health-care providers are usually unable to cure *mal de ojo*. Making a referral to a folk health-care provider is advisable |
| • ***Caida de mollera*** (fallen fontanel) has numerous causes, which may include removing a nursing infant too harshly from the nipple or handling an infant too roughly. Symptoms vary from failure to thrive to irritability. The usual treatment is to hold the infant upside down by the feet. Assess for dehydration. A referral to a folk health-care provider is also recommended |
| • ***Susto*** (magical fright or soul loss) is associated with epilepsy, tuberculosis, and other infectious diseases as well as "overwhelming feelings of loss." Symptoms may be physical or psychological in nature and may also be consistent with depression. Gear treatment to the underlying causes, and treat physical and psychological symptoms |

**Not all Mexicans in the United States have the extended family to care for relatives at home.**

- **Because long-term care facilities are either nonexistent or are of very poor quality in Mexico, family members may be very reluctant to place family members in such a facility. Suggesting a visit to one may be an option.**
- Extraordinary means being used to preserve life are frequently frowned upon and are often determined by finances, education, and availability of services.
- Family members who are documented are more likely to seek and receive health care versus those who are undocumented. In some families, several members may be documented and others are undocumented, potentially resulting in inequitable care.
- Blood transfusions are acceptable, but some may be reluctant to accept transfusion or blood products for fear of HIV/AIDS (*SIDA*). The belief that the body must be buried whole deters organ donation. Organ transplantation may not be acceptable because of the belief that *mal aire*, bad air, will enter the body and increase one's risk for developing cancer. **Dispel myths related to organ donation and organ transplantation. Eliciting the assistance of a priest may be helpful.**

## *27.1.11 Health-Care Providers*

- Many Mexicans use a number of folk health-care providers who are usually well known to the patient. See Table 27.4.
- Health-care providers with the right qualities are well respected by Mexicans.
- **Important qualities include addressing the patient formally unless told to do otherwise, showing respect by asking questions, accepting the patient's**

**Table 27.4** Major Mexican folk health-care providers

| |
|---|
| • ***Curanderos*** receive their gift from God or serve an apprenticeship. Some even prescribe over-the-counter medications. They usually treat traditional illnesses not caused by witchcraft |
| • ***Espiritistas*** (spiritualists) treat conditions caused by witchcraft. Amulets and prayer are a large part of the treatment. Seeing an *espiritista* may carry a stigma among some Hispanics |
| • ***Yerberos*** or ***Jerberos*** use herbs, teas, and roots to prevent or treat illnesses. Patients usually purchase the herbs from a ***botanica***, a specialist herb shop that also sells religious figurines |
| • *Sobadores* treat muscle and joint problems using massage and manipulation; usually they do not have formal training |
| • Ask patients if they are using folk health-care providers and the reasons why they are using them, and have them fully disclose all treatments prescribed |

**ideas, taking an interest in the entire family, and being well groomed. Promoting confianca in the caring relationship. Always ask permission, and explain the reason for the necessity of touching the patient during a physical examination.**

## 27.2 Reflective Exercises

1. Describe the extensive name format for Mexican women. How would you address her? What is her legal name?
2. Describe eye contact protocol among traditional Mexicans.
3. How do you assess for jaundice and cyanosis among dark-skinned Mexicans?
4. List three common health conditions of Mexican that have a lifestyle component.
5. Describe Mexican folk healers and describe their treatment modalities.
6. What are the deciding factors for treating Mexicans with extraordinary illnesses?
7. Describe the "hot" and "cold" theory of illnesses among Mexicans.
8. Describe the "hot" and "cold" theory of food among Mexicans. If you are unfamiliar with foods in each category, what are your resources for finding out more?
9. Two herbs commonly used by Mexican Americans are *azarcon* and *greta*, which are used for colic and stomach conditions in children. What are the hazards of giving them to children?
10. What is the folk illness *ataque de nervios*. What are its symptoms? How would you treat it?

## Bibliography

Benito-Gomez M, Williams KN, McCurdy A, Fletcher AC (2020) Autonomy-supportive parenting in adolescence: cultural variability in the contemporary United States. J Fam Theory Rev 12(1):7–26. https://doi.org/10.1111/jftr.12362

Dame-Griff EC (2022) What do we mean when we say "Latinx?": definitional power, the limits of inclusivity, and the (un/re)constitution of an identity category. J Int Intercult Commun 15(2):119. https://doi.org/10.1080/17513057.2021.1901957

Gonzalez N, Méndez-Pounds J (2017) The impact of acculturation and cultural values on Hispanic immigrants' parenting. Contemp Fam Ther 40(1):56–67. https://doi.org/10.1007/s10591-017-9428-8

Hendy HM, Can SH, Heep H (2022) Machismo and caballerismo linked with perceived social discrimination and powerlessness in U.S. Latino men. J Cross-Cult Psychol 53(1):109–121. https://doi.org/10.1177/00220221211054206

Outright International (2024) Country overview: Mexico. Outrightinternational.org. https://outrightinternational.org/our-work/americas/mexico

Planned Parenthood Global (2024) Mexico | Planned Parenthood Global. https://www.plannedparenthood.org/about-us/planned-parenthood-global/where-we-work/mexico

Romero Morales A, Consoli AJ (2020) Mexican/Mexican-American siblings: the impact of undocumented status on the family, the sibling relationship, and the self. J Latinx Psychol 8(2):112–126. https://doi.org/10.1037/lat0000133

Son D, Updegraff KA, Umaña-Taylor AJ (2022) Familism values and Mexican-origin adolescents' disclosure and secrecy with fathers and mothers. J Fam Psychol 36(8):1296–1305. https://doi.org/10.1037/fam0000986

Updegraff KA, Umaña-Taylor AJ (2015) What can we learn from the study of Mexican-origin families in the United States? Fam Process 54(2):205–216. https://doi.org/10.1111/famp.12135

Zoucha R, Zamarripa-Zoucha A (2020a) People of Mexican heritage. In: Purnell LD, Fenkl EA (eds) Textbook for transcultural health care: a population approach. Springer, Cham, pp 613–636

Zoucha R, Zamarripa-Zoucha A (2020b) Health and health care in Mexico. In: Holtz C (ed) Global health care: issues and policies, 4th edn. Jones and Bartlett, Burlington, pp 599–610

# Chapter 28
# People of Polish Heritage

**Latisha Barfield and Nola Holness**

## 28.1 Overview and Heritage

Poland was conceived near the middle of the tenth century and reached its golden age in the sixteenth century. In a series of agreements between 1772 and 1795, Russia, Prussia, and Austria partitioned Poland among themselves. Poland regained its independence in 1918 only to be overrun by Germany and the Soviet Union in World War II. It became a Soviet satellite state following the war, but its government was comparatively tolerant and progressive. Labor turmoil in 1980 led to the formation of the first independent free trade union in Eastern Europe, *Solidarnosc* (Solidarity) that over time became a political force and by 1990 had swept parliamentary elections and the presidency. Poland borders with neighboring countries Belarus, Czechia, Germany, Lithuania, Russia, Slovakia, and Ukraine.

In 1988, formal negotiations between the Polish Communist Party leaders and the unofficial opposition, called the "Round Table talks," resulted in partially free Parliamentary elections. Solidarity won a landslide victory in the 1989 elections and the newly elected Parliament changed the country's name and constitution establishing the Third Republic of Poland.

Polish and their descendants who have immigrated to the United States and elsewhere for many generations have maintained their ethnic heritage by promoting their culture, attending Catholic churches, attending parades and festivals, maintaining ethnic food traditions, speaking the Polish language, and promoting interest in their home country through media events as well as economic and political channels. For newer immigrant Poles, maintaining ethnic heritage means learning

L. Barfield (✉) · N. Holness
Nicole Wertheim College of Nursing and Health Sciences, Florida International University, Miami, FL, USA
e-mail: lbarfiel@fiu.edu; nholness@fiu.edu



E. A. Fenkl, L. D. Purnell (eds.), *Handbook for Culturally Competent Care*,
https://doi.org/10.1007/978-3-031-70492-5_28

English and obtaining a good job. Newer immigrants are less concerned with raising consciousness over Polish American issues than they are with financially helping families who remain in Poland and raising concerns over the political and economic climate in their homeland.

***Polonia*** is the name given to Polish communities in the United States. Poles are a heterogeneous group. Much of the diversity within this ethnic group is due to the degree of acculturation, individuality, and variant cultural characteristics as described in Chap. 1.

The coronavirus (COVID-19) caused by severe acute respiratory syndrome coronavirus 2 (SARS-CoV-2) spread throughout Europe, reaching Poland in 2020. The pandemic had a significant impact on Poland's health system, the health of its citizens, and social, economic, political, and cultural services even after the disease diminished. Social interactions among Polish citizens and travel restrictions severely affected tourism to the country.

Another humanitarian crisis challenged Poland. The Russian military invaded Ukraine on February 24, 2022, which resulted in an unprecedented influx of Ukrainian war refugees into Poland. As of September 2023, 15.4 million people have fled Ukraine. One major factor for this influx is Poland's geographical location adjacent to Russia. Since the war began, Poland received over 1.7 million refugees, mostly women and children. Poland is a major destination in Western Europe. As of February 2023, there are over eight million Ukrainian refugees residing in Poland. The current refugee crisis caused by this conflict has resulted in additional challenges with Poland's healthcare system and social services.

### *28.1.1 Communications*

- The dominant language of people living in Poland is Polish, although there are some regional dialects and differences.
- The Polish language was influenced by the countries surrounding Poland and by the Latin of eleventh- and twelfth-century kings. Depending on the regional and cultural background of the speaker, Polish may sound German, Russian, or French.
- Poles are an animated group, and facial expressions generally convey the tone of the conversation.
- Poles use touch as a form of personal expression of caring. Touch is common among family members and friends, but Poles may be quite formal with strangers and healthcare providers.
- Handshaking is considered polite. In fact, failing to shake hands with everyone present may be considered rude. **Greet patients formally and with a handshake.**
- Most Poles feel comfortable with close personal space, but distances increase when interacting with strangers.
- First-generation Poles commonly kiss "Polish style," that is once on each cheek and then once again on each cheek.

- Two women may walk together arm in arm, or two men may greet each other with an embrace, a hug, and a kiss on both cheeks.
- Many Polish Americans consider the use of spoken second-person familiarity rude. Polish people speak in the third person. For example, they might ask, "Would Martin like some coffee?" rather than "Would you like some coffee?" Although the first expression might sound awkward, the latter expression may be considered impolite and too informal, especially if the person being asked is older.
- Polish people use direct eye contact when interacting with others. Many Americans and elsewhere may feel uncomfortable with this sustained eye contact and feel it is quite close to staring, but to Poles, it is considered ordinary. **Sustained eye contact does not necessarily mean anger.**
- Poles, as a group, tend to share thoughts and ideas freely.
- Punctuality is important to Polish people. To be late is a sign of bad manners. Even in social situations, people are expected to arrive on time and stay late.
- Polish people are both past and future-oriented. The past is very much a part of Polish culture with the families passing on their memories of World War II, which still haunt them in some way. A strong work ethic encourages Poles to plan for the future.
- Many Polish names are difficult to pronounce to outsiders. Even though a name may be mispronounced, a high value is placed on the attempt to pronounce it correctly.
- Some examples of common Polish names include *Kowal*, meaning "blacksmith." Numerous suffixes, such as "icz," "czyk," "iak," and "czak," which mean "son of," can be added. Two of the most common suffixes are "ski" for male and "ska" for female, which originally were added to many names because they were associated with nobility. The suffix "cki" became the phonetic version of "ski." Surnames ending in "y," "ow," "owo," and "owa" are usually derived from names of places. The "ak" suffix is typical of western Poland, whereas "uk" is found in the east.

### *28.1.2 Family Roles and Organization*

- Polish culture centers on family. Each member has a certain position, role, and related responsibilities. All members are expected to work, make contributions, and strive to enhance the entire family's reputation and social and economic position.
- In most families, the father is perceived as the head of the household. Depending on the degree of assimilation, the father may rule with absolute authority in first-, second-, and even third-generation Polish families. However, among some third- and fourth-generation Polish and second- and third-wave immigrants, more egalitarian gender roles are becoming the norm.
- **Determine the spokesperson and decision maker for healthcare decisions.**

- Historically, large families were expected and common among Poles, with women often experiencing between 5 and 10 pregnancies.
- The most valued behavior for children is obedience. Taboo child behaviors include anything that undermines parental authority. Parents are quite demonstrative with young children, but they resist showing much affection toward them once they are older than toddler age. This is the parents' way of teaching children to be strong and resilient.
- Many parents praise children for self-control and completing chores. Little sympathy is wasted on failure, but doing well is openly praised. Children are taught to resist feelings of helplessness, fragility, or dependence.
- Traditional family values and loyalty are strong in most households. The elderly are highly respected. They attend church regularly and carry on Polish traditions.
- The elderly play an active role in helping grandchildren learn Polish customs and in assisting adult children in their daily routine with families.
- For some families, one of the worst disgraces, as seen through the eyes of the Polish community, is to put an aged family member in a nursing home. Third- and fourth-generation Polish may consider an extended-care or assisted-living facility because of work schedules and demands of care, but first-generation immigrants rarely perceive this as an option.
- Extended family, consisting of aunts, uncles, and godparents, is very important to Poles. Long-time friends become aunts or uncles to Polish children.
- Alternative lifestyles are seen as part of assimilation into the blended cultures. Same-sex couples are frowned upon and may even be ostracized, depending on the level of assimilation. **Do not disclose same-sex relationships to others.** Poland is very conservative. In Poland, there is still no legal acknowledgment of marital or family relationships for same-sex couples. Compared to Western countries, recognition of LGBTQ+ families is faced with many legal and social obstacles that play a role in how same-sex family structures are constructed, viewed, and practiced. Polish law opposes adoption and artificial reproductive technologies for same-sex couples.
- When divorce occurs, single heads of households are accepted in the Polish American community.

### *28.1.3 Workforce Issues*

- Some Polish immigrants are underemployed and may have difficulty working with authority figures who are less educated. Poles quietly comment that they are disrespected for their educational background and that they must endure decreased status to stay in America and elsewhere.
- The Polish characteristic of praising people for their work makes Poles strong managers.

- Because Poles learn deference to authority at home, in the church, and in parochial schools, some may be less well suited for the rigors of a highly individualistic, competitive market.
- The strong Polish work ethic, exhibited as volunteering for overtime, being punctual, and rarely taking sick days, is valued by employers.
- Native-born Polish Americans have little, if any, difficulty with the English language. Foreign-born Poles frequently have some difficulty understanding the subtle nuances of humor. Recent Polish immigrants, who had experience working under a Communist bureaucratic hierarchy, may have some difficulty with the structure, subtleties, and culture of the American workplace. **Carefully explain workforce rules and expectations.**
- The pandemic impacted the workforce in Poland by increasing unemployment rates by 5.8–6.3% from April to December 2020 due to decreased employment opportunities. Employment opportunities decreased by 25% compared to 2019. The main areas impacted by the pandemic were the healthcare industry, hospitality, retail, human services, cultural activities, and religious and family observances. Initially, the Polish government acted on prevention and management strategies to control the transmission of COVID, aiding employees and preventing companies from claiming bankruptcy. The pandemic also resulted in the advancement of technology and working remotely.
- Before the Russian invasion of Ukraine, there were an estimated 1.35 million Ukrainian immigrants (mostly men) working and residing in Poland. Due to the continuing conflict and destruction of Ukraine's infrastructure, 46.7% of Ukrainians came to Poland. More Ukrainians will continue to reside in Poland compared to before the war. The outlook of Poland's economy is expected to be challenged as labor migration will increase. In the long term, tensions may increase between Polish and Ukrainian refugees because of increased use of social services and in the labor market.

### *28.1.4 Biocultural Ecology*

- Most Poles are of medium height with a medium-to-large bone structure. As a result of foreign invasions over the centuries, Polish people may be dark and Mongol-looking or fair with delicate features, blue eyes, and blonde hair. Those with fair complexions are predisposed to skin cancer. **Teach patients about protecting themselves from sun exposure and cancer.**
- In Polish culture, a common belief is that enduring pain without complaining or asking for relief demonstrates virility in men and self-control in women. **Encourage acceptance of pain medication, explaining that it will promote healing.**
- Risk factors for newer Polish immigrants relate to their employment in industries in their homeland. Heavy industry in Poland produced prolonged, significant air pollution and environmental neglect. Living in polluted environments led to an

increase in premature deliveries, low-birth-weight children, diseases of the pulmonary and circulatory systems, and various forms of cancer.

- In Poland, air pollution remains a serious problem because of sulfur dioxide emissions from coal-fired power plants. In 1986, the Chernobyl incident in Russia contaminated the land and water systems of eastern Poland.
- **Assess new immigrant Poles for respiratory conditions and cancer resulting from radiation exposure.** Table 28.1 lists commonly occurring health conditions for Poles in three categories: lifestyle, environment, and genetics.
- The first case of COVID was recorded on March 4, 2020 in Poland. There were 6.4 million confirmed cases of COVID in Poland from 2020 to 2023. The impact of COVID caused the Minister of Health to declare a state of emergency on March 20, 2020, and social restriction protocols were enacted. Mood disorders were on the rise among young Poles because of social isolation. The number of deaths significantly increased. In January 2023, the number of deaths due to COVID was greater than 118,000. COVID restrictions and travel advisories in Poland were lifted in March 2022. The vaccination rate doses are 145 per 100 people in Poland.
- The war began in the third year of the COVID pandemic. The war has caused a humanitarian crisis with the influx of refugees. The increased transmission of infectious diseases places challenges to public health in Poland. Since the war, the COVID vaccination rate among Ukrainians is 38% in comparison to 56% of Polish citizens coming out of the pandemic.
- It is estimated that up to 8000 Ukrainian refugees are unaware of their HIV status and up to 18,000 will need treatment for HIV disease, which will lead to increased HIV cases reported in Poland. Before the war, HIV screening in Poland singled out drug users and men who have sex with men.

### *28.1.5 High-Risk Behaviors*

- Poles have had a long history of excessive smoking. **Encourage anti-smoking beginning in elementary school.**
- In Poland, a high rate of alcoholic psychosis, cirrhosis of the liver, and acute alcohol poisoning exists. **Encourage socially responsible alcohol intake and refer patients to support groups for those with alcohol misuse.**

**Table 28.1** Health conditions common to Polish people

| Health conditions | Causes |
|---|---|
| Cardiovascular disease | Environment, lifestyle |
| Diabetes mellitus | Genetic, lifestyle |
| Alcohol misuse | Environment, lifestyle |
| Cancer | Environment, lifestyle |
| Pulmonary disorders | Environment, lifestyle |

- **Children of immigrants should especially be targeted for counseling regarding the health effects of smoking and alcohol consumption.**
- **Screen Polish immigrants for diseases common in their home country: hypertension, CVDs, respiratory conditions, cancer (particularly leukemia), and thyroid disorders.**

### *28.1.6 Nutrition*

- Meats and vegetables are cooked for a very long time, resulting in destruction of B and other vitamins.
- The strong Catholic influence is evidenced by attending many food-laden celebrations, festivals, and rituals, each of which has its own traditional high-calorie foods.
- **Complete a dietary assessment on admission. Help patients structure a diet that is culturally acceptable, promotes healthy food choices, and is sustainable.**
- Staples of the Polish diet are millet, barley, potatoes, onions, radishes, turnips, beets, beans, cabbage, carrots, cucumbers, tomatoes, apples, and wild mushrooms.
- Common meats are chicken, beef, and pork. Traditional high-fat entrees include pigs' knuckles and organ meats such as liver, tripe, and tongue. *Kapusta* (sauerkraut), *golabki* (stuffed cabbage), *babka* (coffee cake), *pierogi* (dumplings), and *chrusciki* (deep-fried bowtie pastries) are common ethnic foods. Table 28.2 lists a variety of traditional Polish foods.

**Table 28.2** Polish foods

| Common name | Description | Ingredients |
|---|---|---|
| *Babka* | Coffee cake | Yeast bread |
| *Barszcz* | Beet soup | Served plain or with sour cream |
| *Bigos* | Hunter's stew | Stew with game, sausage, sauerkraut |
| *Chrusciki* | Polish bowties | Fried egg dough |
| *Golabki* | Cabbage rolls | Cooked cabbage stuffed with chopped meat and rice in tomato sauce |
| *Kielbasa* | Sausage | Ground pork and spices |
| *Ogorki smietanie* | Sour cream and cucumbers | Sour cream, cucumbers |
| *Pierogi* | Boiled dumplings | Dumplings filled with potatoes, cheese, or sauerkraut |
| *Sledzie* | Herring | Pickled fish |

### 28.1.7 Pregnancy and Childbearing Practices

- Because family is very important, most Poles want children. In Poland, the Catholic Church strongly opposes abortion, which is the prevailing attitude of many Poles in the United States and elsewhere.
- Fertility practices are balanced between the needs of the family and the laws of the Church. Polish women believe in the power of prayer to guarantee a successful pregnancy.
- Pregnant Polish women are expected to seek preventive health care, eat well, and get adequate rest to ensure a healthy pregnancy and baby. Pregnant women usually follow the physician's orders carefully.
- They depend on intuition and believe in "evil eye" powers, causing the mother and baby harm. Polish women hang a red ribbon on the baby carriage to protect the infant.
- The emphasis on food and "eating for two" is a common philosophy. **Counsel pregnant Polish women on weight gain and the long-term hazards.**
- Many consider it bad luck to have a baby shower. Even now, many Polish grandmothers may be reluctant to give gifts until after the baby is born.
- The birthing process is considered the domain of women. During labor, movement is an important method to hasten childbirth. Polish women believe massage is important to deliver the placenta. Newer Polish immigrants may feel uncomfortable with men in the birthing area or with family-centered care. **Do not make men feel guilty if they do not wish to be in the delivery room.**
- Women are expected to rest for the first few weeks after delivery. For many, breast-feeding is important for postpartum recovery. **Lactation education during breast-feeding helps the woman understand the balance between diet, rest, and exercise after delivery.** Polish women believe in traditional personal hygiene practices to prevent infections after childbirth.
- Fertility trends are declining in Poland due to postponed childbearing related to advanced age, low socioeconomic status, scarceness or instability of partners, cost of childrearing, infertility, and uncertain future. Delayed childbearing increases the risk of infertility and other pregnancy complications.

### 28.1.8 Death Rituals

- Most Poles have a stoic acceptance of death as part of the life process.
- Family and friends stay with the dying person to negate any feelings of abandonment.
- Make arrangements for the family to remain with the dying patient.
- Home hospice care is acceptable to most Poles.

- Healthcare providers may encounter difficulty in convincing the family that the dying member may choose to refuse food as a result of the illness. **Provide factual information about end-of-life care.**
- Polish American family members follow a funeral custom of having a wake for 1–3 days followed by a Mass and religious burial.

### *28.1.9 Spirituality*

- The Catholic Church, with its required attendance at Mass on Sundays and holy days, is an integral part of the lives of most Polish people.
- Polish people have had a renewed interest in their ethnic roots. For example, their attendance at language classes, festivals, and Polish Catholic churches has become very widespread.
- Primary spiritual sources are God and Jesus Christ, with many Polish immigrants praying to the Virgin Mary, saints, and angels to ward off evil and danger. **Contact a chaplain if the patient wishes.**
- Honor and special attention are paid to the Black Madonna or Our Lady of Czestochowa. During times of illness and serious family concerns, one might hear a Pole evoking *Matka Boska*, which translates literally to "Mother of God."
- Many older Polish people believe in the special properties of prayer books, rosary beads, medals, and consecrated objects. **Do not remove religious items and amulets from patients.**
- Polish Americans commonly exhibit devotions to God, such as crucifixes and pictures of the Virgin Mary, the Black Madonna, and Pope John Paul II, in their homes.

### *28.1.10 Health-Care Practices*

- Most Poles place a high value on stoicism and doing what needs to be done. Many go to healthcare providers only when symptoms interfere with function; then they may carefully consider the advice provided before complying.
- Many Poles are reluctant to discuss their treatment options and concerns with healthcare providers and routinely accept the proposed care plan.
- If Poles believe they are unable to pay the medical bill, they may refuse treatment unless the condition is life-threatening. Many have a strong fear of becoming dependent and resist relying on charity.
- Because many Poles consider Medicare, Medicaid, and managed care as forms of social charity, they are reluctant to apply for them. **Encourage patients and families and help them access social services when needed.**

- Poles usually look for a physical cause of disease before considering a mental disorder. If mental health problems exist, home visits are preferred. Few Poles turn to psychiatrists or mental health providers for help. Those who seek help from mental health providers do so as a last resort. Many individuals choose their priest or seek assistance from a Polish volunteer-run agency before going to a health provider for psychiatric help.
- Talk-oriented interventions and therapies without pharmaceutical or suitable psychosocial strategies are dismissed unless interventions are action-oriented. In addition, Poles consult other family members and the community to assess the appropriateness of treatments. The role of exercise and increased intake of vegetables have minimal impact on reducing overweight, obesity rates, and lifestyle changes. Many Poles do not make the connection that weight reduction prevents cardiovascular diseases.
- Polish people often seek self-help groups such as Alcoholics Anonymous before seeing a healthcare provider.
- Assimilated Poles respect the healthcare system and tend to seek specialized care when necessary.
- Given the continuation of limited access to care and the strong work population, health promotion practices are often undervalued by Polish.
- Explain the benefits of health promotion activities and wellness checkups.
- Breast self-examination and Pap smear tests are poorly understood by many newer female immigrants. **Explain the benefits of breast self-examination and the necessity of Pap smears.**
- The Polish ethic of stoicism discourages the use of over-the-counter medications unless a symptom persists.
- Poles may use certain remedies to cure an illness, such as tea with honey and spirits to "sweat out" a cold. Herbs and rubbing compounds may also be used for problems associated with aches, pains, and inflammation from overworked joints and muscles. Some additional common cultural practices included treating the symptoms of colds with herbs or poultices made from goose grease or fat. **Specifically, ask about the use of home remedies and over-the-counter medications.**
- When a Pole is asked to undress for a physical examination, pay special attention to any medals pinned to the patient's undergarments. Most of these medals have special religious significance to the wearer. **Do not remove medals from clothing without asking permission from the patient or family.**
- Being unable to speak and understand a foreign language and the cost of healthcare and its complexity are the greatest barriers to health care for newer Polish immigrants. **Obtain interpreters when needed.**
- Poles are polite to authority figures and avoid offending a healthcare worker by disagreeing with them. Thus, they may be reluctant to ask for clarifications on questionable issues.
- Owing to their strong sense of stoicism and fear of being dependent upon others, many Polish Americans use inadequate pain medication and choose distraction as a means of coping with pain and discomfort. Many Poles either deny or mini-

mize their pain or level of discomfort. Poles with chronic illnesses may have similar attitudes; thus, persevering with pain is common. **Explain that pain medicine will promote healing. Use a visual analog scale to assess pain and assist patients with distraction techniques.**

- The ethic of being useful, independent, and a good Catholic influences one to refrain from using extraordinary means to keep people alive. The individual or family determines what means are considered extraordinary. Receiving blood transfusions or undergoing organ transplantation is acceptable.

### *28.1.11 Health-Care Providers*

- Immigrant Poles often assess healthcare providers by their demeanor, warmth, and show of respect.
- Health advice may be sought from chiropractors and local pharmacists as well as neighbors and extended family.
- **When caring for Polish patients, particularly older adults, make every attempt to address individuals by their surname. Attempting to pronounce the name demonstrates respect for the patient.**
- Polish women are modest and self-conscious and may refuse health care when asked to disrobe in front of a male healthcare provider. **Provide a same-sex healthcare provider when possible.**
- Physicians are held in high regard in Polish communities. Poles may change physicians if they believe their recovery is too slow or if a second opinion is needed.
- Immigrant Poles may be unfamiliar with the advanced roles of nurses who are expected to know about, plan, and be directly involved in the patients' care. Thus, many Poles may still want only the physician to explain all aspects of their care.
- In 2020, the number of physician and dental practices was 3.7 per 1000 population, a 46% decrease compared to 2010. The number of nurses increased to 236.4 thousand in 2022 compared to 2021. The number of practicing midwives increased marginally to 29.4 thousand in 2022.
- Since the beginning of the war, the Polish healthcare system faced many challenges. There is an increased need for mental and public health services in Poland. Many of Ukraine's healthcare facilities were destroyed such as hospitals, ambulatory clinics, and pharmacies. As a result, many Ukrainian healthcare providers became refugees and fled to Poland. Poland's Ministry of Health made provisions for Ukrainian healthcare workers to be able to receive temporary medical licenses to practice. There is also a language barrier with Ukrainian healthcare workers. Poland offered a free e-learning course in Polish medical language to remove barriers of communication in the medical community. The Ministry of Health and WHO Europe initiated the Organization of Healthcare course in Poland. Due to the damage to the Ukrainian infrastructure, Polish med-

ical students attending Ukrainian universities returned and continued their education at Polish universities.

## 28.2 Reflective Exercises

1. Describe Polish communication practices as they relate to the practice of using the third person.
2. What do Poles expect in terms of timeliness?
3. Describe taboo behaviors for Polish children.
4. Name and describe at least three diseases common among Poles.
5. What are some of the foods a Pole might eat?
6. Describe some prescriptive and restrictive Polish prenatal/postpartum practices.
7. What might you do to encourage a stoic Pole to accept pain medication?
8. How is mental illness perceived by Poles?
9. Identify at least three common folk remedies commonly used by Polish people.
10. Describe a traditional Polish funeral.
11. Post-COVID, how have health, social, and economic services changed in Poland?
12. What measures will the Ministry of Health implement to improve COVID vaccination rates in Poland with Poles and Ukrainians?
13. Why is Ukraine important to Poland?
14. What is the current impact on Poland because of the Russian-Ukrainian war?
15. What will the Ministry of Health leadership and government officials implement to improve culturally competent care in Poland?

## Bibliography

Adamowicz M (2022) COVID-19 pandemic as a change factor in the labour market in Poland. Sustain For 14(15):9197. https://doi.org/10.3390/su14159197

Chawlowska E, Lipiak A, Krzysztoszek J, Krupa B, Staszewski R (2020) Reproductive health literacy and fertility awareness among Polish female students. Front Public Health 11(8):499. https://doi.org/10.3389/fpubh.2020.00499

CIA (2024) World Factbook: Poland. https://www.cia.gov/the-world-factbook/countries/poland/

Duszczyk M, Kaczmarczyk P (2022) The war in Ukraine and migration to Poland: outlook and challenges. Intereconomics 57(3):164–170. https://doi.org/10.1007/s10272-022-1053-6

Ejder Apay S, Nagórska M, Sonmez T, Yagmur Gur E, Yilmaz F, Zych B, Lewandowska A, Lesińska-Sawicka M (2022) Traditional practices during pregnancy, delivery, and puerperium used by women in Poland and Turkey. Med Sci Pulse 16(2):22–32. https://doi.org/10.5604/01.3001.0015.8819

Elflein J (2023) COVID vaccine dose rate worldwide by select country or territory March 20, 2023. Statista. https://www.statista.com/statistics/1194939/rate-covid-vaccination-by-county-worldwide/

Jankowski M, Lazarus JV, Kuchyn I, Zemskov S, Galazkowski R, Gujski M (2023) One year on: Poland's public health initiatives and national response to millions of refugees from Ukraine. Med Sci Monit 29:e940223-1–e940223-5. https://doi.org/10.12659/MSM.940223

Mendoza SL, Halualani R, Dzewiecka JA (2002) Moving the discourse on identities in intercultural communication: structure, culture, and resignifications. Commun Q 50(3–4):312–327. Update

Mizielińska J, Stasińska A (2017) "There is nothing like a family": discourses on families of choice in Poland. J Homosex 64(13):1793–1815. https://doi.org/10.1080/00918369.2016.1267460

Olszanecka-Glinianowicz M, Chudek J (2013) The level of health education in the Polish population. Ann Agric Environ Med 20(3):559–565

Pietkiewics I (2015) Burial rituals and cultural changes in the Polish community: a qualitative study. http://www.academia.edu/2277040/Burial_rituals_and_cultural_changes_in_the_polish_community_a_qualitative_study

Polish Culture Profile (2006) http://www.diversicare.com.au/wp-content/uploads/2015/10/Polish.pdf

Public Opinion Research Center (2010) Polish values: what is important, what is permitted, what must not be done. http://www.cbos.pl/PL/publikacje/public_opinion/2010/07_2010.pdf

Purnell L (2013) People of Polish heritage. In: Purnell L (ed) Transcultural health care: a culturally competent approach. F.A. Davis, Philadelphia

Rempusheski VF (1998) Caring for self and others: second generation Polish American elders in an ethnic club. J Cross Cult Gerontol 3:223–271

Ryndyk O, Johannesses L (2015) Review of research literature on parenting styles and childrearing practices among poles. Historical and contemporary perspectives. https://core.ac.uk/download/pdf/30919587.pdf

Sas A (2023a) Number of coronavirus (COVID-19) deaths in Poland 2020–2023. Statista. https://www.statista.com/statistics/1110787/poland-coronavirus-mortality/

Sas A (2023b) Number of doctors and dentists practices in Poland 2010–2021. Statista. https://www.statista.com/statistics/1121819/poland-number-of-doctors-and-dentists-practices/

Sas A (2023c) Practicing midwives: number of employees in Poland 2012–2022. Statista. https://www.statista.com/statistics/462782/practising-midwives-employment-in-poland/

Sas A (2023d) Number of Ukrainian refugees in Poland 2022–2023 by date of report. Statista. https://www.statista.com/statistics/1293564/ukrainian-refugees-in-poland/

Stewart C (2023) Number of coronavirus (COVID-19) cases in Europe 2023, by country. Statista. https://www.statista.com/statistics/1104837/coronavirus-cases-europe-by-country/

Szczepańska A, Pietrzyka K (2023) The COVID-19 epidemic in Poland and its influence on the quality of life of university students (young adults) in the context of restricted access to public spaces. Z Gesundh Wiss 31(2):295–305. https://doi.org/10.1007/s10389-020-01456-z

Weissman AL (2017) Repronormativity and the reproduction of the nation-state: the state and sexuality collide. J GBLT Fam Stud 13(3):277–305. https://doi.org/10.1080/1550428X.2016.1210065

# Chapter 29
# People of Puerto Rican Heritage

**Mariceli Comellas-Quinones**

## 29.1 Overview and Heritage

The Island of Puerto Rico is home to a population of approximately 3.5 million persons. The Puerto Rican community, comprising over 5.5 million individuals, represents the second largest demographic group in the continental United States accounting for nearly 10% of the Hispanic communities. They self-identify as *Puertorri-queños* or ***Boricua*** (the Taíno Indian word for Puerto Rican) or ***Niuyoricans*** for those born in New York. Diversity among Puerto Ricans includes assimilation, individuality, and the variant characteristics of culture as presented in Chap. 1.

Puerto Rico is a commonwealth of the US Citizenship status, which has created a controversial *va y ven* (go and come) circular migration in which individuals and families are often caught in a reverse cycle of immigration, alternately living a few months or years in the United States and then returning to Puerto Rico. Some Puerto Ricans fear that the dominant American culture is a potential threat to the Puerto Rican culture, language, and future.

Although the educational system in Puerto Rico is similar to that in the United States, when children emigrate from Puerto Rico to the mainland United States, many educational organizations place them one grade below their previous academic year, sometimes due to language barriers.

M. Comellas-Quinones (✉)
Nicole Wertheim College of Nursing and Health Sciences, Florida International University, Miami, FL, USA
e-mail: mcomella@fiu.edu



E. A. Fenkl, L. D. Purnell (eds.), *Handbook for Culturally Competent Care*,
https://doi.org/10.1007/978-3-031-70492-5_29

### 29.1.1 Communications

- English and Spanish are both official languages.
- Most Puerto Ricans speak with a melodic, high-pitched, fast rhythm maintained when speaking English. **Avoid making comments about accent, use caution when interpreting voice pitch, and seek clarification when in doubt about the content and nature of a conversation that may seem confrontational.**
- Many prefer to read or share sensitive information, options, and decisions with close family members. Some obtain assistance from extended family or community members knowledgeable in health matters. **Seek clarification of the information provided, ask for language preference in verbal and written communication, and allow time to exchange information with questions and answers when critical decisions need to be made.**
- Great value is placed on interpersonal interactions such as ***simpatía*** when an individual is perceived as likable, attractive, and fun-loving.
- Most individuals enjoy sharing information about their families, heritage, thoughts, and feelings and expect the healthcare provider to exchange personal information when beginning a professional relationship.
- If ***confianza*** (trust) is established, healthcare providers can establish open communication channels with individuals and families.
- Most Puerto Ricans readily express their physical ailments and discomforts, except for sexuality, which is often considered taboo. **Set boundaries with discretion, emphasizing personal rather than impersonal and bureaucratic relationships.**
- Among younger generations and those born in the United States, eye contact is maintained and is often encouraged, with the more traditional, limited eye contact being a sign of respect, especially toward older people.
- The culture of Puerto Rico places significant importance on the values of reverence and nurturing toward the elderly population. **Clinicians need to acknowledge and attend to the concerns and preferences of elderly patients while recognizing the significance of involving family members in the decision-making process and incorporating them into healthcare discussions when suitable**.
- Hand, leg, head, and body gestures are commonly used to augment messages expressed by words. Feelings and emotions are also expressed through touch.
- Gesturing includes an affirmative nod with an "Aha!" response which does not necessarily mean agreement or understanding related to the conversation.
- Women greet each other with a robust and familiar hug, and if among family or close friends, a kiss is included. Young women may take offense to verbal and nonverbal communications that portray women as non-assertive and passive. Men may greet other men with a firm handshake and a left hand stroking the greeter's shoulder.
- **Greet Puerto Ricans formally or with a friendly handshake.**
- Most individuals are present-oriented, having a relativistic and serene view and way of life.

- **Carefully explain appointment times and time limits at the beginning of an interview.**
- Respect is reflected in how children talk, look, and refer to adults and older people. Rather than *Señora* (Mrs.) and *Señor* (Mr.), children and adults are expected to use the terms *Doña* (Mrs.) and *Don* (Mr.) for most adults. Aunts and uncles have their names preceded by *tití* or *tío* (auntie/uncle) and *madrina* or *padrino* (godmother or godfather).
- **Address Puerto Ricans as *Sr.*, *Sra.*, *Don*, and *Doña*.**
- A single woman may use her name as Sonia López Mendoza, with López being her father's surname and Mendoza her mother's. When she is married, the husband's last name, Pérez, is added with the word *de* to reflect that she is married. This woman's married name would be Sonia López de Pérez; the mother's surname is eliminated. In business and health-care organizations, Señora López de Pérez is the correct formal title for promoting conversation or building relationships. **Ask patients their complete name as well as their legal name.**
- Some parents, hailing from conservative backgrounds, believe that children do not have to engage in conversations about sex with them. Some women perceive parents lacked awareness of children's sexual activity during adolescence or before marriage.
- Sometimes, parents may struggle to initiate open and candid conversations regarding sexual matters with their children due to cultural norms that deem discussions about sex taboo within the household. Information about sex is predominantly facilitated through the educational system, with teachers and peers playing a crucial role in the education of young individuals. Involving Puerto Rican parents in conversations regarding sexual health education can pose specific difficulties attributed to potential unease surrounding the subject matter. **Establish trust and cultivate an environment conducive to open communication. Offer guidance on engaging in conversations with children about safe sexual practices. Educate against sexually transmitted infections (STIs) and unintended pregnancies while underscoring the importance of privacy and consent. Ensure that their children's health information is kept confidential. When language becomes a barrier, it is advisable to use interpreters to facilitate clear communication.**

### *29.1.2 Family Roles and Organization*

- Traditional families may view women as lenient, submissive, and wanting to please men. **Accept family decision-making styles without judgment.**
- Traditional cultural norms discourage an overt sexual-being image for women, but many of these conventional values disappear with family assimilation. **Topics such as sex, sexually transmitted infections, or other infections establish an environment built on *confianza* and *personalismo*. Employ culturally sensitive terminology and refrain from presumptions or imposing your views.**

**Promote an environment conducive to open communication where clients are encouraged to articulate their opinions and share their concerns. Refer patients to specialty care or community support groups with specialized expertise in working within the Puerto Rican cultural context**.

- Most families expect their children to stay home until they get married or pursue a college education. The mother is expected to actively discipline, guide, and advise children. Most fathers expect to be consulted but mainly see themselves as financial providers. **Address family conflict within the context of the family to promote adolescents' mental health issues rather than using individual approaches.**
- Many traditional families socialize male children to be robust.
- Female children are socialized with a focus on home economics, family dynamics, and motherhood.
- Many Puerto Rican mothers use threats of punishment, guilt, and discipline, which can create stress and difficulties for adolescents as they struggle with the more permissive cultural patterns of the United States such as dating. **Provide counseling and explain US child abuse laws.**
- The family structure may be nuclear or extended. Family members include grandparents, great-grandparents, married children, aunts, uncles, cousins, and even divorced families with their children.
- Dependent older people are expected to live with their children and be cared for. Placements in nursing homes and extended-care facilities may be considered inconsiderate and cause family members to feel guilty and experience depression and distress. **Explore alternatives for long-term care and provide information to all family members involved in this decision-making process—address discharge planning and hospice and palliative care. Develop strategies for providing emotional support and assistance to family members caring for older adults.**
- A family member is expected to be at the sick person's bedside. **Ask for the name of the family spokesperson and document it in the patient's chart.**
- Some families may have difficulty educating young women about sexuality and reproductive health. Thus, adolescents depend on educational organizations to learn about sexual health, including body changes such as menstruation and the reproductive system. **Educating the family about sexuality issues gains respect and entrance into the trusted family environment.**
- Homosexuality is predominantly a taboo topic that carries a tremendous cultural stigma among most Puerto Ricans. Some men may perceive sexual intercourse with men as a sign of virility and sexual power rather than homosexual behavior. **Do not disclose same-sex relationships without the patient's consent. When providing care for individuals who identify as lesbian, gay, bisexual, or transgender, it is vital to inquire about their sexual orientation status, specifically whether they have openly shared this information (disclosed) or have chosen to keep it private (undisclosed)**.

### 29.1.3 Workforce Issues

- Some have a relativistic view of time and may not value regular attendance and punctuality in the workforce. **Explain the timeliness requirement in the dominant American (and other countries) workforce and any repercussions for chronic tardiness.**
- Most Puerto Ricans value personal relationships at work. Work is perceived as a place for social and cultural interactions, including listening to background music while performing job activities. This practice may lead to loud, cheerful, and noisy conversations requiring the employer's attention. **Explain the culture of the dominant American workforce etiquette and the allowable noise level.**
- For many women, family responsibilities, pregnancy, and the health of their children and other family members are prioritized over work. **Negotiate flexible work hours when possible.**
- Although most Puerto Ricans are bilingual, some speak broken or street English, such as ***Spanglish***. **Communicate composed and articulately, employing straightforward vocabulary and expressions that can aid in promoting comprehension**.
- Among younger generations and those born in the mainland United States, eye contact is maintained and is often encouraged. With the more traditional, limited eye contact is a sign of respect, especially with older people. **Ask questions and do not assume that the lack of eye contact means the person is not listening, caring, or being less than truthful.**

### 29.1.4 Biocultural Ecology

- The Puerto Rican population displays diverse skin tones, encompassing individuals with fair, olive, and darker complexions. The diversity observed can be attributed to the prevailing ancestral influences characterized by European, Spanish, African, and Native American physical traits.
- The facial features of individuals of Puerto Rican descent exhibit considerable diversity due to their varied ancestral backgrounds.
- Table 29.1 presents commonly occurring health conditions and diseases for Puerto Ricans in three categories: lifestyle, environment, and genetics. See Chap. 2.
- **Advise patients and families traveling to Puerto Rico to avoid exposure to endemic areas and always to use mosquito repellent and protective clothing. Become familiar with dengue fever's signs, symptoms, and treatment recommendations.**

**Table 29.1** Health conditions common to Puerto Ricans

| Health conditions | Causes |
|---|---|
| Heart disease | Genetic, environment, lifestyle |
| Lung cancer | Environment, lifestyle |
| Ovarian and cervical cancer | Genetic, environment, lifestyle |
| Asthma | Environment, lifestyle |
| Prostate cancer | Environment, lifestyle |
| Stomach and esophageal cancer | Environment, lifestyle |
| Pancreatic cancer | Environment, lifestyle |
| Diabetes mellitus | Genetic, environment, lifestyle |
| Dengue fever | Environment, lifestyle |

### 29.1.5 *High-Risk Health Behaviors*

- Puerto Ricans have shown high rates of alcoholism. **Use validated screening tools such as the brief version of the Alcohol Use Disorders Identification Test (AUDIT), to assess individuals for alcohol misuse. Recommend patients to specialized treatment programs, such as outpatient counseling, inpatient rehabilitation, or mutual support groups. Conduct evaluations on patients to ascertain the coexistence of comorbid mental health disorders that commonly co-occur with alcoholism such as depression or anxiety. Develop a regular schedule of consecutive sessions to methodically assess the progress of the individual, address any potential relapses, and make adjustments to treatment**.
- Puerto Rican men and women have a high smoking prevalence. **Assess smoking habits and promote programs aimed at helping individuals to quit smoking. It is beneficial for young children and adolescents to establish networks with schools, beginning in elementary school, to support healthy behaviors.**
- Lack of condom use may be a significant risk behavior. Some men fear that if they use condoms, they portray a less macho image, have decreased sexual satisfaction, or portray that they have a sexually transmitted infection or HIV. Some women may refuse a condom, believing the man thinks she is "dirty." **Provide choices such as contraceptive methods or pharmaceutical interventions**.

### 29.1.6 *Nutrition*

- Puerto Ricans celebrate, mourn, and socialize around food. Food is used (a) to honor and recognize visitors, friends, family members, and health-care providers, (b) as an escape from everyday pressures, problems, and challenges, and (c) to prevent and treat illnesses. Patients may bring homemade goods to healthcare providers to express appreciation, respect, and gratitude for services rendered. **Refusing food offerings may be interpreted as a personal rejection.**

- Being overweight is a sign of health and wealth. Some individuals eat to excess, believing that if they eat more, their health will be better. **Educate and negotiate a healthy weight with patients.**
- Diets include roots called *viandas*, vitamin-rich vegetables, and starches. The most common are celery roots, sweet potatoes, dasheens, yams, breadfruit, breadnuts, green and ripe plantains, green bananas, tanniers, cassava, and chayote squash or christophines. The dietary habits of Puerto Ricans often consist of processed foods while lacking in fresh fruits, vegetables, and whole foods. This dietary pattern may contribute to the development of many health disorders, including obesity, heart disease, and diabetes. In turn, it is crucial to evaluate Puerto Ricans' access to nutritious food options and consider providing referrals to social workers to address this issue. **Familiarize with some traditional foods and their nutritional content to assist families with dietary practices that integrate their conventional or preferred food selections.** Table 29.2 lists a variety of typical Puerto Rican food and ingredients.
- Many families believe a healthy child is *gordita* or *llenito* (diminutive for fat or overweight) and has red cheeks. **Educate mothers and other family members about healthy nutrition and the health risks for children who are overweight.**
- A traditional breakfast includes hot cereal such as oatmeal, cornmeal, or rice and wheat cereal cooked with vanilla, cinnamon, sugar, salt, and milk. Although less

**Table 29.2** Common Puerto Rican meals and fritters

| Puerto Rican meal | English translation |
|---|---|
| *Alcapurrias* | Green plantain fritters filled with meat or crab |
| *Arepas de maíz y queso* | Cornmeal and cheese fritters |
| *Arroz con pollo* | Rice with chicken |
| *Arroz con gandules* | Rice with pigeon peas |
| *Arroz blanco (con aceite)* | Plain rice (with oil) |
| *Arroz guisado básico* | Plain stewed rice |
| *Asopao de pollo* | Soupy rice with chicken |
| *Bacalaitos* | Codfish fritters |
| *Bocadillo* | Grilled sandwich |
| *Mondongo* | Tripe stew |
| *Lechon* | Roasted pig |
| *Paella de mariscos* | Seafood paella |
| *Pastelillos de carne, queso o pasta de guayaba* | Turnovers filled with meat, cheese, or guava paste |
| *Pasteles* | Traditional dish mainly served during the Christmas season |
| *Pollo en fricase con papas* | Stewed chicken with potatoes |
| *Relleno de papa* | Potato ball filled with meat |
| *Sancocho* | Viandas and meat stew |
| *Sofrito* | Condiment (1 tbsp) |
| *Surullo de queso* | Cornmeal fritters filled with cheese |

common, many eat corn pancakes or fritters for breakfast. Lunch is served by noon, followed by dinner at around 5 or 6 p.m.

- Rice and stew *habichuelas* (beans) are the main dishes. Rice may be served plain or cooked with many starch-rich side dishes. Rice cooked with vegetables or meat is considered a complete meal. *Arroz guisado* (rice stew) is seasoned with *sofrito*, a blend of spices such as cilantro, *recao* (a type of cilantro), onions, green peppers, and other nonspicy ingredients. Rice is cooked with chicken, pork, sausages, codfish, calamari, or shrimp. It is also cooked with corn, several beans, and *gandules* (green pigeon peas), a Puerto Rican bean rich in iron and protein.
- Various pastas, breads, crackers, vegetables, and fruits are eaten. Fritters are also common foods. Fried green or ripe plantains are a favorite side dish.
- Many ascribe to the hot–cold classifications of foods for nutritional balance and dietary practices during menstruation, pregnancy, the postpartum period, infant feeding, lactation, and aging. Table 29.3 identifies many of the foods that are

**Table 29.3** Puerto Rican hot–cold classification of health-illness status, medications, herbs, and selected foods

| Hot–cold classification | Health/illness status | Western medications | Traditional herbs | Foods |
|---|---|---|---|---|
| Hot | G.I. illnesses (constipation, diarrhea, Crohn's colitis, ulcer bleeding)<br>Gynecologic issues (pregnancy, menopause)<br>Skin disorders (rashes, acne)<br>Neurological disorders (headache)<br>Heart disease<br>Urologic illnesses | Syrups<br>Dark-colored pills<br>Aspirin<br>Anti-inflammatory agents<br>Prednisone<br>Antihypertensives<br>Castor oil<br>Cinnamon<br>Vitamins (iron)<br>Antibiotics | Teas:<br>Cinnamon<br>Dark-leaf teas | Cocoa products<br>Alcoholic beverages<br>Caffeine products<br>Hot cereals (wheat, corn)<br>Salt<br>Spices and condiments<br>Beans<br>Nuts and seeds |
| Cold | Osteomuscular illnesses (arthritis, rheumatoid arthritis, multiple sclerosis)<br>Menstruation<br>Respiratory illnesses | Diuretics<br>Bicarbonate of soda<br>Antacids<br>Milk of magnesia | Teas:<br>Orange-lemon, chamomile<br>Linden<br>Anise | Rice<br>Rice and barley water<br>Milk<br>Sugar and sugar products<br>Mint<br>Root vegetables<br>Avocado<br>Fruits<br>Vegetables<br>White meat<br>Honey<br>Onions |

considered either "hot" or "cold." **Becoming familiar with dietary practices facilitates the provision of expert recommendations for developing culturally sensitive food alternatives.**

- Understanding that iron is considered a "hot" food not usually taken during pregnancy can help healthcare providers negotiate approval. **Educate women about the importance of adhering to iron recommendations, even during pregnancy and lactation. Demonstrate recognition and reverence toward these beliefs while delivering evidence-based knowledge regarding the significance of sufficient iron consumption and viable avenues for obtaining iron-rich dietary options. Participating in inclusive and culturally aware dialogues can facilitate the reconciliation of disparities between cultural values and medical recommendations.**
- A summary of food habits, reasons for practices, and recommendations for health-care providers during such developmental stages is included in Table 29.4.
- Black cohosh, evening primrose, St. John's wort, ginkgo, ginseng, valerian root, sarsaparilla, chamomile, red clover, linden leaves, and passionflower are the most common herbs. **Discuss the safety and efficacy of the most frequently used alternatives. Include the use of complementary and alternative therapies in routine health assessments.**
- If the individual is believed to have low blood pressure and is weak or tired, brandy may be added to black coffee to enhance the work of "an old heart."
- **Exercise sensitivity toward folk practices as they can discourage patients from seeking further medical care and undermine their trust and confidence. This inquiry pertains to nutritional practices, focusing on integrating culturally sensitive meal choices or non-conflicting behaviors into dietary regimens.**
- During illness, chicken soups and *caldos* (broth) are used as a hot meal to provide essential nutrients.
- A honey, lemon, and rum mixture is used as an expectorant and antitussive. A malt drink, *malta* (grape juice), or milk is often added to an egg yolk mixed with sugar to increase the hemoglobin level. Ulcers, acid indigestion, and stomach discomfort are treated with warm milk, with or without sugar.
- Herbal teas are used to treat illnesses and to promote health. Most herbal teas do not interfere with medical prescriptions. **Incorporating herbal tea with traditional Western medicine may enhance treatment compliance.**

### *29.1.7 Pregnancy and Childbearing Practices*

- Due to the Catholic Church's endorsement of "natural" family planning methods and sexual abstinence, women often refrain from using contraceptives such as foams, creams, and diaphragms, as they are regarded as morally objectionable. Historically, abortion has been predominantly seen as unacceptable, with limited exceptions granted when the mother's life is at risk. However, it is worth noting

**Table 29.4** Puerto Rican cultural nutrition and health beliefs and practices during health stages

| Behavioral period | Dietary and health practices | Cultural justification | Recommendation for health-care providers |
|---|---|---|---|
| Menstruation | Food taboos: Avoid spices, cold beverages, acidic citric fruits and substances, chocolate, and coffee | It may induce cramps, hemorrhage clots, and physical imbalance. It may produce acne during menstruation | Assess individual beliefs and acknowledge them. Incorporate traditional beliefs with treatments as required in nonsteroidal anti-inflammatories for dysmenorrhea |
| | Foods encouraged: Plenty of hot fluids such as cinnamon tea and milk with cinnamon and sugar. Teas such as chamomile, anise seed, linden tea, mint leaves | Fluids encourage body cleaning of impurities. Hot beverages promote circulation and reduce abdominal colic, cramps, and pain. Teas are soothing to all body systems | Encourage passive exercise. Provide information about the role of exercise in the reduction of menstrual pain. Support other practices |
| | Health practices: Avoid exercise and practice good hygiene. Do not walk barefoot. Avoid wind and rain. Stay as warm as possible | Exercise may increase pain and bleeding. Good hygiene is important for health. Walking barefoot during menstruation may cause rheumatoid arthritis and other inflammatory diseases. Warm temperatures promote circulation and the health of the reproductive system, as well as prevent cramps | |

(continued)

| Behavioral period | Dietary and health practices | Cultural justification | Recommendation for health-care providers |
|---|---|---|---|
| Pregnancy | Food taboos: Hot foods, sauces, condiments, chocolate products, coffee, beans, pork, fritters, oily foods, and citric products | It may cause excess flatus, acid indigestion, bulging, and constipation. Chocolate and coffee may cause darker skin in the fetus. Some believe citric products may be abortive | Encourage healthy food habits. Provide information about chocolate and coffee myths. Encourage fruits<br>Discourage the use of raw eggs in beverages because of the possibility of salmonella poisoning |
| | Food encouraged: Milk, beef, chicken, vegetables, fruits, and ponches | Considered healthy and nutritious. Increase hemoglobin and strengthen and promote good labor | Encourage the use of food recommended for pregnancy. Provide information about sexual activity |
| | Health practices: Rest and get plenty of sleep. Eat plenty of food. Follow diet cautiously. Many avoid sexual intercourse early in pregnancy. Practice good hygiene and take warm showers | It enhances health and prevents problems during birth. Sex may cause issues with the baby or preterm labor | Encourage a balanced plan of exercise with emphasis on weight control and the health of the baby |
| Lactation | Food taboos: Avoid beans, cabbages, lettuce, seeds, nuts, pork, chocolate, coffee, and hot food items at all times | These foods cause stomach illnesses for the infant and mother, including baby colic, diarrhea, and flatus | Include a dietary plan that is balanced with substitute food items. Clarify any myths about infant diarrhea, colic, and flatus |
| | Food encouraged: Milk, water, ponches, chicken soup, chicken, beef, pastas, and hot cereals | Improve health and increase hemoglobin and essential vitamins. Protect mother and infant from illnesses. Fluids and ponches increase milk supply. Red meats reduce cravings | As above with raw eggs. Provide information about reasons for stroke and facial paralysis. Provide time to ask questions and reduce anxiety during winter season deliveries |
| | Health practices: Avoid cold temperatures and wind. A few may avoid showering for several days after birth. Great attention is paid to health of the mother | Cold temperatures and winds are believed to cause stroke and facial paralysis in a new mother. Showering may cause respiratory diseases. Mother is believed to be at risk and fragile | |

(continued)

**Table 29.4** (continued)

| Behavioral period | Dietary and health practices | Cultural justification | Recommendation for health-care providers |
|---|---|---|---|
| Infant feeding | Food taboos: Beans, too much rice, and uncooked vegetables | It is believed to cause stomach colic, flatus, and distended abdomen. Too much rice causes constipation | Provide information about appropriate dietary patterns for infant |
| | Foods encouraged: Hot cereals, ponches, chicken broth, or *caldos*. Fresh fruits, cooked vegetables, *viandas* (raw eggs, cereals, baby food in milk bottles). Fresh fruit juices. Mint, chamomile, and anise tea. Sugar and honey used for hiccups | Believed to be nutritious, healthy, and to decrease hunger. *Caldos* are fortifying and preventing illness. Cooked vegetables are healthy and avoid constipation. Bottle food fills the baby. Fresh juices and fruits refresh the stomach. Teas help babies sleep and cure flatus. Sugar and honey have curing properties | Instruct about infant diet and timely introduction of food items to diet. Explain the consequences of excessive weight in infants. Discourage food in a bottle to prevent choking. Discourage raw eggs because of the risk of salmonella and egg allergies and the use of honey because of the risk of botulism. Teas are harmless and provide additional fluid when used in moderation without sugar |
| | Health practices: Keep baby warm while feeding | Warm babies eat, chew, and digest food better, and choking is decreased | Provide information about choking and babies |
| Physical discomfort or illness | Health practices: Some persons may believe they can alleviate discomfort by relying on faith and engaging in prayer | Praying or invoking the Christian God to seek relief from physical discomfort or illness | Evaluate dietary practices, including habits. Encourage prayer if it brings comfort and peace, emphasize medical treatment's significance, and respect individuals' beliefs. Puerto Ricans' beliefs around worship have the potential to offer emotional and psychological benefits, which can have an indirect positive impact on their overall Well-being |

that this perspective is shifting, with an increasing number of women embracing this option.

- Many women tend to initiate prenatal treatment later in their pregnancies. **Encourage early prenatal visits and explain their importance in having a healthy pregnancy and newborn.**
- Men are socialized to be tolerant, understanding, and patient regarding pregnant women and their preferences. Women are encouraged to rest, consume large quantities of food, and carefully watch what they eat. Some individuals expect women to "get fat" and place little emphasis on weight control. Strenuous physical activity and exercise are discouraged, and lifting heavy objects is prohibited.
- Many women refrain from *tener relaciones* (having sexual intercourse) after the first trimester to avoid hurting the fetus or causing preterm labor.
- Loud and verbally expressive behaviors are culturally accepted and encouraged to cope with pain and discomfort. Pain medications are welcomed. Most women oppose having a cesarean section because it portrays a "weak woman." **Discuss the possibility of a cesarean section early in the pregnancy. Explain reasons for the necessity of invasive interventions during labor. Encourage using social support networks comprised of mothers who have undergone comparable experiences.**
- Women are encouraged to avoid exposure to wind and night lower temperatures, not to lift heavy objects, and not to do housework for 40 days after delivery. Some traditional women do not wash their hair during this time.
- Mothers are encouraged to drink fluids such as milk and chicken soup if breastfeeding. They drink milk or fresh juices mixed with raw egg yolk and sugar if they feel weak or tired. Hot foods such as chocolate, beans, lentils, and coffee are discouraged because they are believed to cause stomach irritability, flatus, and colic for the mother and infant.
- Some women believe that breastfeeding increases their weight, disfigures their breasts, and makes them less sexually attractive and undervalue the benefits of breastfeeding. Puerto Rican women exhibit inferior evidence in several health habits and outcomes, such as lower exclusive breastfeeding rates.
- **Provide information about breastfeeding practices and educate women about myths and misconceptions. The inclusion of grandmothers and significant others in breastfeeding programs is encouraged due to their impact on breastfeeding practices.**

### *29.1.8 Death Rituals*

- Death is perceived as a time of crisis. The body is considered sacred and is guarded with great respect.

- Give news about the deceased to the head of the family first, usually the oldest daughter or son. **Use a private room to communicate such notice and have clergy or ministers present when the news is disclosed. Allow time for the family to view, touch, and stay with the body before it is removed.**
- Burial rituals may be delayed until all close family members can be present. Cremation is rarely practiced.
- Among Catholics, religious ceremonies such as praying, *velorio* (wake), novenas, and 9 days of saying the rosary follow the death.
- Families freely express themselves through loud crying and verbal expressions of grief. Some may talk in a thunderous way to God. Others may express their sorrow through sensitive but continuous crying or sobbing. Some believe that not expressing their feelings could mean a lack of love and respect for the deceased. Some develop psychosomatic symptoms; others may experience nausea, vomiting, or fainting spells due to an *ataque de nervios* (nervous attack). **Be nonjudgmental with patients' expressions of grief. Provide a private environment to help minimize interruptions.**

### *29.1.9 Spirituality*

- Most Puerto Ricans are Catholic; however, many have joined Evangelical churches because they offer a more personal spiritual approach. Many individuals who adhere to religious beliefs often rely on the concept of a higher power and miracles to overcome challenging circumstances.
- A few individuals practice *espiritismo*, a blend of Native Indian, African, and Catholic beliefs that deal with rituals related to spiritual communications with spirits and evil forces. ***Espiritistas*** (individuals who communicate with spirits) may be consulted to promote spiritual wellness and treat mental illnesses. *Espiritistas* are often consulted to determine folk remedies compatible with Western medical treatments.
- Because older persons may consider illness a result of sins, a cure should be sought through prayer or by the "laying on of hands."
- **Inquire about the family's wishes regarding the Sacrament of the Sick. Clergy are a resource for spiritual wisdom and help with various spiritual needs.**
- Although amulets have lost their popularity, some still use them. An *azabache* (tiny black fist) or a rabbit's foot might be used for good luck, to drive away evil spirits, and to protect a child's health. Rosary beads and patron saint figures may be placed at the head or side of the bed or on the patient to protect them from outside evil sources. **Ask permission before removing, cleaning, or moving amulets. A benediction may be requested before removing amulets or religious objects, giving the Sacrament of the Sick, or providing spiritual support.**

### 29.1.10 Health-Care Practices

- Many Puerto Ricans underuse health promotion and preventive services such as regular dental or physical examinations and Pap smears. The variant characteristics of culture (see Chap. 1) influence health-seeking beliefs and behaviors.
- **Integrate individual, family, and community resources to encourage a focus on health promotion and enhance early health screening and disease prevention. Offering weekend, evening, and late-night healthcare services increases the use of preventive services.**
- Most Puerto Ricans believe in "family care" rather than self-care. Women are considered the primary caregivers and promoters of family health and are the source of spiritual and physical strength. The involvement of the family in the provision of care for individuals who are sick is a crucial aspect to consider. Given the cultural values held by Puerto Ricans, which prioritize the significance of family, faith, and religious practices, healthcare providers should exercise care when contemplating the possibility of referring families to social support services. **Acknowledge the importance of family in healthcare-related matters. Involve family members in discussions and decision-making processes.**
- Natural herbs, teas, and over-the-counter medications are often used as initial interventions for symptoms of illness.
- Patients who use folk practices visit ***botánicas*** (folk religious stores) and use natural herbs, aromatic incense, special bathing herbs, prayer books, prayers, and figurines for treating illness and promoting good health. **Refrain from making prejudicial comments that may inhibit collaboration with folk healers.**
- Most families prefer to keep chronically or terminally ill family members at home.
- Some place a damp cloth on the forehead to refresh the "hot" inside the body and relieve nausea. Some put the head between the legs to stop vomiting. Mint, orange, or lemon-tree leaves are boiled and used as tea to relieve nausea and vomiting.
- Barriers to using healthcare services include poor English language skills, low socioeconomic status, lack of insurance, and lack of transportation. Many individuals experience apprehension when requesting time off from their employment concern for childcare.
- **Provide interpreters and offer social work or transportation services if needed.**
- Puerto Rican males may experience stigma and unease when engaging in open discussions about sexual health, primarily due to cultural norms and societal expectations. Consequently, individuals may exhibit more remarkable reticence and less inclination to pursue knowledge and assistance about matters of sexual health. **Foster an environment that is inclusive and free from bias. Certain healthcare providers specialize in sexual health that can result in a higher level of expertise and sensitivity while engaging in discussions about sex-related matters.**

- Most Puerto Ricans tend to be loud and outspoken in expressing pain. Some prefer oral or intravenous medications for pain relief rather than intramuscular injections or rectal medications. Herbal teas, heat, and prayer are often used to manage pain.
- **Do not censure the expression of pain or judge it as an exaggeration. This expressive behavior is a socially learned mechanism to cope with pain. *¡Ay bendito!* is a common verbal moaning expression for *dolor* (pain).**
- **Rural elderly individuals might have difficulty interpreting and quantifying pain. The** use of numerical pain-identifying scales may be inappropriate.
- Because mental illness carries a stigma, obtaining information or talking about mental illness may be difficult. Some families might not disclose the presence or history of mental illnesses, even in a trusting environment. They might use phrases like "*estar bajoneado/*a" (feeling down) or "*tener los nervios mal o de punta*" (having bad nerves) as indirect ways to refer to these conditions. However, it's important to note that this changes as mental health awareness grows and more people openly discuss these issues. **Monitor the psychological well-being considering family dynamics and refer patients to support services as appropriate.**
- Symptoms of mental illness are often perceived as the result of *nervios* (nerves), having done something wrong, or failing God's commandments. When someone is anxious or overcome with emotions or problems, they are just *nervioso*. Similarly, someone experiencing a mental health concern such as melancholy, anxiety, or lack of sleep may be *nervioso(a)* or suffering from an *ataque de nervios* (nervous attack) rather than clinically depressed or mentally ill. **Demonstrate empathy and respect toward the individual's beliefs. Recognize and validate the cultural perspective of the individual while acknowledging and affirming their emotional state and worries. Communicate by adapting to the cultural understanding of the listener. Refrain from using technical terminology commonly used in the medical field. Provide psychoeducational information regarding mental health disorders, emphasizing the multifactorial etiology of these problems, such as biological, psychological, and socio-cultural elements**.
- **Acknowledge confidentiality when obtaining healthcare history. If trust is built, health-care providers may get a more accurate response to their questions**.
- **Promote physical and mental health by encouraging the active involvement of community-based churches, schools, or childcare centers**.
- Organ donation is considered an act of goodwill and a gift of life. Autopsy may be regarded as a violation of the body. **When discussions regarding autopsies and organ donations are necessary, proceed patiently and provide precise and straightforward information. A priest or minister is often helpful and may be expected to be present at the time of death.**

- Many individuals are reluctant to receive or donate blood for fear of contracting HIV.
- **Explore beliefs about blood donation and transfusions to dispel myths.**

### *29.1.11 Health-Care Providers*

- Puerto Ricans hold healthcare providers in high regard. Distrust may develop if the healthcare provider (a) lacks respect for issues related to traditional health practices, (b) ignores *personalism* in the relationship, (c) does not use advanced technological assessment tools, and (d) has a physical or personal image that differs from the traditional "well-groomed white attire" image.
- Puerto Ricans may use traditional and folk healers such as *espiritistas* and *santeros* with Western healthcare providers. Some *espiritismo* practices are used to deal with the power of good and evil spirits in the physical and emotional development of the individual. ***Santeros***, individuals prepared to practice *santería*, are consulted in matters related to the belief of object intrusion, diseases caused by evil spirits, the loss of the soul, the insertion of a spirit, or the anger of God.
- Traditional therapeutic practices like *curanderismo*, which mix herbal medicine, spiritual healing, and energy therapy, are just one example. Natural treatments, such as herbs and plants, are part of the healing practices employed by *curanderos* and *curanderas*. Another way that people deal with the disease is through the spiritual cleansing practice. Many Puerto Ricans put faith in the healing powers of spiritual purification. This may include using holy water or smudging negative energy away with herbs like sage or palo santo. **Encourage patients to disclose any herbal remedies they are using alongside conventional treatments. Monitor the use of herbal remedies. Some herbs or vitamins may interact with prescribed medications or cause side effects: healthcare providers monitoring their use is essential for patient safety.**
- Men prefer male physicians and may feel embarrassed and uncomfortable with a female physician.
- Sometimes, individuals may discount female physicians' academic and intellectual competencies and distrust their judgment and treatment.
- Some women feel uncomfortable with a male physician; a few prefer a male doctor.
- Older people may prefer more aged healthcare providers because they are considered wise and mature in health, life experiences, and folk practices.
- Some people experience adverse societal perceptions when seeking help for mental health issues. **Create a clinical environment that promotes support and inclusivity, acknowledging and valuing Puerto Ricans' unique characteristics and cultural diversity. Use a systematic approach for facilitating**

**appropriate referrals within mental healthcare. Integrate comprehensive health education initiatives that promote the normalization of mental healthcare and well-being, emphasizing the importance of seeking help when experiencing negative emotions.**

## 29.2 Reflective Exercises

1. Identify three common health conditions among Puerto Ricans. What educational recommendation would you make to decrease them?
2. Explain the "hot" and "cold" theory of foods and illnesses. If you are unfamiliar with the foods, what might you do to learn more about them?
3. Describe Puerto Rican communication practices, including temporality and punctuality.
4. What are men's and women's roles in the traditional Puerto Rican culture?
5. What are acceptable fertility control practices for traditional Puerto Ricans? How might they view the use of condoms?
6. Explain Santería. For what conditions would a Puerto Rican cult a practitioner of Santería?
7. What are some traditional practices for pregnancy?
8. Does the sex of the provider raise any concerns? If yes, please explain the stated reasons.
9. What might you do to get allopathic and traditional practitioners to work together?
10. Identify the death and funeral practices of traditional Puerto Ricans.

## Bibliography

American Lung Association (2010) Hispanics. http://www.lungusa.org/stop-smoking/about-smoking/facts-figures/hispanics-and-tobacco-use.html

Barcelona de Mendoza V, Harville E, Theall K, Buekens P, Chasan-Taber L (2016) Acculturation and intention to breastfeed among a population of predominantly Puerto Rican women. Birth 43(1):78–85. https://doi.org/10.1111/birt.12199

Caraballo RS, Yee SL, Gfroerer J, Mirza SA (2008) Adult tobacco use among racial and ethnic groups living in the United States, 2002–2005. Prev Chronic Dis 5(3):1–9

Centers for Disease Control and Prevention (2024a) Fast Stats Hispanic Latino Health. https://www.cdc.gov/nchs/fastats/hispanic-health.htm

Centers for Disease Control and Prevention (2024b) Health disparities in Hispanic or Latino people. https://www.cdc.gov/health-disparities-hiv-std-tb-hepatitis/populations/hispanic-latino.html

CIA (2024) World Factbook: Puerto Rico. https://www.cia.gov/the-world-factbook/countries/puerto-rico/

Diaz L (2012) A structuralist analysis of Puerto Rican Santería. http://future-nonstop.org/c/b684d4ce0746140c5c08714153843d32

Díaz DHS, Garcia G, Clare C, Su J, Friedman E, Williams R, Vazquez J, Sánchez JP (2020) Taking care of the Puerto Rican patient: historical perspectives, health status, and health care access. MedEdPORTAL 16:10984. https://doi.org/10.15766/mep_2374-8265.10984
Dominguez K, Penman-Aguilar A, Chang MH, Moonesinghe R, Castellanos T, Rodriguez-Lainz A, Schieber R, Centers for Disease Control and Prevention (CDC) (2015) Vital signs: leading causes of death, prevalence of diseases and risk factors, and use of health services among Hispanics in the United States—2009-2013. MMWR Morb Mortal Wkly Rep 64(17):469–478
Echenique M, Melgar L (2018) Mapping Puerto Rico's hurricane migration with mobile phone data. Bloomberg. https://www.bloomberg.com/news/articles/2018-05-11/where-puerto-rico-s-residents-migrated-since-maria
Kondo KK, Rossi JS, Schwartz SJ, Zamboanga BL, Scalf CD (2016) Acculturation and cigarette smoking in Hispanic women: a meta-analysis. J Ethn Subst Abus 15(1):46–72. https://doi.org/10.1080/15332640.2014.1002878
Llibre-Guerra JJ, Jiménez-Velázquez IZ, Llibre-Rodriguez JJ, Acosta D (2020) The impact of COVID-19 on mental health in the Hispanic Caribbean region. Int Psychogeriatr 32(10):1143–1146. https://doi.org/10.1017/S1041610220000848
Moslimani M, Lopez MH, Noe-Bustamante L (2023) 11 Facts about Hispanic origin groups in the U.S. https://www.pewresearch.org/short-reads/2023/08/16/11-facts-about-hispanic-origin-groups-in-the-us/
Munet-Vilaro F (1998) Grieving and death rituals of Latinos. Oncol Nurs Forum 25(10):1761–1763
Noland C (2006) Listening to the sound of silence: gender roles and communication about sex in Puerto Rico. Sex Roles 55:283–294. https://doi.org/10.1007/s11199-006-9083-2
Puerto Rico (2016) Countries and their culture. http://www.everyculture.com/No-Sa/Puerto-Rico.html
Puerto Rico (2024) Family culture. http://knowpuertorico.weebly.com/family-culture.html
Puerto Rico (2024) Non-verbal communication. http://knowpuertorico.weebly.com/-non-verbal-communication.html
Purnell L (2013) People of Puerto Rican heritage. In: Purnell L (ed) Transcultural health care: a culturally competent approach, 4th edn. F.A. Davis, Philadelphia
SAMHSA (2024) Alcohol Use Disorders Identification Test (AUDIT). https://www.samhsa.gov/resource/dbhis/alcohol-use-disorders-identification-test-audit. Accessed 29 Aug 2023
Scaramutti C, Salas-Wright CP, Vos SR, Schwartz SJ (2019) The mental health impact of hurricane Maria on Puerto Ricans in Puerto Rico and Florida. Disaster Med Public Health Prep 13(1):24–27. https://doi.org/10.1017/dmp.2018.151
Tang H, Choudhry S, Mei R, Morgan M, Rodriguez-Cintron W, Burchard EG, Risch NJ (2007) Recent genetic selection in the ancestral admixture of Puerto Ricans. Am J Hum Genet 81(3):626–633. https://doi.org/10.1086/520769

# Chapter 30
# People of Russian Heritage

**Tatayana Maltseva and Irina Rozenfeld**

## 30.1 Overview and Heritage

The Russian Federation (Russia), the largest country in the world with a population of 141.5 million, is composed of 21 republics, several regions, and multiple provinces. Russian geography consists of two continents: Asia and Europe. The multicultural diaspora of Russians shows the diversity of acculturation, individuality, and the variant characteristics of culture (see Chap. 1) as well as the physical size of the country. There are approximately around 2.9 million Russian Americans living in the United States. This chapter focuses on Russians who are immigrants to the United States. Many Russians are very well educated. Given the complexity of Russian immigration, most of the information in this chapter pertains to people of Russian heritage who emigrated in the latter part of the twentieth century and at the beginning of the twenty-first century.

Everyone could obtain higher education in the former Soviet Union. However, upon arrival to the United States, many highly qualified and educated Russian immigrants have difficulty in practicing their profession and securing employment in their specialties. This contributes to diminishing their self-esteem.

T. Maltseva (✉)
Nicole Wertheim College of Nursing and Health Sciences, Florida International University, Miami, FL, USA
e-mail: bogopolt@fiu.edu

I. Rozenfeld
Dr. Kiran C. Patel College of Osteopathic Medicine, Nova University, Fort Lauderdale, FL, USA
e-mail: rozenfel@nova.edu



E. A. Fenkl, L. D. Purnell (eds.), *Handbook for Culturally Competent Care*,
https://doi.org/10.1007/978-3-031-70492-5_30

### 30.1.1 Communications

- The official language of Russia is Russian. However, most educated Russians in the United States speak English to some extent because scientific literature in Russia is published in English.
- Many Russians do not understand medical language and have difficulty communicating abstract concepts or understanding American dialects. **It is recommended to use a professional interpreter and speak slowly and clearly with a medium tone of voice and refrain from a loud conversation when communicating with Russian immigrants.**
- Older Jewish immigrants speak **Yiddish**. Younger Jewish immigrants usually do not speak Yiddish because of lack of resources in learning Yiddish in Russia.
- Punctuality is a cultural standard, and many Russians arrive early for their scheduled appointments. It is customary for Russian immigrants to take medications until symptoms disappear and save the remaining for future use. Russians prepare food and essential supplies for current and future usage. It is common for them to have an abundant stock of items for many days and weeks ahead. It is also customary for Russian immigrants to borrow medications from their Russian friends and neighbors. **Explain the necessity of taking all medications on a regular basis as prescribed, emphasize that some prescriptions can be refilled, and use only medications prescribed to them.**
- Direct eye-to-eye contact is maintained among family and friends. Speech is usually clear and concise in most situations. Most Russians prefer a clear action plan instead of an uncertain dialog without a plan or goals.
- **Eye contact is not usually maintained when speaking with government officials or authority figures. Direct eye contact during meetings and communicating with authority figures may be perceived as an invasion of privacy and may feel uncomfortable during the conversation.**
- Russians accept touch in healthcare settings but prefer not to be hugged or kissed by strangers. A handshake is acceptable.
- Russians do not appreciate nonverbal gestures such as standing with hands inserted into pockets or arms crossed over the chest; neither do they appreciate slouching postures when being interviewed. Many Russians prefer to seek care from Russian-speaking healthcare providers. They are prone to seek multiple consultations from different providers until they develop trust with a provider.
- **Address Russians by their last name and title such as Mr., Mrs., Miss, Ms., or Dr. You can greet patients with a handshake.**

### 30.1.2 Family Roles and Organization

- Family, children, and older adults are highly respected and valued. It is customary for Russians to have extended family living together for communication and support.

- Decision-making among current immigrants is usually democratic. They prefer to consult with their parents and oldest children in the decision-making process. **Ask patients who they want to be a healthcare surrogate and include them in medical decisions.**
- It is common in Russian culture for grandparents to take care of their grandchildren while their parents are at work. **Include grandparents and older family members in health education for their grandchildren.**
- Older parents live with their children when self-care is a concern. It is customary for Russians not to place their elderly parents in long-term care facilities. Russians ask for help from friends and other family members to take care of their aging parents.
- **Help families to prevent caregiver burden. Provide resources about long-term care facilities if needed and encourage the family to visit the facility before a decision is made to place a family member there.**
- Children of all ages are expected to succeed in school, obtain a higher education, help care for older family members, and do household chores in a timely matter.
- Adolescents are not expected to engage in sexual activity. Sexual and contraceptive education is still lacking in Russia. Acculturation to the American culture and lack of traditional family roles may predispose young Russian immigrants to high-risk behaviors, unprotected sexual activities, substance usage, and unexpected pregnancy.
- Divorce is accepted without stigma. Same-sex marriages are not supported by Russians even though the Russian penal code was revised in 1997 and homosexuality is no longer considered a crime. However, even now, many hide their homosexuality from the family in fear of rejection.
- **Do not disclose sexual orientation, particularly same-sex orientation, to family members.**

### *30.1.3 Workforce Issues*

- Russians practice positive politeness, respect, and accountability in the workplace. They value time and support the rule of work and society. Not showing up for an appointment or to work without prior notification is considered disrespectful in Russian culture.
- When negotiating work contracts or project implementation, Russians express their passion for completing the project and invest considerable time and effort in supporting decisions.
- With colleagues and friends, Russians communicate directly, which is considered a sign of respect and sincerity.
- Russians may practice non-direct eye contact when speaking with government officials, a practice common in Russia where making eye contact with government officials could lead to questioning, invasion of personal space, or disrespect. **Do not assume that the lack of eye contact means the person is not paying attention, does not care, or not providing correct information.**

- The tone of voice may be loud, as Russians prefer to prove their point of view by articulating loudly. **Do not interpret a loud voice volume as a sign of anger.**

### *30.1.4 Biocultural Ecology*

- Russians who have immigrated to the United States and elsewhere are predominately white Caucasians and have a physical structure like other white Americans, making them prone to skin cancer.
- **Encourage the use of sunblock and instruct all family members about the dangers of skin cancer.**
- Table 30.1 lists commonly occurring health conditions for Russians in three categories: lifestyle, environment, and genetics.

### *30.1.5 High-Risk Health Behaviors*

- Both men and women have high smoking rates. **Encourage smoking cessation and assist with finding smoking cessation programs.**
- Molecular genetic risk factors for alcohol dependence are associated in the Russian population with increased alcohol consumption, particularly in Russian males.
- Intimate partner violence is common in Russia and is related mostly to high rates of alcohol consumption. Russian Americans tend not to report domestic violence to authority figures and are reluctant to seek services for domestic violence.
- **Explain support services available for intimate partner violence and help patients gain access to the services.**
- **Screen all newer immigrants for conditions common in their home country. Provide scientific information and assist in finding services and resources.**

**Table 30.1** Health conditions common to Russians

| Health conditions | Causes |
|---|---|
| Alcoholism | Genetic, environment, lifestyle |
| Depression | Environment, lifestyle |
| Gastrointestinal disorders | Environment, lifestyle |
| Respiratory diseases | Environment, lifestyle |
| Cardiovascular diseases | Genetic, environment, lifestyle |
| Cancer due to radiation | Environment |
| Dental disease | Environment, lifestyle |
| Tuberculosis | Environment, lifestyle |
| Diabetes mellitus | Genetic, environment, lifestyle |
| Hyperlipidemia and obesity | Genetic, environment, lifestyle |

- Many who come from Ukraine, Byelorussia, and the Western part of Russia were exposed to the radiation effects of the Chernobyl disaster, resulting in a high incidence of thyroid cancer, respiratory problems, cardiovascular diseases, depression, and post-traumatic stress disorder (PTSD).

### *30.1.6 Nutrition*

- Common foods in the Russian culture include a variety of dairy products, marinated or pickled vegetables, salads from fresh vegetables with sour cream and mayonnaise, soup made from beets (borscht), cabbage, buckwheat, potatoes, yogurt, soups, stews, and hot tea and milk with honey. Cold and iced drinks are not preferred.
- Meat choices include beef, pickled herring, smoked fish, anchovies, sardines, cold tongue, chicken, ham, sausage, and salami. Bread is a must with every meal. The diet overall is high in fat, salt, and cholesterol.
- **Determine preferred foods and preparation practices before providing dietary counseling. Incorporate traditional foods into dietary modifications. Encourage fruits and vegetables.**

### *30.1.7 Pregnancy and Childbearing Practices*

- Marriage and childbearing are part of the Russian culture. Contraception for Russian women is allowed without sanctions or taboos. Currently, the most popular contraceptive methods among Russian families are condoms, followed by intrauterine devices and birth control pills. Russian males do not consider vasectomies as a method of contraception. Many new immigrants may not be aware of different methods of fertility control. The abortion rate has decreased in Russia since the introduction of family planning programs by the Russian Ministry of Health.
- **Inform patients of the variety of common birth control methods.**
- Pregnant women have regular prenatal checkups, which are mandatory in Russia. During pregnancy, women are encouraged to rest. They do not perform heavy lifting, hard work, and strenuous physical activities. They are also protected from bad news or unpleasant information that can be harmful to the fetus. Pregnant women are encouraged to eat foods that are high in iron, calcium, and vitamins. Strawberries, citrus fruits, peanuts, and chocolate are avoided to prevent allergies in the newborn.
- As labor approaches, women take laxatives and enemas to facilitate delivery. Traditionally in Russia, husbands and relatives do not participate in the delivery or visit the hospital postpartum. There are no cultural restrictions for fathers or

female relatives not to participate in delivery in the United States and elsewhere. **Ask patients who they wish to be involved in the delivery.**
- Breastfeeding is encouraged as Russians believe that breastmilk will increase the immune system of the newborn. Some women breastfeed until the infant reaches the toddler stage.
- Peri care with warm water is important. A binder is worn to help the mother's figure return to normal.
- In Russia, women receive 8 weeks of maternity leave before delivery and up to 3 years of leave after newborn delivery. **Explain family medical leave practices for expectant parents.**

### *30.1.8 Death Rituals*

- Russians' families protect a terminally ill family member. They want to be notified about imminent death first before the patient is told.
- Most families prefer to care for the dying family member at home.
- **Facilitate transferring the dying person to the home if the family wishes and has the means to provide terminal care. Assist with obtaining hospice home care.**
- Russian families believe that everything should be done to save their loved ones and often will not agree to disconnect the life support. They will not discuss advance directives, living will, durable power of attorney, or do not resuscitate procedures. They might not sign these documents. A Russian Orthodox pastor may perform a special prayer, hymn, or vigil over the deceased.
- **Ask permission from the family before contacting clergy.**
- Many Russian families do not believe in cremation; most prefer interment.
- Both men and women may wear black as a sign of mourning. All mirrors are covered in the deceased's home. Expression of grief varies greatly. Russian Orthodox families mourn the death of a family member for at least 40 days. **Recognize a wide variety of grieving and bereavement practices among Russians.**

### *30.1.9 Spirituality*

- Russians are closely associated with the Orthodox Christian Church. The Russian Federation recognizes Orthodox Christianity, Buddhism, Islam, and Judaism as established religions.
- Russians in the United States and elsewhere may or may not practice a religion or call themselves atheists. **Discuss individual religious practices and spirituality with patients and family during the interview.**

### *30.1.10 Health-Care Practices*

- Most Russians define health as the absence of disease.
- Russian immigrants might not be familiar with the healthcare insurance system in the United States because Russia has a public healthcare system. Some clinics and hospitals in Russia are private, where patients pay out of pocket for their treatment. **Explain health reimbursement procedures and elicit assistance from case managers to obtain health insurance coverage as needed.**
- Hospitalization lasts in Russia between 3 and 4 weeks. Some patients may expect the same on immigration. **Explain hospitalization routines. Upon discharge refer patients to home healthcare agencies if needed and teach the family how to care for recuperating family member at home.**
- Single females do not do Pap tests in Russia; only married women perform them in women's center clinics. **Explain the significance of Pap tests and mammography, regardless of the woman's marital status.**
- Many individuals prefer staying warm to prevent colds and other illnesses. Most like to keep the temperature warm in their rooms. Russians may also be unwilling to apply ice or drink cold drinks with ice. **Explain the application of ice compress in exact terms. Ask if the patient wants ice in their drinks.**
- Most Russians consider pain as a sign of weakness and may not ask for pain medicine.
- **Offer and encourage pain medicine and explain that it will help with the healing process.**
- Many Russians fear having X-ray or CT scan because of exposure to radiation. **Explain the necessity of X-rays and CT scan procedures and that healthcare facilities use minimal radiation exposure.**
- Patients are not comfortable being told about terminal diagnoses or cancer because it is believed to make the condition worse. **Inform patients about poor prognosis gradually, on their terms, and preferably in several meetings. Consult with family members first and pay attention to their verbal and nonverbal cues for readiness to hear news about a terminal illness.**
- Cupping is a common procedure in Russia for the treatment of respiratory infections in health care settings and at home. **Do not mistake circular ecchymotic areas on children's or adults' back as abuse. Ask if they have been practicing cupping.** Alcohol-saturated cotton or other materials may be introduced into the cup. Cotton or another material is lighted inside the cup and removed; then, the hot cup is turned upside down on the patient's back. The skin is drawn into the cup by vacuum. The cup is pulled off the skin within 10–15 min, leaving round ecchymotic areas 1–1½ in. wide.
- Common cultural practices include using herbal remedies and tinctures available from pharmacies, soaking one's feet in warm water for a cold, gargling with salt water for a sore throat, aromatherapy for a variety of respiratory illnesses, mud and mineral baths to promote healing, and herbs and teas for fever and colds. Russians are accustomed to using sauna called *banya* and massage therapy to

prevent respiratory illnesses and to facilitate relaxation. **Incorporate nonharmful folk practices into allopathic prescriptions.**

- A national Russian tradition is to collect mushrooms and berries from the forest. Russians practice the preservation of wild mushrooms and berries. This may contribute to botulism from mushroom preservation and diabetes from high sugar concentration in berries preserve. **Assess traditional Russian practices of wild mushrooms and berries. Educate Russians about the prevention of botulism and diabetes.**
- Most Russians are accustomed to seeing multiple healthcare providers. Each provider may give prescriptions. Patients rarely inform the provider about seeing other providers. **Specifically ask patients if they are seeing other healthcare providers and taking any prescription medications, over-the-counter medications, or traditional herbal remedies and teas.**
- Russians will not disclose mental health issues unless they are experiencing a crisis. They might present with vague somatic complaints because mental illness carries a significant stigma and punishment in Russia. Many Russians consider mental illness as a weakness. **Screen for anxiety, depression, and post-traumatic stress disorder in immigrants.**
- Russians may have a great fear of contracting HIV from blood transfusions. Most Russians do not believe in organ donation. **Explain safety procedures with blood transfusions and organ donation.**

### *30.1.11 HealthCare Providers*

- Healthcare providers are very well respected by Russian immigrants. Because nurses function in higher roles in the United States and some other countries, Russian immigrants may view nurses as physicians. **Explain the roles and education of nurses.**
- Men and women are accustomed to living together in very small apartments or communal living; thus, most do not have a problem with privacy. Healthcare professionals' gender is not generally a concern in providing care.
- **Always ask about gender concerns before performing breast pelvic or rectal exams on patients.**

## 30.2 Reflective Exercises

1. What are traditional verbal and nonverbal communication practices in Russian culture?
2. How do Russians value their time and punctuality?
3. What are the decision-making practices of current Russian immigrants?

4. List one health care condition in Russians in each of the three categories: environment, lifestyle, and genetics. What can you do to decrease the incidence and prevalence of these diseases?
5. What are common high-risk health behaviors among Russian immigrants? Include some strategies for reducing them.
6. What are fertility practices among Russians? What education would you provide to the family about childbirth and contraception practices?
7. Many Russians are afraid of having X-rays and CT scans. What is the reason for this? What can you do to decrease their fear?
8. Identify a traditional treatment that Russians use for respiratory symptoms. How would you know that this treatment is being used?
9. Identify some practices women use during pregnancy.
10. Identify some cultural and traditional Russian practices. Which of them might have harmful effects?

## Bibliography

Ameredia Integrated Cultural Market (2023) Russian American demographics. https://www.ameredia.com/resources/demographics/russian.html

Eckemoff EH, Sudha S, Wan D (2018) End of life care for older Russian immigrants—perspectives of Russian immigrants and hospice staff. J Cross Cult Gerontol 33:229–245. https://doi.org/10.1007/s10823-018-9353-9

Handtke O, Schilgen B, Mösko. (2019) Culturally competent healthcare—a scoping review of strategies implemented in healthcare organizations and a model of culturally competent healthcare provision. PLoS One 14:e0219971. https://doi.org/10.1371/journal.pone.0219971

Maltseva T (2021) People of Russian heritage. In: Purnell LD, Fenkl EA (eds) Textbook for transcultural health care: a population approach, 5th edn. Springer, Cham

Nersessova KS, Jurcik T, Hulsey TL (2019) Differences in beliefs and attitudes toward depression and schizophrenia in Russia and the United States. Int J Soc Psychiatry 65(5):388–398. https://doi.org/10.1177/0020764019850220

Orekhovskaya NA, Chistyakov AA, Kryukova NI, Krokhina JA, Ospennikov YV, Makarova EV (2019) Orthodoxy and modernity their contact facets in Russian society. Eur J Sci Theol 15(2):67–77

Stoeckl K (2020) The rise of the Russian Christian right: the case of the world congress of families. Relig State Soc 48(4):223–238. https://doi.org/10.1080/09637494.2020.1796172

U.S. Census Bureau (2023) Russia—demographic data as of July 1, 2023. https://www.census.gov/popclock/world/rs

# Chapter 31
# People of Somali Heritage

**Trudy Gaillard and Tamara LaCroix**

## 31.1 Overview and Heritage

Somalia is approximately 246,200 miles$^2$ (637,658 km$^2$) and has a population of over 17 million. Somalia occupies the "horn" of Africa, a large land mass jutting into the Indian Ocean just south of the Arabian Peninsula, wrapping around Ethiopia and Kenya to the west. With a long coastline on the Indian Ocean and the Gulf of Aden, which leads into the Red Sea and Suez Canal, Somalia has a long trade and interaction with other cultures. Most of the country is arid and sparsely settled by nomadic herders. The southern part has a tropical climate supporting agriculture, and cultivators settled in urban communities, including Mogadishu, the large capital city.

The Somali Democratic Republic is the formal name of this African country, also known as Somalia in short form. Somali is the predominant language with several dialects in different coastal regions. English and Italian are used in small urban areas. Arabic is used in religious contexts. Introduced by the Arabs in the seventeenth century, the official religion is Islamic with an overwhelming majority of Sunni Muslims, less than 1% of the population are Christians. Arabic neighbors greatly influence the culture in Somalia. However, as in many countries, tribal allegiance is primary. Social stratification is based on classes and castes. The Samaal clan-family believes that they are superior to the Saab. In turn, the Saab developed a caste system that awards status to different groups according to their heritage or occupation. In the twentieth century, education, income, and the ability to speak foreign languages were the standards by which status was attained in urban Somalia.

T. Gaillard (✉) · T. LaCroix
Nicole Wertheim College of Nursing and Health Sciences, Florida International University, Miami, FL, USA
e-mail: tgaillar@fiu.edu; tlacroix@fiu.edu



E. A. Fenkl, L. D. Purnell (eds.), *Handbook for Culturally Competent Care*,
https://doi.org/10.1007/978-3-031-70492-5_31

After the collapse of the Somali government in 1991, at least one million Somali people fled to neighboring counties such as Ethiopia, Djibouti, Yemen, and especially Kenya. Many refugees have resettled in the United States (US), Canada, the United Kingdom, Holland, Sweden, Denmark, Germany, Australia, and New Zealand.

Immigrants to Western countries were initially scattered, but many have joined family and clan members in urban centers throughout the US, especially Minneapolis–St. Paul, Minnesota. Many have built prosperous lives, opened businesses, and taken advantage of educational opportunities. In some Minnesota high schools and community colleges, more than half the students are Somali. Residual conflict persists in their new homes between members of different clans still at war in Somalia. Diversity among Somalis varies according to acculturation and the variant characteristics of culture (see Chap. 2).

### *31.1.1 Communications*

- Somali is the universal language understood by all Somali people despite minor differences in dialect. Many people speak Arabic. A few older and educated Somalis speak Italian or English, depending on experiences with former colonial powers. In 1972, during a literacy campaign, the Somali language began to be written using phonetic spelling and the European alphabet and taught in primary schools. Older Somalis are generally not able to read this language. **Upon admission, determine language preference, Somali, Arabic, or English, and obtain an interpreter if necessary.**
- Somalis are polite and appreciate complete introductions of everyone in the room, including interpreters and family members. **Formally introduce yourself by title and purpose.**
- Women and men are not shy about obtaining medical care and often ask detailed questions until satisfied. This may include challenging questions like “Why is this medicine not working?” However, they are almost always respectful and usually conclude a medical or nursing interaction by expressing thanks.
- **Explain the purpose of every intervention.**
- Somalis dress up for medical appointments. Women and female children as young as 7 or 8 generally wear the Muslim headscarf and dress discreetly, often with a strong sense of color and fashion. Men dress in slacks and shirts with good-quality leather belts and shoes. One rarely encounters sloppy or worn clothing. **Sloppy or worn clothing may be a symptom of other problems besides the reason for the appointment.**
- Traditional Somalis follow Muslim traditions regarding social touching. A man or woman may not touch someone of the opposite sex except for spouses or close relatives. However, Somalis are practical people; in the exam room, it is under-

stood that touching is permitted for a specific reason. **Because shaking hands when greeting a patient of the opposite sex is inappropriate, saying hello while putting your hands in your pocket or clasping them together puts a patient at ease.**

- Many Somali patients are uncomfortable having a genital exam by a provider of the opposite sex. **Provide a same-sex provider for intimate exams.**
- Eye contact and voice volume are not noticeably different from the European American cultural norms. If a patient is timid and soft-spoken, the explanation may be depression or post-traumatic stress disorder rather than culture.
- **Smiling, informality, expressing friendship, and engaging in humor by the healthcare provider are acceptable.**
- Naming is by the usual Muslim patronymic system. Both girls and boys are given a specific name at birth. This is followed by the name of their father (for both sexes) and then the name of their grandfather. Jibrill Ibrahim Mohamud would be a boy named Jibrill whose father was Ibrahim and grandfather was Mohamud.
- Punctuality is not expected in a social context. Still, Somali people are generally quick to adapt to new cultures and learn that they are expected to be on time for appointments. **Explain the necessity of being on time for appointments.**

### *31.1.2 Family Roles and Organization*

- A Somali nuclear family includes a father, mother, children, and grandchildren. The father heads the family and makes the final decision in family affairs. Family relations are based on hierarchal respect, as the **Qur'an** expects children to obey their parents unconditionally. Children are also likely to care for their parents when needed, especially during sickness, aging, or financial hardship. There was no social security or pension back home in Somalia. **Accept family dynamics and hierarchy without judgment.**
- The father is the leading income producer and manager of family finances. Other family members, including his wife, their grown children, and sometimes grandchildren work and help as they can. The mother is a homemaker in cooking, cleaning, shopping, and caring for children. She is revered and respected in this role, even more so than the father. **Accept gender roles without judgment.**
- When children reach the age of marriage, the father has the final say regarding whom and when his daughters marry. When a son matches, his father participates in decision-making and the marriage ceremony but does not have the same level of authority as the bride's father.
- Marriage in Somalia was an entirely religious ceremony conducted by a religious official versed in *Sharia* law with two male witnesses and male members of the bride's family. Civil authorities sometimes issued certificates of marriage, but these were not essential.

- The husband can declare divorce under Sharia in the presence of two male witnesses. A wife must apply to a court for judgment if she wishes to divorce. In rural areas, she may be represented by her male relatives instead. Grounds for divorce include lack of financial support or sexual impotence. These traditions do not coincide with Western civil law. Confusion and misunderstanding can occur, especially among newly arrived immigrants. **Carefully explain the legalities of divorce in the US and elsewhere.**
- Extended families are critical. Most Somalis in the US help support their family members in Somalia or refugee camps in neighboring countries.
- **Include extended family in health teaching.**
- Clan relationships are essential. Wealthy and productive clan members are expected to help others, especially in the event of significant illness or legal trouble. The clan leader administers it and persists in Western countries.
- The family is becoming less cohesive in the US. Adolescents are less obedient to their parents. Fathers may believe they are losing their traditional positions as providers and heads of the family when two incomes are needed to survive. Mothers may feel long work hours interfere with their role as homemakers. Large families add to financial stress. These factors and the American culture of individualism are loosening the tight relationships and traditional roles characteristic of the collectivistic Somali family for generations. **Assist the family in understanding the dominant culture and refer for appropriate parenting skills.**
- Same-sex relationships are hidden and not discussed. Marrying and producing children is the expectation. **Do not disclose same-sex relationships to family or outsiders.**

### *31.1.3 Workforce Issues*

- In general, Somali culture values practicality and assimilation.
- Sometimes Somali employees will join each other and speak Somali, but they are generally polite about switching to English to comfort English-speaking coworkers. **Explain organizational policies as to when English only is required.**
- Somali culture also values hard work, self-reliance, and entrepreneurship. Conflict may ensue when employers fail to permit prayer at times prescribed by the Qur'an (imagine an assembly line). **Employers may not be aware that worship at these times is mandatory and not something that can be done when convenient.**
- Conflict between Somali and African American teenagers occurs in some high schools and may carry over into the workplace.
- In the US, Somali immigrants are generally quick to learn American culture and the English language. Keeping apart from mainstream culture and language, as seen in some immigrant communities, is distinctly unusual.

### 31.1.4 Biocultural Ecology

- Somali people tend to have thin frames and light skin. Men and women tend to vary in appearance once settled in America. Table 31.1 lists common Somali health conditions in three categories: lifestyle, environment, and genetics.
- Tuberculosis is by far the most important infectious disease among Somali immigrants. Providers should not assume that testing for this has been accomplished during the immigration process. Extrapulmonary tuberculosis, especially in lymph nodes, is more common than pulmonary tuberculosis and often presents with fever alone or vague other symptoms. **A tuberculin skin test should be performed unless the patient gives a history of a previously positive test. Screen every Somali patient for tuberculosis.**
- Other common infectious diseases are hepatitis B, *Helicobacter pylori*, intestinal parasites, and malaria. Many arrivals come from refugee camps in countries other than Somalia. HIV and hepatitis C are relatively uncommon. Almost all Somali immigrants have antibodies to hepatitis A and do not need vaccination against this disease.
- *H. pylori* infection is common, although it has an uncertain relationship to common upper gastrointestinal symptoms. **Antibodies to *H. pylori* in the blood do not indicate current condition. A better indicator is the presence of *H. pylori* DNA in the stool. Screen Somali refugees for hepatitis B, *H. pylori*, intestinal parasites, and malaria.**
- Intestinal parasites may persist for years after immigration. Usually, these do not cause symptoms but should be treated for public health reasons. Common infections involve *Trichuris trichiura, Enterobius vermicularis, Entamoeba histolytica, Dientamoeba fragilis, Ascaris lumbricoides,* and *Schistosoma mansoni.* Stool tests will also reveal many nonpathogenic organisms, such as *Endolimax nana, Entamoeba coli, Entamoeba hartmanni, Iodamoeba buetschlii,* and *Chilomastix mesnili.* These do not need treatment. There is controversy about

**Table 31.1** Health conditions common to Somalis

| Health condition | Causes |
|---|---|
| Depression, post-traumatic stress disorder | Environment, lifestyle |
| Tuberculosis | Environment, lifestyle |
| Hepatitis B | Environment, lifestyle |
| *Helicobacter pylori,* intestinal parasites, and malaria | Environment, lifestyle |
| *Trichuris trichuriasis*, *Enterobius vermicularis*, *Entamoeba histolytica*, *Dientamoeba fragilis*, *Ascaris lumbricoides*, and *Schistosoma mansoni* | Environment, lifestyle |
| Lactase deficiency | Genetic |

whether *Blastocystis hominis* is pathogenic; generally, treatment is recommended only if the patient is symptomatic or immunosuppressed.

- Little is written about drug metabolism specific to Somali. Lactase deficiency occurs in adulthood. **Assess for lactose intolerance on intake interviews.**

### *31.1.5 High-Risk Health Behaviors*

- Rates of obesity and diabetes have increased dramatically, especially among women adjusting to food availability in Western countries and decreased physical activity climates. **Encourage indoor physical activities such as swimming, working out at gyms, and dancing when appropriate.**
- The Muslim religion prohibits alcohol consumption and few Somali immigrants use alcohol. **If you ask about alcohol, explain, "We ask these questions of everybody."**
- More of a problem is *khat or qat*, a vegetable stimulant chewed like tobacco or drunk as an infusion that produces euphoria. *Khat* can be obtained from leaves, shoots, or twigs of an evergreen shrub found in Arabia, Ethiopia, Kenya, and Yemen but not in Somalia. It can cause psychosis and is classified as a controlled substance by the United Nations. Possession in the US is a felony. It loses potency within days after harvest, so imported *khat* must be used promptly. People who use *khat* regularly tend to gather in houses where they sit together for many hours, talking with each other, abandoning family and work obligations. *Khat* can cause loss of appetite and erectile dysfunction. In pregnancy, it is associated with low birth weight.
- Promiscuity, intravenous drug use, and HIV are relatively uncommon. **If you ask about promiscuity, *khat*, intravenous drug use, or HIV, explain, "We ask these questions of everybody."**
- Teenagers are more oriented toward school and family life and less likely to engage in premarital sex, gangs, weapons, and drugs than their American counterparts.

### *31.1.6 Nutrition*

- As in many cultures, the day's main meal is an essential opportunity for a family to gather and talk. In urban areas, the main meal is usually in the early afternoon. In rural areas, it's in the evening after work.
- In Somalia, everyday foods are rice, pasta, bread, potatoes, corn, beans, and sugar. Meat products include goat, lamb, beef, camel, and fish. Fruits include bananas, mango, papaya, grapefruit, oranges, lemons, grapes, and dates. Beverages include milk from goats, cows, and camels; tea and coffee; and soda.

- A typical breakfast includes *angela*, a flour product resembling pizza, to which oil or sauce can be added, and tea. At mid-day, the meal is rice or pasta with sauce and vegetables, including onions and potatoes, often with goat meat and bananas. A mixture of beans, corn, oil, and sugar, with milk, is consumed for dinner.
- Somali food is less spicy than Ethiopian, Middle Eastern, or Indian cuisine.
- Adapting to American culture and the ready availability of food has produced an epidemic of obesity and diabetes, especially in Somali women who were physically active in their home country and now live indoors in cold northern climates.
- **Initiate nutritional assessments and food preferences on intake assessments.**
- Vitamin D deficiency may occur in winter due to lack of sunlight from indoors and wearing a headscarf.
- Everyone fasts from sunup to sundown during the lunar month of **Ramadan**. Some religious individuals may select additional days of fasting at any time. **Adjust medication schedules to coincide with nonfasting hours. See Chap. 8 for Ramadan schedules.**
- Meat slaughtered by a butcher of faith is called ***halal***. The butcher is ideally Muslim, but Christianity and Judaism are respected religions, and butchers from these faiths are acceptable. Pork of any kind is forbidden. **Provide *halal* meats when possible; otherwise, ask the patient's family to bring food from home. Some online services have frozen *halal* foods available.**

### *31.1.7 Pregnancy and Childbearing Practices*

- Somali families tend to be prominent by American and European standards. Over 60% of Somalia's population is younger than 25 as of 2020, and the fertility rate is among the world's highest at almost 5.5 children per woman—a rate that has decreased little since the 1970s.
- Somalia has the highest rate of female genital mutilation (FGM) or infibulation worldwide. Infibulation is attributed to gender inequality. Prepubertal girls aged 5–11 years can be circumcised by removing the clitoris, labia minora, and labia majora. In another variety of this practice, only the clitoris is removed. The vaginal orifice is then narrowed by sutures, leaving little room for menstruation. At the time of marriage, these sutures are removed. After childbirth, infibulation may be done. In the US, and some other countries, this practice is illegal. In response, some families arrange for their children to travel to Africa for FGM.
- Somali women who have undergone FGM have many health concerns that should not be overlooked. These issues include pain, urinary dysfunction, sexual dysfunction, and scarring. During the birthing process, Somali women with FGM are at risk for difficult labor and neonatal complications secondary to birth. In addition, the mental health of Somali women with FGM may be compromised. Many Somali women may have depression, low self-esteem, and low self-confidence due to FGM.

- **Carefully explain that female circumcision is illegal in the US and most other countries. Obtain permission for students, residents, and fellows' presence during exams and birthing experiences**.
- Children and pregnancy are invariably considered precious gifts. Infertility is a psychological burden that causes marked suffering and elicits sympathy. Abortion is prohibited.
- Contraception to space children is allowed by Islam. Cesarean sections are avoided because of the fear of preventing subsequent normal pregnancy and delivery.
- During pregnancy, women must eat well and avoid heavy exertion or lifting. The custom of working until the due date seems strange to many Somali women. However, Somalis tend to be practical and adapt to new situations and expectations.
- **Incorporate healthy traditional practices into care.**
- After childbirth, ***afartanbah*** is observed, a period of 40 days during which the woman stays home, eats well, is assisted by her family and neighbors, and abstains from sexual intercourse.
- Breastfeeding for as long as possible, up to 2 years, is a religious obligation. Supplementation with milk, formula, and other food as the child grows is allowed.
- A naming ceremony is held between 2 weeks and 40 days.

### 31.1.8 Death Rituals

- Upon learning a family member is dying, Somali families come together to support each other and the dying person. This may include many members of the extended family. A religious person, such as the ***imam*** from the local mosque, may visit the person in the hospital, offer prayers, and help mobilize community support. **Assist the family to obtain an *imam* and make arrangements for the family to be at the dying patient's bedside.**
- As faithful Muslims, Somalis believe that Allah will determine how long a person will live. In addition, Muslim Somalis believe the soul returns to God upon death. For this reason, discussing end-of-life care, advance directives, or efforts to prolong life may be misunderstood. In general, impending death is not concerned directly with the patient. "Allah will decide, and I will follow" is common. **When the need arises to discuss end-of-life care, hospice, palliative care, or life-prolonging measures, approach the nearest relative to the patient. Be patient and allow time for the family to discuss this among the elders of the family.**
- After death, the funeral home cooperates with the mosque to prepare the body, wash it, wrap it in white cloth, and deliver it to the mosque for a prayer service. Bodies are never cremated, always buried on the right side facing Mecca. In

Somalia, bodies are buried without coffins, as the belief is that being "in a box" may be uncomfortable.

### 31.1.9 Spirituality

- Essentially, all Somalis are devout Sunni Muslims who believe that Allah cares for them in their daily lives and follow instructions in the Qur'an regarding prayer and fasting, caring for the poor, and kindness to strangers. They have a great interest in Sufi spiritualism, characterized by chanting, whirling, chewing *khat,* and falling into a trance to communicate with Allah. Somalis also include the veneration of Somali saints in religious worship.
- In Islam, no priests mediate between the believer and God, but there are religious leaders, including teachers, preachers, and mosque officials. Somalis believe that both their religious and secular leaders have the power given by Allah to bless and curse people. This power, baraka, is believed to linger on tombs of Somali saints to help cure illness and resolve other troubles upon visit to the tomb. Somalis believe that some kinds of illnesses are caused by the body being possessed by spirits. Among several religious practitioners, the Yibir from the clan of Saab is called on to exorcise spirits and restore health, good fortune, or prosperity to individuals through prayers and ceremonies, including animal sacrifice.
- In the US and elsewhere, Somalis attend formal services at mosques, which serve as centers of community life and religion. Prayers (*salat*) are performed five times daily in any location facing Mecca. These prayers take about 3 min and involve kneeling and touching the forehead to the ground. On Fridays, men must attend the local mosque to listen to the *khadi,* who gives a sermon called *khudba*. Women may go to the mosque to pray and study the Qur'an separately from men, but attendance is not required. **Plan care around prayer times if at all possible. Assist patients and families with prayer as necessary. Contact a chaplain for assistance.**
- Strict fasting occurs during the lunar month of Ramadan. Ramadan advances each year within the Western calendar. During this time, all people older than 14 refrain from taking food or drink from sunrise to sundown. Sometimes, this is interpreted to include medication, including insulin. Exceptions can occur during illness, travel, menstruation, pregnancy, etc. At the end of this month comes a feast called ***Eid al-Fidri***, in which everyone joins to give thanks, make a special prayer, wear their best clothes, and dine out. **Adjust meal and medication schedules to accommodate for Ramadan.**
- Another feast day, *Eid al-Adha*, coincides with the annual pilgrimage to Mecca (*haj*). Every Muslim is expected to make *haj* once in their life.
- There is a strong sense of sharing and being part of a faithful community. Individualism is regarded as self-centered. Somali culture does not encourage celebrating birthdays. Many refugees will give January 1 of the year they were

born when confronted by immigration authorities or Western medical records systems.

### *31.1.10 Healthcare Practices*

- Somalis sometimes seek healing from traditional and religious healers first. However, they do not hesitate to go to a healthcare provider when symptoms persist or are alarming. This includes seeking care for mental health problems. Older individuals who feel well may not understand the concept of preventive medicine but usually accept advice when the reasons are explained to them. **Integrate nonharmful traditional healthcare practices with allopathic practices and treatments.**
- Somali patients listen, ask questions, and value medical advice. Seeking care and keeping appointments are usually not barriers to care; lack of insurance may be a barrier.
- Affording prescription medication is a problem for many patients; prescribing less expensive generic drugs is a discreet way to help. Newly arrived persons will appreciate assistance in navigating the complex American healthcare system. **Recommend generic medications when feasible and assist patients and families in navigating the healthcare system. Refer to social services and community agencies as necessary.**
- Traditional Somali healing practices include ritual reading or reciting of verses from the Qur'an; creating superficial skin burns on the abdomen, chest, or head; external cutting of the skin (primarily used for jaundice in children suffering from hepatitis A); *minis* (or *saar*), which consists of ceremonial drumming, singing, and dancing around the sick person, and herbs that may be consumed or used as a poultice. Such practices are guided by a traditional healer, usually an older man called an ***alaqad***, and may include participation by a large group of people. The patient is expected to provide food and clothing for the gathering and to pay the *lead* a sum of money. Some traditional healers specialize in setting fractures or trephine procedures for subdural bleeding.
- In Somalia, prepubertal girls are circumcised by removing the clitoris, labia minora, and labia majora. The vaginal orifice is then narrowed by sutures, leaving room for menstruation. At the time of marriage, these sutures are removed. In another variety of this practice, only the clitoris is removed. While these practices seem extreme to Western medical providers, they are accepted as expected by many Somali women who value purity and tradition. In the US and many other countries, this practice is illegal. In response, some families arrange for their children to travel to Africa for circumcision. **Carefully explain that female circumcision is illegal in the US and most other countries.**
- There are no restrictions on accepting blood products when needed.
- Pain is a common complaint, including the unusual pain pattern on one half of the body. Pain may be a culturally acceptable way to express sadness or social

and psychological discomfort. **Ascertain if the pain is physical or emotional in origin.**

- Healthcare providers should be informed of the cultural idioms of distress Somali immigrants express. One such idiom is *jinn*. *Jinn* is believed to be a spirit that causes many mental health experiences. A sudden unknown reason for the change in personality or behavior may be caused by *jinn* possession.
- Many Somali immigrants have significant depression or post-traumatic stress disorder related to war, family disruption, and long years in refugee camps. This cultural term for mental distress is referred to as ***qaracan***.
- Symptoms include depressed mood, insomnia, rumination, social isolation, lack of trust, and a host of somatic symptoms for which investigations do not reveal a specific cause. **Antidepressants are generally more helpful than insight psychotherapy.**
- **There is a significant stigma attached to mental health among Somalis.** Somalis may respond better to more straightforward explanations like troubled thinking, too much stress, insufficient sleep, or a reaction to the past. There is a sense that people should bear up under their symptoms, that Allah would not ask a person to bear more than they can tolerate. Poor mental health can be perceived as a weakness or lack of religious participation.
- **Avoid using "depression" on the first encounter because it may elicit feelings of weakness or guilt. Approach mental health issues cautiously and indirectly until trust is established, unless the situation is an emergency.**

### *31.1.11 Healthcare Providers*

- In urban Mogadishu, Western medicine was practiced by European standards before the civil war, supplemented by traditional practices. The breakdown of the government and the destruction of most hospitals and clinics have declined Somalia's healthcare system. This also caused the availability of few doctors and hospitals, and many unqualified persons practice a form of medicine. Most healthcare is free in Somalia and is provided by the World Health Organization and nongovernmental organizations (NGOs). Folk medicine is practiced by nomads and farmers who have no immediate access to healthcare. Somali Western medicine and classic methods are complementary. **Integrate traditional healthcare providers and traditional treatments with Western healthcare providers and treatments.**
- Respect for medical education and expertise is the rule. Doctors and nurses are valued as professionals. A healthcare career is a common choice for young people. Somalis integrate easily into Western medical systems, displaying medical values such as patience, tactfulness, and kindness. **Encourage Somali children and older adults seeking a second career to think about the health professions.**

## 31.2 Reflective Exercises

1. What do Somali people speak as their primary language? Where might you find an interpreter for this group?
2. How would you greet a Somali male patient? A Somali female patient?
3. Explain decision-making practices among the Somali people. What implications does this have for your nursing care?
4. What are the primary non-communicable health conditions among Somalis? What might you do to help decrease some of them?
5. Name at least three infectious conditions common among Somali immigrants. What are the implications for your initial assessment and follow-up care?
6. Explain the name format for Somali men. For Somali Women.
7. Identify infant feeding beliefs among Somali families. What educational advice would you offer?
8. What is the dominant religious practice among Somalis? How does this affect your care?
9. How is depression seen among Somalis? What is your approach to addressing depression?
10. Identify several traditional healthcare practices. How might you incorporate traditional healers with allopathic treatments?

## Bibliography

Adair R, Jama Y (2013) People of Somali heritage. In: Purnell L (ed) Transcultural health care: a culturally competent approach, 4th edn. F.A. Davis, Philadelphia

Adair R, Nwaneri O (1999) Communicable disease in African immigrants in Minneapolis. Arch Intern Med 159:83–85

Adair R, Greminger A, Post B (2006) Differences between insured and uninsured patients in South Minneapolis. Minn Med 89:46–47

Chung S (2015) Refugees from Somalia in Buffalo—background: Somali and Somali-Bantu refugees, partnership for the public good. United States of America. https://policycommons.net/artifacts/3802519/refugees-from-somalia-in-buffalo-background/4608347/. Accessed 4 Oct 2023. CID: 20.500.12592/spfv8b

CIA (2024) World Factbook: Somalia. https://www.cia.gov/the-world-factbook/countries/somalia/

Epstein R, Fiscella K, Gipson T, Volpe E, Jean-Pierre P (2007) Caring for Somali women. Implications for clinician-patient communication. Patient Educ Couns 66(3):337–345

Gerritsen AAM, Bramsen I, Deville W, van Willigen LHM, Joven JE, Ploeg HM (2006) Physical and mental health of Afghan, Iranian, and Somali asylum seekers and refugees living in the Netherlands. Soc Psychiatry Med 41(1):18–26

Im H, Ferguson A, Hunter M (2017) Cultural translation of refugee trauma: cultural idioms of distress among Somali refugees in displacement. Transcult Psychiatry 54(5–6):626–652

Lewis T (2016) Somali cultural profile. Ethnomed. https://ethnomed.org/culture/somali/somali-cultural-profile

Markova V, Sandal GM (2016) Lay explanatory models of depression and preferred coping strategies among Somali refugees in Norway. A mixed-method study. Front Psychol 7:1435

Mohamed AH, Mohamud RYH, Mohamud HA et al (2022) Somalian women with female genital mutilation had increased risk of female sexual dysfunction: a cross-sectional observational study. Sci Rep 12:15633. https://doi.org/10.1038/s41598-022-19949-0

Nelson HD (ed) (1982) Somalia, a country study (Vol. 550, No. 86). US Government Printing Office

Njeru JW, Wieland ML, Okamoto JM et al (2020) Social networks and obesity among Somali immigrants and refugees. BMC Public Health 20:238. https://doi.org/10.1186/s12889-020-8315-7

Pavlish CL, Noor S, Brandt J (2010) Somali immigrant women and the American health care system: discordant beliefs, divergent expectations and silent worries. Soc Sci Med 71(2):353–361

Somalis in Minnesota (2016) Culture care connection. http://www.culturecareconnection.org/matters/diversity/somali.html

Steinman L, Doescher M, Keppel G, Pak-Gorstein S, Graham E, Haq A, Johnson DB, Spicer P (2009) Understanding infant feeding beliefs and practices and preferred nutrition education and health provider approaches an exploratory study with Somali mothers in the US. Matern Child Nutr 6(1):67–88

Wallmo S, Allgurin K, Berterö C (2020) The lived experience among Somali women of giving birth in Sweden: an interpretive phenomenological study. BMC Pregnancy Childbirth 20:262. https://doi.org/10.1186/s12884-020-02933-9

# Chapter 32
# People of Thai Heritage

**Ratchneewan Ross**

## 32.1 Overview and Heritage

Cities in the United States with sizable Thai populations include Los Angeles, New York City, Houston, and Chicago. Many Thais with graduate degrees work in the United States in professional fields such as medicine, nursing, and engineering. Others own Thai restaurants or grocery stores and provide work for other Thais. Diversity among Thais is significant according to acculturation, individuality, and the variant characteristics of culture as described in Chap. 1 and the region in Thailand from which they come.

Some of the earliest Thai immigrants in the United States were *Eng* and *Chang*, the famous Siamese twins who captured the world's attention because of their conjoined chests and whose career was as a public exhibition. They settled in North Carolina in 1839, later changed their Thai last name to Bunker, and married American women. However, prior to 1960, Thai immigration to the United States was largely non-existent and didn't become more prevalent until during and after the Vietnam War. Today, it is estimated that approximately a quarter of a million Thai immigrants reside in the United States.

R. Ross (✉)
School of Nursing, University of Louisville, Louisville, KY, USA
e-mail: ratchneewan.ross@louisville.edu





E. A. Fenkl, L. D. Purnell (eds.), *Handbook for Culturally Competent Care*,
https://doi.org/10.1007/978-3-031-70492-5_32

### 32.1.1 Communications

- The official language in Thailand is Thai with English as a secondary language of the elite. Thai is derived from *Pali* and *Sanskrit* (ancient South Asian languages) and has several dialects. The three main dialects are ***pasha dai***, ***pasha issam***, and ***pasha nua***.
- The Thai language is a fixed tonal language having five tones. Thus, the same phonetic sound can have different meanings depending on the tone. The written alphabet is a complicated system of 44 letters with over 33 vowels or vowel combinations.
- Age and status contribute greatly to how Thais communicate with one another. A younger person is expected to show respect for an older person through his or her gestures and language. A Thai female uses the word *kah* and a Thai male uses *kraab* at the end of a sentence to add politeness in a conversation.
- Looking into a person's eyes and conversing quietly reflects respect and politeness. A distance of 1½–2 ft between two speakers is preferable.
- Thais usually greet each other with the *wai* motion—putting the palms of both hands together in a prayer-like gesture and bowing the head slightly. This gesture is used by both men and women of all age groups. **Health-care providers are not expected to greet patients with the *wai* motion.**
- Most Thais have long first and long last names. A Thai is usually referred to by his or her first name, even in an official setting like school or work. The names usually have clear meanings in Thai. The first name is often given by a Buddhist monk or fortune-teller based on the date, day of the week, and time of a newborn's birth. Often, parents name their children themselves. More recently, some parents have begun to give their children Western first names.
- When a woman marries, she usually takes her husband's last name, although this has become less prevalent in recent years. A couple's children also take their father's last name.
- When Thai names are transcribed in English, the spelling is merely a phonetic translation from its spelling in the Thai alphabet. Because Thai is a tonal language, the correct pronunciation of names cannot be ascertained from their spelling in English. **Ask Thai patients how to pronounce their names and do the best you can in approximating them.**
- Almost all Thais have a short nickname used by their family and close friends and often by colleagues at work. Nicknames normally have no relationship with first names. They are often humorous to Thais themselves. Nicknames are usually either Thai or English words. They might be derived from names of colors, body types, fruits, or any number of other things. **Feel free to ask patients if they wish to be called by their nickname. The patient may well prefer it. Maintain the legal name for medical record keeping.**

### 32.1.2 Family Roles and Organization

- In a traditional Thai family, the man is usually the breadwinner and manages important tasks such as managing finances, making decisions on major purchases, and deciding where the family will live. Household chores and taking care of children are the woman's responsibilities. However, more Thai families today have begun to divide household chores between men and women.
- Older Thais generally value tranquil relationships with family and friends.
- Thai children are taught to respect elders. Talking back to elders is discouraged.
- Thai female adolescents have traditionally been expected to protect their virginity until marriage. Dating with a chaperone is preferable. However, more and more Thai adolescents date on their own today.
- Children are the center of the family. Many Thai children sleep with their parents from birth until some point in time before they reach adolescence.
- Often, children are spoon-fed by adults until they are 4–5 years old.
- Marriages used to be mainly arranged by the parents. Today, young Thais have more freedom to select a spouse. Nevertheless, sometimes parents may make the final decision as to whether or not a bride or groom is acceptable.
- Gays and lesbians in Thailand are more accepted today than in the past. However, for some, being gay or lesbian continues to carry a stigma. **Do not disclose a patient's same-sex relationships to the family or others.**

### 32.1.3 Workforce Issues

- Most Thais usually try to avoid personal conflicts at work and are hard workers.
- Although the family is deemed very important, in many circumstances, especially for economic reasons, work comes before family.
- In general, Thai Americans tend to socialize among themselves rather than mix with Americans or people from other cultures. Therefore, some Thai Americans may not fully understand American culture.
- Thai respect their supervisors because seniority is strongly valued in their culture. Thus, they might not be assertive at work. **Supervisors may be wise to provide open discussions and expression of opportunities for their Thai American colleagues.**
- English proficiency among some Thais is low, especially among newer immigrants and older people who did not learn English in Thailand. **The language used in the workplace should be clear.** Avoid **slang expressions**.

### 32.1.4 Biocultural Ecology

- An estimated 75% of the population in Thailand are pure "Thai"; 14% are Chinese; and the rest (11%) are Malay, Lao, Mon, Cambodian, Vietnamese, Asian Indian, Caucasian, or hill-dweller tribes—Karen, Lisu, Ahka, Lahu, Mien, and Hmong.
- Thais from northeast Thailand (*Isaan*) emigrated from Laos or Cambodia and have dark brown skin.
- Some Thais in the north immigrated to Thailand from China or Burma and tend to have finer skin texture and lighter skin color. Their noses are a little longer and their cheekbones narrower than those of *Isaan* Thais.
- Central Thais generally have medium skin color compared with those in the rest of the country. Their facial profile is a mixture of *Isaan* Thais and northern Thais.
- Southern Thais, some of whom migrated from Malaysia, are likely to have a darker skin color.
- Regardless of skin color or facial profile, the Thais' size and body structure are usually much smaller than those of Caucasians.
- Common genetic conditions among Thais are glucose-6-phosphate dehydrogenase deficiency (G-6-PD) and thalassemia. Different ethnic groups may have different pharmacokinetic functions. Table 32.1 lists commonly occurring health conditions for Thais in three categories: lifestyle, environment, and genetics.

### 32.1.5 High-Risk Behaviors

- Major health problems among Thais in Thailand include low birth weight and perinatal asphyxia, HIV/AIDS, road traffic injuries, drug and alcohol abuse, schizophrenia, diabetes, cerebrovascular disease, emphysema, and liver cancer. It is unknown to what degree these conditions are common among Thais in the United States.
- Sexual promiscuity among Thai youth has also contributed to an increased risk for other sexually transmitted infections (STIs) and unwanted pregnancies. **Teach safe sex practices at every opportunity.**

**Table 32.1** Health conditions common to Thailanders

| Health conditions | Causes |
|---|---|
| Glucose-6-phosphate dehydrogenase deficiency | Genetic |
| Thalassemia | Genetic |
| Lactase deficiency | Genetic |

### 32.1.6 Nutrition

- In general, an individual portion of a Thai dish is about one-third to one-fifth of a typical US dish in terms of volume. Most Thais are slim owing to these smaller portions and also the types of food they eat.
- A Thai balanced diet usually includes low-fat/low-meat dishes with a large percentage of vegetables and legumes.
- Rice and fish are the main staples. Rice is the main source of carbohydrates in Thai dishes, but noodles are also found in many favorite recipes.
- Vegetables and meats are usually fried or grilled and prepared in many combined variations to supplement rice.
- Pork and chicken are eaten more than beef. All meats are consumed more sparingly in proportion to vegetables when compared with a Western diet.
- Fish and seafood are also regularly enjoyed.
- Noodle recipes are much loved by Thais and are prepared with the noodles already mixed in with meats and vegetables.
- For all foods, seasonings are critical to the Thai artistry of accommodating different palettes.
- Fish and oyster sauces are very often combined with soy sauce as a basic starting point for many recipes.
- Thai chili pepper is the basic ingredient added to control the degree of spiciness in foods. Many, but not all, Thais love very spicy food.
- *Tom-Yum* is a traditional spicy Thai soup that is gaining popularity worldwide. It has been found to have positive effects on people's health because of its ingredients, which include lemongrass, galangal roots, kaffir lime leaves, hot chilies, red onions, and garlic. The soup's antimicrobial effects come from its chilies, onions, and garlic. Onions and garlic can function against diabetes and hypercholesterolemia.
- *Som-tum* is a famous spicy Thai salad. Its ingredients include fresh shredded papaya, cut tomatoes, tamarind juice, fish sauce, salt, sugar, fresh crushed garlic, and hot chilies. Sometimes, cooked or raw fermented fish is added. *Som-tum* is usually served with hot *sticky* (sweet) rice. Fresh garlic, used as an ingredient in *Som-tum* and many other Thai dishes, has been identified as an antifungal, antiparasitic, and antiviral agent.
- Thais in the United States have changed their food intake habits. They skip more meals and consume more Western foods and snacks such as white bread, salty items, fruit juice, soft drinks, and sweets. **Assess Thai patients' food intake habits and encourage more fruits and vegetables.**
- Hot or warm foods or drinks are considered healthier than cold ones.

### *32.1.7 Pregnancy and Childbearing Practices*

- Thai women view pregnancy as a special time in their lives when they need extra care, both physically and emotionally. They acknowledge that this is a time when their moods can be unstable.
- Owing to modesty, especially during a vaginal examination, Thai women prefer female health-care providers. **Provide female health-care providers for intimate care during labor and delivery and the postpartum period.**
- Women often receive advice from their mothers and/or mothers-in-law about what to do or not do during childbearing. Pregnant women are advised not to complain or get upset so that newborns will be happy and stay happy for the rest of their lives. They may also be advised not to sit on stairs or doorsills to avoid difficult labor and delivery. It is believed that when a pregnant mother blocks other people from going up and down stairs or in and out of a doorway, the unborn baby could be blocked inside the mother's uterus.
- Astrology and animism play major roles in many Thais' lives. In general, pregnant women are discouraged from visiting a hospitalized person, attending a funeral, or visiting a house where a death has occurred. Such practices prevent the woman and her unborn baby from catching any illness or getting haunted by a spirit or ghost.
- Some avoid drinking coconut juice, believing that it can cause too much vernix caseosa (fat on the newborn's skin). Others drink a lot of the juice, believing that it will help their newborns to have smooth and beautiful skin texture.
- Most Thais view lighter skin as more favorable and that drinking chocolate milk, eating chocolate, or drinking coffee will cause newborns to have a darker skin color.
- Pregnant women from the central region of Thailand are often seen with a safety pin on their outfit over their belly. The pin works against a kind of ghost who always wants to steal the unborn baby from a mother's womb. **Do not remove pins or amulets of any kind unless the patient gives permission.**
- After a child is born, the mother is left cold and wet. Therefore, the mother should gain some heat to dry out her body, especially her uterus. To gain heat, some Thai mothers practice *yue fai*, which literally means "being with fire," a complicated process that is rarely practiced in the United States and is practiced less in Thailand.
- Mothers are advised to use a perineal heat light, a hot sitz bath, sauna heat belts, and warm showers. Warm and hot drinks and foods are consumed; ice chips or ice cubes are avoided.
- Some postpartum mothers are advised to not eat eggs, chicken, or buffalo meat, believing that the new mothers' perineum may not heal. Some believe chicken is a taboo food for women after delivery because a chicken likes to scratch the ground to look for food. The chicken meat, therefore, could scratch open the perineum. Eggs are avoided by some, believing that they could cause a big scar on the perineum.

- Vegetables include lettuce, banana flower, lemongrass, onion, ginger, cabbage, hairy melon, snake beans, chili, peppers, and bamboo shoots.
- Acceptable fruits after the postpartum period include oranges, bananas, tamarind, watermelon, jack fruit, and durian, an oval fruit with a hard spiny rind.
- Some women drink *ya dong*, a Thai nonalcoholic or alcoholic drink herbs such as *ya dong,* ginseng, galangal, peppermint, cinnamon, spirulina, and plant roots. As perceived by many Thais, *ya dong* is famous for its medicinal qualities. The drink helps with blood production and drying out the uterus quickly.
- Fathers have begun to participate more in the childbearing experience.

### 32.1.8 Death Rituals

- Most Thais are Buddhists. Buddhists believe that after a person dies, the person will be reborn somewhere else based on that person's *karma*, the process by which a person's moral behavior or actions have consequences for the person's future, either in the present or later life.
- In general, Thai Buddhists follow the custom of cremating the bodies of the deceased, because when the Buddha passed away, his body was cremated.
- In Buddha's teaching, a funeral ceremony should be simple. In the funeral ceremony, often Buddhist monks are invited to chant verses to the dead and the family.
- The ashes from the cremation are buried at a cemetery. Sometimes, a portion of the ashes is sprinkled in a river. If possible, when the family of the dead returns home after a sojourn away from Thailand, some of the ashes may be sprinkled again in a river or near the deceased's hometown.
- Female relatives normally wail quietly. The family members pray quietly to the dead before the cremation to ask for forgiveness and wish the dead to be reborn in a happy and peaceful home. **Expect and accept a wide range of behaviors to occur with grieving.**

### 32.1.9 Spirituality

- Most Thais are Buddhist and are free to incorporate any other religious values and animism into their beliefs and practices when deemed good. Thais in all socioeconomic strata incorporate animism, fortune-telling, and astrology to some degree into their religious practice.
- For most Thais, family support along with Buddhism is a crucial source of strength.
- Meditation and prayer are used by many Thais to cope with an illness. Meditation is a means for Thai older people to enhance their self-awareness, peace of mind, sleep, and physical health.

- For older people in the United States, meditation and prayer also help them feel peaceful, perceive life as valuable, and experience meaning and confidence in death. For Thais, health and spirituality are intertwined and are important aspects of life. **Seek assistance from chaplain services when requested.**

### *32.1.10 Health-Care Practices*

- Health promotion and disease prevention practices among the Thais are very limited. Folk practices are common among the less educated. **Promote healthy behaviors at every opportunity.**
- Many believe that bad *karma* and negative supernatural power cause mental illness. Therefore, folk therapies from traditional healers are the first resource for many families. When such therapies do not seem to work, they go to contemporary medical facilities. Folk therapies may include healing ceremonies, using shamans (as a mediator) to converse with supernatural beings (such as black magic, evil beings, and ancient/natural spirits), and negotiating with them that the sick person might be released from his or her illness.
- Stigmatization attached to mental illness and beliefs in animism and *Karma* tend to prevent some Thais from seeking health-care providers' help when mental health problems arise.
- Many Thais may appear stoic in trying to withhold expressions of pain or suffering.
- **Rely more on nonverbal clues for pain or psychological-emotional distress when assessing Buddhist Thai patients.**
- **When clinical depression is diagnosed, make extra efforts to encourage patients to get help and treatment, along with assuring them that depression and psychosis are different disorders.**
- No religious beliefs against blood transfusion exist for Thais.
- Thais who believe in rebirth might object to organ donation, believing that they might not have the organ when needed in the next life.

### *32.1.11 Health-Care Providers*

- Thais in the United States and elsewhere tend to consult their family and friends first when they feel ill or have medical problems.
- Thai women usually seek female health-care providers for childbearing care and gynecological problems owing to their modesty and their culture. If female providers are not available, they are generally willing to accept male health-care providers.

- **Provide a female health-care provider for female intimate care whenever possible.**
- Respect for seniority is a strong cultural value among Thais. In general, when comparing physicians, head nurses, and junior nurses, physicians receive the most respect, followed by head nurses and junior nurses. In some cases, very senior head nurses receive the same level of respect as physicians.

## 32.2 Reflective Exercises

1. What are the three main dialects spoken in Thailand? How might you get an interpreter for each dialect?
2. What are nonverbal practices used by Thais?
3. Identify at least two genetic conditions common among Thais. What implications do they have for health and health teaching?
4. Identify common health conditions among Thais.
5. What is *Tom-Yum*? What are the ingredients and what are its health properties? For what conditions is it advantageous?
6. *Som-tum*? What are the ingredients and what are its health properties? For what conditions is it advantageous?
7. What foods might be avoided among some Thai postpartum women? Why are these foods avoided?
8. What is the dominant religion of Thais? What are its principles?
9. Identify several Thai folk practices. For what conditions are they used?
10. What are the roles of men and women in traditional Thai culture?

## Bibliography

Asian Immigrants in the United States (2016) http://www.migrationpolicy.org/article/asian-immigrants-united-states#Distribution%20by%20Region%20and%20Country%20of%20Origin

Chotchoungchatchai S, Saralamp P, Jenittikul J, Pornsiripongase S, Prathanturarug S (2012) Medicinal plants used with Thai traditional medicine in modern healthcare services: a case study in Kabchoeng Hospital, Surin Province, Thailand. J Ethnopharmacol 141(1):193–205

CIA (2024) World Factbook: Thailand. https://www.cia.gov/the-world-factbook/countries/thailand/

Hays J (2013) Folk beliefs (superstitions), astrology, fortunetellers, and beliefs about the supernatural in Thailand. http://factsanddetails.com/southeast-asia/Thailand/sub5_8b/entry-3216.html#chapter-0

König A, Nabieva J, Manssouri A, Antia K, Dambach P, Deckert A, Horstick O, Kohler S, Winkler VA (2022) Systematic scoping review on migrant health coverage in Thailand. Trop Med Infect Dis. 7(8):166. https://doi.org/10.3390/tropicalmed7080166. PMID: 36006258; PMCID: PMC9415742. https://www.ncbi.nlm.nih.gov/pmc/articles/PMC9415742/

Liamputtong P (2009) Pregnancy, childbirth, and traditional beliefs and practices in Chiang Mai, Northern Thailand. In: Science across cultures, pp 175–184. https://link.springer.com/chapter/10.1007/978-90-481-2599-9_15

Ross R, Ross J (2013) People of Thai heritage. In: Purnell L (ed) Transcultural health care: a culturally competent approach. F.A. Davis, Philadelphia

Thailand's Spiritual Fusion (2016) http://www.pattayaunlimited.com/thailands-spiritual-fusion/

# Chapter 33
# People of Turkish Heritage

**Marshelle Thobaben and Sema Kuguoglu**

## 33.1 Overview and Heritage

The Turkish diaspora in the United States is primarily concentrated in states such as New York, California, New Jersey, and Florida. The community comprises individuals from diverse backgrounds, with many coming from elite and upper-middle-class segments. Education holds considerable value in Turkish culture, and a substantial number of Turkish Americans hold advanced degrees, reflecting their commitment to academic pursuits. The Turkish community in the United States exhibits a wide array of beliefs and practices influenced by factors such as acculturation, individuality, and varying cultural characteristics (discussed in Chap. 1).

Prior to its independence, Türkiye had already initiated a process of Westernization. However, it was under the leadership of its former president, Mustafa Kemal Atatürk, that Westernization and secularism became synonymous with the country's transformation. From 1923 to 1938, Atatürk implemented numerous progressive reforms, which included the ban on the traditional fez, the prohibition of polygamy, the establishment of civil marriage contracts, the abolition of communal law for ethnic minorities, the transition from Islamic law to a civil law legal system, the promotion of nationalism and pride, educational and cultural reforms, the compulsory adoption of surnames, the shift of the weekly day of rest from Friday to Sunday, and the election of 17 female deputies to the National Assembly.

M. Thobaben (✉)
Nursing, Cal Poly Humboldt, Arcata, CA, USA
e-mail: Marshelle.Thobaben@humboldt.edu

S. Kuguoglu
Nursing, Medipol University, Istanbul, Turkey



E. A. Fenkl, L. D. Purnell (eds.), *Handbook for Culturally Competent Care*,
https://doi.org/10.1007/978-3-031-70492-5_33

Even after his passing on November 10, 1938, Mustafa Kemal Atatürk remains highly revered as the founding father of Türkiye, given his transformative impact on the nation's social, political, and cultural landscape.

### 33.1.1 Communications

- Turkish, spoken by approximately 90% of the population, employs a linguistic structure that does not differentiate between gender-specific pronouns, such as "he" and "she," or possessive forms like "her" and "his." However, it is noteworthy that Turkish does maintain a distinction between formal and informal forms of the second-person singular pronoun "you." This linguistic characteristic reflects the societal emphasis on hierarchical status within Turkish culture.
- The pronunciation of Turkish names can pose challenges for Americans, and others, due to historical and script-based differences. Up until 1928, the Turkish language was transcribed using the Arabic script. However, guided by Atatürk's initiatives, a distinct Turkish alphabet based on the Latin script was introduced. This modern Turkish alphabet bears similarities to the English counterpart but lacks the letters "w" and "x." Furthermore, the Turkish alphabet accommodates certain distinctive phonetic nuances. For instance, certain sounds are represented by an undotted "i," as well as accented "ö" and "ü," while a "ç" with a cedilla signifies sounds akin to "sh" and "ch." These nuanced aspects of the Turkish alphabet mirror the phonetic intricacies of the language and contribute to the distinctive pronunciation of Turkish names.
- Elevated vocal tones are common in Turkish communication, indicating enthusiasm rather than anger. Simultaneous speech or interruptions during discussions are not inherently rude, except when a person of lower status interrupts a higher-status individual. **Do not interpret speaking in a loud voice as anger.**
- In Turkish society, there is a prevailing emphasis on group affiliation over individualism. Identity is often shaped by factors such as familial ties, group memberships, educational institutions, and professional associations. The notion of privacy holds less significance, as Turks commonly lean toward collaborative dynamics within their social circles encompassing family and friends. However, it's worth noting that inter-group competition can be notably intense despite the cooperative tendencies observed.
- In the context of Turkish society, women typically find avenues for expressing anger within the confines of same-sex friendships and familial bonds. These channels are considered acceptable, as is the case when addressing individuals of lower social strata. However, it is important to acknowledge that women often encounter constraints in openly venting their frustrations toward their spouses or individuals occupying higher positions of influence.
- Engaging in physical contact such as touching, hand-holding, and gentle patting is considered socially acceptable among individuals of the same sex in friend-

ships, as well as between individuals of opposite sexes who are in romantic relationships. Similarly, there exists a closer personal space between same-sex friends compared to opposite-sex partners, with physical proximity being esteemed as an indicator of emotional intimacy. **Touch, when necessary, is allowed and expected from healthcare providers. However, strict Muslims may not shake hands or touch members of the opposite sex, especially if they are not related.**

- Eye contact serves as a means of conveying respect in Turkish culture. In the presence of individuals holding higher social status, it is customary to maintain intermittent eye contact to signal attentiveness. Nonetheless, it is important to exercise discretion, as extended or sustained eye contact can be construed as impolite or even suggestive of flirtatious behavior. **Maintain eye contact with patients, demonstrating respect and truthfulness.**
- In Turkish culture, a prevailing inclination toward formal attire is evident. On social occasions, men often opt for suits instead of more casual sports jackets and slacks. Women, however, tend to embrace modest clothing choices, favoring skirts and dresses over trousers. It is noteworthy that more conservative Muslim women might adhere to particularly modest clothing and may use black or colorful printed scarves to cover their heads. However, it's important to acknowledge the evolving nature of fashion trends. Among the younger demographic, denim jeans and casual attire are progressively gaining traction, particularly for less formal gatherings. This shift underscores the dynamic interplay between tradition and contemporary preferences within Turkish dress codes. **Expect a wide variety of dress among Turks, depending on religiosity, acculturation, and age.**
- Turkish individuals commonly exhibit a proclivity for openly expressing a range of emotions, including happiness, disgust, approval, disapproval, and sadness, through readily discernible facial expressions and gestures. The negation "no" is frequently conveyed through the raising of eyebrows or a subtle lifting of the chin, often accompanied by a sharp snapping or "tsk" sound. An emblem of appreciation involves bringing together the fingertips and thumb, followed by a kissing gesture, which is frequently employed to convey gratitude for food. These distinctive nonverbal cues and gestures form an integral aspect of Turkish communication norms.
- Maintaining a meticulously clean home is a source of pride within Turkish culture, underscored by the customary practice of removing shoes upon entering. It is commonplace for Turkish hosts in both Türkiye and, to a significant extent, the United States and some other countries, to provide guests with slippers. Irrespective of footwear, revealing the sole of one's foot is deemed disrespectful. Additionally, women are expected to observe modest sitting postures, typically maintaining knees together and refraining from crossing them. These cultural norms reflect the value placed on cleanliness, hospitality, and decorum in Turkish society. **Home healthcare providers should look for cues and ask if they should remove their shoes when entering the home of a Turkish family.**

- In the realm of business relationships and healthcare appointments, the significance of punctuality is steadily growing within the Turkish American community. **Explain the necessity of being punctual for healthcare appointments.**
- Titles serve as essential markers of respect and social status within Turkish culture. When addressing individuals, even those unknown, it is customary to employ their relevant titles, such as Bey (Mr.), Hanim (Mrs., Miss, or Ms.), *Doktor*, or *Profesör*. The same principle extends to familial interactions, where specific titles reflecting relationships are employed. For instance, terms like agbi (older brother or older close male friend), amca (uncle or elderly male relative or stranger), abla (older sister or older close female friend), teyze (maternal aunt or older female relative or older female stranger), and yenge (wife of a brother or paternal uncle) are utilized. This practice underscores the importance of demonstrating respect and acknowledging social dynamics through appropriate address. **Greet patients formally with their title or Mr., Mrs., or Ms.**

## *33.1.2 Family Roles and Organization*

- In a highly traditional Turkish household, the father assumes a role of unequivocal authority. However, within families embracing less traditional values, a notable shift toward spousal equality becomes evident, particularly in nuclear family structures where the wife possesses a strong educational background and is engaged in professional pursuits outside the home. Notably, contemporary women in Türkiye tend to exhibit a greater degree of Westernization in comparison to certain counterparts within Middle Eastern or Muslim societies. This variance underscores the evolving dynamics shaping gender roles and cultural influences in modern Türkiye. **Ascertain the family spokesperson before eliciting health-care decisions. Accept family decision-making patterns without judgment.**
- Upon entering the school system, children are met with expectations to dedicate themselves to their studies, demonstrate respect, and adhere to the guidance of their elders, including older siblings. Conventional norms dictate that children refrain from engaging in disruptive behavior or expressing defiance toward their superiors. It's worth noting that traditional views permit a degree of light corporal punishment as a means of discipline, although practices in this regard are undergoing change due to evolving societal perspectives. **Explain the legalities of child abuse in the United States and elsewhere.**
- Circumcision stands as a significant rite of passage for male children within Turkish culture. This occasion is marked by familial celebration involving the extended family and includes the bestowing of gifts upon the newly circumcised boys. While the traditional practice allows for circumcision up to approximately

12 years of age, contemporary customs are shifting toward hospital-based procedures shortly after birth. This evolution reflects the intersection of tradition and modern medical practices in the Turkish context.

- Sexual interaction is notably discouraged among unmarried youth in Turkish society, with a particular emphasis on preserving chastity, especially in the case of women for whom virginity holds substantial cultural significance.
- While financial self-sufficiency holds significance within Turkish culture, the pursuit of independence from the family unit is not strongly promoted. Adult offspring, particularly sons, continue to play a vital role in their parent's lives, and it is anticipated that they will assume caregiving responsibilities for their elders in later years. This emphasis underscores the enduring familial bonds and intergenerational obligations within Turkish society.
- Parents or other family members are often consulted before major decisions are made.
- **Facilitating compliance with health prescriptions and health education can be effectively achieved through the strategic organization of a family conference led by a designated family spokesperson. This approach proves advantageous in fostering understanding and garnering support for recommended health practices within the family unit.**
- Among young individuals, residing within their parental home is customary until marriage unless educational pursuits or professional obligations necessitate alternative living arrangements. While arranged marriages persist within certain segments of Turkish society, this practice is comparatively less prevalent among Turkish individuals residing in Western countries.
- Individuals are ingrained with a sense of responsibility to care for their parents when the need for assistance arises. The active involvement of grandparents in the upbringing of their grandchildren is pronounced, particularly when they share a household. This intergenerational dynamic underscores the pivotal role of family support in Turkish culture.
- Within numerous Turkish families, the extended family encompasses aunts, uncles, cousins, and in-laws. This expansive network fosters a cooperative dynamic that extends to diverse aspects of life, encompassing the sharing of responsibilities such as child care, labor, and sustenance as circumstances require. This collaborative ethos underscores the significance of familial bonds in Turkish cultural traditions. **Use the extended family arrangement for health teaching and home care.**
- The prevalence of divorce is on the rise, although it continues to carry a certain social stigma, particularly for women. The act of divorcing remains culturally unfavorable. In cases of widowhood, there exists a customary practice of extended family support, where the deceased husband's family assumes responsibility for the care of the widow. Depending on factors such as age and socioeconomic status, widows may have the opportunity to remarry, reflecting the flexibility within this traditional framework.

- The acceptance of homosexuality in Turkish society is gradually emerging, albeit with a certain degree of reservation. Many Turks exhibit a reluctance to align themselves with the gay community openly. There exists a noticeable hesitancy among individuals to establish affiliations with this demographic, reflecting the evolving but still somewhat distant reception of homosexuality within the cultural context. **Do not disclose same-sex relationships to family members or others.**

### *33.1.3 Workforce Issues*

- Türkiye is characterized by a culture that prioritizes group dynamics, and this orientation often extends to the workplace, which tends to emphasize a team-oriented approach.
- The Turkish cultural inclination toward relationship-oriented interactions can result in a reliance on personal connections and networks to achieve tasks. The process of cultivating relationships and networks, while intrinsic to Turkish norms, might be perceived as nepotism or excessive socializing from an American standpoint. **Orient nurses to the Turkish perspective on relationships and the Turkish perspective of relationship to American nurses.**
- Within the Turkish work environment, there exists an anticipation for a hierarchical relationship between supervisors and subordinates. The practice of delivering criticism indirectly is not only customary but also valued, as it serves to "save face." It's important to note that openly critiquing a Turkish individual, especially in a public setting, can be deeply offensive and detrimental to interpersonal dynamics. **Do not criticize Turks in front of other employees.**
- They may be reticent about asking questions for fear of exposing a lack of knowledge. **Ask employees to demonstrate a procedure to ensure understanding.**
- Turks handle differences of opinion in a brisk and clear-cut manner.
- Because most Turkish immigrants speak English, language barriers in the workplace may be only subtle.
- Turkish cultural norms dictate that confrontational interactions conducted in an assertive manner have the potential to erode relationships. Consequently, conflict resolution predominantly revolves around collaborative approaches, supported by elements of compromise and assertiveness. Compromise and avoidance tend to manifest among peers, whereas accommodation behaviors are more commonly employed when interacting with superiors. This nuanced conflict resolution framework reflects the cultural emphasis on maintaining harmonious interpersonal connections.
- Türkiye is recognized for its significant power distance, reflecting the substantial psychological and emotional gap that often exists between superiors and subordinates. The culture places a strong emphasis on respecting authority, resulting in a centralized administrative structure and an authoritarian leadership approach that characterizes various facets of Turkish society.

### 33.1.4 Biocultural Ecology

- The Turkish population is a mosaic in terms of appearance, complexion, and coloration. Appearances range from light-skinned with blue or green eyes to olive or darker skin tones with brown eyes. Mongolian spots, usually found at or near the sacrum, are common among Turkish babies and should not be confused with bruising.
- Table 33.1 lists commonly occurring health conditions for Turks in three categories: lifestyle, environment, and genetics. See Chap. 2.
- Endemic goiter associated with iodine deficiency is a major health problem in Türkiye despite iodine prophylaxis. **Screen newer immigrants for helminthiasis and goiter.**
- **Assess for lactose intolerance as part of the intake interviews, and help patients identify alternative sources of calcium. Assess Turkish patients for β-thalassemia before prescribing medications.**
- Turks in Türkiye have high rates of occupational injuries. **Provide education regarding safety issues in occupational health.**
- Turks are at high-risk for multiple cancers because of the Chernobyl accident in the Soviet Union in 1986; this includes children born after this period as well.

### 33.1.5 High-Risk Health Behaviors

- Cigarette smoking is widespread in Türkiye and tends to start at an early age. Türkiye is a major producer of tobacco in the world. **Encourage patients to cease smoking and help them access smoking cessation programs. Start teaching the hazards of smoking in elementary school.**

**Table 33.1** Health conditions common to Turks

| Health conditions | Causes |
|---|---|
| Helminthiasis | Environment, lifestyle |
| Goiter | Environment, lifestyle |
| Behçet's disease | Genetic |
| Thalassemia | Genetic |
| Cardiovascular disease | Genetic, environment, lifestyle |
| Diabetes mellitus | Genetic, environment, lifestyle |
| Hepatitis | Environment, lifestyle |
| Parasitic diseases | Environment, lifestyle |
| Malaria | Environment, lifestyle |
| Tuberculosis | Environment, lifestyle |
| Rickets | Environment, lifestyle |
| Lactase deficiency | Genetic, environment, lifestyle |
| Sickle cell disease | Genetic, environment |

- Turks tend to consume less alcohol than Americans or Europeans, perhaps as a result of the Muslim culture that discourages more than moderate alcohol use.

### 33.1.6 Nutrition

- Food preferences in Türkiye encompass a diverse range of options, often contributing to a well-rounded and health-conscious diet. Turkish cuisine, while not excessively spicy, is distinguished by meticulous and skillful preparation. The culinary tradition places great value on both the aesthetic presentation and the nutritional quality of dishes.
- Although a typical family dinner in Türkiye might exhibit simplicity, guests are commonly treated to an abundant selection of dishes. The prevailing sentiment is that more food signifies generosity leading to instances where dinner guests are presented with a profusion of culinary offerings. This approach sometimes results in guests finding it challenging to consume everything on their plates, as Turkish hostesses are inclined to offer replacements for consumed items repeatedly. In accordance with etiquette, polite guests typically decline the initial offer, signifying a cultural practice rooted in hospitality and respect.
- Breakfast is typically a simple meal of *beyaz peynir* (white feta cheese), olives, tomatoes, eggs, cucumbers, toast, jam, honey, and Turkish tea. Turks typically eat dinner at about 8 p.m. **Take mealtimes into consideration when teaching Turkish patients about medication therapies.**
- The spectrum of soups in Turkish cuisine spans from delicately flavored options to more substantial choices. Hors d'oeuvres encompass a wide array of petite dishes, ranging from hot to cold, including items such as olive oil-stuffed grape leaves, olives, Circassian chicken with walnut sauce, dried mackerel, roasted chickpeas, and crispy fried savory cheese pastries. These appetizers are frequently complemented by *rakí*, a traditional anisette liquor derived from grapes, typically served over ice with water and sipped at a leisurely pace. This culinary tradition reflects the rich diversity and meticulous attention to detail inherent in Turkish gastronomy.
- Salads in Turkish cuisine consist of a medley of fresh ingredients such as lettuce, tomatoes, cucumbers, and onions, often combined with raw vegetables. These salads are typically dressed with olive oil and either lemon juice or vinegar. Turkish culinary practices often involve the preparation of meat in modest portions, often accompanied by various vegetables, potatoes, or rice. Notable dishes within Turkish gastronomy encompass *köfte*, which are small, well-seasoned meatballs, as well as *kebabs* consisting of skewered beef or lamb complemented by an assortment of vegetables. This culinary selection is emblematic of the meticulous blending of flavors and textures that characterizes Turkish cuisine.
- Yogurt, a culinary creation attributed to the Turkish people, holds a fundamental position within the Turkish diet. Its consumption is often integrated into meals

served at elevated temperatures rather than being relegated to a cold breakfast option. This cultural approach to yogurt underscores its versatile role as a complement to various dishes, emblematic of the inventive and diverse culinary traditions of Türkiye.

- Rice and *börek* are important parts of Turkish culinary tradition. *Börek* is made by wrapping *yufka* (thin sheets of flour-based dough) around meat, cheese, or spinach and then frying or baking until the dough is flaky.
- Turkish desserts can be classified into four distinct categories: opulent and sugary pastries, exemplified by *baklava*; creamy and indulgent puddings; cooked fruits offering a sweet profile; and the inclusion of fresh fruits. A customary practice involves concluding meals with a selection of fresh fruits, often accompanied by coffee or tea. This dessert tradition exemplifies the multifaceted nature of Turkish cuisine, where a diversity of flavors and textures are harmoniously integrated into culinary offerings.
- Renowned for its robust and sweet profile, Turkish *kahve*, a dark and dense coffee, holds a distinguished place in the culinary landscape. *Ayran*, a blend of yogurt and milk, is the prevailing cold beverage in Türkiye, enjoyed by individuals of all ages. It's notable that the Islamic faith mandates abstaining from pork consumption and alcohol consumption, but adherence varies based on individual levels of religious observance. The diversity in practice reflects the nuanced relationship between faith and cultural practices within the Muslim community.
- **Given the diversity among food options in the United States and elsewhere, provide dietary counseling according to the individual's unique food choices and practices.**
- **Ramazan**, the Islamic tradition known as **Ramadan** in Arabic countries, entails a month-long fasting practice observed by devout Muslims across the globe. Throughout the course of Ramazan, individuals refrain from eating or drinking from sunrise until sunset. The extent of adherence to this practice varies considerably, ranging from those who do not observe it to those who meticulously adhere to its tenets. This variance reflects the diverse spectrum of observance within the global Muslim community.
- Within the Sunni Muslim tradition, the commencement of Ramazan observance occurs at the onset of puberty. Some adherents hold the belief that women have an obligation to fast even during pregnancy and the postnatal phase. In contrast, Alevi Muslims do not impose fasting obligations on either gender. Typically, pregnant and postpartum women, individuals traveling, and those dealing with illness are exempt from fasting, although there may be an expectation to compensate for missed days at a later point, thereby illustrating the nuanced approach to fasting regulations across different situations within the Islamic framework. **Health teaching strategies for Turks in America and elsewhere should include the recognition and prevention of dehydration, bloating, constipation, and fatigue during periods of Ramazan fasting (see Chap. 8).**
- Fasting can potentially lead to an array of digestive issues and pose health risks, particularly in the case of pregnant or postnatal women and their infants. The physiological changes associated with fasting may jeopardize the well-being of

both the mother and the child, underscoring the importance of considering individual health conditions and circumstances when observing fasting practices within specific religious contexts. **Provide factual information regarding fasting and the potential effect on some health conditions.**

- Molasses, *baklava*, *lokum* (Turkish delight), tahini, honey and nuts, and raisins are believed to increase strength and sexual vigor.
- Fruits, especially bananas, oranges, tangerines, and apples, are brought to convalescing people, helping them to regain their strength and aiding in the healing process.
- Chicken soup is a common remedy for cold and flu symptoms. An *ebe* (a traditional Türkiye midwife) relies on various herbs and home remedies to heal patients. *Ebegömeci* (a spinach-like leaf or herb) may be prepared for topical or oral use to treat inflammation, infection, and sometimes infertility.
- Tea, cinnamon, hot water with sugar, ginger, mint, and various roots are used separately or in various combinations to treat rheumatism, low blood pressure, intestinal gas, colds, and flu. Nettles may be used topically for rheumatism, arthritis, and varicose veins.
- A traditional folk remedy for diabetes involves the process of boiling olive leaves and subsequently refrigerating the resulting infusion before consuming the extracted juice.
- *Lapa*, a watery rice mixture with a gruel-like texture, or a boiled potato may be used to treat diarrhea and is followed by yogurt to replace the natural flora of the intestines.
- **Ask Turkish patients if they are using traditional dietary practices and incorporate them into prescription therapies. Because malnutrition may be a significant problem among economically disadvantaged Turks immigrating to the United States and elsewhere, consider extensive nutritional assessments for more recent immigrants.**

### *33.1.7 Pregnancy and Childbearing Practices*

- Until the 1960s, the sale of contraceptives and the dissemination of birth control education were prohibited in Türkiye. The legality of abortion and sterilization, with exceptions for medical necessities, remained restricted until 1983. Presently, prevalent contraception methods among Turkish women encompass withdrawal, utilization of intrauterine devices, and the use of birth control pills. This historical progression reflects the evolving landscape of reproductive health practices within Turkish society. **Given the history of family planning in Türkiye, be open to culturally congruent ways to discussions of family planning with Turkish people in the United States and elsewhere.**
- Motherhood holds a position of considerable reverence within Turkish culture, and pregnant women are afforded extensive care and comfort, often accommodating their cravings. Pregnant individuals are commonly encouraged to maintain their routine activities or work as long as they feel at ease. This approach

underscores the cultural esteem for motherhood and the collective effort to ensure the well-being of expectant mothers.

- Many pregnant women take prenatal vitamins, drink a lot of milk, and apply salves such as Vaseline to avoid stretch marks. Light exercise, such as walking, is encouraged, but weather conditions often hamper such efforts because Turks generally avoid wet or cold weather, fearing its ill effects on one's health.
- Most Turks prefer hospitals for physician-assisted child delivery. The husband and the birth mother's father may be present during the birthing process. Expressions of discomfort and pain are quite acceptable.
- The postpartum period can last up to 40 days. Light exercise is encouraged during this period, and bathing, an important part of the Muslim tradition, is strongly encouraged.
- A distinct dietary item known as "*lousalik*" is customarily offered to postpartum women. *Lousalik* is a sweet, sherbet-like substance created by dissolving *lousalik* beads in hot water. This carbohydrate-rich preparation is designed to enhance the woman's vitality during the postpartum phase. Hot soups and fluid consumption, including milk, feature prominently in the diets of postpartum women, particularly those breastfeeding. The practice of breastfeeding is highly esteemed among the majority of Turkish individuals, and while it is commonly embraced, it is typically approached with modesty. This cultural perspective reflects the significance attributed to maternal health and child-rearing practices. **Encourage the incorporation of nonharmful postpartum practices with Western interventions.**
- A customary practice involves the placement of a small blue bead, known as a "nazar boncuk," on the left shoulder of a child. This bead is believed to offer protection against the perceived malevolent influence of the "evil eye," specifically guarding against the influence of the evil angel purportedly whispering into the child's left ear. Additionally, other traditional safeguards include positioning iron beneath the baby's mattress as a precaution against anemia. Likewise, tying a yellow ribbon to the crib is thought to avert jaundice, while affixing a red bow serves as a symbolic countermeasure to deflect any potential envious or negative energies. These practices reflect the cultural significance attributed to safeguarding infants from potential harm and ill intentions. **Do not remove protective beads or ribbons from the infant's bedside.**
- Swaddling, a common practice, has been linked to congenital hip dislocation, pneumonia, and upper respiratory infections among young infants. **Explain the harmful effect of swaddling.**

### *33.1.8 Death Rituals*

- Cremation is not commonly practiced among Turkish Muslims due to their religious conviction that the body should remain intact. Following the passing of an individual, a ritualistic washing of the body takes place, after which it is enshrouded in a white sheet. This practice adheres to the cultural and religious importance placed on preserving the sanctity of the deceased individual's physical form.

- Muslim Turks are placed in the grave, with the right side of the body facing Mecca.
- The traditional mourning period is 40 days, during which traditionally, those attending the ceremony wear black clothes or a black scarf.
- While Muslim Turks hold a belief in the afterlife, the passing of an individual is consistently marked by profound sorrow and mourning. A common expression of condolences to those bereaved is "*Basınız sağ olsun*," which translates to "May your head be healthy." This sentiment carries the hope that the mourning individual will not be overcome by grief. This cultural practice reflects the cultural and emotional sensitivity that surrounds the topic of death within Turkish society.

### *33.1.9 Spirituality*

- The predominant religious affiliation among Turks is Sunni Islam, while a minority belongs to the Alevi Muslim group. Notably, a substantial proportion of Turkish immigrants to Western countries tend to adhere to a more moderate interpretation of Islam. This religious diversity among Turkish expatriates underscores the varying degrees of religious observance and beliefs within this demographic, reflecting the broader spectrum of religious practices observed by Turks residing outside their homeland.
- Traditional prayer is practiced five times each day and can take place anywhere, as long as one is facing the holy city of Mecca. A special small rug called *seccade* is used for praying in places other than the ***cami***, mosque. When entering the *cami*, shoes are always removed, and women must cover their heads. Men and women go to separate parts of the *cami* for prayer. One prepares for prayer by ritual cleansing, which, at minimum, includes washing the face, ears, nostrils, neck, hands to the elbow, and feet and legs to the knee, three times each.
- **Caregivers may need to make special arrangements so that Muslims can practice their religious obligations when they are in a healthcare facility.**
- Spiritual leaders or healers are sought for assistance with relationships or emotional problems. A *muska*, a paper inscribed by a *hoca* (an alderman) with a prayer in Arabic, is wrapped in fabric and hidden in the home or worn by the person seeking help. *Turbe* and *yatir* are the practice of going to the graves of scholars to pray about wishes, mental or emotional problems, or fertility problems. ***Tesbih*** (small beads traditionally used for praying) now take a more secular meaning and are often referred to as "worry beads." **Religious or folk items should not be removed because they provide comfort for the patient and removal may increase anxiety.**

### 33.1.10 Health-Care Practices

- While the majority of Turks predominantly seek health and curative care from Western medicine provided by well-trained professionals, vestiges of traditional beliefs persist in parallel. A prevalent viewpoint attributing the origins of illness to imbalances of hot and cold elements endures within the cultural framework. For instance, ailments like diarrhea are ascribed to an excess of cold or heat, whereas conditions such as pneumonia are associated with extreme cold. This coexistence of modern medical practices and enduring traditional beliefs underscores the complex healthcare landscape in Turkish society.
- Many Turks have a traditional fatalistic worldview, which is more common among those living in extended households who believe "You cannot change what God has written." **Incorporate factual information for the entire family regarding disease causation and treatment in patient education.**
- In the context of terminal illness, it is customary in Turkish medical practices to withhold the full severity of the condition from patients. This approach stems from the belief that imparting such information could potentially diminish hope, motivation, and the vital energy required for healing. Additionally, the disclosure of a terminal diagnosis might intensify patient anxieties pertaining to mortality and the impact on their loved ones. This approach is aligned with the perspective that the ultimate outcome is beyond human control and rests with divine will. This blend of cultural and spiritual considerations underscores the nuanced manner in which medical information is conveyed within the Turkish healthcare framework. **Take cues from the patient and discuss the patient's condition with the family before disclosing the severity to the patient.**
- Within Turkish households, women typically assume the role of primary caregivers for the ailing. Moreover, pivotal healthcare decisions often rest in the hands of the individual held in the highest esteem for their level of education. This arrangement underscores the gender-specific care-taking dynamics and hierarchical decision-making process within the context of health and well-being in Turkish society. **Determine the family spokesperson for decision-making on health issues and incorporate group interventions whenever possible.**
- Türkiye demonstrates elevated rates of over-the-counter antibiotic and painkiller utilization, with aspirin often being embraced as a universal remedy for diverse afflictions, including gastric discomfort. A notable trend involves individuals, particularly those encountering financial constraints with regard to accessing medical services, frequently seeking guidance from pharmacists prior to consulting a physician. This practice reflects the significant role of pharmacists in the healthcare field and underscores the broader healthcare dynamics, including self-care and affordability considerations, prevalent within Turkish society. **Assess Turkish patients for their use of over-the-counter medications to prevent conflicting or potentiating effects with prescription medications.**
- The belief in the "evil eye" holds significant cultural prominence within Turkish society. An inherent caution exists against excessively extolling one's state of

health, as it is thought that doing so may inadvertently attract misfortune through others' envious gazes. This sentiment is so deeply ingrained that even taxi drivers and medical practitioners revere the protective power of the "*nazar boncuk*," a blue bead utilized as a shield against the malevolent effects of the evil eye. When discussing ailments, individuals consciously avoid employing themselves or others as examples, lest such descriptions inadvertently invite affliction upon the person being discussed. This intricate interplay between cultural beliefs and health-related conversations demonstrates the enduring influence of traditional superstitions within Turkish cultural norms.

- The practice of using "kolonya" (cologne) is a cherished tradition that transcends religious and secular boundaries within Turkish culture. Stemming from a foundation of reverence for cleanliness, this practice involves applying cologne to the hands of guests both before and after meals, offering a dual benefit of sanitization and infusing a pleasant lemon fragrance. Additionally, inhaling the aroma from a cloth or handkerchief imbued with a cologne is employed as a remedy for alleviating motion sickness. Notably, within medical settings, patients may extend the gesture of offering cologne to physicians or nurses prior to undergoing an examination. This multifaceted utilization of cologne underscores its integral role in Turkish customs, blending hygiene, comfort, and cultural expressions of hospitality. **Respect this custom as appropriate.**
- Other home remedies include using rubbing alcohol or a wet cloth to bring down a fever and warming the back to treat coughing.
- Turkish culture allows freedom to express pain through either emotional outbursts or verbal complaints. **Accept a wide range of pain expression among Turkish patients.**
- Although stigma is attached to mental illness, many families seek treatment or care for the patient at home. The most common reasons given for mental illness include discord among family members, marital and love problems, gossip, and other familial problems. Social causes include financial inadequacies, societal disorders, and a "struggle with life." Men more frequently blame social causes for mental illness, and women often give psychological causes. **Address somatization among Turkish patients with mental health concerns, and encourage the expression of their feelings. Incorporate group interventions whenever possible.**
- Within the context of hospitalization, the term "*refakatçi*" pertains to the individual who remains overnight with the patient, offering indispensable emotional and physical support and solace. Demonstrating care and compassion serves to alleviate apprehensions and diminish feelings of isolation experienced by the patient. Additionally, family members may take on responsibilities such as aiding with bathing and tending to other physical requirements. Emphasizing a well-rounded, nutritious diet is deemed crucial for the process of recuperation, with a common practice being the provision of home-cooked meals brought from outside the hospital setting. This comprehensive approach underscores the multifaceted nature of Turkish healthcare customs, intertwining emotional well-

being, familial involvement, and dietary considerations in the realm of patient care. **Encourage family-assisted care, and allow families to provide homemade foods.**
- While blood transfusions are gaining acceptance, Turkish people usually prefer to receive blood from family members. Muslims traditionally prefer that the body remain intact after death, a belief that may conflict with organ donation.

### *33.1.11 Health-Care Providers*

- Although Turkish people are inclined toward Western health-seeking behaviors, medical care tends to be holistic. Great value is placed on emotional well-being, especially as it affects physical well-being. Physicians may be "adopted" as members of their patient's families, and it is common to give gifts (usually food) to the physician as an expression of gratitude.
- Generally, treating someone of the opposite sex is not an issue among Turks. **Ask each patient for preference on gender issues in health care.**
- Physicians, and to a lesser extent, nurses and midwives, have historically been held in very high esteem. Patients rarely question the authority of physicians, but the notion of obtaining a second opinion is gaining popularity.

## 33.2 Reflective Exercises

1. Explain the Turkish language and how it might be difficult for Americans to pronounce Turkish names. What might you do to ensure you pronounce a name correctly?
2. Give some examples of the collectivistic Turkish family versus the individualistic culture of European American and other nationalities. What implications does this have for patient education?
3. Identify communication practices related to touch among Turkish people. How does this vary among women, men, their religion, and tradition?
4. Identify genetic conditions common among Turks. Which of them has a lifestyle component?
5. Describe the practices conducted during the holy month of *Ramazan*? How would you change your practice to accommodate a person who is conducting these practices during *Ramazan*?
6. What function does somatization have in Turkish culture? How would you address a patient who is somaticizing?
7. What is a *nazar boncuk*, and what is its function?
8. Identify two folk remedies and the conditions they are used for.

9. Explain why newer Turkish immigrants may easily assimilate into the dominant European American culture?
10. As a nurse, you are visiting a Turkish patient in their home. What do you need to consider before entering their home?

## Bibliography


Bosrock MM (2006) European business customs & manners: a country-by-country guide to European customs and manners. Meadowbrook, Minnetonka

CIA (2024) World Factbook: Turkey Turkiiye. https://www.cia.gov/the-world-factbook/countries/turkey-turkiye/

Encyclopedia Britannica (2023) https://www.britannica.com/place/Turkey

Fassaert T, Hesselink AE, Verhoeff AP (2009) Acculturation and use of health care services by Turkish and Moroccan migrants: a cross-sectional population-based study. BMC Public Health 9:332. http://bmcpublichealth.biomedcentral.com/articles/10.1186/1471-2458-9-332

Hendrich B (2012) Introduction—beyond state Islam: religiosity and spirituality in contemporary Turkey. Eur J Turkish Stud 13 | 2011. Online since 17 Dec 2012, connection on 10 Jul 2023. http://journals.openedition.org/ejts/4527. https://doi.org/10.4000/ejts.4527

Kara B (2017) Self-related health and associated factors in older Turkish adults with type 2 diabetes: a pilot study. J Transcult Nurs 28(1):40–48

Karakas F, Sarigollu E, Kavas MJ (2015) Discourses of collective spirituality and Turkish Islamic ethics: an inquiry into transcendence, connectedness, and virtuousness in Anatolian tigers. J Bus Ethics 129(4):811–822

Ozbassaran F, Ergul S, Temel AB, Aslan GG, Coban A (2011) Turkish nurses' perceptions of spirituality and spiritual care. J Clin Nurs 20:21–22

Senay B (2010) State of origin: Turkish migration and multiculturalism in Australia. In: Erdogan MM (ed) Yurtdışındaki Türkler 50 : Yılında Göç ve Uyum = Turks abroad : migration and integration in its 50th year. Orion, pp 287–296

Serdar U (2011) Traditional fold medicine in the Turkish folk culture. Turk Stud 6(4):317–327

Thobaben M, Kuguoglu S (2013) People of Turkish heritage. In: Purnell L (ed) Transcultural health care: a culturally competent approach (chapter on DavisPlus web site), 4th edn. F.A. Davis Company, Philadelphia

Thobaben M, Kuguoglu S (2021) People of Turkish heritage. In: Purnell L, Fenkl E (eds) Textbook for transcultural health care: a population approach. Springer, Cham. https://doi.org/10.1007/978-3-030-51399-3_28

Yaman A, Mesman J, van Ijzendoom MH, Bakermans-Kranenbutg MJ, Linting M (2010) Parenting in an individualistic culture with a collectivistic cultural background: the case of Turkish immigrant families with toddlers in the Netherlands. J Child Fam Stud 19(5):617–628

# Chapter 34
# People of Ukrainian Heritage

**Tatayana Maltseva and Irina Rozenfeld**

## 34.1 Overview and Heritage

Ukraine is the second-largest European country by area in Eastern Europe. Ethnic Ukrainians belong to the Slavic nationality group. The capital of Ukraine is Kyiv. The national groups residing in Ukraine include Ukrainians, Russians, Belarusians, Bulgarians, Crimean Tatars, Moldovans, Jews, and Hungarians. Russians account for about a sixth of Ukraine's population. Most of the people in the Crimean Peninsula are of Russian ethnicity. Ukraine is one of the world's leading farming regions. During A.D. 800s, Kyiv became the center of a Slavic state that modern historians call Kievan Rus (also spelled *Kyivan Rus)*. Parts of Ukraine were under Polish, Lithuanian, and Austrian control until the late 1700s, when almost all of Ukraine came under Russian control. In 1922, Ukraine became one of the four original republics of the Soviet Union (USSR), but in 1991, Ukraine declared its political independence from the USSR. Ukraine has a significant history of movement for national independence with strong aspirations to join the European Union in recent years. During the post-Soviet period, some Ukrainian citizens left the country for economic reasons and have been working abroad and supporting families at home. In 2014, Maidan Revolution in Kyiv led to the removal of the Ukrainian president from office and the beginning of civilian unrest in the Eastern part of the country and Crimea. In February 2022, the Russia–Ukraine war started,

T. Maltseva (✉)
Nicole Wertheim College of Nursing and Health Sciences, Florida International University, Miami, FL, USA
e-mail: bogopolt@fiu.edu

I. Rozenfeld
Dr. Kiran C. Patel College of Osteopathic Medicine, Nova University, Fort Lauderdale, FL, USA
e-mail: rozenfel@nova.edu



E. A. Fenkl, L. D. Purnell (eds.), *Handbook for Culturally Competent Care*,
https://doi.org/10.1007/978-3-031-70492-5_34

contributing to millions of Ukrainian citizens escaping war zone and significant number of citizens volunteered joining military units on the front line or at the local territorial defense units. A new wave of Ukrainian immigrants to the United States comprised of younger generation Ukrainians who are usually well-educated or have good vocational skills. Many are proficient in English and can find gainful employment upon arrival in the United States. Recently immigrated Ukrainians may have higher level of medical and psychological needs due to exposure to war (Popova and Shevel, 2024).

### *34.1.1 Communications*

- Ukrainians tend to be relatively straightforward and use a direct communication style. They expect others to be clear and open when expressing their opinions. They may not be overly polite or use innocent words discussing sensitive topics. While being direct, they may in some cases use indirect communication. When discussing sensitive or confrontational topics, they may use euphemisms instead of addressing the issue.
- Ukrainians may use **nonverbal communications** via facial expressions, gestures, and eye contact to convey emotions and meanings. **Maintaining eye contact during a conversation is generally seen as a sign of attentiveness and sincerity**. Ukrainians may avoid direct eye contact in communication with authorities.
- Ukrainians often show respect for authority figures. They may use titles and honorifics when addressing superiors in hierarchical organizational settings. It is a sign of respect to address older person by first and middle name.
  - **Personal Relationships**: Ukrainians often prefer to do business with people they know and trust. "***Small Talk***" or polite inquiries about one's health, family, and general well-being are common to build rapport.
  - **Language**: Ukrainian is the official language of Ukraine, although Russian was also widely spoken in the eastern and southern regions. **Language choice can be a sensitive issue, so it's important to use the language preferred by your communication partner**. Many Ukrainians do not understand medical language and have difficulty communicating abstract concepts or understanding American dialects. **It is recommended to use a professional interpreter, speak slowly and clearly with a medium tone of voice. Refrain from a loud conversation when communicating with Ukrainian immigrants.**
  - Ukrainians appreciate active listening. **Provide full attention when someone is talking with you. Nodding and making affirmative sounds may show that you are engaged in the conversation and appreciate the speaker**.
  - Ukrainians may prefer to avoid confrontational or aggressive communication styles. **It's essential to be aware of regional and individual differences**

**when interacting with Ukrainians** in any setting to ensure effective communication.
- Punctuality is a cultural standard. Many Ukrainians arrive early for their scheduled appointments.
- Ukrainians accept touch in the healthcare settings but prefer not to be hugged or kissed by strangers. A handshake is acceptable. They do not appreciate nonverbal gestures such as standing with hands inserted into pockets or arms crossed over the chest. Neither do they appreciate slouching postures when being interviewed. Ukrainians are likely to seek and follow the healthcare recommendation which may appear paternalistic in the Western healthcare system.
- **Address Ukrainians by their last name and title such as Mr., Mrs., Miss, and Ms. You can greet patients with a handshake.**

### *34.1.2 Family Roles and Organization*

Ukrainian family roles are influenced by traditional and cultural norms, but there can be variation in family structures and roles depending on factors such as urban or rural living, generational differences, and individual preferences, there are some common characteristics of Ukrainian family life.

- Extended families with multiple generations living together were common in Ukraine. Extended family arrangement may be less common in urban areas. Extended families provide emotional and financial support and play a role in child-rearing. **Ask patients who they want to be a healthcare surrogate and to include in the medical decisions making process. Do not disclose sexual same-sex orientation to family members.**
- **Traditional gender roles** are still prevalent in Ukraine. Men may be seen as the primary breadwinners, while women are expected to take care of the household and children. However, women have equal decision-making power and some families may be described as maternalistic. In urban areas among younger generations, gender roles are more equitable with women pursuing careers and men taking more domestic responsibilities.
- **Child-rearing**: Children's education is highly valued in the Ukrainian culture. Parents, grandparents, and all members of the extended family may play roles in the care and education of children. There is a strong emphasis on education. Parents invest in children's academic success. **Include grandparents and older family members in health education for their grandchildren.**
- **Respect for elders** is a fundamental aspect of Ukrainian family life. Grandparents are highly respected and are often involved in the lives of their grandchildren. Elders are often consulted for advice and play a significant role in decision-making within the family. **Help families prevent caregiver burden. Provide**

**resources about long-term care facilities if needed and encourage family to visit the facility before a decision is made to place a family member there.**

- Marriage and family values are highly valued. Many Ukrainians place a strong emphasis on the importance of a stable and loving family unit. Divorce rates, while rising, are generally lower compared to some Western countries. Families often provide financial support to their members including assistance with education, housing, or other major life events. Family dynamics are evolving in Ukraine among younger generations. There is a growing emphasis on individualism, career aspirations, and personal choices. Ongoing political and war crisis led to changes in family roles and organization as well as family structures and gender roles.
- **Religious institutions** have a meaningful influence on family life. Religious ceremonies and traditions are often integrated into family events such as weddings and baptisms.
- **Hospitality**: Ukrainians are known for their cheerful family gatherings with extended family members coming together for celebrations and holidays. Food plays a significant role in these gatherings with elaborate traditional meals shared with loved ones.
- **Emigration**: During the last 30 years, economic and political instability have led to a significant emigration of Ukrainians in search of better opportunities abroad. This has sometimes resulted in family members being separated across borders adding emotional challenges to families.

### *34.1.3 Workforce Issues*

Before emigration to the United States, about 70% of the Ukrainians were living in big cities residing in apartments. About 30% of Ukrainians live in rural areas and work on farms. Western Ukraine is more rural, while Eastern Ukraine is heavily industrialized. Nevertheless, many young people leave the countryside to live and work in the cities.

- Ukrainians have a strong **work ethic** and dedication to their jobs. Ukrainians value professionalism in the workplace. This includes dressing appropriately and being respectful to colleagues. Ukrainians communicate directly with colleagues and friends, and this is considered a sign of respect and sincerity. Not showing up for work without prior notification is considered disrespectful. Ukrainians practice politeness, respect, and accountability in the workplace. They value time and support the rule of work and society.
- **Punctuality** is generally expected in the workplace. Being late for meetings or appointments is considered disrespectful so it's important to arrive on time.
- Ukrainians may practice non-direct eye contact when speaking with government officials, a practice common in Ukraine where making eye contact with govern-

ment officials could lead to questioning or even invasion of personal space or disrespect. **Do not assume that the lack of eye contact means the person is not paying attention, does not care, or not providing correct information.**
- The tone of voice may be loud as Ukrainians prefer to prove their point of view by articulating loudly. **Do not interpret a loud voice volume as a sign of anger.**

### *34.1.4 Biocultural Ecology*

- Ukrainians who have immigrated to the United States are predominately white Caucasians and have a physical structure like other white Americans, making them prone to skin cancer.
- **Encourage the use of sunblock and instruct all family members about the dangers of skin cancer.**
- Ukrainian immigrant may have prior exposure to tuberculosis. The burden of tuberculosis (TB) remains significant according to WHO estimates. Ukraine has the fourth highest TB incidence (new cases and relapses) among the 53 countries of the WHO European Region. Other chronic infectious diseases are hepatitis and HIV.
- Ukraine's population faces health risks from a poor diet, smoking, alcohol consumption, and lack of regular physical activity.
- Men are more likely to be exposed to these behavioral risk factors than women.
- Mental health remains a significant challenge. The adult male suicide rate in Ukraine is the highest in the WHO European Region (WHO 2021).
- Physical, psychiatric trauma, and post-traumatic stress disorder can be prevalent conditions in new Ukrainian immigrants due to war exposure and economic and healthcare destruction of the country. Fewer than 2% of Ukrainians seek specialized psychological or psychiatric help and may have impaired social functioning. **Evaluate patients' risk factors for depression, suicide, and post-traumatic stress disorder. Younger age, lack of social support, psychiatric comorbidities, and substance use disorder are significant.**
- The government spending on health in Ukraine is the lowest compared to the European countries, leaving vulnerable groups at risk of impoverishing payments and undermining access to health services. Over half (51%) of all health spending in Ukraine in 2018 was private almost entirely consisting of out-of-pocket payments. **Explain patient governmental resources and structure of healthcare to facilitate obtaining insurance and using healthcare services.**

Table 34.1 lists commonly occurring health conditions for Ukrainians in three categories: lifestyle, environment, and genetics.

**Table 34.1** Top 10 health risk factors common to Ukrainians

| Health condition | Cause |
|---|---|
| 1. High systolic blood pressure | Genetics, environment, lifestyle |
| 2. Cardiovascular disease | Genetics, environment, lifestyle |
| 3. Diabetes mellitus | Genetics, environment, lifestyle |
| 4. Hyperlipidemia and obesity | Genetics, environment, lifestyle |
| 5. Depression and anxiety | Genetics, environment, lifestyle |
| 6. Gastrointestinal disorders | Genetics, environment, lifestyle |
| 7. Respiratory diseases | Genetics, environment, lifestyle |
| 8. Tuberculosis | Environmental, lifestyle |
| 9. Infectious diseases | Environmental, lifestyle |
| 10. Cancer | Genetics, environmental, lifestyle |

### *34.1.5 High-Risk Health Behaviors*

- Smoking is a prevalent health risk behavior in Ukraine with a high rate among men. Smoking is a major contributor to respiratory diseases, cardiovascular diseases, and cancer. **Encourage smoking cessation and assist with finding smoking cessation programs.**
- Ukraine has one of the highest alcohol consumption rates in the world. High levels of alcohol consumption contribute to a range of health problems, including liver disease, alcohol poisoning, accidents, and violence.
- Ukrainian diets are high in salt, saturated fats, and sugar, which can lead to obesity, hypertension, and other diet-related diseases. Insufficient consumption of fruits and vegetables is also a concern.
- A sedentary lifestyle, particularly in urban areas, can contribute to obesity and high levels of diabetes and cardiovascular diseases.
- Unprotected sex and a lack of access to oral contraception as well as awareness about sexually transmitted infections (STIs) lead to the spread of STIs, including HIV/AIDS. **Explain support services in the United States for intimate partner violence. Help patients gain an access to the services.**
- **Injection drug use** is a significant risk factor for the spread of HIV/AIDS and hepatitis C in Ukraine.
- Ukrainians may have **low vaccination rates** for preventable diseases. Inquire about prior vaccinations, and advise about possible health risks, especially among children.
- Some Ukrainian cities, particularly Kyiv and industrial areas, have **high level of air pollution** which contribute to respiratory problems and cardiovascular diseases in Ukrainians.
- Lack of awareness about mental health issues can deter individuals from seeking help and leading to potentially risky behaviors. **Screen all newer immigrants for conditions common in their home country. Provide scientific information and assist in finding appropriate services and resources.**

### 34.1.6 Nutrition

- The Ukrainian diet includes pork, beef, goat, lamb, rabbits, chicken, duck, fish, and various cold-cut products made from pork such as ham, sausage, and bacon. Ukrainians eat large amounts of vegetables such as potatoes, cabbage, tomatoes, cucumbers, and eggplants, cooked and pickled. Breads are among staple foods: white, sour dough, dark rye, and numerous sweetened choices. Grains are popular as a side dish. A prevalent grain is buckwheat, called *kasha.* Popular drinks include tea, coffee, kvas, cocoa, kefir (special soured milk drink), honey and cherry liquor, and vodka with pepper (horilka).
- Traditional Ukrainian dishes include *varenyky, borsch,* holubtsi, and *holushki.* Varenyky consists of boiled dumplings filled with potatoes, sauerkraut, cheese, or cherries. Ukrainians eat dumplings with sour cream-fried onions or bacon bits. Borsch is a soup of meat, beets, carrots, tomatoes, and cabbage served with bread and sour cream. Holubtsi are stuffed cabbage rolls filled with rice, buckwheat, and meat.

### 34.1.7 Pregnancy and Childbearing Practices

Pregnancy and childbearing practices are similar to the Russian culture.

- Contraception for Ukrainian women is allowed without sanctions or taboos. Condoms, followed by intrauterine devices and birth control pills are available. Ukrainian males do not consider vasectomies as a method of contraception. Many new immigrants may not be aware of different methods of fertility control.
- **Inform patients of the variety of birth control methods common in the United States.**
- Pregnant women have regular prenatal checkups, which are mandatory in Ukraine. During pregnancy, women are encouraged to rest. They do not perform heavy lifting, hard work, and strenuous physical activities; they are also protected from bad news or unpleasant information that can be harmful to the fetus. Pregnant women are encouraged to eat foods that are high in iron, calcium, and vitamins. Strawberries, citrus fruits, peanuts, and chocolate are avoided to prevent allergies in the newborn.
- As labor approaches, women take laxatives and enemas to facilitate delivery. Recently husbands in Ukraine are allowed to participate in the delivery or visit the hospital postpartum. There are no cultural restrictions for fathers or female relatives not to participate in delivery in United States. **Ask patients who they wish to be involved in the delivery.**
- Breastfeeding is encouraged as Ukrainians believe that breast milk will increase immune system of the newborn. Some women breast-feed until the infant reaches the toddler stage.

- Peri-care with warm water is important, and a binder is worn to help the mother's figure return to normal.
- In Ukraine, women receive maternity leave before delivery and after newborn delivery up to the 3 years. **Explain family medical leave practices for expectant parents in the United States.**

### 34.1.8 Death Rituals

- Ukrainian' families protect the terminally ill family member. They want to be notified about imminent death first before the patient is told.
- Most families prefer to care for the dying family member at home.
- **Facilitate transferring the dying person to the home if the family wishes and has the means to provide terminal care. Assist with obtaining hospice home care.**
- Ukrainian families believe that everything should be done to save a loved one and often will not agree to disconnect the life support. They will not discuss advance directive, a living will, durable power of attorney, do not resuscitate procedures, and might not sign these documents. Ukrainian Orthodox pastor may perform and institute a special prayer, hymn, and vigil over the deceased.
- **Ask permission from the family before contacting clergy for their preferences.**
- Ukrainian rural families may not accept cremation; most prefer interment. Ukrainians living in the major cities may consider cremation.
- Both men and women may wear black as a sign of mourning and mirrors are covered of the deceased's home. Expression of grief varies greatly. Ukrainian Orthodox family mourns the death of a family member for at least 40 days. **Recognize a wide variety of grieving and bereavement practices in Ukrainian culture.**

### 34.1.9 Spirituality

- Ukrainians are closely associated with the Ukrainian Orthodox Church and Catholic Church. Faith plays a role in internal spirituality. The Ukrainian government recognizes Orthodox Christianity, Buddhism, Islam, and Judaism as established religions.
- Ukrainian Americans often practice religion and attend services in Ukrainian Churches. **Discuss individual religious practices and spirituality with patients and family during the interview.**

### 34.1.10 Healthcare Practices

- Most Ukrainians define *health* as the absence of disease.
- Ukrainian immigrants will need to become more familiar with the healthcare insurance system in the United States as Ukraine has a public healthcare system. Some clinics and hospitals in Ukraine are private, and patients pay out of pocket for their treatment. **Explain health reimbursement procedures in the United States and elicit assistance from case managers to obtain health insurance coverage as needed.**
- Hospitalization in Ukraine lasts between 3 and 4 weeks. Some patients may expect the same in the United States. Explain the hospitalization stays in the United States. **Refer patient to the home health care agencies if needed upon discharge and teach family how to care for recuperating family member at home.**
- Single females do not do Pap tests in Ukraine; only, married women perform them in women center clinics. Explain the significance of Pap tests and mammography, regardless of the woman's marital status.
- Many individuals prefer staying warm to prevent colds and other illnesses. Most like to keep the temperature warm in their rooms. They may also be unwilling to add ice or drink cold drinks with ice. Explain the application of ice compress in exact terms. **Ask if a patient wants ice in their drinks.**
- Most Ukrainians consider pain a sign of weakness or fear of side effects of opioid medications and may not ask for pain medicine. **Offer and encourage pain medicine and explain that it will help with the healing process.**
- Patients are not comfortable to being told about terminal diagnoses or cancer because it is believed to make the condition worse. **Inform patients about poor prognosis gradually on their terms and preferably in several meetings. Consult with family members first and pay attention on their verbal and nonverbal cues for readiness to hear news about a terminal illness.**
- Cupping is a common procedure in Ukraine for treatment of a respiratory infection in healthcare settings and at home. Do not mistake circular ecchymotic areas on children or adults back as abuse. **Ask if they have been practicing cupping.** Alcohol-saturated cotton or other materials is introduced to a cup. The cotton or material is lighted inside the cup and removed; then, the hot cup is turned upside down on the patient's back. The skin is drawn into the cup by vacuum. The cup is pulled off the skin within 10–15 min leaving round ecchymotic areas 1–1½ in. in diameter.
- Common cultural practices include using herbal remedies and tinctures available from pharmacies, soaking one's feet in warm water for a cold, gargling with salty water for sore throat, aromatherapy for a variety of respiratory illnesses, mud and mineral baths to promote healing, and herbs and teas for fever and colds. **Incorporate nonharmful folk practices into allopathic prescriptions.**

- Ukrainians are accustomed to seeing multiple healthcare providers. Each provider may give prescriptions. **Specifically ask patients if they are seeing other healthcare providers and taking any prescription medications, over-the-counter medications, or traditional herbal remedies and teas.**
- Ukrainians may not disclose any mental health issues unless they are experiencing crisis. They might present with vague somatic complaints because mental illness carries a significant stigma and is considered as a weakness. **Screen for anxiety, depression, and post-traumatic stress disorder in immigrants.**
- Ukrainians may have a great fear of contracting HIV from blood transfusions. **Explain US safety procedures with blood transfusions and organ donation.**

### *34.1.11 Healthcare Providers*

- Healthcare providers are very well respected by Ukrainian immigrants. Because nurses' function in higher roles in the United States, Ukrainian immigrants may not understand their independent roles. **Explain the roles and education of nurses in the United States.**
- Men and women are accustomed to living together in very small apartments or in communal living; thus, most do not have a problem with privacy. Healthcare professionals' gender is not generally a concern in providing care.
- **Always ask about gender concerns before performing breast pelvic or rectal exams on patients.**

## 34.2 Reflective Exercises

1. What are traditional verbal and nonverbal communications practices in Ukrainian culture?
2. How do Ukrainians value their time and punctuality?
3. What are the decision-making practices in current Ukrainian immigrants?
4. List one healthcare condition in Ukrainians in each of the three categories: environment, lifestyle, and genetics. What can you do to decrease the incidence and prevalence of these diseases?
5. What are common high-risk health behaviors in Ukrainian immigrants? Include some strategies for reducing them.
6. What are fertility control practices in Ukrainians? What education would you provide to the family about childbirth and contraception practices?
7. Many Ukrainians are not seeking psychiatric care. What is a reason for this? What can you do to decrease their fear?
8. Identify traditional folk treatments. How would you know that this treatment is being used?

9. Identify some practices women use during pregnancy.
10. Identify some cultural traditional Ukrainian practices. Which of them might have harmful effects?

## Bibliography


BBC News (2023) Ukraine country profile. https://www.bbc.com/news/world-europe-18018002

CIA (2024) World Factbook: Ukraine. https://www.cia.gov/the-world-factbook/countries/ukraine/

Eckemoff EH, Sudha S, Wan D (2018) End of life care for older Russian immigrants—perspectives of Russian immigrants and hospice staff. J Cross Cult Gerontol 33:229–245. https://doi.org/10.1007/s10823-018-9353-9

Handtke O, Schilgen B, Mösko (2019) Culturally competent healthcare—a scoping review of strategies implemented in healthcare organizations and a model of culturally competent healthcare provision. PLoS One 14:e0219971. https://doi.org/10.1371/journal.pone.0219971

Maltseva T (2021) People of Russian heritage. In: Purnell LD, Fenkl EA (eds) Textbook for transcultural health care: a population approach, 5th edn. Springer, Cham

Nersessova KS, Jurcik T, Hulsey TL (2019) Differences in beliefs and attitudes toward depression and schizophrenia in Russia and the United States. Int J Soc Psychiatry 65(5):388–398. https://doi.org/10.1177/0020764019850220

Orekhovskaya NA, Chistyakov AA, Kryukova NI, Krokhina JA, Ospennikov YV, Makarova EV (2019) Orthodoxy and modernity their contact facets in Russian society. Eur J Sci Theol 15(2):67–77

Popova M, Shevel O (2024) Russia and Ukraine. Entangled histories, diverging States. Cambridge, London: Polity Press.

Slovo i dilo (2022) ПТСР и обострение хронических болезней: Как война повлияет на здоровье украинцев. https://www.slovoidilo.ua/2022/11/01/infografika/suspilstvo/ptsr-ta-zahostrennya-xronichnyx-xvorob-yak-vijna-vplyne-zdorovya-ukrayincziv

Slovo i dilo. Analytic portal (2023) Mortality rates: most causes of death in Ukraine. www.slovoidilo.ua

Stoeckl K (2020) The rise of the Russian Christian right: the case of the World Congress of Families. Relig State Soc 48(4):223–238. https://doi.org/10.1080/09637494.2020.1796172

U.S. Census Bureau (2023) Russia—demographic data as of July 1, 2023. https://www.census.gov/popclock/world/rs

WHO (2021) European observatory on health and policies (2021). Health systems in action: Ukraine. https://iris.who.int/bitstream/handle/10665/349235/9789289051859-eng.pdf?sequence=1

WHO (2022) Assessing health care in Ukraine after 8 months of war: the health system remains resilient, but key health services and medicine are increasingly unaffordable. News release. https://www.who.int/europe/news/item/24-10-2022-accessing-health-care-in-ukraine-after-8-months-of-war%2D%2Dthe-health-system-remains-resilient%2D%2Dbut-key-health-services-and-medicine-are-increasingly-unaffordable

# Chapter 35
# People of Vietnamese Heritage

**Carol O. Long, Tracy Tram Tran, and Timothy Minh Tran**

## 35.1 Overview and Heritage

Most Vietnamese living in the United States have arrived since 1975. Their departures from Vietnam were often precipitous and tragic. Escape attempts were long, harrowing, and, for many, fatal. Survivors were often placed in squalid refugee camps for years. Many departed Vietnam in small, often unseaworthy and overcrowded vessels in hopes of reaching a non-Communist port. Half died during their journey. Many were forcibly repatriated to Vietnam or eventually returned voluntarily; others continued to languish in camps. The 1979 Orderly Departure Program provided safe and legal exit for the Vietnamese to reunite with family members already in the United States (referring specifically to the United States and not North, South, or Central America). In 1987, The Amerasian Homecoming Act provided for the entry of former South Vietnamese military officers, other political detainees, children of American servicemen, and Vietnamese women and their close relatives. The Vietnamese in America differ substantially, depending on acculturation, individuality, and the variant characteristics of culture as described in Chap. 1. Vietnamese place a high value on education and accord scholars an honored place in society. The teacher is highly respected as a symbol of learning and culture.

C. O. Long (✉)
Transcultural and Palliative Care Consultant, Fredericksburg, VA, USA

T. T. Tran
Banner Desert Medical Center, Mesa, AZ, USA

T. M. Tran
Gila River Health Care, Sacaton, AZ, USA





E. A. Fenkl, L. D. Purnell (eds.), *Handbook for Culturally Competent Care*,
https://doi.org/10.1007/978-3-031-70492-5_35

### *35.1.1 Communication*

- Vietnamese is a single distinctive language with northern, central, and southern dialects, all of which can be understood by individuals speaking any one of these dialects. Vietnamese is the only language of the Asian mainland that is regularly written in the Roman alphabet.
- All words in Vietnamese consist of a single syllable, although two words are commonly joined with a hyphen to form a new word. Each vowel can be spoken in five or six tones that may completely change the meaning of the word. Vietnamese language also tends to be more descriptive. For example, if you say *mau xanh*, it could either be "blue" or "green." However, Vietnamese would describe it as *mau xanh da troi* which means "color of the sky" for blue, whereas *mau xanh la cay* which means "color of a leaf." **To avoid misunderstandings, do not give directions such as take one "blue" pill at a specified time. Instead, provide the name and dosage of the medication.**
- In the same context, Vietnamese tend to pretend to understand the instructions given or what was told to them to avoid creating disharmony. Ask the patient to repeat the instruction or their understanding of what was said back to you. **Ask open-ended questions, and have the patients demonstrate rather than verbalize their understanding of treatments.**
- *Am* and *duong* translate to *yin* and *yang*. This is used in a different context such as when you are talking about the effect (hot or cold) (*duong* or *am*) certain types of food have on the body. Whereas *nong* (hot) or *lanh* (cold) are used to express physical feelings associated with fever and chills. **See Sect. 35.1.6 in this chapter for a fuller understanding about hot and cold.**
- English skills may not be sufficient to communicate in psychiatric interviews, which are usually carried out at a highly abstract level. **Observe patients for behavioral cues, use simple sentences, paraphrase words that may have multiple meanings, and avoid metaphors and idiomatic expressions. Explain all points carefully. Approach patients in a quiet, unhurried manner. Provide an interpreter as needed.**
- Self-control, another traditional value, encourages keeping to oneself. Expressions of disagreement that may irritate or offend another person are avoided. Individuals may be in pain, distraught, or unhappy, yet rarely complain to health-care providers. Some Vietnamese tend to not like to impose on other people for health care. Thus, they tend to use the self-care method to take care of themselves. Going to see a health-care provider is usually a last resort.
- Expressions of emotion are considered a weakness that may interfere with self-control. Negative emotions and expressions may be conveyed by silence or a reluctant smile. A smile may express joy or convey stoicism in the face of difficulty or may indicate an apology for a minor social offense. A smile may also be a response to a scolding, as a way to show sincere acknowledgment for the wrongdoing. **Clarify the meaning of a smile.**

- The head is a sacred part of the body that should not be touched. To place one's feet on a desk is considered offensive. **If it is medically necessary to touch the head, provide an explanation and ask permission. Do not place feet on a table or desk.**
- Looking another person directly in the eyes may be deemed disrespectful.
- To signal for someone to come closer using an upturned finger is considered a provocation, usually used to call a dog. Waving the hand is considered more proper. **Do not point when beckoning patients; call them by name.**
- Hugging and kissing are not seen outside the privacy of the home. Men greet each other with a handshake, but they do not shake hands with a woman unless she offers her hand first. **Greet men with a handshake. Men should wait for the woman to extend her hand for a greeting.**
- Two men or two women can walk hand in hand without implying sexual connotations.
- Women may be reluctant to discuss sex, childbearing, or contraception when men are present.
- Traditional Vietnamese prefer more distance during personal and social relationships than the European American culture.
- Not keeping appointments may relate to not understanding oral or written instructions or not knowing how to use the telephone. More acculturated patients understand the significance of punctuality.
- Little attention is given to one's precise age. Birth dates may pass unnoticed, with everyone celebrating birthdays together during the Lunar New Year (*Tet*) in January or February. A person's age is calculated from the time of conception; children are considered to be a year old at birth and gain a year each *Tet*. A child born just before *Tet* could be regarded as 2 years old when only a few days old. Because of the difficulty in determining age, many immigrants use January 1 as a date of birth for official records.
- Most names consist of a family name, a middle name, and a given name of one or two words, always written in that order. There are relatively few family names, with Nguyen (pronounced "nwin") and Tran accounting for more than half of all Vietnamese names. Other common family names are Cao, Dinh, Hoang, Le, Ly, Ngo, Phan, and Pho. There are relatively few middle names, with Van being used regularly for men and Thi (pronounced "tee") for women. A typical woman's name is Tran Thi Thu. That is how she would write or give her name if requested. She would expect to be called simply Thu or sometimes *Chi* (sister) Thu by friends and family. In other situations, she would expect to be addressed as *Cô* (Miss) or *Ba* (Mrs.) Thu. If married to a man named Nguyen Van Kha, the proper way to address her would be as Mrs. Kha, but she would retain her full three-part maiden name for formal purposes. The man would always be known as *Kha* or *Ong* (Mr.) Kha. Some Vietnamese American women have adopted their husband's family name. Children always take the father's family name.
- Vietnamese are very hierarchical. Depending on where they are in age, you would use a certain term to denote where they are in the hierarchy as a form of respect. For example, if they are similar in age to your parents but a little younger,

you would call them *Co* or *Chu* as in *Co Thu* or *Chu Kha*. If they are older than your parents, you would call them *Bac Thu* or *Bac Kha*. If they are similar in age as you, you would call them either *Em* if they are younger or *Anh* if they are older and male or *Chi* if they are older and female. If you don't know their age, out of respect always assume they are older than you. If they are similar to your children age, you would call them either *Chau* or *Con*. **Ask women their preferred name as well as their legal name.**

### *35.1.2 Family Roles and Organization*

- The family is the main reference point for the individual throughout life, superseding obligations to country, religion, and self. The traditional family is patriarchal with an extended family structure. Men deal with matters outside the home. A wife is expected to be dutiful and respectful toward her husband and his parents. Women often make family health-care decisions.
- Children are expected to be obedient and devoted to their parents and to worship their memory after death. The eldest son is usually responsible for rituals honoring the memory and invoking the blessings of departed ancestors. A son's obligations and duties to his parents may assume a higher value than those to his wife or children.
- For the first 2 years of life, mothers primarily care for their children; thereafter, grandmothers and others take on much of this responsibility.
- Exposure of the younger generation to American culture can become a source of conflict with considerable family strain when adolescents are influenced by the perceived American values of individuality, independence, self-assertion, and egalitarian relationships. **Counsel with parents and children together and discuss cultural values to help them reach satisfactory outcomes.**
- Unattached male refugees may join ***pseudofamilies***, households made up of close and distant relatives and friends who share accommodations, finances, and companionship.
- ***Pseudofamilies* form an important source of social support in the refugee communities.**
- Differing sexual orientation is difficult for Vietnamese to face because of the shame it can bring to the family. When questioning a lesbian, gay, bisexual or transgender (a more inclusive LGBT initialism), queer or questioning, intersex or asexual individuals, as well as those who identify beyond these commonly used sexuality and gender expressions their sexual orientation, an interpreter unknown to the family is an absolute requirement. **Do not disclose these conversations or relationships to family members**.
- Traditional Vietnamese are class-conscious and rarely associate with individuals at different levels of society. Respect is accorded to people in authoritative positions who are well educated or otherwise successful or who have professional titles.

### 35.1.3 Workforce Issues

- Priority is given to the concerns of the family, rather than to those of the employer.
- **Explain the necessity of punctuality in the workforce.**
- The Vietnamese are highly adaptable and adjust their work habits to meet requirements.
- Most Vietnamese respect authority figures with impressive titles, achievements, education, and a harmonious work environment.
- Confucian cultural background results in conformity and reluctance to undertake independent action.
- Many fear losing their job if they speak out about inequities; they are likely to be taken advantage of by some more unscrupulous employers. **Allow extra time, provide visually oriented instructions, and provide programs that enhance communications to promote increased harmony in the workplace.**
- Many individuals cannot easily give a blunt "no" as an answer because they feel that such an answer may create disharmony. A "yes" response, rather than expressing a positive answer or agreement, may simply reflect an avoidance of confrontation or a desire to please the other person. **Ask open-ended questions and have employees demonstrate rather than verbalize their understanding of procedures.**
- Looking another person directly in the eyes may be deemed disrespectful. **Do not assume that lack of eye contact means the person is not listening, does not care, or is being less than truthful.**
- **Do not take offense if the Vietnamese employee stands at a greater distance away from how you are accustomed.**

### 35.1.4 Biocultural Ecology

- Skin color ranges from pale ivory to dark brown. Mongolian spots, bluish discolorations on the lower back of a newborn child, are normal hyperpigmented areas in many Asians. **Assess for oxygenation and cyanosis in dark-skinned Vietnamese by examining the sclera, conjunctiva, buccal mucosa, tongue, lips, nail beds, palms of the hands, and soles of the feet. Assess for patches of melanin in the buccal mucosa and the conjunctiva for petechiae and rashes. Assess for jaundice by observing for yellow discoloration of the conjunctiva.**
- Most Vietnamese are small in physical stature. Adult women average 5 ft tall and weigh 80–100 lb. Men average a few inches taller and weigh 110–130 lb.
- Typical physical features include inner eye folds that make the eyes look almond shaped, sparse body hair, and coarse head hair. They also have dry earwax, which is gray and brittle. People with dry earwax have few apocrine glands, especially in the underarm area, and thus produce less sweat and associated body odor.

- Children are small by US standards. American growth charts do not provide adequate assessments for evaluating the physical development of Vietnamese children. Standing, walking, and language skills begin at a slightly later age in Vietnamese children, but they rapidly catch up with European American norms by the age of 1½–2 years.
- Table 35.1 lists commonly occurring health conditions for Vietnamese in three categories: lifestyle, environment, and genetics. See Chap. 2.
- Betel-nut pigmentation may be found in some adults, resulting from the practice of chewing *chau* (betel leaves) that is common among older women for its narcotic effect on diseased gums. Some elderly women lacquer their teeth, believing that it strengthens the teeth and symbolizes beauty and wealth.
- Cancer and other problems common to this group may be associated with the widespread application of chemical agents during the Vietnam War. Explain the importance of cancer screening. **For recent immigrants or older adults, assess patients for tuberculosis, parasitism, anemia, malaria, and hepatitis B.**
- Melioidosis and paragonimiasis may mimic tuberculosis. Microcytosis may be misdiagnosed as iron deficiency. Inappropriate treatment with iron may be initiated. Erythrocytic microcytosis is a reflection of the presence of thalassemia or of hemoglobin E trait. **Screen for these disorders in people with findings consistent with tuberculosis but with a negative purified protein derivative response.**

**Table 35.1** Vietnamese health conditions

| Health conditions | Causes |
|---|---|
| Cervical cancer | Environment, lifestyle |
| Depression and post-traumatic stress syndrome | Environment, lifestyle |
| Tuberculosis | Environment, lifestyle |
| Hepatitis | Environment, lifestyle |
| Malaria | Environment, lifestyle |
| Parasitosis | Environment, lifestyle |
| Thalassemia | Genetic |
| Liver cancer | Environment, lifestyle |
| Trichinosis | Environment, lifestyle |
| Nasopharyngeal cancer | Environment, lifestyle |
| Lactase deficiency | Genetic |
| Leprosy | Genetic |

- Most Vietnamese people are slow metabolizers of alcohol, making them more sensitive to the adverse effects of alcohol as expressed by facial flushing, palpitation, and tachycardia. They are twice as sensitive to the effects of propranolol on blood pressure and heart rate, experience a greater increase in heart rate from atropine, and require lower doses of benzodiazepines, diazepam, and alprazolam. Because of their increased sensitivity to the sedative effects of these drugs, many require lower doses of imipramine, desipramine, amitriptyline, and clomipramine. They are less sensitive to cardiovascular and respiratory side effects of analgesics (for example, morphine) but are more sensitive to their gastrointestinal side effects.
- **Most Asians require lower doses of neuroleptics. Because Vietnamese people are considerably smaller than most European Americans and many other groups, medication dosages may need to be reduced.**

### *35.1.5 High-Risk Health Behaviors*

- Some newer immigrants have never heard of cancer; among those who have heard of it, some believe it is contagious. Many women have never heard of or had a Pap test. Many have never performed breast self-examination or had a mammogram. **Explain the necessity of Pap tests, mammograms, and breast self-exams.**
- Vietnamese drink a lot of alcohol, mainly beer, wine, and cognac, especially at parties and weddings. Consider screening for liver diseases such as fatty liver and cirrhosis.
- High rates of gastrointestinal cancer may be due to asbestos that, in some parts of the world, is used in the process of "polishing" rice. **Remind patients that polished, imported rice should always be washed.**
- Many do not know that cigarette smoking can cause cancer. **Explain the adverse effects of cigarette smoking and encourage cessation, starting as early as in elementary school.**
- Trichinosis risk is 25 times greater in Southeast Asian refugees than in the general population. This increased risk is related to undercooking pork and purchasing pigs directly from farms. **Screen newer immigrants for the possibility of trichinosis if symptomatology warrants.**

### *35.1.6 Nutrition*

- Meal preparation is precise and may occupy much of the day. Celebrations and holidays involve elaborately prepared meals.

- The normal daily caloric intake of the Vietnamese is approximately two-thirds that of average Americans. White or polished rice is the main staple in the diet, providing up to 80% of daily calories. Other common foods are fish (including shellfish), pork, chicken, soybean curd (tofu), noodles, various soups, and green vegetables. Preferred fruits are bananas, mangoes, papayas, oranges, coconuts, pineapples, and grapefruit. Soy sauce, garlic, onions, ginger root, lemon, and chili peppers are used as seasoning. Rice and other foods are commonly served with *nuoc mam* (a salty, marinated fish sauce). A meal typically consists of rice, *nuoc mam*, and a variety of other seasonings, green vegetables, and sometimes meat cut into slivers. Chicken and duck eggs may be used.
- Most Vietnamese prefer white bread, particularly French loaves and rolls, and pastry. A regular dish is *pho*, a soup containing rice noodles, thinly sliced beef or chicken.
- Food preparation includes *com chien* (fried rice) and *thit bo xao ca chua* (beef fried with tomatoes). *Cha gio* (pronounced "cha-yuh") is a combination of finely chopped vegetables, mushrooms, meat, or bean curd that is rolled into delicate rice paper and deep-fried. Stir-frying, steaming, roasting, and boiling are the preferred methods of cooking. Hot tea is the usual beverage.
- A predominant aspect of the traditional Asian system of health maintenance is the principle of balance between two opposing natural forces, known as *am* (cold) and *duong* (hot) in Vietnamese. The terms have nothing to do with temperature and are only partly associated with seasoning. Rice, flour, potatoes, most fruits and vegetables, fish, duck, and other things that grow in water are considered cold. Most other meats, fish sauce, eggs, spices, peppers, onions, candies, and sweets are hot. Tea is cold, coffee is hot, water is cold, and ice is hot.
- Illness or trauma may require therapeutic adjustment of hot–cold balance to restore equilibrium. Hot foods and beverages, used to replace and strengthen the blood, are preferred after surgery or childbirth. During illness, certain foods, such as *chao* (light rice gruel or porridge) and a few pieces of salty pork cooked with fish sauce, are consumed in greater quantity. Chao is the preferred food because it is easy on the palette, requires minimal effort to chew, and is easy to digest when you don't feel good and tired. Fresh fruits and vegetables are usually avoided, being considered too cold. Water, juices, and other cold drinks are restricted. **Nutritional counseling should take into consideration these factors and other aspects of the usual Vietnamese diet because professional advice to eat certain kinds of American foods may be ignored.**
- Most adults and many children suffer from lactose intolerance, which may cause problems in schools, other institutional settings, and adoptive families. **Consider the use of substitute milk products that use soybeans for those who are lactose intolerant.**

### 35.1.7 Pregnancy and Childbearing Practices

- Women have high fertility rates (average of six pregnancies, with four live births) and commonly have children until their early to middle 40s. Many know little about contraception. Abortions are commonly performed in their homeland because pregnancy outside of marriage is considered a disgrace to the family. More acculturated individuals practice some form of birth control. **Avoid family planning issues on the first encounter; but such information is usually well received on subsequent visits.**
- Prescriptive food practices for a healthy pregnancy include noodles, sweets, sour foods, and fruit and exclude fish, salty foods, and rice. After birth, to restore equilibrium and provide adequate warmth to the breast milk, women consume soups with chili peppers, salty fish and meat dishes, and wine steeped with herbs.
- Foods are also classified as *tonic* and *wind*. Tonic foods include animal protein, fat, sugar, and carbohydrates; they are usually also hot and sweet. Sour and sometimes raw and cold foods are classified as *antitonic*. Wind foods, often classified as cold, include leafy vegetables, fruit, beef, mutton, fowl, fish, and glutinous rice. It is considered critical to increase or decrease foods in various categories to restore bodily balances upset by unusual or stressful conditions such as pregnancy. While the balance of foods may be followed, the terminology is not used consistently.
- Until now, older generations such as grandmothers, mothers, and aunties will restrict certain foods for pregnant women such as watermelon (red inside/green outside is considered not good for the baby). A lot of the restrictions are based on superstitions rather than health based. For example: "don't drink coffee because it will give baby dark skin."
- During the first trimester, the expectant mother is considered to be in a weak, cold, and antitonic state. Therefore, she should correct the imbalance by eating and drinking hot foods, such as ripe mangoes, grapes, ginger, peppers, alcohol, and coffee. To provide energy and food for the fetus, she is prescribed tonic foods, including a basic diet of steamed rice and pork.
- Cold foods, including mung beans, green coconut, spinach, and melon, and antitonic foods, such as vinegar, pineapple, and lemon, are avoided during the first trimester. In the second trimester, the pregnant woman is considered to be in a neutral state. Cold foods are introduced, and the tonic diet is continued.
- During the third trimester, when the woman may feel hot and suffer from indigestion and constipation, cold foods are prescribed, and hot foods are avoided or strictly limited. Tonic foods, which are believed to increase birth weight, are restricted to reduce the chances of a large baby. Wind foods are generally avoided throughout pregnancy, as they are associated with convulsions, allergic reactions, asthma, and other problems. This regimen may appear more complex and restrictive than it actually is in practice. Most women use it only as a general

guide, commonly restricting, rather than totally abstaining from the restricted foods. A great variety of foods, including rice, many kinds of vegetables and fruits, various seasonings, and certain meats and fish are generally permissible throughout pregnancy. **Ask women at each stage of their pregnancy about preferred foods before initiating dietary recommendations. Dispel any myths and recommend foods that are congruent with beliefs.**

- Many women do not seek medical attention until the third trimester because of cost, fear, or lack of perceived need. Women who are generally better educated seek early prenatal care. For obstetric and gynecologic matters, they tend to feel more comfortable with a female physician or midwife.
- Women maintain physical activity to keep the fetus moving and to prevent edema, miscarriage, or premature delivery. It is believed that prolonged labor may result from idleness, and an undesirably large baby may result from afternoon napping.
- Restrictive beliefs include avoiding heavy lifting and strenuous work and raising the arms above the head, which pulls on the placenta causing it to break. Sexual relations late in pregnancy may cause respiratory stress in the infant.
- Many consider it taboo for pregnant women to attend weddings or funerals.
- Some pregnant women look at pictures of happy families and healthy children, believing that it helps give birth to healthy babies.
- Anesthesia during labor and delivery is usually accepted. Once in labor, the woman tries to maintain self-control and may even smile continuously. Her period of labor may be usually short, and there may be no warning of impending delivery.
- **Because the head is considered sacred, neither the mother nor the infant's head should be touched or stroked. Removal of vernix from the infant's head can cause distress. If it is necessary to touch the head for medical treatment, an explanation is essential. Stress the importance and necessity of any invasive procedures that are required.**
- Because body heat is lost during delivery, women avoid cold foods and beverages and increase the consumption of hot foods to replace and strengthen their blood. Ice water and other cold drinks are usually not welcome, and most raw vegetables, fruits, and sour items are consumed in lesser amounts. Prescriptive foods include steamed rice, fish sauce, pork, chicken, eggs, soups with chili or black peppers, other highly seasoned and salty items, wine, and sweets.
- Because water is *am* (cold), women traditionally do not fully bathe, shower, or wash their hair for a month after delivery. This is still in practice, even for Vietnamese Americans and is totally opposite of Western medicine, which is moving and showering right after birth. Some women have complained that they were adversely affected by showering shortly after delivery in American hospitals; others however welcomed the opportunity to shower and seemed willing to give up some traditional practices. Be mindful of individual choice with these decisions.
- Postpartum women avoid drafts and strenuous activity; wear warm clothing; stay in bed, indoors, or both for about a month; and some avoid sexual intercourse for

months. Many use hot water bottles or electric blankets to combat the cold forces of childbirth.

- Other women in the family assume responsibility for the baby's care. In Vietnam, husbands would never be present at their child's delivery; this may also occur in the United States and elsewhere. **Do not interpret the mother's inactivity and dependence on others as apathy or depression. Do not praise the baby, lest jealous spirits steal the infant. Do not cut the child's hair or nails for fear that this might cause illness.**
- The infant is generally maintained on a diet of milk for the first year, with the introduction of rice gruel at around 6 months.
- Some women discard colostrum and feed the baby rice paste or boiled sugar water for several days. This does not indicate a decision against breast-feeding. **Explain the immune properties of colostrum.**
- After the milk comes in, the mother and baby benefit from the hot foods consumed by the mother for the first month. Then, however, a conflict arises: the mother believes that hot foods benefit her health, but that cold foods ensure healthy breast milk. Having the mother change from breast-feeding to formula can easily solve this dilemma. However, if the mother cannot afford formula, she may use fresh milk or rice boiled with water, which may result in anemia and growth retardation. **Reinforce the benefits of colostrum.**

### *35.1.8 Death Rituals*

- Most Vietnamese accept death as a normal part of life. Ancestors are commonly honored and worshipped and are believed to bestow protection on the living.
- Most individuals have an aversion to hospitals and prefer to die at home. Some believe that a person who dies outside the home becomes a wandering soul.
- Many do not want to artificially prolong life and suffering, but it may still be difficult for relatives to consent to terminating active intervention, which might be considered contributing to the death of an ancestor who shapes the fate of the living. **Having a family conference and providing factual information is appreciated.**
- Clergy visitation is usually associated with Sacrament of the Sick; rites influenced by Catholicism can actually be upsetting to hospitalized patients. Sending flowers may be startling, as flowers are usually reserved for the rites of the dead. **Do not place flowers in the patient's room until getting permission from the patient or family. Contact religious leaders only with the permission or request of the patient or family.**
- Families gather around the body of a deceased relative and express great emotion. Traditional mourning practices include the wearing of white clothes for 14 days, the subsequent wearing of black armbands by men and white headbands by women, and the yearly celebration of the anniversary of a person's death.

- Few families consent to autopsy unless they know and agree with the reasons for it. Cremation is an acceptable practice to some families. **Explain the legalities of an autopsy.**

### *35.1.9 Spirituality*

- Most practice Buddhism, either solely or combined with other religions. Other major religions practiced are Confucianism and Taoism; a few are Christians, most of whom are Catholic. Animism is found mainly among people from the highland tribes. Many believe that deities and spirits control the universe and that the spirits of dead relatives continue to dwell in the home.
- Spirituality is also gained by participating in ethnic community activities. **Learn more about individual spiritual practices as a means to gain trust.**

### *35.1.10 Health-Care Practices*

- Good health is achieved by having harmony and balance with the two basic opposing forces, *am* (cold, dark, female) and *duong* (hot, light, male). An excess of either force may lead to discomfort or illness. *Am* (cold) represents factors that are considered negative, feminine, dark, and empty, whereas *duong* (hot) represents those that are positive, masculine, light, and full. Proper balance of these two life forces ensures the correct circulation of blood and good health. If the balance is not proper, life is short.
- Diseases and debilitating conditions result from either cold or hot influences. For example, diarrhea and some febrile diseases are due to an excess of cold, whereas pimples and other skin problems result from an excess of hot. Countermeasures involve using foods, medications, and treatments that have properties opposite those of the problem and avoiding foods that would intensify the problem. Asian herbs are cold, and Western medicines are hot. **Incorporate traditional practices into allopathic or Western care.**
- Common Vietnamese medical practices are discussed in Table 35.2.
- Many Vietnamese try home remedies, allowing the condition to become serious before seeking professional assistance. Once a physician or nurse has been consulted, the Vietnamese are usually quite cooperative and respect the wisdom and experience of health-care providers.
- The belief that life is predetermined is a deterrent to seeking health care. Diagnostic tests may be baffling, and invasive procedures are frightening. The prospect of surgery can be terrifying. **Explain procedures in simple, realistic terms, and reinforce information as necessary.**
- Hospitalization is considered a last resort and is acceptable only in case of emergency when everything else has failed.

**Table 35.2** Traditional Vietnamese medicine

| |
|---|
| • ***Cao gio*** (literally, "rubbing out the wind") is used for treating colds, sore throats, flu, sinusitis, and similar ailments. An ointment or hot balm oil is spread across the back, chest, or shoulders and rubbed with the edge of a coin (preferably silver) in short, firm strokes. This technique brings blood under the skin, resulting in dark ecchymotic stripes, so the offending wind can escape. Dermabrasion may provide a portal for infection. Do not interpret these ecchymotic areas as evidence of abuse |
| • ***Be bao*** or *bar gio* (skin pinching) is a treatment for headache or sore throat. The skin of the affected area is repeatedly squeezed between the thumb and forefinger of both hands, as the hands converge toward the center of the face. The objective is to produce ecchymoses or petechiae. Do not interpret these ecchymotic areas as evidence of abuse |
| • ***Giac*** (cup suctioning), another dermabrasive procedure, is used to relieve stress, headaches, and joint and muscle pain. A small cup is heated and placed on the skin with the open side down. As the cup cools, it contracts the skin and draws unwanted hot energy into the cup. This treatment leaves marks that may appear as large bruises. Do not interpret these ecchymotic areas as evidence of abuse |
| • ***Xong*** (an herbal preparation) relieves motion sickness or cold-related problems. Herbs or an agent such as Vicks VapoRub is put into boiling water, and the vapor is inhaled. Small containers of aromatic oils or liniments are sometimes carried and inhaled directly |
| • **Moxibustion** is used to counter conditions associated with excess cold, including labor and delivery. Pulverized wormwood or incense is heated and placed directly on the skin at certain meridians |
| • Acupuncture, acupressure, and acumassage relieve symptomatic stress and pain (see Chap. 10 for a description of these healing practices) |
| • Balms and oils, such as Red Tiger Balm, available in Asian shops, are applied to affected areas for relief of bone and muscle ailments |
| • Herbal teas, soups, and other concoctions are taken for various problems, generally in the sense of using cold measures to overcome hot illnesses. Eating organ meats such as liver, kidneys, testes, brains, and bones of an animal is said to increase the strength of the corresponding human part. Two additional practices are consuming gelatinized tiger bones to gain strength and taking powdered rhinoceros horn to reduce fever |

- Naturalistic explanations for poor health include eating spoiled food and exposure to inclement weather. Natural elements are associated with bad weather and cold drafts and cause problems such as the common cold, mild fever, and headache. Countermeasures involve dietary, herbal, hygienic, and simple medical practices. Collectively, these measures are categorized as *thuoc nam*, the traditional southern medicine of Vietnam, and *thuoc bac*, the more formal northern medicine of Vietnam or Chinese medicine.
- Supernaturalistic causes, such as gods, spirits, or demons, may cause illness. Illness may be considered a punishment for offending such an entity or violating some religious or moral code.
- A "weak heart" may refer to palpitations or dizziness, a "weak kidney" to sexual dysfunction, a "weak nervous system" to headaches, and a "weak stomach or liver" to indigestion.
- Loss of blood from any route is feared, and some may refuse to have blood drawn for laboratory tests. The patient may complain, although not to the health-care provider, of feeling weak for months. A patient may fear that any body tissue or

fluid removed cannot be replaced and that the body suffers the loss in this life as well as into the next life. **Explain that blood is naturally replenished with good nutrition.**

- Many believe that Western medicine is very powerful and cures quickly, but few understand the risks of over- or underdosages. The concept of long-term medication for chronic illnesses and the acceptance of unpleasant side effects and increased autonomic symptoms are not congruent with the traditional notions of safe and effective treatment of illnesses. Some people politely accept a prescription but may not fill it. Even if they have filled it, they may not take the medicine: they may adjust the dosage without telling the health-care provider. **Extensive education, the repetition of instructions, and home visitations are necessary to promote adherence.**
- **Persistent reminding, as part of an overall effort to improve communication and information dissemination, has been suggested as the best way to encourage Vietnamese women to undergo regular cancer screening and follow-up treatment.**
- **Include family members in all major treatment decisions regarding physical and mental health.**
- Common problems that pose barriers to health care are discussed in Table 35.3.
- Fatalistic attitudes and the belief that problems are punishment may reduce the amount of complaining and expression of pain among those who view endurance as an indicator of strong character. Many Vietnamese accept pain as part of life and attempt to maintain self-control as a means of relief. **Explain that accepting pain medication will hasten the healing process.**
- In times of distress or loss, they often complain of physical discomforts such as headaches, backaches, or insomnia.
- Mental illness is believed to result from offending a deity and brings disgrace to the family. A shaman may be enlisted to help: additional therapy is sought only with the greatest discretion and often after a dangerous delay. **Partner with traditional healers to refer patients early in the diagnostic or treatment stages.**

**Table 35.3** Barriers to health care for Vietnamese people

| |
|---|
| • Subjective beliefs and the cost of health care |
| • Lack of a primary care provider |
| • Differences between Western and Asian health-care practices |
| • Caregivers' judgment of Vietnamese as deviant and unmotivated because of noncompliance with medication schedules, diagnostic tests, follow-up care, and failure to keep appointments |
| • Inability to communicate effectively in the English language with recent immigrants, who lack confidence in their ability to communicate their needs; failure of providers to communicate |
| • Avoidance of Western practitioners out of fear that traditional methods will be criticized |
| • Fear of conflicts and ridicule, resulting in loss of face |
| • Lack of knowledge of the available resources |
| • Fear of stigmatization, difficulty in locating agencies that can provide assistance without distorted professional and cultural communication, and unwillingness to express inner feelings |

- Emotional disturbance is usually attributed to the possession by malicious spirits, bad luck of familial inheritance, or, for Buddhists, bad karma accumulated by misdeeds in past lives.
- The term *psychiatrist* has no direct translation in Vietnamese and may be interpreted to mean nerve physician or specialist who treats crazy people. The nervous system is sometimes considered the source of mental problems, with neurosis being thought of as "weakness of the nerves" and psychosis as "turmoil of the nerves." A person with a mental disability may be stigmatized by the family and society, which can jeopardize the ability of relatives to find marriage partners. People with mental disabilities are usually harbored within their families unless they become destructive: then they may be admitted to a hospital. **Do not disclose mental health conditions outside the family.**
- People with a physical disability are commonly accepted and are treated well and cared for by their families.
- Many women who have abnormal Pap smears fail to return for follow-up care, thereby contributing to the shockingly high incidence of cervical cancer among Vietnamese women. Many women fail to comprehend the severity of the situation and the potential for recovery if regular treatment begins early enough. **To increase follow-up visits and care, carefully explain the problems that may result if they do not follow up after an abnormal Pap smear. Explain that lack of symptoms or pain may be only temporary. Persistent reminding, as part of an overall effort to improve communication and information dissemination, has been suggested as the best way to encourage Vietnamese women to undergo regular cancer screening and follow-up treatment.**
- Because many believe that the body must be kept intact even after death, they are averse to blood transfusions and organ donation. Those who prefer cremation will donate body parts under certain circumstances.
- Women may not want to discuss sexual problems, reproductive matters, and birth control techniques until after an initial visit and confidence has been established in the health-care provider. **Pelvic examinations on unmarried women should not be conducted on the first visit or without careful advance explanation and preparation. When such an examination is necessary, the woman may want her husband present. If possible, the health-care provider and an interpreter should both be female.**

### *35.1.11 Health-Care Providers*

- Four kinds of traditional and folk healers exist in Vietnam, as shown in Table 35.4. Many consult one or more of these healers in an attempt to find a cure. **Partner with traditional healers as a means for social support and treatment adherence.**

**Table 35.4** Folk practitioners

| |
|---|
| • Asian physicians who are learned individuals who use herbal medication and acupuncture |
| • Informal folk healers who use special herbs and diets as cures based on natural or pragmatic approaches. The secrets of folk medicine are passed down through the generations |
| • Spiritual healers, some of whom have a specific religious outlook, and others who have powers to drive away malevolent spirits |
| • Magicians or sorcerers who have magical, curative powers but no communication with the spirits |

- The acknowledgment and support of traditional belief systems are important in building a trusting relationship. Traditional healers often provide Vietnamese people with necessary social support.
- Adults, particularly young and unmarried women, are more comfortable with health-care providers of the same gender, especially for obstetrical and gynecological conditions. **Provide female healthcare providers whenever possible for gynecological care.**
- Some Vietnamese are not accustomed to female authority figures and may have difficulty relating to women as professional health-care providers.
- While many individuals have great respect for professional well-educated people, they may be distrustful of outside authority figures. Most have come to the United States and elsewhere to escape oppressive authority. A common suspicion is that divulging personal information for a medical history may jeopardize their legal rights. Respect and mistrust are not mutually exclusive concepts.

## 35.2 Reflective Exercises

1. Explain the naming format for Vietnamese men and women.
2. When does a traditional Vietnamese have a birthday? What might this mean for child development in accordance with the US standards and expectations?
3. What are traditional Vietnamese family roles and decision-making?
4. Describe two high-risk health behaviors that may require additional screening during a health care intervention.
5. List five ideologies or ideals that Vietnamese people value as part of the culture.
6. Explain the *am* and *duong* concepts of nutrition.
7. How do *am* and *duong* practices vary with pregnancy and postpartum practices?
8. How are the principles of *am* and *duong* displayed in illnesses?
9. How are mental illnesses seen among the Vietnamese? As a health-care provider, what can you do to improve Vietnamese's' access to mental health professionals?
10. Explain the five different traditional Vietnamese medical practices.

## Bibliography

Budiman A (2021) Vietnamese in the US fact sheet. Pew Research Center. https://www.pewresearch.org/social-trends/fact-sheet/asian-americans-vietnamese-in-the-u-s-fact-sheet/

CIA (2024) World Factbook—Vietnam. https://www.cia.gov/the-world-factbook/countries/vietnam/

Global Health: Vietnam (2016) Centers for Disease Control and Prevention. https://www.cdc.gov/globalhealth/countries/vietnam/

Gordon S, Bernadett M, Evans D, Shapiro NB, Dang L (2006) Vietnamese culture: influences and implications for health care. Molina Institute for Cultural Competency. https://www.molinahealthcare.com/~/media/Molina/PublicWebsite/PDF/providers/nm/medicaid/resource/health_nm_vietnameseculture-influencesandimplicationsforhealthcare_materialandtest.pdf

Harjanto L, Batalova J (2021) Vietnamese immigrants in the United States. Migration Policy Institute. https://www.migrationpolicy.org/article/vietnamese-immigrants-united-states

LaBorde P, Duong B (2010) Vietnamese cultural profile. EthnoMed. https://ethnomed.org/culture/vietnamese/vietnamese-cultural-profile

Long CO (2021) People of Vietnamese heritage. In: Purnell L, Fenkl EA (eds) Textbook for transcultural health care: a population approach, 5th edn. Springer, Cham

Nguyen DS, Giap N (2019) Death and dying: belief, fear and ritual in Vietnamese culture. In: Selin H, Rakoff RM (eds) Death across cultures. Death and dying in non-western cultures. Springer, Cham. https://doi.org/10.1007/978-3-030-18826-9_5

Nguyen VVT, Zheng MY, Liu SM, Kallen MA, Kay K, Ivey SL (2022) Prevalence of traditional Asian postpartum practices at a federally qualified health center. J Immig Minor Health 24(5):1251–1260. https://doi.org/10.1007/s10903-021-01299-0

Thai C (2003) Traditional Vietnamese medicine: historical perspective and current usage. EthnoMed. https://ethnomed.org/clinical/traditional-medicine/traditional-vietnamese-medicine-historical-perspective-and-current-usage

Vietnam—Culture, etiquette and business practice (2022) Commisceo Global. http://www.commisceo-global.com/country-guides/vietnam-guide

Vietnamese pregnancy and birth practices (2024) http://www.vietnam-culture.com/zones-28-1/Pregnancy-and-Birth-Customs.aspx

# Correction to: People of Jewish Heritage

**Teri Aronowitz and Polly Zavadivker**

## Correction to:
## Chapter 25 in: Eric A. Fenkl et al. (eds.), *Handbook for Culturally Competent Care*, https://doi.org/10.1007/978-3-031-70492-5_25

Author's name was inadvertently published as "Polly Zavadivken" in the Table of Contents and in Chapter 25. It has been corrected as "Polly Zavadivker" in this version.

The updated version of this chapter can be found at
https://doi.org/10.1007/978-3-031-70492-5_25


E. A. Fenkl, L. D. Purnell (eds.), *Handbook for Culturally Competent Care*,
https://doi.org/10.1007/978-3-031-70492-5_36

# Glossary

**aagwachse** Amish folk illness, referred to in English as livergrown, with symptoms of abdominal distress believed to be caused by too much jostling, especially occurring in infants during buggy rides.
**Abnemme** Amish folk illness characterized by "wasting away"; usually affects infants or young children who seem to be too lean and not active.
**Abwaarde** Amish term for ministering to someone by being present and serving when someone is sick in bed.
**Acadia** Part of the Canadian Maritime Provinces.
**adab** Egyptian term for politeness.
**afatanbah** Somali term for after childbirth.
**alaqad** Somali traditional healer.
**Allah** Greatest and most inclusive of the names of God. Arabic word used to describe the God worshipped by Muslims, Christians, and Jews.
**am** Pervasive force in Vietnamese traditional medicine, associated with cold conditions and things that are dark, negative, feminine, and empty.
**Anabaptist** Adherent of the radical wing of the Protestant Reformation who espouses baptism of adult believers.
**antyesti** Hindu equivalent of last rites.
**Arabic** Semitic language of the Arabs.
**arwah** Egyptian term for spirits.
**Ashkenazi** Descended from Eastern Europe and Russia.
**!Ay bendito!** Frequently used Puerto Rican phrase expressing astonishment, surprise, lament, or pain.
**Ayurveda** Traditional Asian Indian medicine.
**baten** Iranian term for inner self.
**be bao or bat gio** Vietnamese folk practice in which the skin is pinched in order to produce ecchymosis and petechiae; practiced to relieve sore throats and headaches.



E. A. Fenkl, L. D. Purnell (eds.), *Handbook for Culturally Competent Care*, https://doi.org/10.1007/978-3-031-70492-5

**Boat people** Haitian or Cuban immigrants who arrive in small boats; usually of undocumented status.

**Boricua** Puerto Rican term used with great pride; name given to Puerto Rico by the Taino Indians.

**botanica** Traditional Cuban or other Spanish store selling a variety of herbs, ointments, oils, powders, incenses, and religious figurines used in Santería.

**Brauche** Folk healing art common among Pennsylvania Germans.

**Braucher** Amish practitioner of brauche, a folk healer.

**bris or brit milah** Ritual circumcision of a male Jewish child.

**caida de la mollera** Condition of fallen fontanelle, believed to occur because the infant was withdrawn too harshly from the nipple; common among some Spanish-speaking populations.

**Cami** Turkish word for mosque.

**cao gio** Vietnamese practice of placing ointments or hot balm oil across the chest, back, or shoulders and rubbing with a coin; used to treat colds, sore throats, flu, and sinusitis.

**Catimbozeiros** Portuguese word for sorcerer; can be a folk practitioner.

**Chasidic (or Hasidic)** Ultra-Orthodox Jewish sect.

**Chiac or ciac** French dialect used in New Brunswick.

**cho** Haitian word for cold.

**Chondo-Kyo** Korean naturalistic religion that combines Confucianism, Buddhism, and Daoism.

**choteo** Cuban term for a lighthearted attitude, involving teasing, bantering, and exaggeration.

**comadre** Portuguese term for godmother.

**compadre** Portuguese term for godfather.

**confianza** Hispanic term for trust developed between individuals; essential for effective communication and interpersonal interactions in health-care settings.

**Conservative** Jewish term for the religious group between Reform and Orthodox in terms of religious practice.

**contadini** Italian term for peasants.

**cornicelli** Italian charm with little red horns worn for good luck.

**Cultural awareness** Appreciation of the external signs of diversity such as the arts, music, dress, and physical characteristics.

**Cultural competence** Having the knowledge, abilities, and skills to deliver care congruent with the patient's cultural beliefs and practices. (See Chap. 1 for a more extensive definition.)

**Cultural sensitivity** Having to do with personal attitudes and not saying things that may be offensive to someone from a cultural or ethnic background different from the health-care provider's background.

**Culture** Totality of socially transmitted behavior patterns, arts, beliefs, values, customs, lifeways, and all other products of human work and thought characteristics of a population of people that guides their worldview and decision-making. Patterns may be explicit or implicit, are primarily learned and transmitted within the family, and are shared by the majority of the culture.

**curandeiro** Portuguese folk practitioner whose healing powers are divinely given.

**curandero** Traditional folk practitioner common in Spanish-speaking communities; treats traditional illness not caused by witchcraft.

**Deitsch/Duetsch** Pennsylvania German (sometimes incorrectly anglicized as Pennsylvania Dutch); American dialect derived from several uplands and Alemannic German dialects, with an admixture of American English vocabulary.

**demut** German term for humility, a priority value for the Amish, the effects of which may be seen in details such as the height of the crown of an Amish man's hat, as well as in very general features such as the modest and unassuming bearing and demeanor usually shown by Amish in public. This behavior is reinforced by frequent verbal warnings against its opposite, hochmut, pride, or arrogance, which is to be avoided.

**descensos** Spanish term for fainting spells.

**doule** Haitian word for pain.

**duong** Vietnamese force used in traditional health practice, associated with things positive, masculine, light, and full.

**Eid** Arabic, Iranian, and Somali term for celebration of a feast—for example, Eid Gorgan (day/feast ending pilgrimage to Mecca); Eid Fetr (last day of the month of Ramadan).

**el ataque/ataque de nervios** Hyperkinetic spasmodic activity common in Spanish-speaking groups. The purpose is to release strong feelings or emotions. The person requires no treatment, and the condition subsides spontaneously. It is an expression of deep anger or depression.

**endropi** Greek term for shame.

**espiritista (espiritualista)** Spanish or Portuguese folk practitioners who receive their talent from "God"; treat conditions believed to be caused by witchcraft.

**Falasha** Black Jews originating from Africa.

**Farsi** The national language of Iran.

**Fatalism** Acceptance that occurrences in life are predetermined by fate and cannot be changed by human beings.

**fret** Haitian word for cold.

**Gaelic** The language spoken in Ireland.

**garm** Iranian term for hot.

**Generalization** Reducing numerous characteristics of an individual or group of people to a general form that renders them indistinguishable. Generalizations have to be validated by the individual.

**geophagia** The practice of eating clay and nonfood substances.

**giac** Vietnamese dermabrasive procedure performed with cup suctioning.

**giagia** Greek term for grandma.

**Gullah** Creole language spoken by African Americans who reside on or near the islands off the coasts of Georgia and the Carolinas.

**halal** The lawful—that which is permitted by Allah; also, the term used to describe ritual slaughter of meat.

**haram** The unlawful—that which is prohibited by Allah; anyone who engages in what is prohibited is liable to incur punishment in the hereafter (as well as legal punishment in countries that incorporate Islamic law into legal codes).

**Hasidic** Jewish ultra-Orthodox sect.

**Hasidic (or Chasidic)** Ultra-Orthodox Jewish sect.

**hijab** Modest covering of a Muslim woman; conceals the head and the body, except for the hands and face, with loosely fitting, nontransparent clothing.

**hilot** Filipino folk healer and massage therapist.

**Hochmut** Amish term for pride and arrogance.

**honor** Spanish term for goodness or virtue; can be diminished or lost by an immoral or unworthy act.

**Hot-and-cold theory** Hispanic concept that illness is caused when the body is exposed to an imbalance of hot and cold; foods are also classified as hot or cold.

**hwa-byung** Korean traditional illness that occurs from repressing anger or other strong emotions.

**Ihteram** Egyptian Arabic word for respect.

**il mal occhio** Italian term for the evil eye.

**imam** Muslim leader of the prayer; usually the most learned member of the local Islamic community.

**Indian Health Service** Federal agency that has the responsibility for providing health services to Native Americans.

**Indochinese** Individuals originating from Vietnam, Cambodia, or Laos.

**Islam** Monotheistic religion in which the supreme deity is Allah; according to Muslim belief, God imparted his final revelations—the Holy Qur'an—through his last prophet, Mohammed, thereby completing Judaism and Christianity.

**issei** First-generation Japanese immigrant.

**itami** Japanese term for pain.

**jenn** Egyptian term for the devil.

**Jinn** Arabic term for demons.

**kaddish** Jewish prayer said for the dead.

**kampo** Japanese term for East Asian or Chinese medical practices and botanical therapies.

**karma** Hindu term for actions performed in the present life and the accumulated effects from past lives.

**kashrut or kashrus** Jewish laws that dictate which foods are permissible under religious law.

**Koran** See Qur'an.

**kosher** Kashrut laws in the Jewish religion.

**Ladino** Guatemalan term for Mestizo (mixed American Indian and Spanish).

**Latino(a)** Person from Latin America.

**laub** Hmong dish made with raw pork and vegetables and spices.

**lien** Vietnamese concept that represents control over and responsibility for moral character.

**maghi** Italian word for witch.

**magissa** Greek folk healer.

**mal ojo or mal de ojo** Spanish term for the evil eye, a hex condition with unspecific signs and symptoms believed to be caused by an older person admiring a younger person; condition can be reversed if the person doing the admiring touches the person being admired.

**matiasma** Greek term for the evil eye.

**Métis** People of mixed Native American and European, especially French Canadian, heritage.

**mien** Vietnamese concept based on wealth and power.

**mohel** Ritual circumciser in the Jewish faith.

**morita therapy** Indigenous Japanese school of psychotherapy.

**Moslem** See Muslim.

**moxibustion** Vietnamese health-care practice in which pulverized wormwood is heated and placed directly on the skin at specified meridians to counter conditions associated with excess cold.

**Muslim** Person who follows the Islamic faith, the world's second-largest religion.

**naharati** Iranian term for generalized distress.

**Naikan therapy** Japanese indigenous therapy focusing on bodily illnesses that are emotionally induced.

**nervioso(a)** Hispanic term used to describe signs and symptoms of nervousness, anxiety, sadness, and grief.

**nevra** Greek folk illness.

**Nihon/Nippon** Japanese name for Japan.

**nisei** Japanese term for the second generation of an immigrant family.

**Nuyoricans** Puerto Ricans born in New York.

**Old Order Amish** Most conservative and traditionalist group among the followers of Jacob Ammann; today simply called Amish, but technically known as Old Order Amish Mennonite to distinguish them from other related Amish and Mennonite groups.

**Oppression** Haitian ailment related to asthma; describes a state of anxiety and hyperventilation.

**oreo** Derogatory term for "black on the outside and white on the inside."

**Paj ntaub (pan dow)** Form of embroidery that Hmong women do to decorate their clothing and make historical story cloths.

**pappou** Greek word for grandfather.

**parve** Jewish term used for foods that are neutral and can be eaten with meat or milk products.

**Pasah Dai** Dialect in southern Thailand.

**Pasah Isaan** Dialect in northeastern Thailand.

**Pasah Nua** Dialect in northern Thailand.

**philptimo** Greek term for respect.

**Polonia** Communities heavily occupied by Polish immigrants and descendants of Polish nationals. Also the medieval name for Poland.

**Ponos** Greek word for pain.

**practika** Greek herbal remedies.

**Pseudofamilies** Vietnamese households made up of close and distant relatives and friends that share accommodations, finances, and fellowship.
**pu tong hua** Recognized language of China.
**qi** One of the five substances or elements of traditional Chinese medicine; encompasses the foundation of the energy of the body, environment, and universe; includes all sources and expenditures of energy.
**Qur'an or Koran** Muslim holy book; believed by Muslims to contain God's final revelations to humankind.
**rabbi** Jewish religious leader.
**Ramadan or Ramazan** The ninth month of the Islamic year during which Muslims are required to fast during daylight hours for 30 days.
**Reconstructionism** Mosaic of the three main branches of Judaism; is an evolving religion of the Jewish people; seeks to adapt Jewish beliefs and practices to the needs of the contemporary world.
**Reform** Liberal or Progressive Judaism.
**respeto** Hispanic term denoting respect; refers to the qualities developed toward others such as parents, the elderly, and educated people who are expected to be honored, admired, and respected.
**Sabra** Jew who was born in Israel.
**sansei** Japanese term for the third generation of an immigrant family.
**Santería** A 300-year-old Afro-Cuban religion that syncretizes Roman Catholic elements with ancient Yoruba tribal beliefs and practices.
**santero** Practitioner of Santería.
**sard** Iranian term for cold.
**sensei** Japanese term for master; used to address teachers, physicians, or those in seniority in a corporate setting.
**Sephardic** Jewish term for being descended from Spain, Portugal, the Mediterranean, Africa, or Central or South America.
**sheikhs** The most learned individuals in an Islamic community.
**Shinto** Indigenous religion of Japan.
**simpatia** Spanish term for smooth interpersonal relationships; characterized by courtesy, respect, and the absence of harsh criticism or confrontation.
**sobador** Spanish folk practitioner, similar to a chiropractor, who treats illnesses and conditions affecting the joints and musculoskeletal system.
**Solidarnosc "Solidarity"** Union of interests, purposes, and sympathies promoting fellowship with Polish nationals.
**Stereotype** Oversimplified conception, opinion, or belief about some aspect of an individual or group of people.
**Synagogue, temple, or shul** Jewish house of worship.
**ta'arof** Iranian ritual expressing courtesy.
**tae-kyo** Korean term, literally fetus education, with the objective being health and well-being of the fetus and the mother through art, beautiful objects, and a serene environment.
**Tagalog** Filipino national language.

**Tesbih** Turkish small beads traditionally used for praying, now take a more secular meaning and are often referred to as worry beads.

**Tet** Asian Lunar New Year; celebrated in January or February.

**Torah** Five books of Moses; referred to in the Jewish faith.

**treyf** Jewish term for forbidden or unclean.

**Tribe** Native American social organization comprising several local villages, bands, districts, lineages, or other groups who share a common ancestry, language, and culture.

**Unani** Traditional medicine practiced in Arab countries and Indochinese such as Vietnam.

**Variant cultural characteristics** Determine a person's adherence to beliefs and values of his or her dominant culture. Includes nationality, race, color, gender, age, religious affiliation, educational status, socioeconomic status, occupation, military experience, political beliefs, urban or rural residence, enclave identity, marital status, parental status, physical characteristics, sexual orientation, gender issues, and reason for migration (sojourner, immigrant, or undocumented status).

**velorio** Spanish term for a wake; a festive occasion following the burial of a person.

**vendouses** Greek practice of cupping.

**voudou or voodoo** Vibrant religion born from slavery and revolt; the term means sacred in the African language of Fon.

**yang** In Chinese belief system, one of the two opposing principles of the balance of life; can be either a single phenomenon or a state of being of a phenomenon. See yin.

**yarmulke** Jewish head covering worn by men.

**yerbero** See jerbero.

**Yiddish** Language often spoken by elderly Jews.

**yin** In Chinese belief system, one of the two opposing principles of the balance of life; can be either a single phenomenon or a state of being of a phenomenon. See yang.

**zaher** Iranian term for public persona.

**zar** Egyptian transmeditative ceremony.

**Zhong guo** The Chinese name for China and means "middle kingdom."